AF615648

SURFACE MINING EQUIPMENT

James W. Martin, P.E.
Professor, Colorado School of Mines
President, Martin Consultants, Inc.

Thomas J. Martin, P.E.
Sr. Project Engineer, Martin Consultants, Inc.

Thomas P. Bennett
Sr. Technical Assistant, Martin Consultants, Inc.

Katherine M. Martin
Illustrator, Martin Consultants, Inc.

FIRST EDITION
1982

MARTIN CONSULTANTS, INC.
P.O. BOX 1076, GOLDEN, COLORADO 80402

Printed in the United States of America

Printing (last digit) 9 8 7 6 5 4 3 2 1

Library of Congress Catalog Card Number: 82-81951

ISBN 0-9609060-0-2

Copies available from publisher: Martin Consultants, Inc.
P.O. Box 1076
Golden, Colorado 80402

TABLE OF CONTENTS

INTRODUCTION

Surface mining is a major source of the world's supply of metal and non-metal minerals. In the U.S. it is broadly applied in the development of deposits of:

- Coal/lignite
- Iron/taconite
- Copper
- Phosphate
- Bauxite
- Uranium
- Gypsum
- Molybdenum
- Limestone
- Sand/gravel
- Kaolin

Surface mining offers the potential for high recovery ratios and high production levels. It is, in addition, generally safer than underground operations, but careful planning is necessary to minimize accompanying reclamation costs.

Unfortunately, surface mineral reserves are decreasing gradually in quality and must be mined at increasing depths, which means ever increasing mining costs. This must be offset by technological improvements in the mining methods and equipment.

Technology has met the challenge to-date but each year the challenge is greater. The future will not be as dependent on the introduction of ever larger equipment but on the more astute selection and application of the units and their design refinement. More combined systems using an array of different types of machines would also appear to be the pattern for the future. This approach maximizes the opportunity for planning and operational flexibility. Further, it permits using each piece of equipment under its most efficient operating conditions.

The tendency in looking at mining systems has been to follow the patterns of the past and then to concentrate on calculations related to machine production rates, matching units, and ownership and operating costs. Prior to making these decisions and calculations, however, it is important that one has a broad basic understanding of what options are, in fact, available and why and when each should be considered. The start of actual mining trails initial planning by a number of years, making it essential that the systems and equipment selected are optimal for that future period.

This manual provides an overview of equipment selection considerations based on equipment design as well as mine plan and geological constraints. Individual machine types are considered in terms of their applications, characteristics, operational practices and basic design features. Machine selection criteria are reviewed as well as future development trends. Detailed machine specifications are provided and comparative charts are presented to facilitate preliminary machine evaluation. It is assumed that data on specific machine production capabilities and the related ownership and operating costs are available from other sources. This latter information must be carefully correlated with specific site conditions, the mine location and targeted economic goals.

Emphasis is placed on the major pieces of mobile production equipment currently available and employed in direct mining operations. All processing and transport equipment employed outside the mining area are omitted. Because of limited current usage in the U.S., in-pit conveyor systems have not been treated. Some pieces of equipment are only briefly referenced because they are not commonly considered economically competitive or are of customized design, such as large stripping shovels, percussion and jet piercing drills, horizontal drills, mobile hopper-crushers, etc. The mines also utilize an array of mobile support equipment such as water trucks, lube trucks, service trucks, powder trucks, graders, cable cars, cranes, pick-up trucks, etc., which are considered as beyond this book's intended scope.

To increase its value as a reference source and include as much information as possible in a concise manner, the text employs an outline type format when practical. Further, the chapters treating the various types of equipment are divided into the same categories to facilitate a unified presentation and comparison of different units. Emphasis has been placed on the comparative performance of different types of equipment to provide added insight for the selection process. An attempt has been made to minimize direct references to proprietary features and avoid any recommendations with respect to specific products or manufacturers. The subject is approached in terms of its present technical status, rather than a historical review of the past.

The first chapter introduces the comparative capabilities of the equipment as they relate to surface mining operations. The mine tasks of excavation, transportation and in-bank material preparation can each be accomplished by more than one type of machine; this section reviews these alternative methods and some of the competitive considerations between different types of equipment.

Each subsequent chapter deals with an individual type of equipment. The first three sections in each subsequent equipment chapter are designed to introduce the machine, provide some background, discuss its capabilities, uses, etc. Included are:

- Typical Units: Describes in general the typical units available. Manufacturers and the range of their product lines are listed.
- Basic Machine Operations: Gives a general introduction of how the machine works and illustrates a typical operating cycle.
- Applications: This section briefly describes the alternative uses for the equipment. It does not go into details on overall mine planning.

The remaining sections of each equipment chapter provide additional data on machine operation and design.

- General Characteristics: Presents the key features and capabilities of each machine as it relates to the unit's application. Comparative graphs of machine specifications are presented. All the key powered functions of the machine are illustrated and discussed.
- Operational Practices: The procedures and factors affecting performance and operating techniques are discussed.
- Design Features: The key design considerations of the machine are illustated and analyzed and, where appropriate, differing design concepts are discussed.
- Selection Considerations: Mine plan geometry, equipment limitations, equipment matching and operational considerations involved in machine selection are reviewed. Standard and optional machine features are tabulated.
- New Developments: Covers the current direction of machine design evolution — factors such as size, component design and options. Also treated are innovative applications.
- Machine Specifications: Tabulates the specifications for each make and model of machine currently available in the United States. A representative sketch of machine dimensions is provided for reference.

Numerous sketches, tables, graphs, and photographs are included to summarize and highlight specific concepts. The specification tables were generated from data contained in manufacturers' brochures. The graphs of machine specifications use this same data; the plotted curves are the result of statistical analysis of the actual data points and not the theoretical relationship.

The two appendices on tires and diesel engines elaborate on these two components common to many types of mining equipment. The tire section covers the types and uses of off-the-road tires, as well as presenting operating guidelines.

The final section of the manual is a list of selected references. These references have been chosen for the reader who would like additional information on such topics as safety, equipment maintenance, design developments, production estimating, cost estimating, etc.

This manual has been in development for over five years. It was originally prepared as a set of notes for a short course offered to the mining community and as a senior level course presented at the Colorado School of Mines. The authors gratefully acknowledge the cooperation of mine personnel and equipment manufacturers who, throughout the years, have provided extensive information on machine operation and design. We hope this manual will aid those in the mining industry and serve as a valuable reference source for equipment manufacturers and students of the mining industry.

Comments and suggestions from readers are welcomed.

Chapter 1

EQUIPMENT SELECTION CONSIDERATIONS

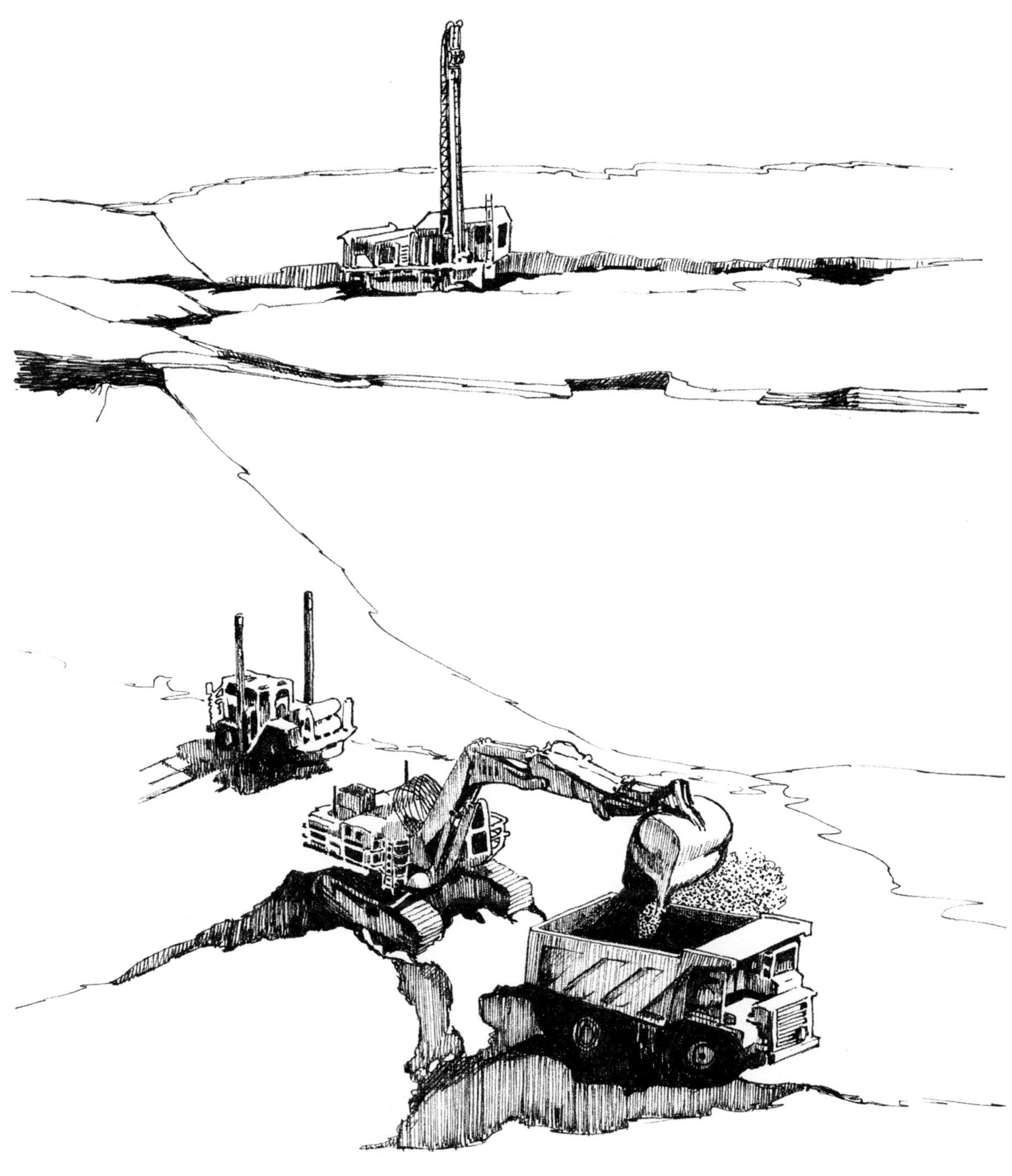

Figure 1.1 A Variety of Equipment Options Are Available

MINE PLAN OPTIONS

Mine planning is complex because of its very site specific nature and the high inter-dependence of the decisions required. Table 1.1. depicts some of the broad inter-relationships that exist. Overlaying the entire review analysis process is the goal of optimizing unit costs to assure a viable economic operation. The general approach to mine planning is one of progressive plan refinement as alternatives are evaluated.

The equipment selection process begins with the initial conception of mine development. Detailed analysis starts after the exploration has located a deposit and preliminary analysis has established that development is feasible and financially justified. Three sets of essentially fixed constraints are identified at this point which define, to a large degree, the criteria for selecting the mining equipment to be utilized.

General site conditions essentially set the equipment operating environment:

- Altitude
- Temperature range
- Rainfall
- Type of terrain
- Power availability
- Site assessibility
- Skilled labor availability
- Convenience to manufacturers' support facilities

The first group of conditions must be incorporated into equipment design considerations. The second group is a little more subtle in its impact, but are always reflected in long term machine productivity.

Formation and deposit characteristics broadly set the type and ruggedness of the primary production equipment, and the type of mining. These characteristics are:

- Overburden
  - — Depth
  - — Nature and degree of consolidation
  - — Highwall stability
  - — Spoil angle of repose
- Coal/ore
  - — Thickness
  - — Pitch
  - — Physical properties
- Partings
  - — Thickness
  - — Nature and degree of consolidation
  - — Underclays
- Hydrology
- Material properties
  - — Abrasiveness
  - — Stickiness
  - — Unit weights
  - — Swell

The mine parameters define the scheduling, production and investment targets established for the operation.

- Property limits
- Production rates
- Product quality
- Mine life
- Reclamation requirements
- Return on investment target
- Cash flow availability

When the site data has been accumulated and preliminary agreement has been reached on these mine parameters, the next level of decisions relate to the basic type of mining, unit operations required and the mining equipment systems to be employed. As noted earlier, these are all closely inter-related decisions.

Table 1.1

BASIC INPUT DATA

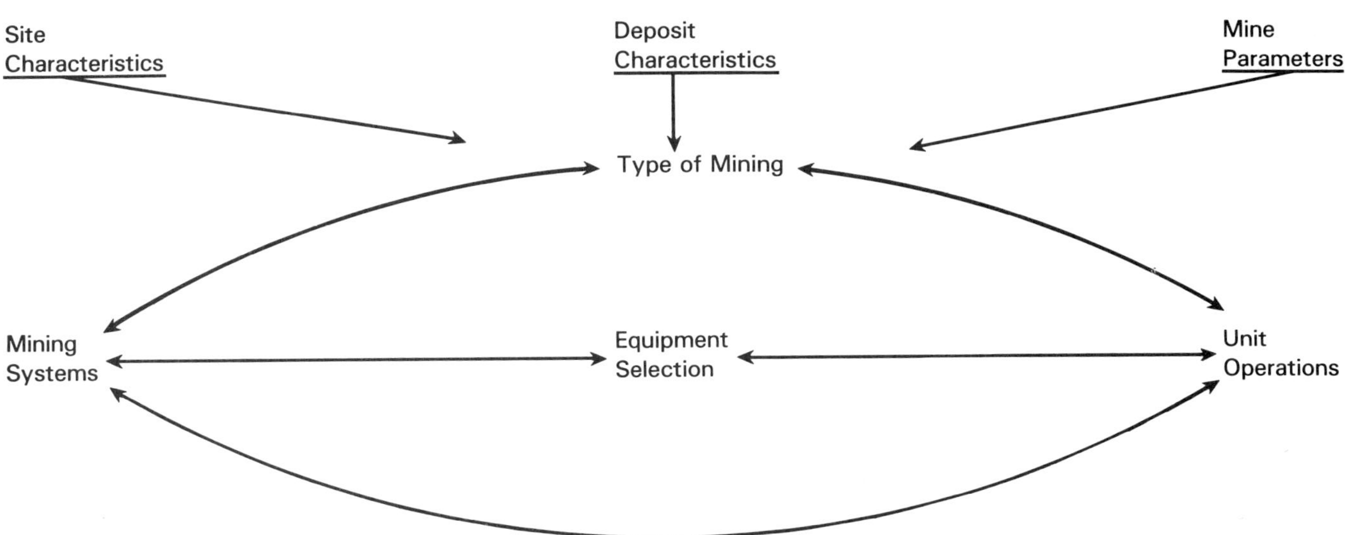

The general type of mining is controlled primarily by the terrain and the nature and size of the deposit.

- Pit/quarry
- Area (modified open pit)
- Strip
- Contour
- Mountain top removal
- Placer

Pit/quarry and area mines are associated with thick deposits. These justify mining of deeper overburden so that total depths can, in the extreme, exceed 1,000 feet. They are typically multiple bench type operations with special consideration required for the vertical lift requirements of the haulage system. Strip mines are commonly associated with shallow overburden conditions in hilly or relatively flat terrain. Single or multiple seams may be involved with present maximum depths of about 200 feet. Direct overburden disposal techniques are utilized as much as possible to minimize rehandling. Contour and mountain top removal permit mining in very hilly or mountainous terrain. Benching and direct disposal techniques are used individually or in combination for overburden removal and replacement. Maximum mining depth is variable and may be dictated by reclamation considerations. Placer mining, while a recognized approach, is not commonly utilized in the United States. The equipment associated with such operations will not be discussed.

Overall mine size has a significant effect on equipment selection. As the size decreases, the volumes of materials to be excavated and transported are reduced, often with a corresponding reduction in the size of the machines required or economically justifiable. Less capital is generally available and the planning tends to be of shorter range. There is less specialization in the equipment operator, and service and maintenance personnel. Maintenance facilities are more limited.

Paralleling the initial concepts on possible mining methods, it is necessary to establish the unit operations. Table 1.2 sequentially summarizes the operations which might commonly be anticipated. This has been arranged as a flow chart, the multiple branches in each case indicating alternative approaches where selection is based on specific site and mine requirements. Some mines may omit indicated unit operations because of unique site conditions. For example, with unconsolidated materials there may be neither blasting

Table 1.2

UNIT OPERATIONS

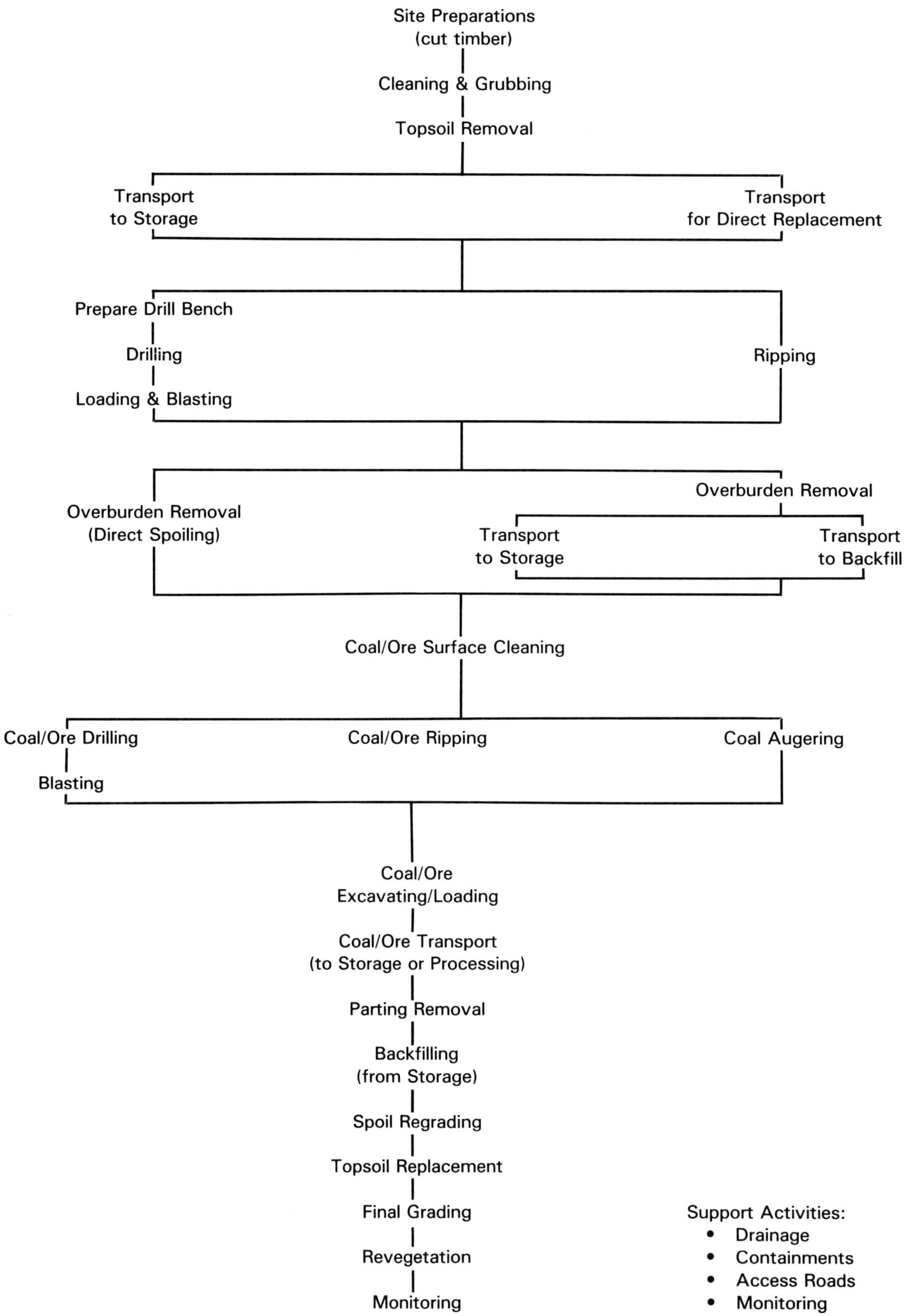

nor ripping. Note also that some of these unit operations could be repeated. For example, in a multiple seam situation, the parting removal might require a repeat of operations similar to that for handling the overburden and this would be followed again by coal/ore preparation and removal operations. Some of the initial operations may be performed on an intermittent basis as mining progresses.

Typical mining system options are summarized in Table 1.3. Essentially three functions are involved in each, preparation, excavating and transport. For simplicity the excavators have been grouped into two classes, cyclic and continuous. Each class includes several types of equipment which will be discussed later. Rail and pipeline transport have been omitted from discussion on in-mine transport because of their specialized applications. They may be effective and economically competitive in some situations.

Judging the relative effectiveness of alternate systems involves both qualitative and quantitative considerations. Direct numerical estimates can be developed for production, and ownership and operating costs. They are based on a broad variety of assumptions with respect to actual site conditions, formation characteristics, machine performance, equipment support, operator and management skills. Everything is essentially related to "average" or "typical" conditions which, of course, occur rarely. Broad qualitative factors, such as the following are difficult to evaluate.

- System flexibility (adaptability to change)
- Equipment utilization
- Excavating and transport efficiency
- Mechanical/electrical equipment availability
- Conservation of energy
- Conservation of resources
- Environmental consequences
- Safety performance

Because of the problems associated with accessing their operational and, in particular, economic impact, the above have in the past been frequently given limited consideration. Undoubtedly, in the decades ahead considerable emphasis will be placed on development of analytical tools together with acceptable criteria and standards for comprehensive evaluation of these areas.

If the mine activities are limited to coal/ore transport to an intermediate stockpile (omitting further preparation, load out facilities, etc.) then the operations consist of:

- Clearing
- Bank preparation
- Excavating/loading
- Transport (hauling)
- Grading and revegetation

Table 1.3

BASIC MINING SYSTEMS

Drill & Blast — Cyclic Excavator (direct spoiling)
Drill & Blast — Continuous Excavator (direct spoiling)

Rip — Excavator/Transport (combination machines)
Rip — Cyclic Excavator — Trucks

Drill & Blast — Cyclic Excavator — Trucks
Drill & Blast — Continuous Excavator — Belt Conveyors
Drill & Blast — Cyclic Excavator — Hopper/Crusher — Belt Conveyors

The first and last of these activities are generally considered secondary in the sense that the time and cost are small in comparison with the overall mining commitment. The clearing is part of the initial site preparation. It is commonly performed with dozer type equipment and, in the case of heavily wooded areas, supplemented with specialized equipment from the lumbering industry. Clearing of wooded areas is often subcontracted to organizations familiar with these activities and the lumber marketed. Spreading and grading are a part of the reclamation phase. The work is presently performed by scrapers, dozers and off-highway type graders. Equipment fleet requirements vary significantly with mine type, size, and whether the work can be scheduled continuously to keep pace with the mining.

Equipment employed for clearing, grading, revegetation, and (in addition) for building and maintaining haul roads has largely been adopted from the heavy construction industry or from farming. These units have been refined and, to a great degree, standardized through the years. Applications, machine sizes and types, general design, and manufacturing sources are changing relatively little. Components and detail designs are being continually modified to incorporate evolving technology and assure compliance with new codes and regulations.

In contrast, equipment for the other direct mining operations seems to be in a more rapid state of change as thinking on the optimum mine plan is revised and analytical techniques become more sophisticated. Units are larger, more expensive and more complex. Matching of production requirements with site conditions dictated by economics has resulted in some customizing of machines for specific applications. Meeting world-wide mining requirements has necessitated considerable flexibility in machine specifications and available optional accessories. Production requirements have led to larger and larger sizes. The need for around-the-clock operation has resulted in a continued search for simple, rugged designs with maximum reliability.

The high investment related to the direct production equipment, as well as the developing technology mentioned above, suggests that attention be concentrated on this equipment area. The scope, at first glance, seems relatively narrow:

- Blast hole drills and/or rippers
- Excavators/loaders
- Transport units

Unfortunately, most equipment falls into a "combination" category. Further complicating the issue is the fact that similar functions can be performed by different types of equipment. This manual will concentrate on these machines.

GENERAL MACHINE CHARACTERISTICS

While some surface mines utilize special equipment in portions of their operations, the bulk of surface mining production equipment is commercially available world-wide from multiple manufacturers. The commercial units are, to a degree, standardized in product lines, providing a graduation of sizes to meet anticipated needs. Such units offer inherent advantages in terms of product support but, most significantly, provide an accumulated background of performance experience in a wide range of applications. Restricting initial mine planning to commerical units still necessitates a great deal of analysis because of the range of equipment available and the number of manufacturers.

The nine types of mobile production equipment which will be considered in this manual are listed in Table 1.4. Limited reference will be made to the special variations that are indicated. The basic machine types have major variations which have a significant effect on their application so that seventeen different configurations result. A comprehensive discussion of each is provided in subsequent chapters.

It is of interest to note some rough statistics which shed light on the scope of the equipment selection problem.

- Major configurations of mobile production equipment–17

Table 1.4

TYPES OF MOBILE SURFACE MINING EQUIPMENT

Standard	Special Variations
Track dozers Wheel dozers	Paired units
Standard scrapers Standard tandem scrapers Elevating scrapers Elevating tandem scrapers	Push-pull combinations
Rear dump trucks Bottom dump trucks	
Front-end loaders	
Hydraulic shovels Hydraulic hoes	
Electric loading shovels	Electric stripping shovels Diesel mechanical shovels
Crawler draglines Walking draglines	
Bucket wheel excavators (medium sizes)	Stripping BWE Surface miner type wheel excavators
Crawler rotary blast hole drills Truck mounted rotary blast hole drills	Horizontal drills Jet piercing machines Pneumatic drills

- Average number of sizes (capacities)–5
- Average number of manufacturers–8

The above is limited to those machines employed in surface mining and to the major suppliers of the equipment-domestic and foreign-for North American operations.

Table 1.5 shows the size, weight and horsepower range for representative types of equipment. The spread in these parameters indicates size ranges which vary roughly from two to one, to twenty to one. Weight and horsepower reflect the ruggedness and the productivity of the machine. The bucket wheel excavator classification has been restricted to two crawler, medium sized machines for service comparable to the hydraulic excavators and the electric shovel. The total power available on the electric shovels cannot be simply related to the other machines because of the characteristics of the static power supply. Therefore, this has been omitted. Similarly, power on rotary drills, which is a mixture of electric and diesel units cannot be meaningfully correlated to provide comparable data. The small machines in these categories are the dozers, and the largest are the draglines. The biggest dragline weighs approximately 90 times that of the biggest dozer. Graph 1.1 further illustrates this range of sizes for some of the excavating equipment.

The major configuration differences between the various machine types are outlined in Table 1.6. Where electric power is indicated it means that an eletric power distribution system is

Table 1.5

COMPARATIVE EQUIPMENT SIZE

	Size Range	Weight Range (1000lbs)	Total Power Range (HP)
Dozers (crawler)	—	30–190	140–700
Scrapers (tandem)	24–52 cubic yards	60–140	288–950
Front-end Loader	3.5–30 cubic yards	33–389	180–1,270
Hydraulic Excavator (shovel)	4–30 cubic yards	114–930	300–2,350
Electric Shovel	6–75 cubic yards	471–3,350	—
Draglines (walking)	9–180 cubic yards	900–17,500	1,080–18,000
Wheel Excavators (medium size)	to 3,500 cubic yard/hr	—	—
Trucks	50–350 tons	58–520	420–3,000
Rotary Drills (crawler)	3.0–17.5 in. hole diameter	40–282	—

required with the capability of being readily extended or contracted as the machine is moved. The technology for the hydraulic drive systems, in the large sizes employed, is relatively new; this means that these designs are less proven at this time. The propel type is an indication of machine mobility and the accompanying time loss associated with relocating the equipment around or between pits. Torque converters, gears, cylinders, linkages, cables, etc., which transmit the power for the prime machine functions, are not easy to assess but are an indication of machine complexity, reliability and maintenance requirements.

While the functions performed are quite similar, the various types of equipment are significantly different in design and capability. This is brought out rather strikingly by the comparisons in Table 1.7. These are approximate figures because of the variations with machine size. Relationships tied to machine selling price can be readily challenged because they are constantly changing and because of the strong part negotiations play in many actual purchases. No attempt has been made to estimate average operating costs which vary greatly with the location and application. Some of the previously noted deficiencies in data resulted in the gaps in the table.

The differences, for example, in machine weight and price per cubic yard of bucket capacity, emphasize the critical nature of the selection process. The configuration as well as the physical size of the machine has a dramatic effect on these factors. The configurations differ in the efficiency

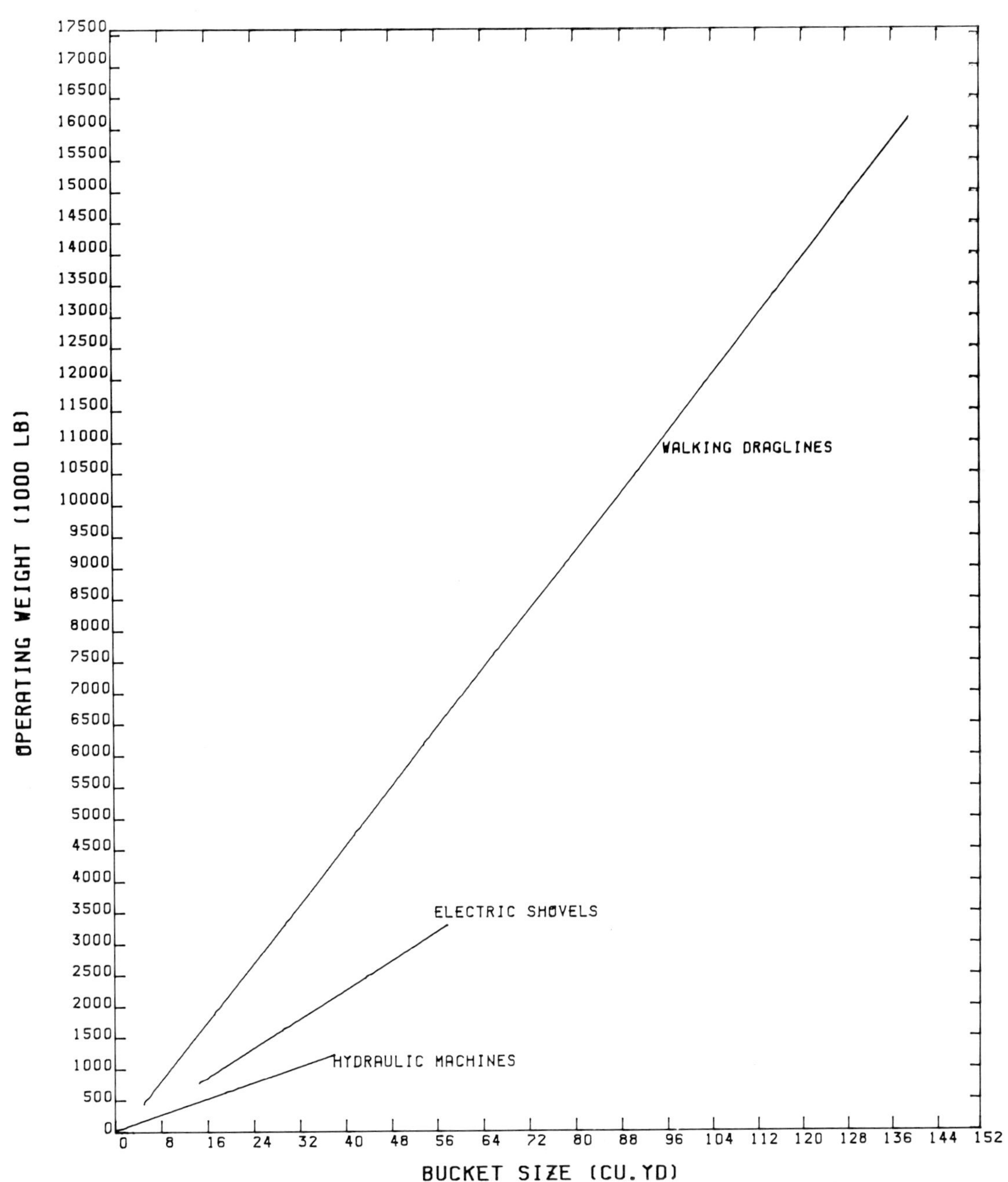

Graph 1.1 Operating weight/nominal bucket size (excavating machines)

Table 1.6

COMPARATIVE FEATURES

	Power	Primary Drive System	Type Propel	Powered Function Design
Dozers	diesel	torque converter/ transmission	crawler or wheel	hydraulic cylinders/ linkages
Scrapers	diesel	torque converter/ transmission	wheel	hydraulic cylinders/ linkages
Front-end Loaders	diesel	converter or electric motor	wheel	hydraulic cylinders/ linkages
Hydraulic Excavators	diesel	hydraulic	crawler	hydraulic cylinders & gears
Electric Shovels	electric	electric	crawler	cable & drums/ gears
Draglines	diesel or electric	electric	crawler or shoes	cable & drums/ gears
Wheel Excavators	diesel	hydraulic and/or electric	crawler	hydraulic cylinders & gears
Trucks	diesel	converter or electric motor	wheel	converter or motors & gears
Rotary Drill	diesel or electric	hydraulic/ electric	crawler or wheel	chain & gears

with which they can perform their functions. As the size increases, the relative cost of manufacturing rises and auxiliary equipment must be added to permit component handling, access, etc. The machines with more range and stability necessitate added strength and weight.

The relationships of operating weight and horsepower for alternate types of loading equipment are further illustrated in Graphs 1.2 and 1.3. These graphs show some of the differences between types of equipment that should be recognized and understood so that proper equipment/application matches can be made.

To expand on the machine requirements for the primary production functions, it is helpful to tabulate the factors which must be considered. These can be conveniently broken down into four categories:

- Performance factors
- Design factors
- Support factors
- Cost factors

Table 1.7

COMPARATIVE MACHINE RATIOS

(Approximate Averages)

	Machine Weight (lbs) per cubic yard of capacity[1]	Total Horsepower[3] per cubic yard of capacity[1]	Machine Weight (lbs) per HP[3]	Price ($) per cubic yard of capacity[1]	Service Life (hrs)
Dozers (crawlers)	—	—	235	—	12,000
Scrapers (standard)	2,900	14	200	9,500	12,000
Front-end Loaders	12,000	52	235	36,000	12,000
Hydraulic Excavators (shovels)	30,000	72	425	85,000	30,000
Electric Shovels	54,000	41	1,300	110,000	75,000
Draglines[2] (walking)	114,000	102	1,100	210,000	100,000
Wheel Excavators (medium size)	—	—	—	—	30,000
Trucks (rear dump)	1,800	12	150	—	20,000
Rotary Drills	—	—	—	—	30,000

[1]bucket capacity (3000 lb./cu.yd. material)
[2]short boom, large bucket machine
[3]diesel horsepower and electric horsepower ratings are not equivalent figures

Summary tables of key comparisons for each of the above are provided. They are, in effect, check lists which summarize the machine features to be considered and evaluated. There is obviously a great deal of inter-dependency among these factors with the ultimate acceptability of any machine dependent on the combined effect of all. Despite the variety of equipment available, many of the essential characteristics are common.

Performance factors (Table 1.8) are directly tied to the productivity of the unit. Most are included in the manufacturer's specifications; others must be determined from supplementary data available on request. In this case, the detailed requirements are different for bank preparation, excavation/loading, and transport. For simplicity, drills are assumed to be representative of the preparation units and trucks of the transport options. Similar reasoning can be readily applied to the alternative machines for these operations. The influence of specific site conditions and the selected mining system will be discussed further, under machine applications in a later section of this chapter.

In very broad terms, performance is a function of cycle speed, available force (power), machine range and reliability. The efficiency with which the power can be utilized is dependent on the design of the machine and is, of course, different for each type of equipment. Range, generally

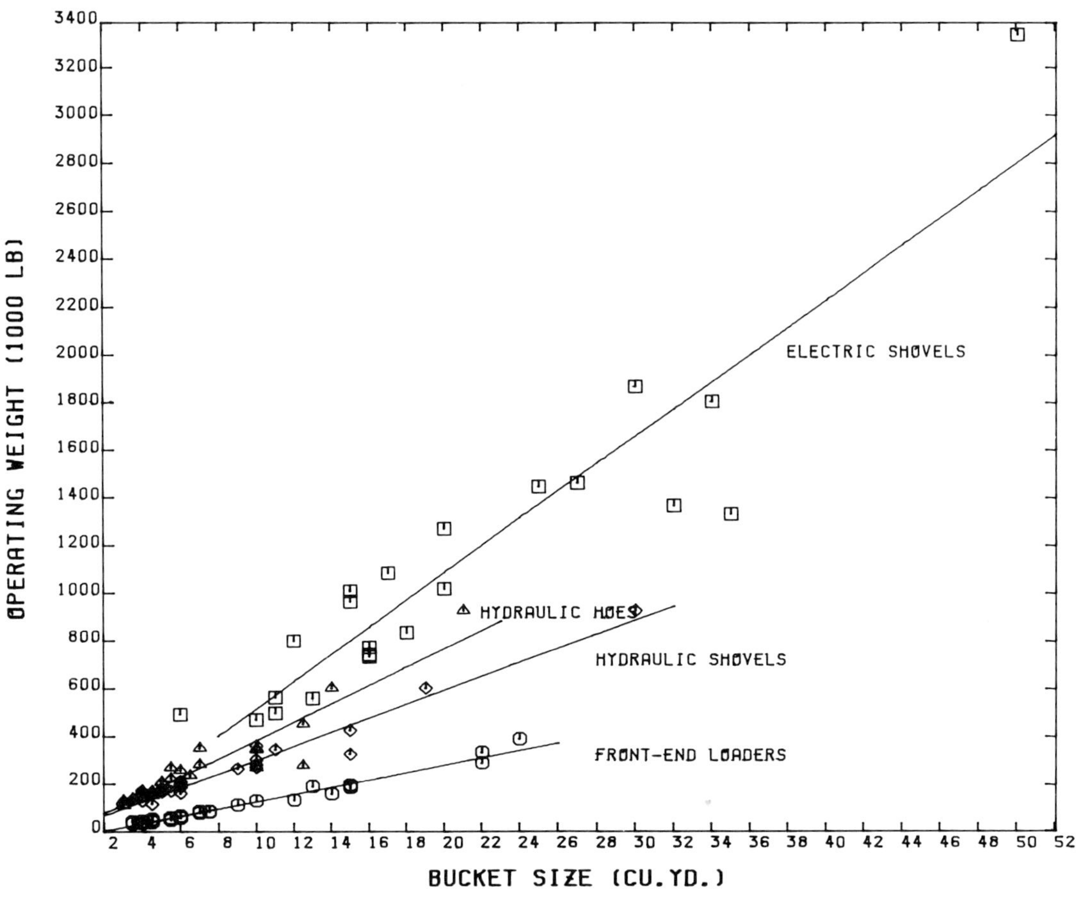

Graph 1.2 Operating weight/nominal bucket size (loading machines)

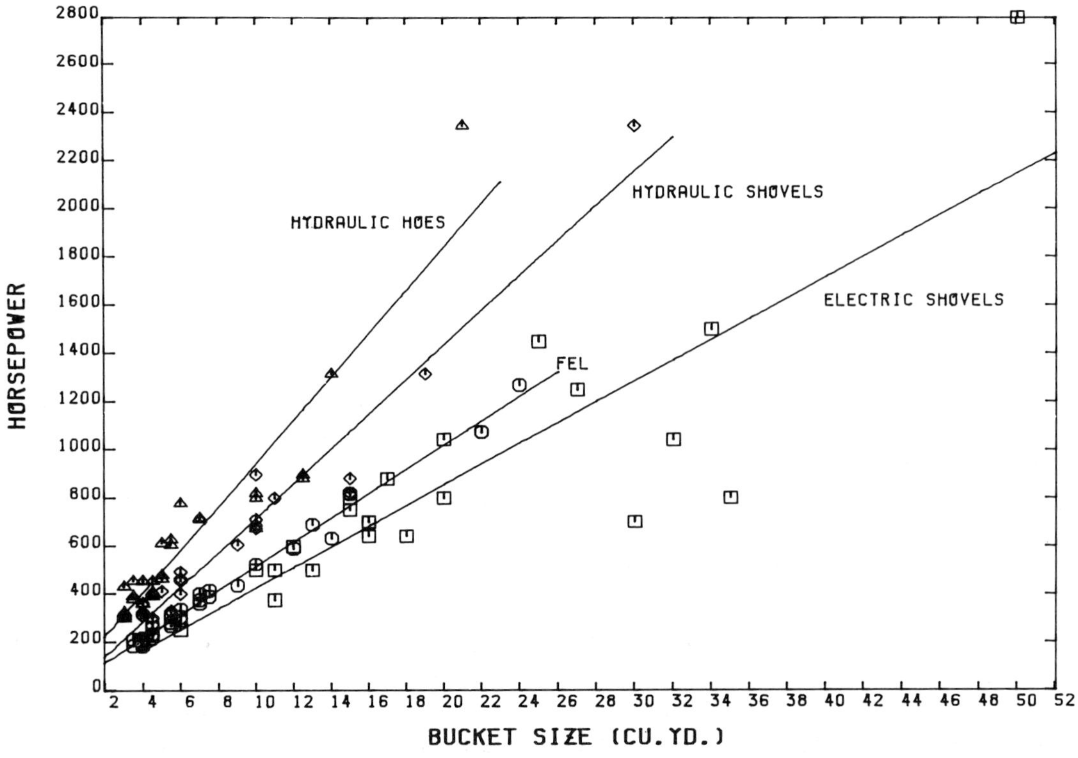

Graph 1.3 HP/nominal bucket size (loading machines)

Table 1.8

PERFORMANCE FACTORS

Bank Preparation Drills	Excavators/Loaders	Transport Units Trucks
Hole size	Bucket capacity	Payload
Drilling rate: Down pressure Rotary torque Rotary speed Compressor size	Digging capabilities: Cutting forces Cutting speeds Cutting path/action Cycle time	Gross weight Power
. . . .	. . . .	Gradeability
Mast height		Speed ranges
	Digging height	
Max. drilling depth	Dumping height	Acceleration
Cycle time: Pipe changes Machine relocation	Reach	Retardation/braking
. . . .	. . . .	. . . .
Maneuverability	Ground pressures	Traction
Travel speeds	Swing speed/turning radius	Turning radius
Gradeability	Propel speeds	Dump speed
Degree of automation	Gradeability	. . . .
. . . .	Tractive effort	Reliability
Reliability	. . . .	
	Reliability	

associated with excavators, is directly tied to size, reflecting the physical dimensions of the machine and its capabilities in terms of its travel limits in the dig-dump cycle.

Capabilities of individual machines have not changed materially in recent years. The introduction of new models of larger size has, however, expanded the power and range available in some of the equipment types. Reliability is the "bottom line", in a sense, because it is a measure of the portion of scheduled production time that the machine is mechanically/electrically operable. This consideration often overshadows many other minor performance differences and, therefore, is reviewed in some detail at the end of this chapter.

The **design factors** (Table 1.9) provide critical insight into the quality and effectiveness of the detail design. They determine the full scope of the man-machine interface for both the operators and those who work on the machines in terms of comfort, effort, visibility, noise levels and safety. They reflect the level of technology employed, the effort directed towards assuring machine reliability, and the overall complexity.

The machine geometry and drive system design establishes the efficiency of power distribution and its application. Machine and component life are tied to the proper matching of design criteria and field requirements. The skill, knowledge and experience applied in every aspect of the design is

Table 1.9

DESIGN FACTORS

Total power
Service life
Weight
Physical size
Stability
Ground clearance

Basic configuration
- Geometry and linkages
- Complexity
- Modular construction

Componentry
- Proven components
- Interchangability units
- Critical component life

Appearance

Serviceability
Maintainability
Transportability
Ruggedness

Altitude limitations
Temperature limitations
Noise levels
Crew size
Dust generation

Operator skill required
Operator effort required
Operator safety
Operator visibility

Control response
Power rating and characteristics
Hydraulic vs. electric vs. mechanical drive system
Diesel vs. electric power

On-board diagnostic and testing devices
Back-up systems
Overload protection devices
Fire suppression equipment
Accessory options
Optimal specifications

ultimately the key to the value received for the investment. While the performance factors have not changed significantly over recent years, the designs have increased in sophistication, with improved automated functions and feed-back control systems. These factors will most likely become increasingly important in machine selection.

The **support factors** (Table 1.10) are often overlooked in machine evaluation. Each machine brought into the mine requires that a unique spectrum of activities, skills, inventory and facilities be available. This is not only an additional financial commitment but, in many cases, poses a special problem with finding and retaining personnel with the necessary qualifications. Maintenance and servicing performed in-house, in a well equipped shop, is more efficient, economical and more readily scheduled than work performed off-site or in the pit. Note that "well equipped" means facilities tailored and staffed to handle the specific type of equipment.

The severe operational demands placed on mining equipment require that the manufacturer also play an active back-up role. It is essential that the manufacturer be realistically committed to promptly providing, as needed, technical expertise, training, replacement parts and assemblies. The expertise must include trouble shooting, failure correction, special modifications, accessory and application guidance. Parts standardization across product lines can facilitate service and maintenance work, reduce inventory investment and unit costs. With the growth in size of the manufacturing organizations and extensive world wide marketing programs, the degree of support available has been broadening and is now a major consideration.

Cost factors (Table 1.11) are possibly the best understood because a variety of set procedures are available in the literature which permit plugging numbers into formulae and getting numerical final answers. The numbers can readily be compared with alternate equipment and provide a neat decision justification. Unfortunately, as with most such calculations, it is relatively easy to make them come out any way desired. High inflation in recent years has increased the difficulty in estimating ownership costs. The cost components have been affected by different inflation rates,

Table 1.10

SUPPORT FACTORS

Auxiliary machines required

Frequency of servicing required
Shop vs. field maintenance
Maintenance skills required
Maintenance facilities required
Portion maintainable in-house

Power distribution system requirements
(Fuel storage facilities)
Component standardization

Manufacturers' support available
Manufacturing quality
Parts availability

Table 1.11

COST FACTORS

Ownership costs
— Depreciation
— Interest
— Insurances
— Taxes

Initial investment includes:
— Basic machine
— Accessories and options
— Special features
— Freight
— Ballast
— Field assembly
(minus salvage value)

Operating costs
— Labor
— Supplies
— Fuel/power
— Lubricants
— Repairs, maintenance

Optional
— General supervision
— Storage and warehousing

and fluctuating rates have made projecting future values risky. Interest rates and salvage values have cycled with the changing economy. Taxes have been significantly restructured to meet revised governmental goals and, unfortunately, will be subject to progressive clarification in the courts for a number of years. Operating costs are commonly derived from past experience in other operations. Inflation, compounded by unusually large changes in fuel and parts costs, have distorted these values and complicated the extrapolation of available data. The ownership and operating cost determination provides necessary input in the mine planning process but it is important to recognize that it represents only a limited evaluation of the equipment and is based on rather broad assumptions of future site conditions and requirements.

MACHINE APPLICATIONS

Matching equipment capabilities to site conditions requires considerable experience and knowledge. Each application is usually unique and each equipment fleet must be matched and balanced.

The material to be excavated falls into four broad categories:

- Topsoil
- Overburden (waste)
- Partings
- Coal/ore

They differ in terms of the degree of difficulty in excavating. The size, shape and basic design of the digging tool must be matched with the unit weight, swelling, stickiness and abrasive characteristics of each material. Blocky material or the presence of large boulders places special constraints on the size and type of equipment that can be employed. Coal/ore presents additional unique demands in terms of minimizing the dilution with waste materials and matching minimum/maximum lump size to subsequent handling and processing requirements.

Successful penetration of a bank by the cutting edge of a bucket or dipper is a function of its shape, applied force and orientation, combined with the resistance offered by the formation. Numerous attempts have been made to correlate the measurable properties of material formations with the machine design requirements. The material properties, regrettably, often vary significantly with faults, climatic conditions, depth, and across the planned mining areas. Both operator skill and cutting geometry of the machine are changing variables which can fluctuate through wide ranges. These factors have so far frustrated most efforts to develop quantitative approaches to digging condition analysis. The only success to-date with this type of correlation appears to be with wheel excavator applications in thick uniform deposits. Actual experience in similar formations has been the most popular practical approach to predicting digging ability; some cases are confirmed with small scale test operations.

Each type of excavator has distinctly different digging characteristics and capabilities. When all factors are fully considered, none can be classed as a universal tool, although the more rugged configurations that can dig the most difficult formations are often treated as such. The different excavators have overlapping digging application zones, limited primarily by machine abuse which, in turn, is reflected in operating availability and costs. In the following chapters for each type of machine, sketches are included of the digging actions, forces applied, cut profiles, typical production cycles, and common operating patterns.

Since the degree of consolidation in the digging face restricts the equipment selection and, in all cases, establishes the level of machine abuse, it is logical to consider auxiliary means for fragmenting or breaking up the formation to reduce the demands on the excavator. Alternatively, drilling and blasting and/or ripping can be employed to prepare the material; Table 1.12 list some characteristics of these methods. Blasting can additionally provide, in some cases, lateral displacement; this can facilitate subsequent operations. How much material preparation is desirable is basically an economic trade-off type of evaluation. Unfortunately, direct correlation of machine maintenance costs vs. preparation costs is difficult and hard to substantiate because of the detailed operational records required. Estimates based on experience in similar formations are usually employed for preliminary estimating and the final practice is established by experimentation.

In drilling and blasting, the extent of fragmentation is dependent on the type of explosives used, the hole spacing, the diameter of the hole, and the distribution and amount of explosive in each hole. (See Photographs 1.1, 1.2, and 1.3) Requirements

Photograph 1.1 Holes drilled in a pattern ready for charging

Table 1.12

APPLICATION GUIDELINES–PREPARATION

	Ripping	Rotary Drilling
Hole sizes (in)	—	4–17
Formation hardness	soft-medium	medium-hard
Degree of consolidation	unconsolidated to moderately consolidated	moderate to highly consolidated
Drilling rate (ft/hr)	—	to 120
Single pass depth (ft)	to 11	to 65
Operating depth (ft)	—	to 250
Mobility	high	limited
Advance preparation	none	prebenching
Auxiliary equipment	none	explosive trucks, etc.
Special considerations	high power tractor	holes located, shock/noise control, operations interrupted during blasting
Ownership cost	low	high
Operating cost	medium	high

and results vary with the nature of the formation, the layout of the holes and the sequencing of the hole firing. This technique permits in one operation the preparation of a block of material to the full depth planned for subsequent excavation (such as a mining bench).

Ripping involves the use of large single or multiple teeth attached to the back of a dozer; these teeth are essentially used to scarify the material surface to depths of from 2 to 11 feet. The spacing between passes and/or fixed teeth determines the degree of fragmentation. If employed in conjunction with scrapers or dozers, the ripping operation is performed intermittently as the surface is removed. The feasibility of ripping is determined from field tests of the seismic wave velocities in the material to be prepared.

There are two basic approaches used in excavating:

- Dozers and scrapers make progressive, long, skimming cuts off the top surface;
- Front-end loaders, shovels, hoes, draglines and conventional bucket wheel excavators generally take short, upward cuts from an inclined face (30° to 80° above horizontal).

The digging characteristics of the dozer and scraper (also the grader) are ideal for grading and leveling (see Table 1.13) and shallow layer removal of material. Hence they are broadly applied in topsoil removal and reclamation work. They are combination machines in that they are used both to excavate and transport material.

Photograph 1.2 Explosives truck charging drilled holes

Photograph 1.3 Blasting as final step in preparation for excavating

Table 1.13

APPLICATION GUIDELINES–LEVELING & GRADING

	Dozer	Scraper	Grader	Dragline
Rough leveling	excellent	fair	poor	good
Rough grading	excellent	good	good	fair
Fine grading	good	fair	excellent	poor
Lateral material displacement (ft)	to 300	to 5000	to 15	to 500
Surface compaction	medium	high	medium	low
Special considerations	super wide & angle blades, paired/tandem units		high flotation tires	Sauerman type buckets, long booms, low ground pressure shoes
Ownership costs	low	low	low	high
Operating costs	high	high	medium	low

Note: Gradations are relative to other units of the same general size.

This, together with their special digging characteristics, make them unique and they, therefore, can be treated separately from the other excavators.

Draglines are unique in their ability to discharge material at a considerable radius. This permits direct disposal of the excavated material without the use of haulage equipment in some mine applications. (See Figure 1.2) These units have very deep digging capabilities below the level of the machine. While their excavating efficiency is poor when digging much above their floor in consolidated material, they are effective in shallow level cuts in looser formations, such as encountered in spoil pile leveling. For these reasons, they are included in Table 1.13.

The remaining units can be conveniently classified as excavator-loaders. While they do transport the material some distance, because of their size this is often not a primary consideration. The material is dumped or disposed of by discharging onto a pile, hopper, or into haulage equipment. Some of their basic selection considerations are depicted in Figure 1.3. These range from digging face conditions, floor conditions and production requirements, to discharge constraints.

The degree of consolidation and abrasiveness must match machine ruggedness and power in the digging face. Secondarily, the nature and control of the machine's digging path comes into play, together with operating range limitations on cutting height. Only the front-end loader does any significant amount of maneuvering on the working floor. This characteristic, added to its rubber tired mounting, results in high operating costs if floor conditions are wet and/or consists of sharp abrasive material which would reduce tire life.

Under any specific face conditions, production rate is primarily a function of machine size and operating cycle time. Cycle time increases with

Figure 1.2 Walking Dragline Direct Spoiling

machine size, depth of cut, and the arc (swing angle, in most cases) through which the material is transported to dump.

Discharge constraints are:

- Adequate height
- Reach
- Discharge severity.

Height and reach, for example, define the maximum truck or hopper size that the unit can effectively dump into. Discharge severity is a measure of the size of material and level of impact that is characteristic of the discharge action. The importance of this consideration is tied to operational practices. Minimal effect is typified by operational techniques that release the load with limited free fall height. The dumping action of the front-end loader and the hydraulic excavator, resulting from a tilting of the bucket, is generally less severe than the opening trap door arrangement on an electric shovel. The wheel excavator, which uses a number of small buckets (rather than a single large one), cannot dig and discharge large boulders, etc.

Table 1.14 expands on this reasoning to provide some broad application guidelines for excavators. It attempts to summarize past performance requirements/expectations and highlight some of the characteristics important to the evaluation process. All indicated gradations are relative to other units of the same overall size. The spread in

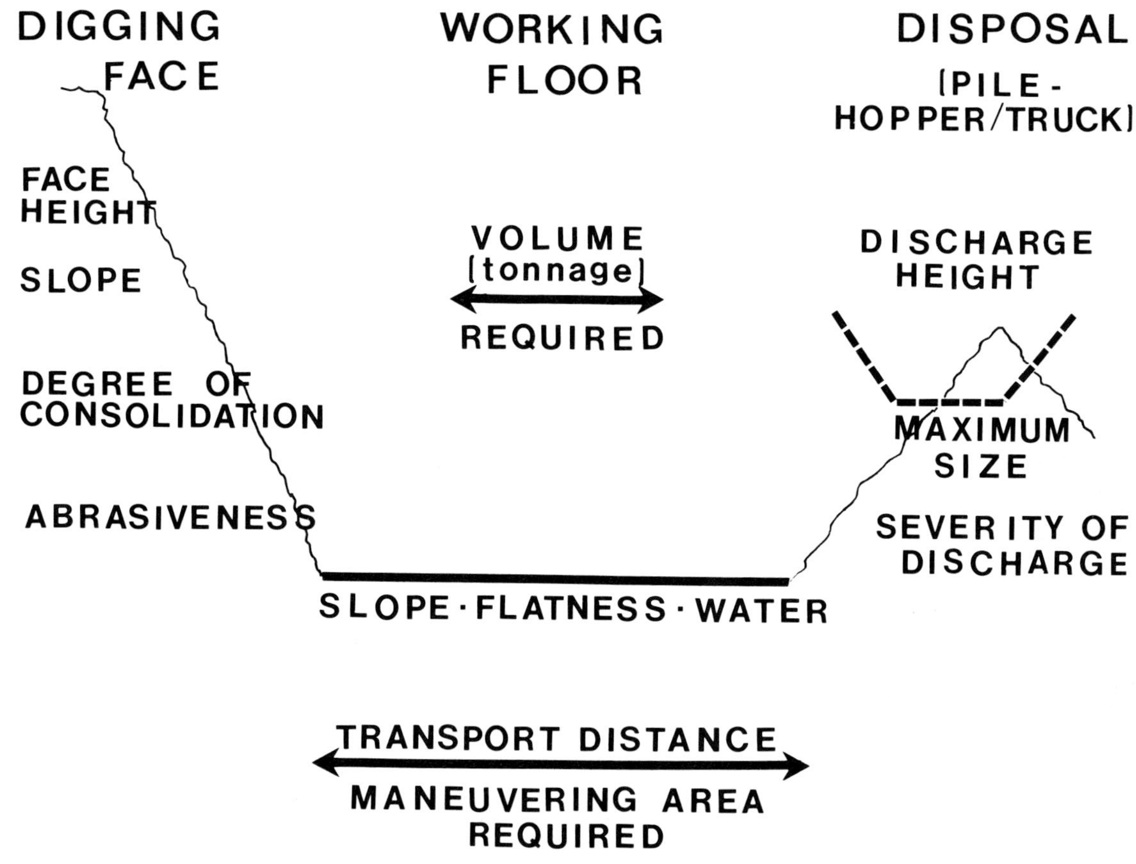

Figure 1.3 Excavating Constraints

the general gradations varies widely, dependent on the characteristic being considered. For example, the range for expected reliability might be from a low of 60% to a high of 90%, but delivery times might span from 3 weeks to 3 years. More specific numerical comparisons are only feasible when detailed site conditions and requirements are known. Graph 1.4 illustrates one of these important operational dimensions that can vary with equipment size and type.

The bulk of the FEL's, hydraulic excavators and electric shovels are dumping into trucks and/or hoppers and, occasionally, into railcars. The wheel excavators can discharge to trucks and/or hoppers, but more commonly are feeding belt conveyors. Draglines can be, and are presently, used only on a limited basis to load trucks and/or hoppers.

The dragline has been included in this machine grouping for convenience. It is, in many ways, an independent category because of its superior direct spoiling capabilities. While not at first apparent, the dragline, due to its great reach, is also a transport machine. Transport is accomplished in the air without any limitations imposed by the terrain. Transport speeds and accelerations/decelerations are high, without restrictions imposed by safety and traffic. The inherent dump height permits stacking loose material in windrows of sufficient storage volume to permit deep digging with simple intermittent machine moves. The walking device (propel system), combined with the fully rotating upper works, provides excellent maneuvering. The dragline production rate indicated in Table 1.14 is misleading and understates its capabilities unless it is recognized that the distances the material is moved are substantially higher than for the other machines.

Material handling in the mine involves excavating and transport. As noted above in the case of the dragline, some units transport the material

Table 1.14

APPLICATION GUIDELINES–EXCAVATORS

	FEL	Hydraulic Excavator	Electric Shovel	Diesel BWE	Dragline
Sizes available (cu. yd. of 3000 lb/LCY material)	3 to 24	3 to 30	15 to 50	—	10 to 180
Avg. production (cu. yd./hr. of 3000 lb/LCY material)	to 2000	to 1800 (hoe) to 2400 (shovel)	to 5500	to 3000	to 7500
Hard digging capability	fair	good	excellent	limited (no boulders)	good
Digging path	semi-fixed	variable	semi-fixed	fixed	variable
Maximum face height	low	high	high	medium	high
Dump height	marginal	good	good	good	excellent
Severity of discharge	easy	easy	can be severe	easy	can be severe
Segregation capability	good	excellent	fair	poor	fair
Mobility	excellent	good	poor	poor	poor
Required floor conditions	good	not critical	not critical	not critical	good (flat)
Influence of weather	high	low	low	low	low
Reliability	medium	medium	high	low	high
Crew size	one	one	1–2	1–2	1–3
Auxiliary equipment	none	none	dozer required	dozer required	dozer required
Operating cost	high	medium	low	low	low
Delivery time	short	medium	medium	long	long

Note: Gradations are relative to other units of the same general size.

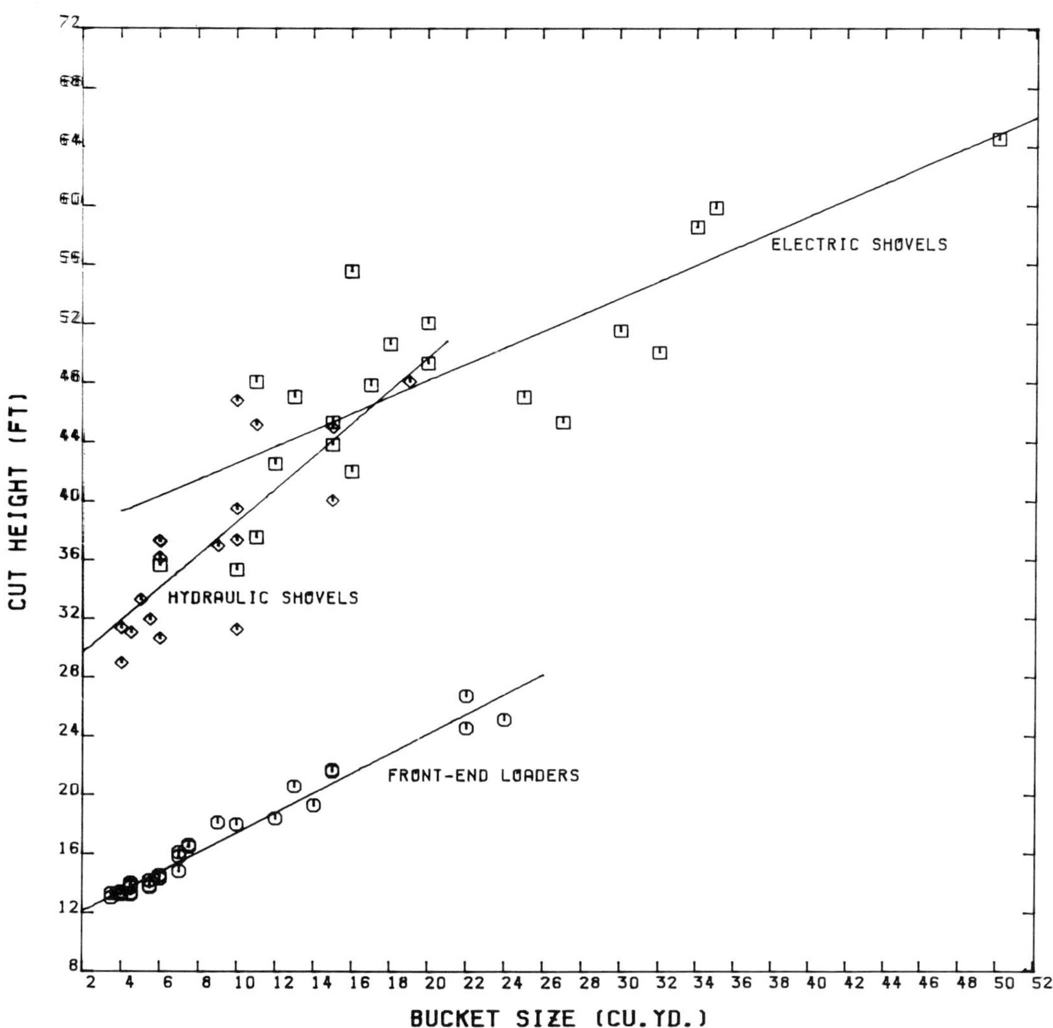

Graph 1.4 Cut height/nominal bucket size (loading machines)

in addition to excavating it. There are other types of equipment which only transport.

Transportation requirements involve the movement of material either to storage, processing or spoiling (waste dumps). It can be either continuous with conveyors or pipelines, intermittent with units such as trucks, or cyclic with direct disposal from a non-continuous excavator. The transport route can be through, around or across the pit. It is two dimensional in that both horizontal (lateral) and vertical (lift) distances must be considered.

The necessary characteristics of the transportation system are defined by the method of loading, the nature of the material, volumes, distances and grades, and the disposal method. Each excavator provides some material movement which must be recognized; in some cases, it will be adequate for the application. This is the simple solution since there is no rehandling required and the single machine offers the optimum in production availability. For greater distances, a combination of equipment is necessary. Table 1.15 lists the available transport options. Rail has been omitted, as noted earlier, because of its limited

Table 1.15

TRANSPORT DISTANCES

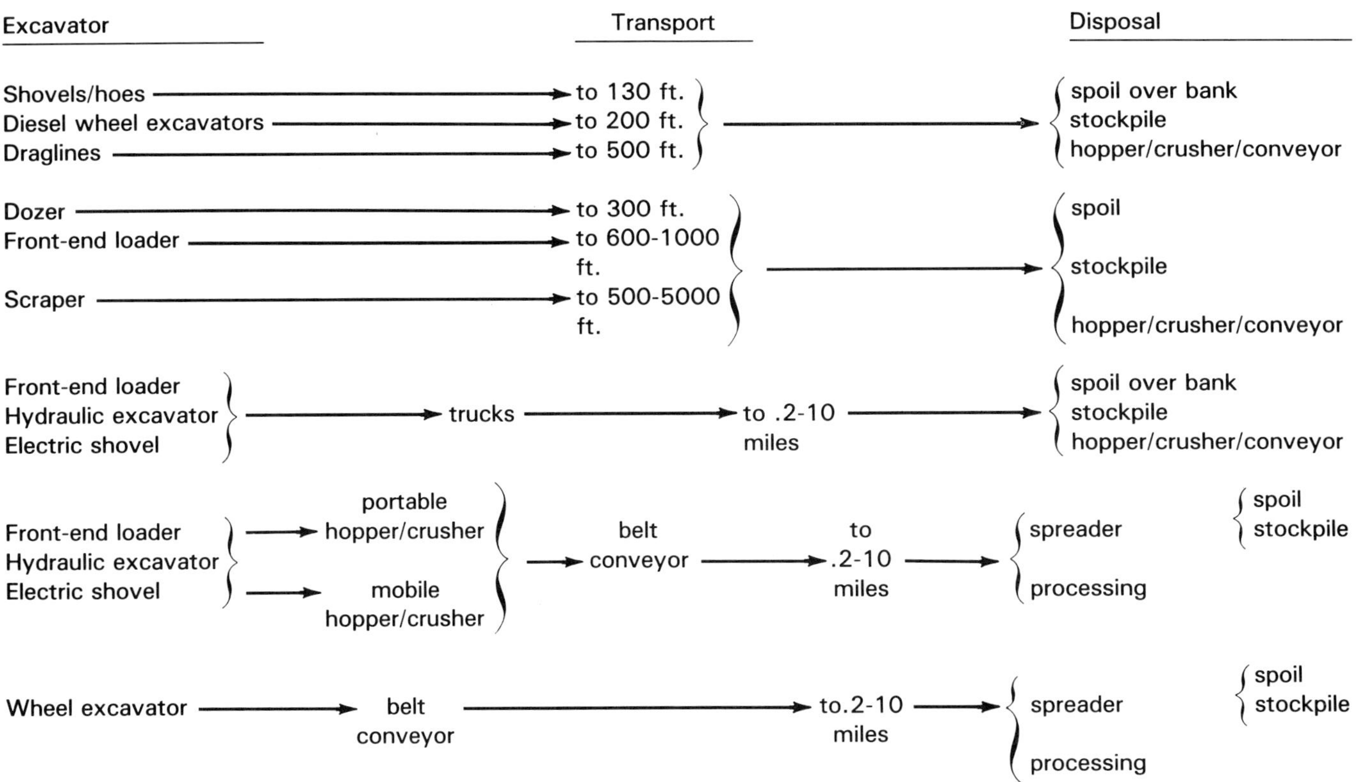

use in the newer mines. The trucks considered are the large off-road models typically used in surface mining.

The maximum transport distances indicated in the table are representative of mine "in pit" applications. There are no technical limitations for either the trucks or conveyors to preclude lengths double, or more, than the ten miles indicated. Mine layout and economics establish the maximum feasible distances.

Individual machine productivity for the excavator-loaders is keyed to the angle of swing (rotation) of the upper works. A 180 degree swing would maximize the distance the material is moved but this also maximizes the machine cycle time. Therefore, decreasing the distances increases productivity. In the typical application of this type of machine, the emphasis is placed on production. Note that only in the case of hoes, draglines and/or special wheel excavator designs is there any appreciable elevating of the material.

The combination machines — dozers, FEL's, scrapers — provide intermediate haulage capabilities. Speeds are moderate but, because of the power required for digging, they have high gradeability. Dozers have very limited range but can operate on very steep slopes. FEL's and scrapers can prepare and maintain their own haulroads, performing these operations on the empty-return portion of their cycle. Two engine scrapers have increased acceleration and hill climbing capabilities.

Truck haulage is common because of the mine planning flexibility it provides. High vertical lifts are possible but maximum grades must be restricted by truck capability and the accompanying reduction in travel speeds. Rear dump units can negotiate steeper slopes than the bottom dump models. Long uphill grades can materially increase fuel consumption. Haulroad maintenance is critical to controlling operating cost and productivity.

To-date, belt conveyors have had limited application in the larger U.S. surface mines. There is, however, growing interest because they offer the potential for high volume continuous haulage systems. Since they are not discussed in detail in any other chapter of this book, it is appropriate to summarize briefly some of their features.

- Can provide continuous haulage
- Capable of high capacities
- Can transport up or down 27% slopes
- High unit availability
- Quiet operation
- Can be automated
- Relatively safe
- Simple maintenance procedures
- Low electric power consumption, uniform demand
- Adaptable to rolling terrain
- Simple crossing of roads/railways with bridges or tunnels
- Individual flights must be relatively straight
- Maximum lump size restricted (1/4 to 1/3 belt width)
- Sticky materials can collect on belts and chutes
- Covers required in high wind areas
- Dust control equipment may be required at belt transfer points
- High initial investment

Individually powered conveyor sections (50 to 110 feet long) can be mounted on wheels and arranged so that the material cascades from one section to another. (See Figure 1.4) Skid mounted units made up of 15 to 20 foot modules with a continuous belt can be shifted laterally in steps to

Figure 1.4 Conveyor Modules

allow a segment of the belt to be advanced parallel to the digging face as the mining advances. (See Photograph 1.4) For sections remaining in a fixed location for a period of years, the structure is supported on permanent foundations.

Conveyor installations are uniquely tailored to the individual mine requirements and site conditions. At present, belt conveyors are not readily available as packaged units for general mining application, except for small mobile modules of low capacity that are utilized in sand and gravel processing and stockpiling activities. The basic components necessary to assemble units are, however, broadly available in a wide range of design configurations. A number of large consulting firms, together with some excavator manufacturers, will, on request, design a belt conveyor and provide all the specifications and supervise the erection.

The comparative characteristics of the four transport modes for distances beyond, roughly, six hundred feet are indicated in Table 1.16. The specific configurations selected for the scraper and truck would affect, primarily, the maximum grades and the materials that could be handled. A single engine scraper has less gradeability; the elevating type has a lower tolerance for large blocky material. The bottom dump truck is designed for less abusive, smaller sized, free flowing materials. For economic performance, all rubber tired equipment requires a haulroad surface free from obstructions and sharp abrasive rocks. Special tire guards, chains, etc. are available at extra initial cost to extend tire life under very severe conditions. Wet weather, icy conditions, sharp turns, and restricted visibility reduce maximum travel speeds.

So far the focus has been on individual pieces of equipment, with each assessed in terms of its application for excavating and/or transport. In most cases, the interest is in meshing a number of different types of equipment to form a mining system which performs a unit operation. Table 1.17 outlines in some detail the machines used in all of the common mining systems.

The system concept includes consideration of multiple machines being required for each phase of the operation. Equipment fleets introduce additional factors to be evaluated. Scheduling is required to coordinate, sequence and minimize delays and maximize equipment utilization. Adjacent activities must not interfere or block each other, or produce unsafe conditions. Traffic and congestion must not reduce productivity.

Photograph 1.4 Shiftable conveyor

Table 1.16

APPLICATION GUIDELINES–TRANSPORT

	FEL	Scraper	Truck	Belt Conveyor*
Avg. tonnages (T/hr): single unit	to 800	to 1200	to 2500	to 5000
Distances (ft)	50–1000	500–5000	1000 & up	6000 & up
Grades (%) loaded	to 12	to 12	5 to 10	to 27
Avg. speed (mph): loaded	5–10	15–25	20–35	6–14
Handling large blocky material	good	poor	good	poor
Haulroad requirements	prepared by loader	prepared by scraper	graded	none (service rd.)
Flexibility for length change	excellent	excellent	excellent	fair
Flexibility for route change	excellent	excellent	excellent	poor
Flexibility for production change	good (adjust fleet or schedule)	good (adjust fleet or schedule)	good (adjust fleet or schedule)	limited to scheduling
Wet weather constraints	reduced production or shutdown	reduced production or shutdown	reduced production or shutdown	minimal
Reliability	medium	medium	medium	high
Lost time (exc. service and repair)	road maintenance	road maintenance	—	conveyor relocation
Auxiliary equipment	none	push dozer in some cases	road maintenance equipment	equipment for periodic relocations
Ownership cost	medium	medium	medium	medium/high
Operating cost	high	high	high	low

*Shiftable conveyeyors and/or a series of modules (72 inch and smaller)
Note: Gradations are relative to other units of the same general size.

Table 1.17

PRIMARY MINING SYSTEMS

Front-end loaders loading trucks, with or without blasting or ripping

Hoes loading trucks, with or without blasting or ripping

Shovels loading trucks, with or without blasting or ripping

Draglines loading trucks, with or without blasting or ripping

Bucket wheel excavators loading trucks, with or without blasting or ripping

Front-end loaders loading hopper/crusher/conveyor,
with or without blasting or ripping

Hoes loading hopper/crusher/conveyor,
with or without blasting or ripping

Shovels loading hopper/crusher/conveyor,
with or without blasting or ripping

Draglines loading hopper/crusher/conveyor,
with or without blasting or ripping

Bucket wheel excavators loading conveyor,
with or without blasting

Dozer — Front-end loader loading trucks,
with or without blasting or ripping

Dozer — Front-end loader loading hopper/crusher/conveyor,
with or without blasting or ripping

Dozer — scraper, with or without ripping

Dozer — belt loader/conveyor, with or without ripping

Stripping dragline (direct disposal), with or without blasting

Stripping shovel (direct disposal), with or without blasting

Stripping bucket wheel excavator (direct disposal),
with or without blasting

Ideally, the units in a system should be matched so that there is a uniform, uninterrupted flow of material from the excavating face to disposal. This infers that all the units have the same production capacity, are dimensionally compatible, will handle the face materials and will function within the operating environment. Balancing production capacity is often possible by varying the work schedule of units that are not interdependent in their production operation. For example, drilling and blasting may be a single or double shift operation, while a dragline in the system might be scheduled for three shifts.

Machine range limits and output information is readily available to determine the optimum match of specific excavator-loaders with trucks and/or hoppers. Graphs 1.5 and 1.6 illustrate excavator dump height and truck loading height, two of the factors that must be considered in this analysis. Once individual machine sizes are established the number of each required can be found for the desired mine production level.

It is important to recognize that any system that is made up of a number of units which perform sequencial (or stepped) operations has a lower production availability than the individual units. If three machines, each with a mechanical availability of 90%, are employed in an integrated system, the total system availability could be as low as 90% × 90% × 90% = 73%. Similarly, the system can be paced by the slowest operator or slowest machine. Delays which throw part of the system out of balance ultimately impact on total system performance.

There are three approaches to compensating for these possibilities. The first is incorporating excess capacity into the system by oversizing all the

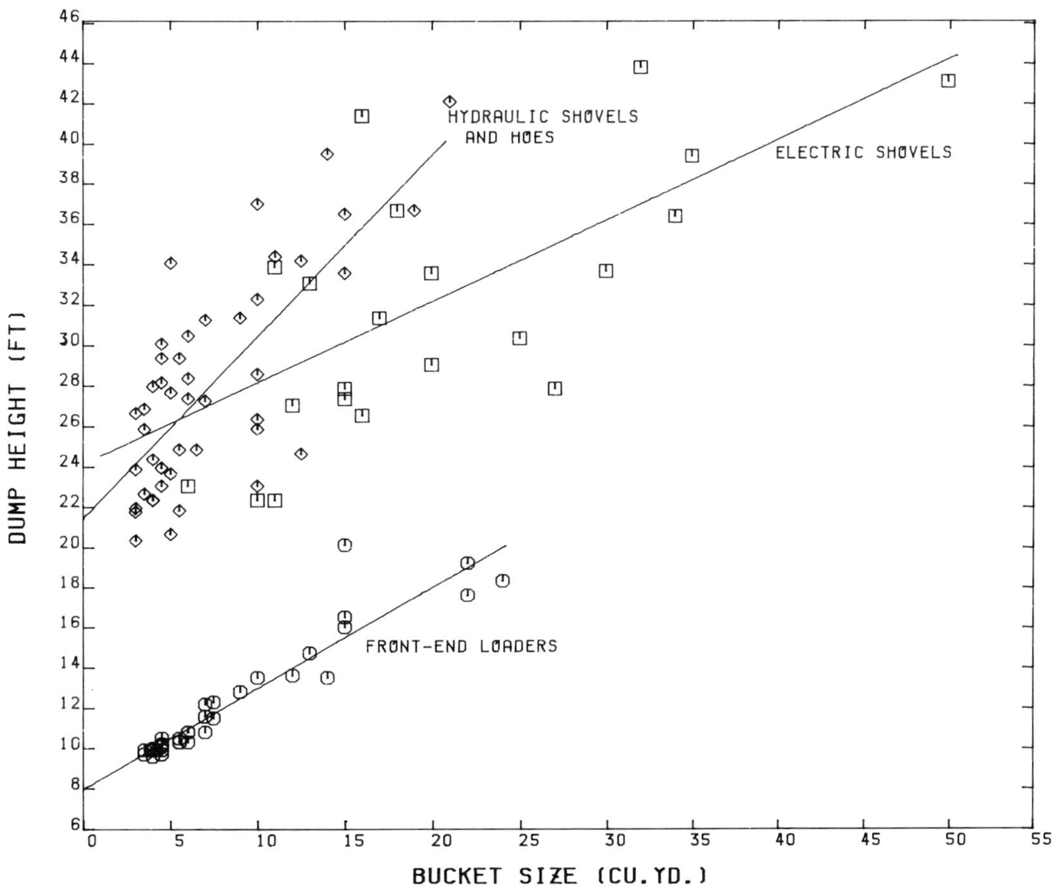

Graph 1.5 Dumping height/nominal bucket size (loading machines)

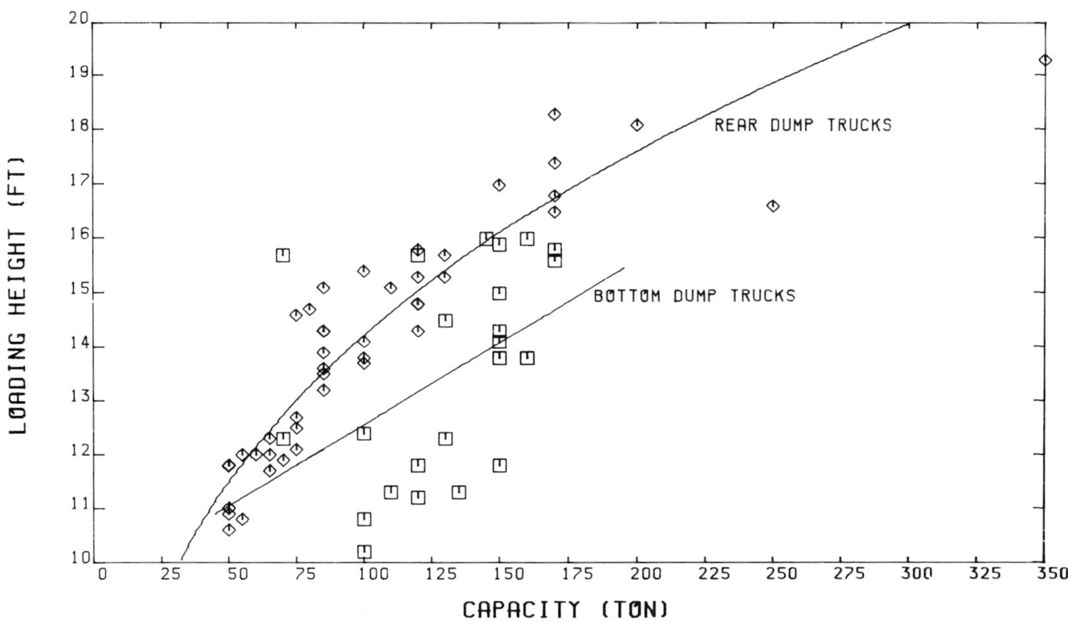

Graph 1.6 Loading height/ton of capacity (trucks)

units, so that anticipated lost time is offset by higher production during operating periods. The second is to provide back-up units for the most critical machines so that units out of service for extended intervals can be replaced. Such unit interchange is fairly simple with highly mobile equipment, but may involve shifting to alternate face locations if the equipment cannot be readily relocated. Back-up units simplify rotation of machines for major maintenance. The third technique is to incorporate surge piles wherever possible between each operation so that this temporary storage can take up the slack when units get out of phase. If this is possible, balancing can also be achieved with unit scheduling. The penalty in this case is the extra rehandling created when material is withdrawn from the surge pile. All of these corrective measures add to the initial investment and/or operating costs.

Detailed review of each of the factors influencing system selection is beyond the scope of this book. The interactions are numerous, complex and interwoven with site conditions and mining objectives. Equipment limitations and capabilities, however, play a key role. Some insight into the scope of the problem can be gained by identifying the basic systems and their characteristics. Table 1.18 attempts to provide an overview and, in particular, to point out some of the considerations in the decision process. The gradations indicated between alternative systems is subjective and can be challenged but are intended to stimulate thinking.

A number of systems are often employed simultaneously to perform different unit operations at the mine. In addition, several systems may be combined in a single operation. The best example of this latter is the current trend to use a separate system to prepare a working bench for a dragline when the overburden depth exceeds the machine's capabilities. The material removed in the benching must be overlaid on the dragline spoil piles. All of the other systems can meet these requirements, depending on the materials to be handled and the preferred method for transporting and disposing of it.

There are special problems related to the disposal of overburden/waste. Ultimately it is anticipated that mining areas will be reclaimed and revegetated. The recontouring of the overburden/waste and subsequent planting is a significant mining cost. To minimize this outlay, the overburden/waste should be initially deposited as close as possible to the planned final contour, with any toxic materials buried. Ideally, the

Table 1.18

SELECTION GUIDELINES (systems)

	Dozer—Front-end Loader	Dozer—Scraper	Dragline (direct casting)	Excavator—Truck	Excavator—Hopper—Crusher—Conveyor	Wheel Excavator—Conveyor
Maximum production	medium	medium	high	high	high	high
Production rate	medium	low	high	medium	medium	high
Pit life	short	short	long	medium	long	long
Pit depth	medium	flat and shallow	medium	deep	deep	medium
Deposit	unconsolidated	unconsolidated	consolidated	consolidated	consolidated	uniform no large boulders
Preparation (if required)	ripping	ripping	drill and blast	drill and blast	drill and blast	drill and blast
System complexity	low	medium	low	medium	high	high
Operational flexibility	high	medium	low	high	low	low
Blending capability	high	high	low	medium	low	low
Selective placement (disposal)	good	excellent	poor	good	medium	medium
Wet weather impact	high	high	low	medium	low	low
Scheduling requirements	low	high	low	high	medium	medium
System availability	medium	medium	high	medium	low	low
Support equipment	low	low	medium	medium	high	high
Ease of start-up	simple	simple	moderate	simple	comples	complex
Investment	low	low	medium	medium	high	high

material would be deposited in horizons duplicating the original formation. The different systems meet these objectives to varying degrees, depending on the mine plan and the sequencing of the operation. Scrapers can excavate and spread horizon by horizon. Excavator-loaders, combined with trucks, can deposit the material in thick layers with the detrimental portions placed the deepest. Direct spoiling draglines, digging deep, have limited ability to segregate or deposit in the preferred manner.

A dozer/ripper, working in combination with a front-end loader (see Photograph 1.5), is a low investment, simple system with a great deal of flexibility. The FEL can load directly into trucks/hoppers or carry for a short distance to spoil directly or dump into a centrally located hopper. The dozer performs two functions: it intermittently rips to prepare the material; and it pushes the material to the loader. The "pushing", usually down grade, further breaks up the material and reduces the height of the loader digging face, improving the loader's productivity. In many cases, multiple dozer/rippers can be teamed with a loader if the conditions are adverse. Such practices are commonly found in sand and gravel operations and are particularly attractive if the material leaves the pit on a belt conveyor fed by a common, conveniently located hopper.

The primary production application of the dozer/ripper and scraper system is a fairly recent transplant from heavy construction to mining. (See Photograph 1.6) The key was the development of operating techniques which made deeper removal economical. Area mining methods have been outlined in recent publications which attack the problem by coupling the efficient short downhill material movement of the dozer with a very short circular dig-dump scraper circuit. The dozer periodically rips and the scraper digs and dumps on downhill legs of the route. The sequencing gets progressively more complex with increasing mining depth. Relatively large fleets of equipment are required.

Photograph 1.5 Dozer — front-end loader — truck operation

Photograph 1.6 Dozer — scraper fleet removing overburden

One of the most broadly used systems for overburden removal in the U.S. in past years has been the dragline. It is particularly popular because of its simplicity, high volume output, and deep mining capability. A wide variety of dragline side casting techniques have been developed to fit mining depths, pit widths, single or multiple seams, haulroad requirements. etc. These necessitate rather sophisticated application studies and careful planning of machine maneuvers and pit dimensions. Drilling and blasting generally precede the dragline sufficiently to permit independent scheduling and avoid delays. The entire dig-transport-dump cycle is combined in a single machine, maximizing reliability. The risk associated with dependence on a single machine is reduced by a basic design which incorporates multiple parallel drives.

In benching type operations, excavator-truck systems offer a high degree of mining flexibility (Photographs 1.7, 1.8, and 1.9), together with a wide selection of equipment to fit site conditions that range from easy to tough digging. Blending is achieved by mixing the output from selected faces. Three to five trucks are normally assigned to each excavator so that relatively large truck fleets are involved. Careful scheduling and traffic control are essential for optimum machine utilization. Operations are shut down during blasting. Truck haulage efficiency is reduced on steep grades and during bad weather, night time or periods of limited visibility.

Systems in which the excavated material is transported continuously to final disposal have been applied on a limited basis in the United States, but extensively overseas. The overseas arrangement combines a bucket wheel excavator with a conveyor through or around the pit. These systems are subject to the limitations of the wheel excavator, which are, primarily, the inability to

Photograph 1.7 Hydraulic hoe and front-end loader — team loading coal into rear dump truck

Photograph 1.8 Hydraulic shovel loading rear dump trucks

Photograph 1.9 Electric shovel loading rear dump trucks

handle large boulders and/or blocky materials in the digging face. The major components in the total system are:

- Wheel excavator
- Mobile bridge conveyor
- Hopper car (riding on belt structure)
- Belt conveyors
- Tripper car
- Spreader (stacker)

The belt conveyor consists of one to three independently powered flights, some or all shiftable so that they can be periodically relocated as mining progresses. For systems of higher capacity, additional special transport units are needed to assist in the shifting of the heavy combination belt drive and transfer stations. Since the wheel is propelling intermittently as the digging face is advanced and the belt conveyor can only be practically shifted periodically (every few months), additional equipment is required to provide an extending link between the excavator and the pit belt. The mobile bridge conveyor performs this function. (See sketches in chapter on wheel excavators and Photograph 1.10) Material on the belt is removed by the tripper car which can travel (similar to the hopper car) on the conveyor supporting structure. A mobile spreader takes the material from the tripper car and discharges it directly to spoil or into stockpiles. Generally, the spreader for waste materials is capable of distributing or spreading the material in layers around an arc to facilitate subsequent leveling. System availability is directly tied to the reliability of all of the units in the chain. There is no intermediate surge capacity at any stage of the process. All of the units are relatively complex and expensive. (See Photograph 1.11)

Photograph 1.10 Medium size bucket wheel excavator working with a mobile transfer conveyor and a shiftable conveyor

Photograph 1.11 Medium size bucket wheel excavator removing top layer of overburden in advance of large stripping shovel.

There has been growing interest in continuous haulage arrangements which would incorporate the more versatile excavator-loaders. The large buckets on these units allow them to handle boulders, frozen lumps, and blocky materials of a size which could damage (shorten the life of) a matching belt conveyor. Likewise, the cyclic discharge has to be smoothed out before it can be loaded onto the belt. These factors suggest that some form of a hopper and a crusher (if oversized pieces are anticipated) are necessary as an interface between the two. (See Photograph 1.12 and 1.13) This unit must be mobile so that it can follow the excavator, continuously maintaining an optimum position for fast excavator dumping. Big buckets and large blocks of material necessitate large hoppers and crushers. Crushing should be limited only to the oversized material. The units connecting the hopper/crusher into the rest of the haulage system are the same as those described for the wheel excavator. Some design concepts incorporate a swingable discharge conveyor boom as a part of the hopper/crusher unit, which eliminates the need for a mobile bridge conveyor. A limited amount of surge capacity is built into the system with the material volume in the hopper.

It should be noted that there are other variations of the continuous haulage system which require customized equipment beyond the scope of this book. Falling into this category are mobile cross-pit conveyors and steep angle conveyor units.

Photograph 1.12 Front-end loader — portable crusher — conveyor system

Photograph 1.13 Walking dragline — mobile hopper — conveyor system

MACHINE RELIABILITY

The evaluation process this presentation has been following is outlined in Table 1.19. The next aspect to consider is machine reliability.

Judgement decisions on the reliability of each piece of equipment are vital to the equipment selection process input. They significantly impact both unit and system productivity. Further and perhaps of equal importance, however, is the insight they give of the design quality of the machine. As existing basic designs are progressively refined with new technology, this could well be one of the future key determinants in equipment selection.

Reliability as an evaluation yardstick is growing in importance because of:

- Increasing machine complexity
- Increasing maintenance skill requirements
- High cost of the equipment
- High cost of the loss time and/or standby units
- Variations between alternate types of units
- Actual performance below expectations
- Broader recognition of its economic impact on the overall operations

The equipment manufacturers have recognized this expanding role and have taken some steps in response to this emphasis.

- Increased on-board diagnostic equipment
- Maximizing service accessibility
- Greater use of modular and interchangeable assemblies
- Providing quick inspection and replacement capabilities

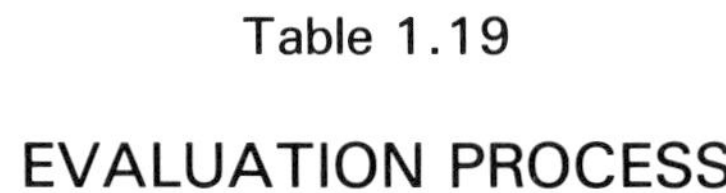
Table 1.19

EVALUATION PROCESS

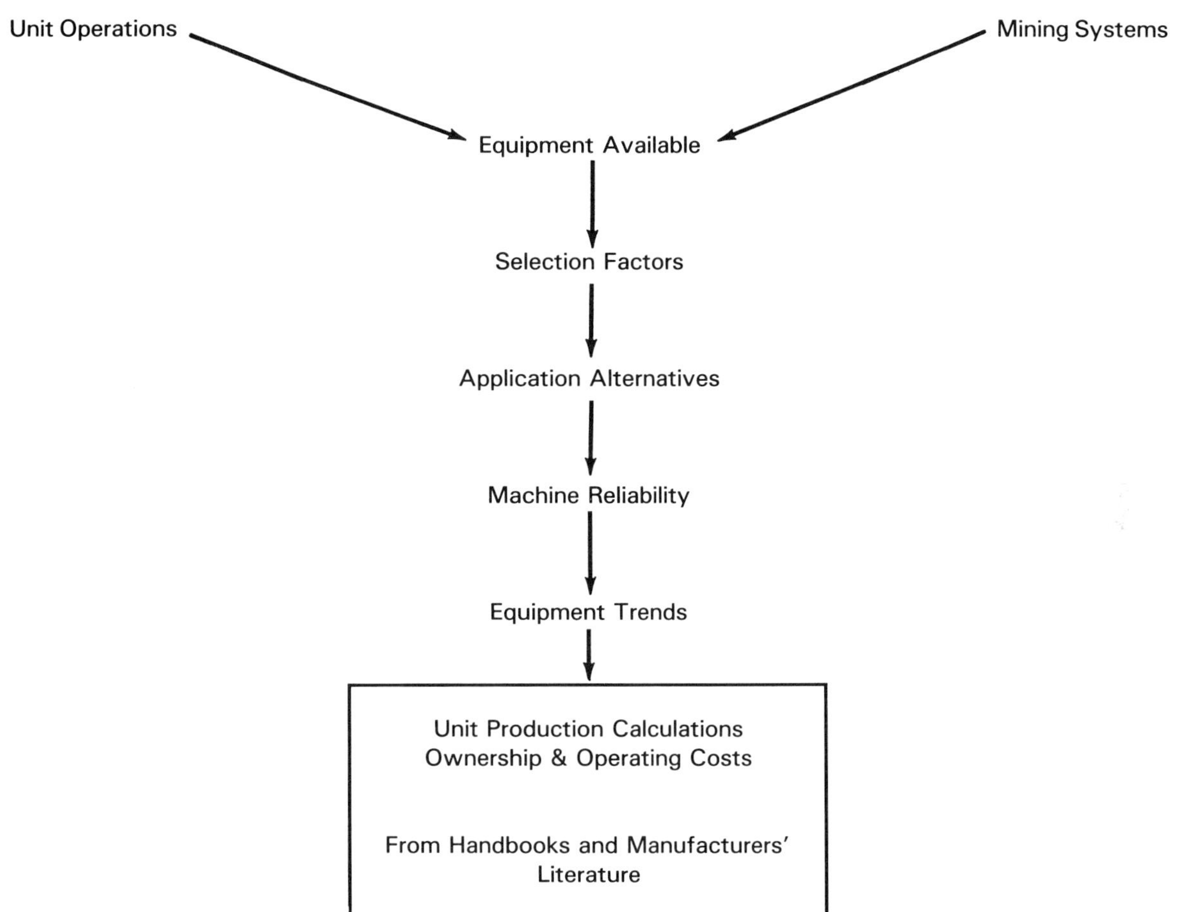

- Incorporating failure check systems
- Expanding manufacturer/dealer training aids and support programs

Reliability estimates have been based primarily on historical experience. To be meaningful, this experience must be well documented and based on the same equipment operating under similar conditions for a reasonable period of time. Rarely can these stipulations be fulfilled. In addition, a significant proportion of the current models incorporate major new technology and componentry with insufficient field service to verify actual performance. Some organizations have insufficient historical records to project performance for the new, larger equipment now available. Others are considering mining systems and equipment for which they have no prior experience. For these situations, broad industry averages must be used or a rating scheme devised.

Table 1.20 illustrates typical data at one mine for three similar front-end loaders. From this information, some justification could be found for picking availability (reliability) values ranging from 63% to 85%. More detailed statistical data often seems only to further complicate this problem. If similar data is used for different types of machines for comparative purposes, the possibility for error increases.

There are many important considerations which affect past performance that are not reflected in most statistical data. Further, the historical machine records seldom include adequate back-up information to permit extrapolating the indicated performance into the future. Some feel of the value of the information available can be obtained by attempting to answer questions such as the following.

- Is it accurate?

- Does it cover an equivalent machine to that under consideration?
- How does one recognize deterioration with age?
- How does one identify technical obsolescence?
- How does one identify the nature and frequency of abuse?
- How does one identify the nature and frequency of unusual operating requirements?
- How does one identify frequency of servicing and maintenance?
- How does one identify design deficiencies?
- How does one identify loss time due to external factors (such as vendors)?
- How does one identify delays, waiting for parts, maintenance equipment or maintenance personnel?
- How does one identify delays resulting from operator training?
- How does one identify reduced performance during operations at partial power, etc.?

One of the most difficult areas to assess is the extent of machine abuse. Frequently extreme cases are documented, but accumulative damage from severe service conditions is not. It is apparent that, at best, this overall approach provides a very rough estimate of limited value for comparative evaluation of alternative equipment.

With rapidly advancing technology and the broader range of equipment being utilized, it is increasingly difficult to equate the various types

Table 1.20

FRONT-END LOADER–AVAILABILITY STATISTICS

Machine Purchased April 1972

	Operator hrs	Repair hrs	Mechancial Availability
Day	14	2	87.5
Month	59	64	48.0
Year	2,350	794	74.7
Life to date	23,553	10,156	69.9

Machine Purchased March 1974

	Operator hrs	Repair hrs	Mechanical Availability
Day	0	14	0.0
Month	7	35	16.7
Year	1,472	872	62.8
Life to date	21,910	6,852	76.2

Machine Purchased May 1977

	Operator hrs	Repair hrs	Mechanical Availability
Day	15	1	93.7
Month	106	15	87.6
Year	2,499	536	82.3
Life to date	6,831	1,241	84.6

of equipment. As noted earlier, the equipment varies substantially in configurations. The extent of this can be shown using as an example the powered functions on all machines.

(2) DOZER—Propel, blade positioning

(3) TRUCK—Propel, dump, steering

(4) SCRAPER—Propel, bowl positioning, ejecting, steering

(4) FRONT-END LOADER—Propel, bucket positioning, wristing, steering

(4) SHOVEL—Propel, swing, hoist, crowd

(4) DRAGLINE—Propel, swing, hoist, drag

(6) HYDRAULIC EXCAVATOR (shovel)— Propel, swing, boom positioning, bucket arm positioning, wristing, dumping

(8) WHEEL EXCAVATOR—Propel, swing, ladder positioning, wheel drive, ladder conveyor, discharge conveyor, discharge swing, discharge boom positioning

(8) DRILL—Propel, mast positioning, jacks, rotary drive, pull down, bailing, dust control, pipe handling (rack & wrench)

There is a four to one range in the number of powered functions. Similar spreads would be found if other design features were summarized.

The quality of the reliability decisions would be enhanced if the estimate for a new unit were based on present and future equipment capabilities and site requirements. Relative ratings can be established by totaling points assigned to critical elements with a grading system. Each contributing factor can be assigned a proportionate value relative to the other factors under consideration. This is a subjective approach so that the absolute numerical values are meaningless. But if the procedure were applied consistently, they might be used effectively to compare different types of equipment.

The two predominant factors which appear to dictate reliability are application severity and machine complexity. If these are reduced to a numerical rating, the application severity would provide an index to compare operating conditions and requirements. The machine complexity would provide an index to compare basic design reliability/maintainability. Elements to consider for rating application severity are outlined in Table 1.21. Point assignments to these would vary dependent on the range of conditions at the mine site.

Table 1.22 illustrates a rating scale for machine complexity. Eight design features were selected for evaluation with a low (1 point), medium (2 points), or high (3 points) designation assigned to each of the common variations. Adding up the points for any specific type of machine established the rating. This scale, if applied to a typical dozer and a wheel excavator, would provide the following ratings shown in Table 1.23.

It is debatable whether or not a useful analytical approach such as this can be developed. The brief discussion, however, points up the level of understanding required to upgrade present practices.

Table 1.21

MACHINE APPLICATION SEVERITY

Environment
- Temperature range
- Rain & snow conditions
- Altitude
- Wind velocities

Floor & haulroad conditions
- Preparation
- Abrasiveness
- Grades
- Traction/rolling resistance

Face conditions
- Degree of fragmentation
- Hardness/abrasiveness
- Percent oversize

Move frequencies–distance

Operating schedule (shifts & days per week)

Maintenance/servicing
- Schedules
- Quality
- Facilities
- Personnel

Operation
- Supervision
- Operators

Table 1.22

COMPLEXITY RATING–SCALES

Power source	
• Diesel	High
• Electric	High
Primary power distribution	
• Hydraulic cylinders, pumps (low pressure)	Low
• Hydraulic cylinders, motors, pumps (high pressure)	Medium
• Torque converter/transmission	High
• D.C. motors	High
• A.C. motors	High
• Hydraulic and electric motors	High
Number of powered functions	
• Less than three	Low
• Four	Medium
• More than four	High
Control system	
• Mechanical linkage, simple hydraulic valve	Low
• Static electric	Medium
• Ward Leonard electric	High
• Variable displacement hydraulic	High
Basic frame or structure	
• Single unit	Low
• Two units articulated	Low
• Two units rotatable	Medium
• Three units rotatable	High
• Four units rotatable	High
Excavating linkage system	
• Cable and drum	Low
• Cylinders and arms	Medium
• Rack and pinion	High
• Digging and discharge booms	High
Excavating means	
• Blade	Low
• Bucket	Low
• Fixed dipper	Low
• Wristing dipper	Medium
• Bucket wheel	High
Propel mechanism	
• Wheel	Low
• Walking device	Low
• High speed crawler	Medium
• Independent track drive	High

Table 1.23

ILLUSTRATIVE EXAMPLE

Dozer Design Complexity

	Type	Relative Complexity
Power source	Diesel engine	High (3)
Primary power distribution	Torque converter-transmission	High (3)
Number of powered functions	Three	Low (1)
Control system	Mechanical linkage, low pressure hydraulics	Low (1)
Basic frame	Single unit	Low (1)
Excavating linkage system	Cylinder & arms	Medium (2)
Excavating means	Blade	Low (1)
Propel mechanism	High speed crawler	Medium (2)
		Rating–14

Medium Size (2 Crawler) Bucket Wheel Excavator Design Complexity

	Type	Relative Complexity
Power source	Diesel	High (3)
Primary power distribution	Hydraulic cylinder, hydraulic and electric motors	High (3)
Number of powered functions	Eight	High (3)
Control system	Variable displacement hydraulics	High (3)
Basic frame	3 units rotatable	High (3)
Excavating linkage system	Digging & discharge booms	High (3)
Excavating means	Bucket wheel	High (3)
Propel mechanism	Independent track	High (3)
		Rating–24

EQUIPMENT TRENDS

Selecting equipment for mine start-up two to five years later, and with anticipated service extending from five to twenty five years into the future, requires careful assessment of the possible impact of new concepts and technology. Mining costs must remain competitive throughout the life of the equipment. This translates into the need for competitive machine productivity and owning and operating costs during this same period.

Projected machine life is bracketed between two extremes: one, the time until the unit will no longer function properly or is impractical to maintain; the other, the time until the ownership and operating cost per unit of production is no longer equal to or less than that of any other machine which can satisfy the mining requirements. The latter assumes that an economic justification can be made for replacement of the existing unit and, of course, that the mine management will make the required investment at that time. The better capabilities of the replacement unit result either from deterioration of the existing machine, changed mining conditions and/or new product technology. While the technology factor has been recognized, it has been largely ignored to date because of the difficulty in forecasting future developments. Service life estimates have been based on tax regulations and/or past general industry experience.

Design of mining equipment has always been a difficult task. The markets, in many cases, are relatively small so that only a limited amount of research can be justified. Site conditions vary so greatly that a considerable amount of customized design has been required, particularly for the larger, high investment units. The machine population is widely dispersed and in remote areas, making engineering follow-up expensive and time consuming. The critical need for reliability has forced conservative approaches, and some new technology is accepted only after it has been proven for extended periods. The physical size of the componentry necessitates major investments in production facilities with each design change. To these limitations must be added the general design criteria that is required in all product development. (See Table 1.24)

Table 1.24

GENERAL MACHINE DESIGN CRITERIA

Performance specifications
- Size
- Range
- Cycle time
- Mobility
- Weight
- Power source
- Crew size
- Operating environment

Innovative

Maximize new technology

High reliability

Maximize productivity (include automated functions)

Comply with codes and standards

Avoid patent infringement

Effective man-machine interface

Safe operation — servicing

Serviceable (include diagnostic elements)

Maintainable (accessible)

Permit future update modules

Clean appearance

Competitive product life

Minimum machine cost

Minimum cost for delivery and assembly

Minimum power consumption — operating cost

Matched to manufacturing facilities & capabilities

Match to distribution requirements

Requirements for surface mining equipment have been changing rapidly as increased effort has been directed towards energy and natural resource development. The inflow of new people into mine management has resulted in more challenging of past practices and a greater willingness to try new techniques. External factors, such as the following, are re-shaping future equipment needs.

- Larger mines handling substantially higher volumes of material
- Thicker coal seams which drastically shift the ratio of ore to overburden
- Deeper deposits which frequently necessitate combined equipment systems
- Lower quality ore/coal reserves that require blending and minimal contamination while maximizing recovery
- The necessity for moving large volumes of material for final recontouring
- Increased emphasis in overburden removal planning to use systems which minimize later rehandling for reclamation
- Special procedures required for the removal, storage, and replacement of topsoil
- Special procedures for fertilizing, planting, and watering restored areas
- Energy conservation — fuel or power consumption
- Availability of diesel fuels
- Voluminous, ever-changing federal and state governmental regulations, constraining operational practices

Responding to the new equipment demands is not as easy as in the past. The manufacturer does not have the same flexibility; the risks and product support requirements are substantially expanded. The pressure for sophisticated new products is across entire product lines, increasing the total commitment needed. Current conditions have further compounded these problems.

- Increasing number of suppliers, international in scope, involving small to huge organizations
- Substantially increased competition between different types of equipment which will perform the same function
- New resource/energy developments which have significantly different machine application requirements
- Energy conservation must be incorporated wherever possible
- Government regulations introduce severe design and application restraints, particularly with respect to safety
- Excessive costs of new model introduction and product liability
- Ever increasing capital investment requirements for major equipment is placing renewed emphasis on feedback control systems aimed at optimizing and monitoring machine performance
- Increasing labor costs and the difficulty in finding skilled operators is forcing a progressively higher degree of automation in equipment operation
- Performance refinements are resulting in increased machine complexity and a corresponding need for serviceable/maintainable configurations with high operational availability
- Increasing prices are forcing greater consideration of means to reduce the lost time associated with routine servicing/maintenance activities
- There is an expanding range of design alternatives
  - Diesel vs. electric power
  - Mechanical vs. hydraulic vs. D.C. electric vs. A.C. electric drives
  - Track vs. wheel propels
  - Bucket vs. wheel
  - Cable vs. hydraulic linkages
  - Metal vs. fiberglass

Recent design trends of specific types of equipment are discussed in subsequent chapters. There is some overlap so it is helpful to bracket the different areas of activity and identify the basic component areas under development.

- Buckets
  - — Ejector and bottom dump
- Propel systems
  - — Bogies for crawlers
  - — Independent crawler drives
  - — Larger tires
- Drive systems
  - — Larger mechanical transmissions
  - — Static A.C. drive systems
  - — High pressure, multi pump/motor hydraulic systems
- Control systems
  - — Automated systems
  - — Summation and constant horsepower hydraulic circuits
  - — Diagnostic systems using microprocessors

One continuing trend has been the application of multiple drive systems, so that many units have a combination of mechanical, electrical and hydraulic drives. There has been a great deal of up grading in the operator environment (including noise reduction) and physical arrangements to speed all servicing activities. Considerable progress has been made in reducing the frequency of servicing and providing module interchanges for major repairs.

New models have been introduced which overlap current units—but with basic configuration changes, such as:

- Articulated rear dump trucks
- Two axle bottom dump trucks
- Cross body bottom dump trucks
- Superfront electric shovels

These are essentially the repackaging of known technology which provides some operational improvements. Into this category falls the vertical auger-type scraper now being evaluated.

The prime development emphasis over the last twenty years has been for larger sized machines. "Big was beautiful". Experience has shown that this is not an assured path to reducing cost per unit of productivity. High investments, reduced reliability and support requirements restrict the supersized machines to very special mining situations. Table 1.25 attempts to summarize the probable future developments in machine size in the next five years. New, larger models of dozers and hydraulic shovels have been introduced within the last year.

The role continuous excavators will have in future mining projects is particularly difficult to project. Much discussed for many years, they have made only a small inroad into the United States mining scene. The medium sized units on two crawlers appear to be attracting more interest, but few proven standardized designs are available in mining sizes.

Small "surface miner" units which take a 1/2 to 12 foot cut with a drum or wheel cutting head, while propelling, are being used in coal. The available machines are produced by small manufacturers and differ significantly in their configurations. The U.S. Bureau of Mines has supported some research to advance this art.

Those making large investments in equipment must consider the potential future impact of new technology from other fields. Of keen interest are any new concepts, machines, or componentry which could substantially upgrade equipment operational capability and performance. Major innovations are infrequent but the combinations of new components and systems often can produce the same results. The data available for making an assessment in this area is fragmented. Little has been published that consolidates the many technological areas and relates them to the unique requirements of surface mining equipment.

A rough overview of possible developments over the next five years can be made by identifying deficient areas, advances in other fields, and promising studies and tests. This approach suggests the following:

- Gaps in componentry
  - — Larger mechanical transmissions
  - — Larger hydraulic pumps and motors

Table 1.25

SIZE TRENDS–NEXT FIVE YEARS

	Probable Future Development of Larger Models	Notes
Trucks Draglines Stripping BWE	None	Supersized units in service
Scrapers Blast hole drills	Marginal	Present maximum size adequate
Dozers (track) Front-end loaders Hydraulic shovels Electric shovels	Low	New larger models: accumulating operating experience
Surface miner type wheels Loading BWE (two crawler machines)	Good	Applications developing: few available machines

Additional new larger models as U.S. manufacturers top out product lines

- More rugged television equipment

- Areas where a need is recognized
  - Improved fuel economy
  - Improved cable life
  - Improved lubricants
  - Improved tire life
  - Improved wear-resistant materials

- Applications of new electronic and computer technology
  - Performance monitoring systems
  - Load sensing feedback control systems
  - Data recording systems

- Areas where studies and tests indicate that a potential exists
  - Trolley assist systems for electric trucks
  - Buried cable guidance systems for vehicles
  - Mobile hopper/crushers
  - Steep angle conveyors
  - Cross pit conveyors
  - Vibratory assist for cutting edges

Many of these are continuations of current efforts. Some concepts which have been suggested in the past are omitted because of the lack of interest shown by the industry.

Chapter 2

DOZERS

Figure 2.1 Typical Crawler Dozer With Ripper

The dozer, or bulldozer, is a crawler or wheel driven tractor with a front mounted blade for digging and pushing material. The dozer is one of the old-timers in terms of equipment employed in mining and on large construction jobs around the world. The concept and utilization have not changed greatly through the years but there have been marked refinements in the design and significant increases in the size of the units available.

Track type tractors with numerous attachments are extremely versatile and can be incorporated into virtually every phase of surface mining. While these units retain their dominant position, the large wheel units are making inroads. Wheeled machines historically have been used in less severe applications.

TYPICAL UNITS

As illustrated in Figure 2.2, there are two basic types of dozers; the crawler dozer and the wheeled dozer. While they are distinguished by the difference in propel system (crawler vs. wheel), there are usually other major design differences. For example, the crawler unit has a heavy rigid frame and the wheeled unit has an articulated frame (often a variation of the manufacturer's front-end loader product line). The primary operating difference between these two types is the force they can apply to digging, their ability to maneuver under sloppy ground conditions, and overall mobility to move between job sites.

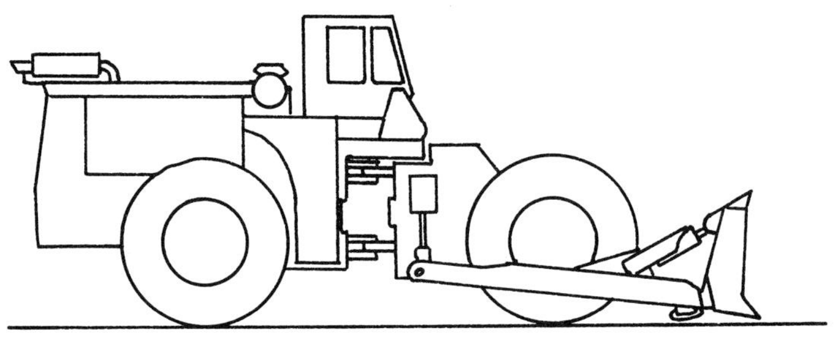

WHEEL DOZER

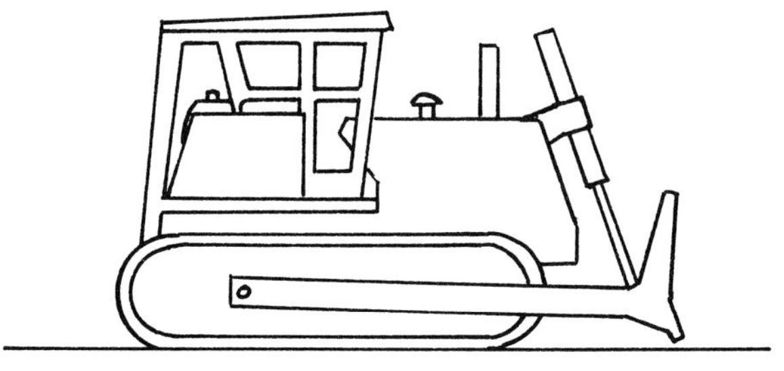

CRAWLER DOZER

Figure 2.2 Dozer Types

Crawler Mounted

- Rigid frame
- Maximum speed from 5 to 9 mph
- Horsepower from 140 to 700, diesel powered
- Mechanical drive system
- Machine weights from 30,000 to 190,000 lb.
- Ground bearing pressures 7 to 22 psi
- Drawbar pulls from 47,000 to 242,000 lb.
- Gradeability to approximately 45 degrees

Wheel mounted

- All wheel drive
- Articulated frame with 40 to 45 degrees steering
- Maximum speeds from 10 to 40 mph
- Horsepower from 170 to 820 with diesel engine
- Mechanical or electric drive system
- Machine weights from 41,000 to 214,000 lb.
- Ground bearing pressures 30 to 50 psi
- Rimpull from 40,000 to 182,000 lb.

Manufacturers

The market is primarily orientated about the U.S. manufacturers, but Komatsu has provided effective competition. (See Table 2.1) Most of the companies also market other mobile equipment in the same class for the mining/construction market. Sales and service are conducted through a dealer organization.

Table 2.1
LIST OF DOZER MANUFACTURERS

J.I. Case Company Construction Equipment Div. 700 State Street Racine, Wisconsin 53404	Crawler mounted 40–140 FWHP 7,500–30,000 lbs. wt.
Caterpillar Tractor Co. 100 NE Adams Street Peoria, Illinois 61629	Wheel mounted 170–300 FWHP 40,000–70,000 lbs. wt. Crawler mounted 60–700 FWHP 14,000–190,000 lbs. wt.
Clark Equipment Company Construction Machinery Div. Pipestone Rd., P.O. Box 547 Benton Harbor, MI 49022	Wheel mounted 270–570 FWHP 68,000–130,000 lbs. wt.
Deere & Company John Deere Road Moline, Illinois 61265	Crawler mounted 40–150 FWHP 14,000–37,000 lbs. wt.

Table 2.1 (Continued)

DOZER MANUFACTURERS

Fiat-Allis Construction Machinery, Inc. Box F, 106 Wilmot Road Deerfield, Illinois 60015	Crawler mounted 90–520 FWHP 20,000–115,000 lbs. wt.
International Harvester Construction Equipment Group 600 Woodfield Avenue Schaumburg, Illinois 60196	Wheel mounted 600 FWHP 140,000 lbs. wt. Crawler mounted 100–300 FWHP 28,000–72,000 lbs. wt.
Komatsu Limited Komatsu Building 2-3-6 Akasaka Minato-Ku Tokyo, Japan 107	Crawler mounted 40–1000 FWHP 7,500–265,000 lbs. wt.
Marathon LeTourneau Co. Longview Division P.O. Box 2307 Longview, Texas 75606	Wheel mounted 800–820 FWHP 193,000–215,000 lbs. wt.
Melroe Multi-Wheel P.O. Box 1059 Longmont, Colorado 80501	Wheel mounted 670–750 max HP 165,000–195,000 lbs. wt.
Terex, Corp. IBH Group Hudson, Ohio 44236	Wheel mounted 280 FWHP 45,000–50,000 lbs. wt. Crawler mounted 90–370 FWHP 23,000–82,000 lbs. wt.

BASIC MACHINE OPERATION

The dozer is one of the simplest pieces of equipment in terms of its basic operations and operator training. It is used to both excavate and transport material over short distances.

- The relatively high speed crawlers or rubber tired mounting generally provide good mobility for propelling the machine to sites anywhere on the property. Auxiliary carriers may be required for long distance or on highway transport.

- Grading and/or excavating is accomplished by forcing the dozer blade into the ground surface and using the propel power to excavate and push the material.

- Dependent on the size and configuration of the blade, a quantity of material can be transported or spread by adjusting the vertical height of the blade with respect to the ground while propelling.

- During excavation, the machine makes a short forward cut, pushes the material a typical distance of 50 to 300 feet, then backs up and positions for succeeding, generally parallel, cuts. This typical production dozing cycle is illustrated in Figure 2.3

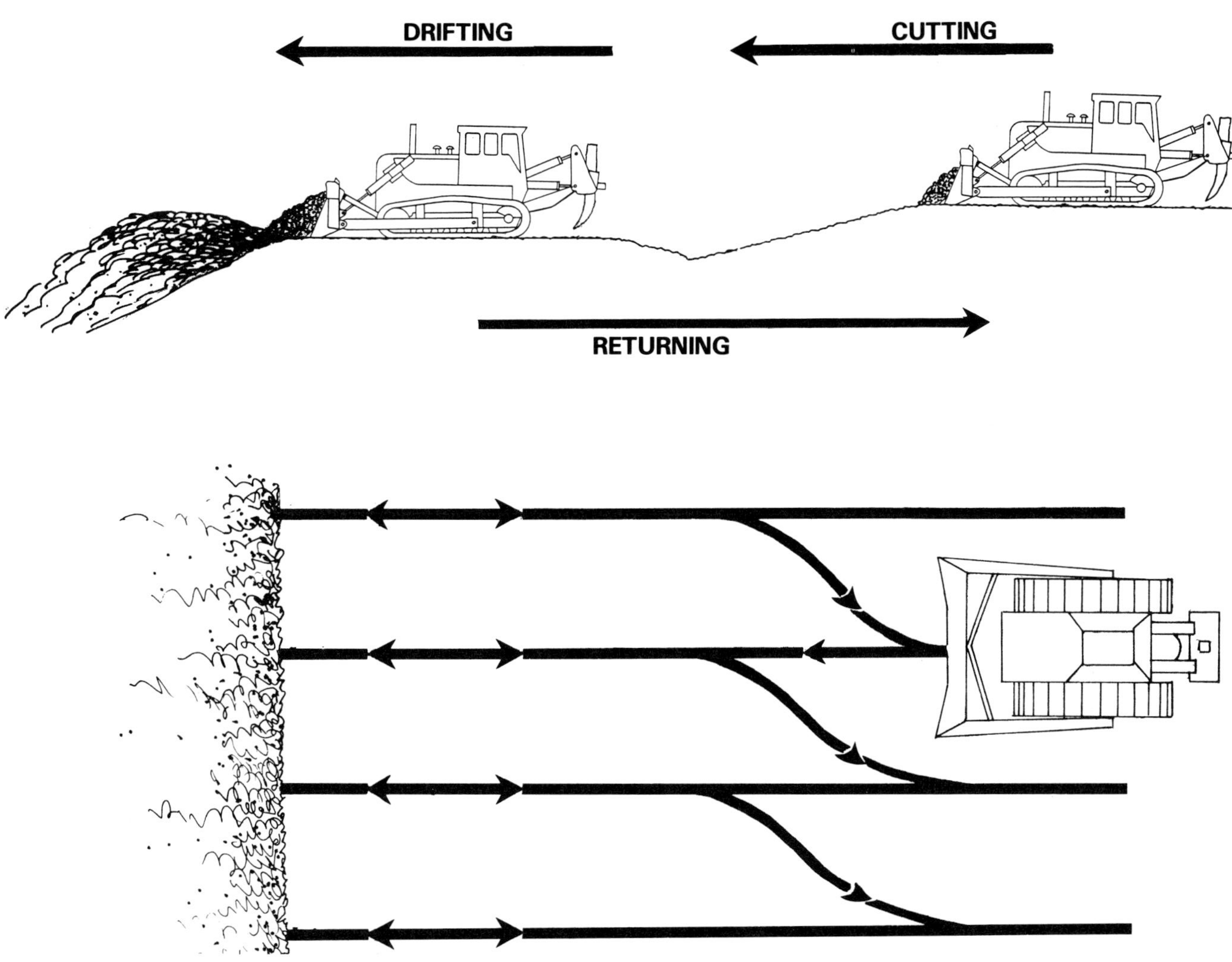

Figure 2.3 Typical Dozer Production Cycle

- Blades can be angled or tipped forward or back, to further enhance both the digging capabilities and/or transport efficiency dependent on the materials.

- The dozers have ample power to slip the wheels/tracks so that the ultimate cutting or pushing forces are generally limited by traction conditions.

- The blade is carried high during reverse travel or placed in a "float" control position which permits it to drag on the surface without any applied down pressure to produce a rough grading action.

- The high traction power of the dozer can also be utilized in pushing scrapers to assist in their loading cycle and for dragging ripper teeth across the surface to break-up the material for subsequent removal.

APPLICATIONS

As indicated, the dozer is a very versatile piece of equipment. It is essentially a rugged high powered traction device capable of applying brute force to pushing and pulling, under a variety of ground conditions. Its applications can be divided into four basic categories.

Earthmoving (Dozing)

The front mounted blade digs a slice of material as the dozer propels forward; the material is then pushed to the disposal area. The operation can be used in many diverse mining situations:

- Pioneering and land clearance: The dozer can be sized to provide sufficient power, and with proper operating techniques can move most obstacles in its path, including boulders, trees, etc. This makes it the primary tool in clearing land prior to mining. Special blades are available for this application.
- Stripping overburden: Some mine plans utilize scrapers and dozers for overburden removal. The dozer, in these operations, moves a portion of the overburden by pushing it over the highwall.
- Grading and leveling mining benches: Draglines, electric shovels and wheel excavators require a flat work surface free of boulders; dozers are commonly used in this clean-up operation.
- Feeding a belt conveyor: The dozer can be effectively employed to push material into a "belt loader" which in turn feeds a belt conveyor. See Figure 2.4.
- Trapping for loaders: The efficiency of small to medium sized loading equipment can be improved by using a dozer to rip and position material to be loaded. (See Figure 2.4)

Photograph 2.1 Terex 82-50 crawler dozer pushing overburden over the highwall

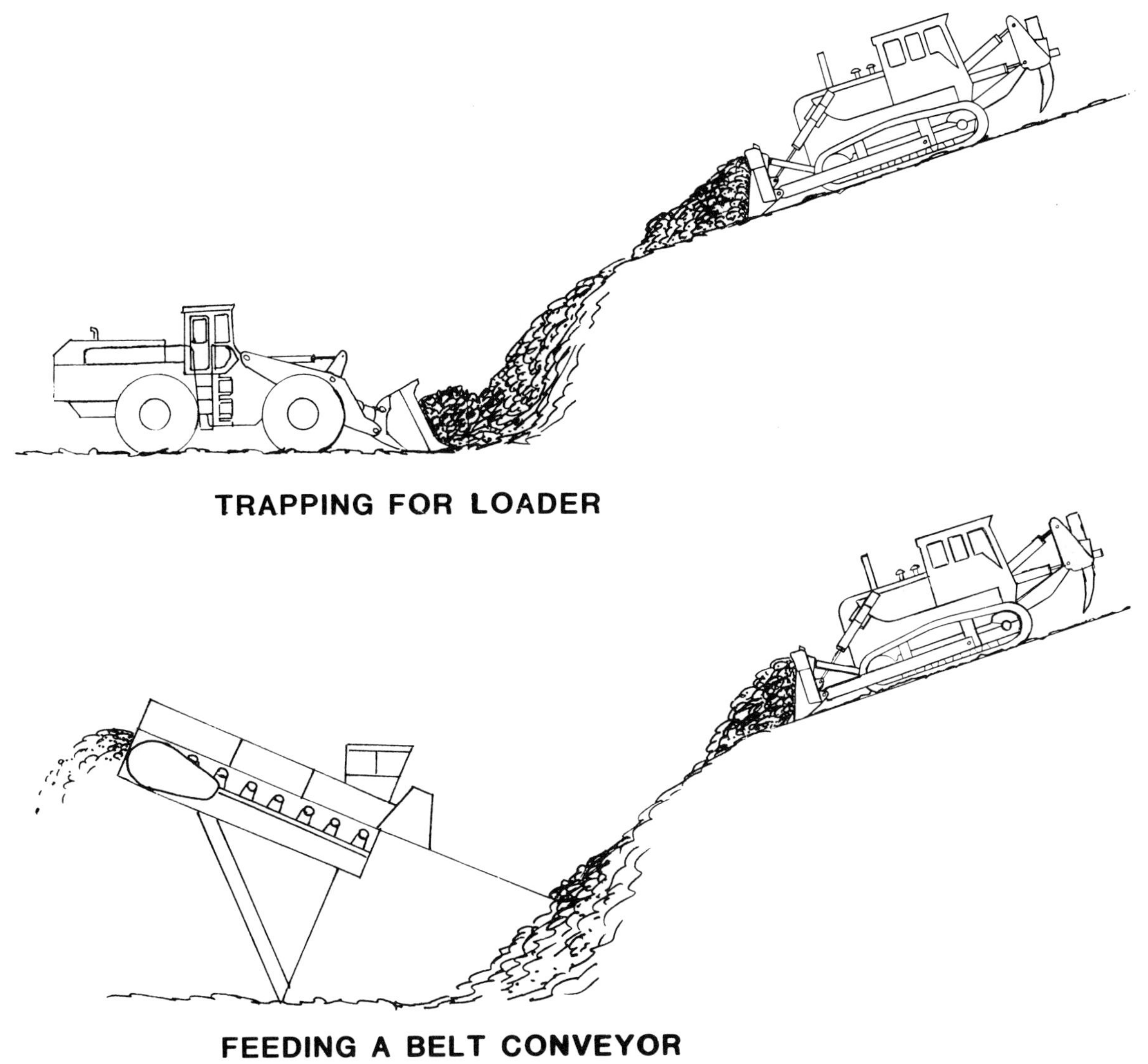

Figure 2.4 Trapping For Front-End Loader & Feeding Belt Conveyor

- Reclamation: Dozers are a basic tool for leveling and recontouring mined out land. Special blades and special wheel models are available for this type of work.

Ripping

Ripping is a method of fragmenting and loosening consolidated material. The ripper is simply an adjustable heavy duty shank hinged to the back of the dozer and pulled through the earth. The advantages of ripping are:

- Relatively safe and simple operation
- Minimal equipment and expertise
- Avoids noise and shock waves (blasting regulations)
- Generally cheaper than drilling and blasting if applied in suitable material

Ripping depth is normally limited by available dozer power with maximum depths of about eleven feet per pass. As the surface layers are removed, ripping can continue to any required depth. It is a convenient approach for scraper and/or front-end loader operations where the top layers are progressively removed in relatively thin lifts.

Pushing

Dozers are used to push other equipment. This is a desirable operational procedure when the pushed vehicle needs a significant increase in propel power during a short part of its work cycle. By using the dozer to provide this increased power, the pushed vehicle's design is simplified and power reduced. Overall results can be reduced investment and operation costs. When pushing is their primary function, the dozers are often equipped with special narrow cushioned pushing blades to minimize impact and tire damage to the other vehicle. The primary applications are:

- Pushing scrapers while loading. Some common dozer paths while pushing consecutive scrapers are illustrated in Figure 2.5. Significant increases in scraper productivity are possible with push loading, especially in difficult digging conditions.
- Pushing scrapers and trucks while transporting up a steep incline
- Pushing other dozers while ripping

Utility work

Because of the dozer's overall versatility, it is frequently assigned to perform special tasks or maintenance activities such as:

- Stockpiling and blending
- Drainage work–ditches and ponds
- Snow removal
- Road maintenance

Photograph 2.2 Clark Michigan 380B wheel dozer leveling and cleaning up mining bench

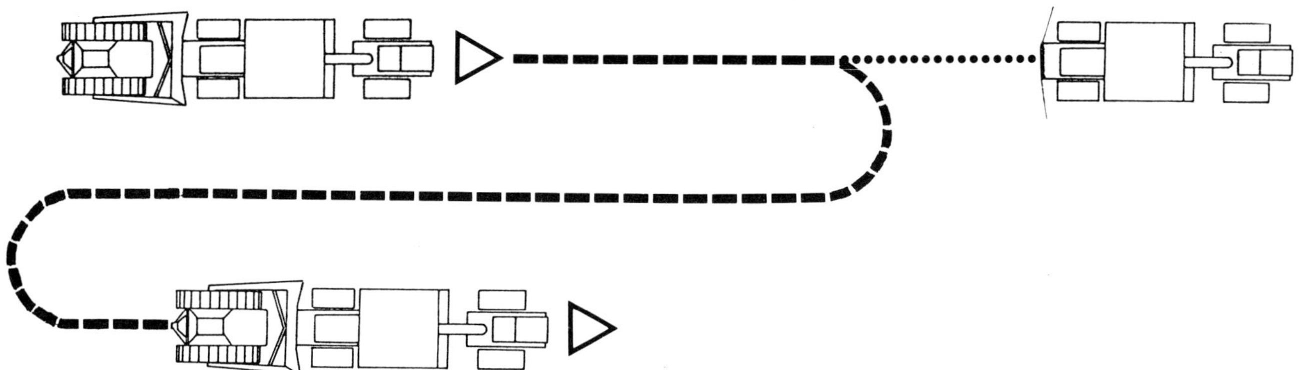

BACK TRACK LOADING

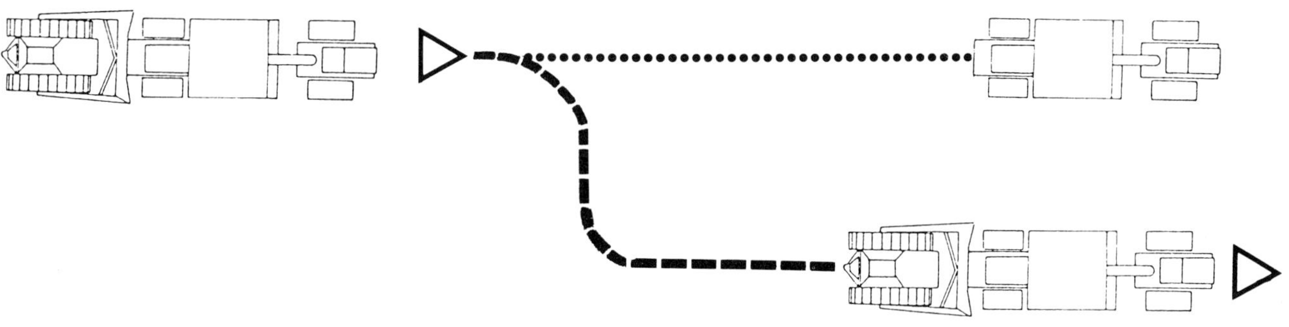

CHAIN LOADING

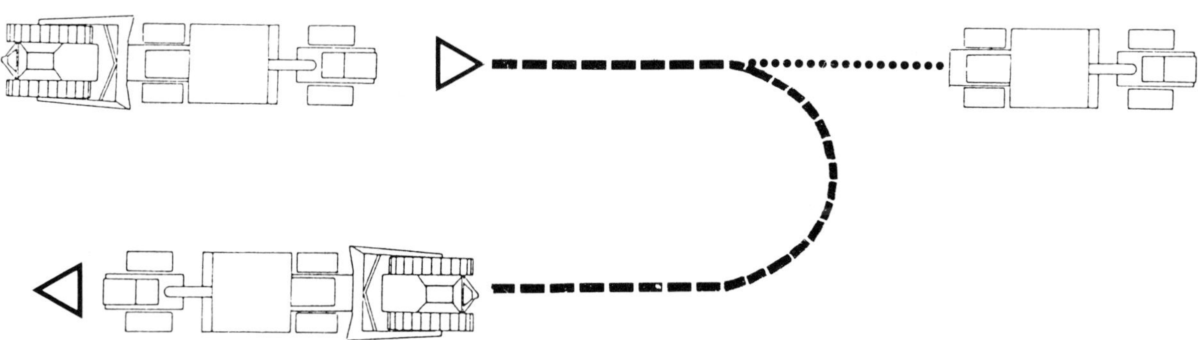

SHUTTLE LOADING

Figure 2.5 Scraper Push Loading Configurations

Photograph 2.3 Crawler dozer trapping for a front-end loader loading a rear dump truck

GENERAL CHARACTERISTICS

In the size range used in surface mining, all dozers utilize diesel engines as the primary power source. The powered functions are illustrated in Figure 2.6.

- Propel: direct mechanical with torque converter or electric wheel
- Vertical blade positioning: hydraulic cylinders
- Blade tilting: hydraulic cylinders
- Ripper tooth positioning: hydraulic cylinders

Note that there are relatively few powered functions and other than the main propel drive, they utilize proven hydraulic cylinder systems. This suggests a simple overall design with a potential for high reliability.

The relationship between machine weights and horsepower available are summarized in Graphs 2.1 and 2.2. These do not show any significant differences between crawler and wheel dozers. The ratio of operating weight to horsepower is fairly consistent for all units.

The relationship between drawbar pull and horsepower is summarized in Graph 2.3. Drawbar pull represents the theoretical capability of a crawler/dozer in terms of maximum forward thrust or push force that is possible. Rimpull is the equivalent figure for wheel dozer; these numbers are not readily available and are not plotted. The relationship between drawbar pull per unit of blade width per machine horsepower is shown in Graph 2.4. This graph implies that the larger machines have superior digging ability since the force per blade width is larger.

Dozers will handle unconsolidated materials, boulders, fragmented rock, soft shales and clays, spoil, topsoil and coals. Consolidated materials must be ripped or blasted prior to bulldozing.

Dozing capability is dependent on machine weight and power, limited by traction. Machine weight is influenced by attachments such as winch or ripper, these also shift machine balance.

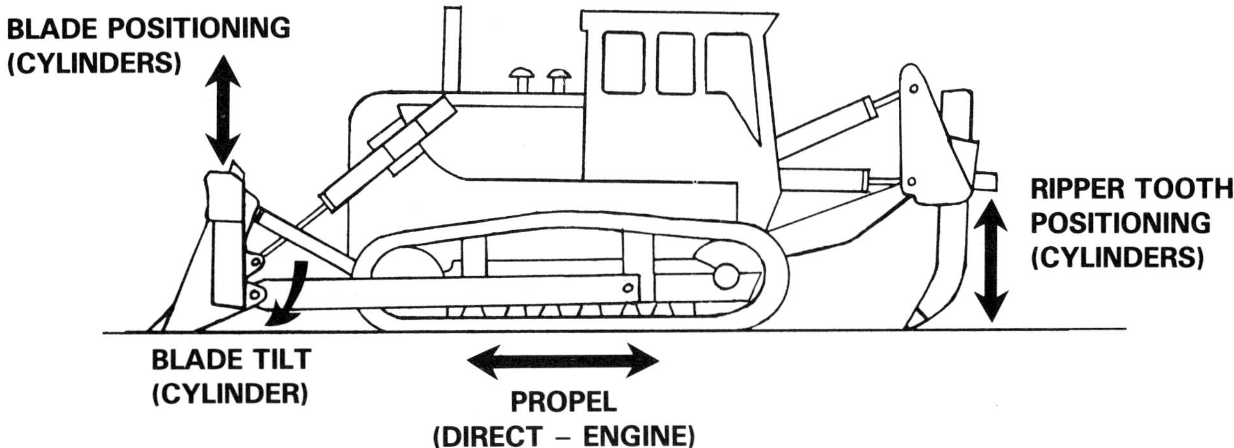

Figure 2.6 Dozer Powered Functions

Photograph 2.4 Fiat-Allis 41B with single shank ripper leveling dragline spoil piles

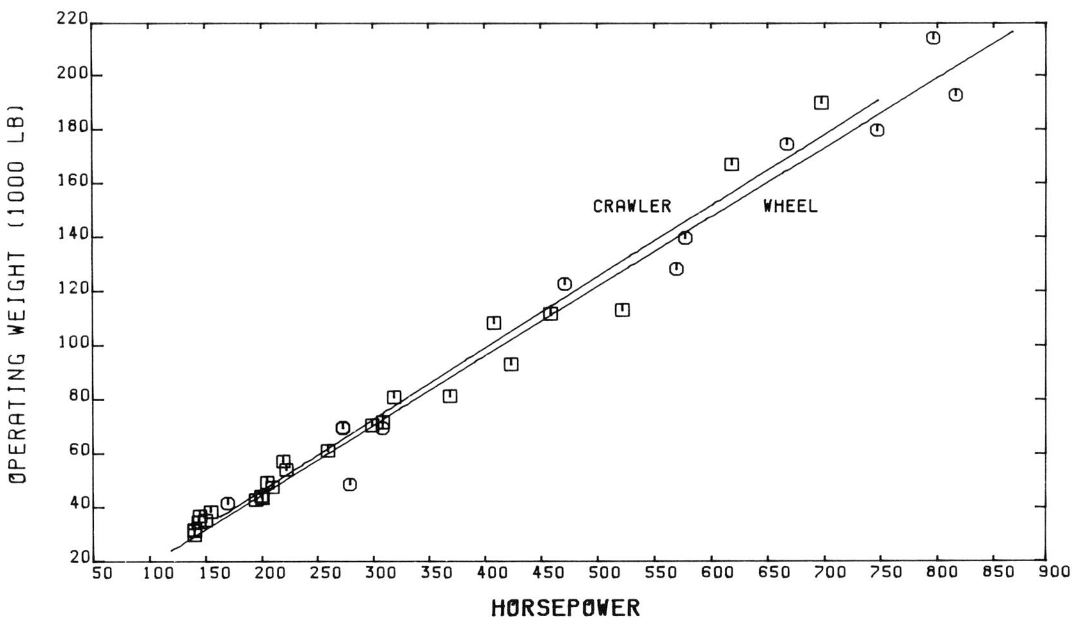

Graph 2.1 Operating weight/HP

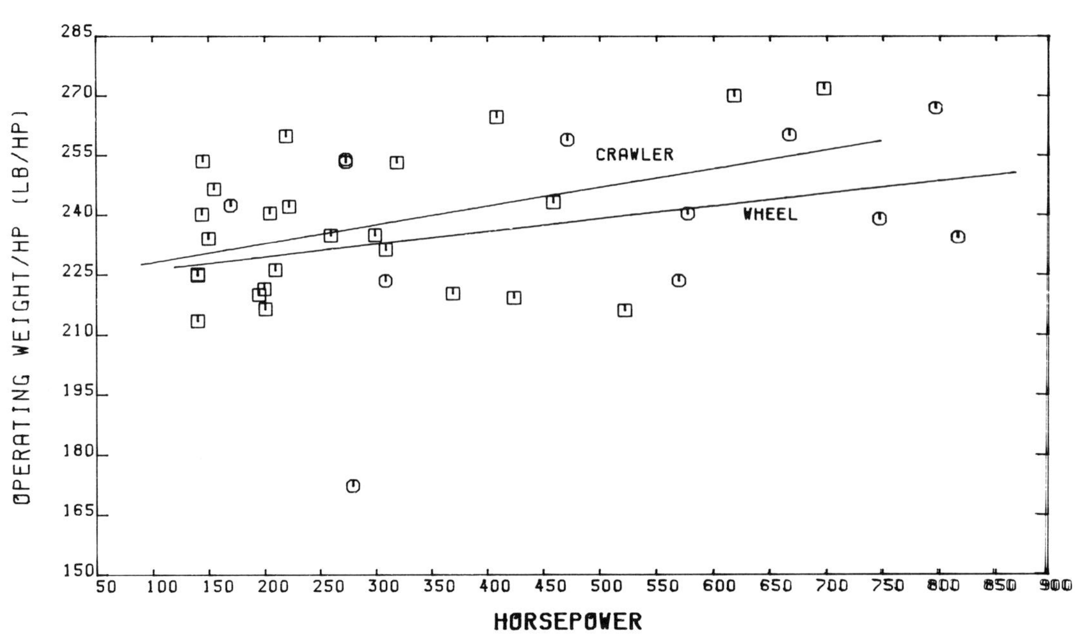

Graph 2.2 Operating weight per HP/HP

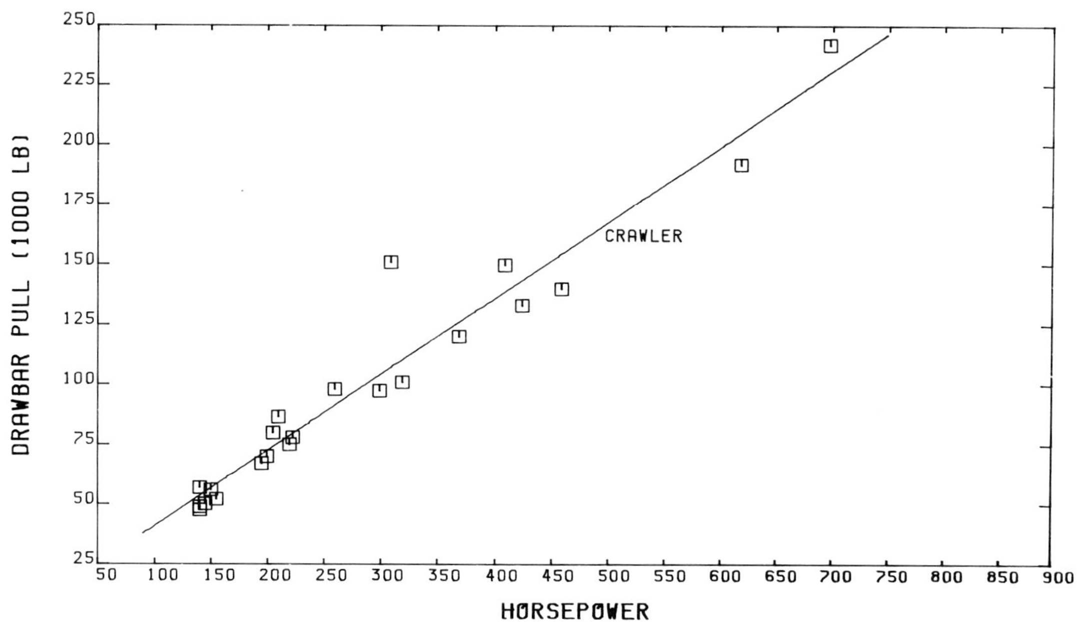

Graph 2.3 Drawbar pull/HP

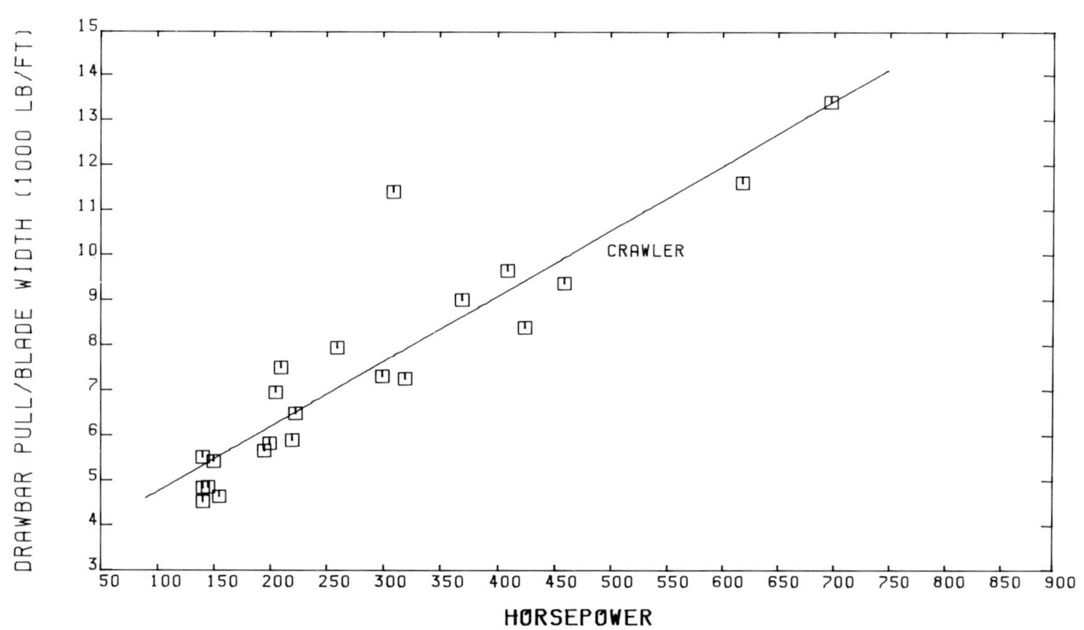

Graph 2.4 Drawbar per blade width/HP

These machines are effective large volume material transporters for short distances of 50 to 300 feet. Larger and more powerful tractors can generally move more material at a lower cost per cubic yard.

Operation of a dozer is fairly straight forward and requires limited operator training. One operator is required per machine.

The machines have:

- High operational versatility
- Excellent gradeability
- Good maneuverability and mobility
- Good stability
- High reliability

The crawler units are commonly used for deep cutting, working on sharp abrasive materials and on the steeper grades.

They require a low capital investment, and because they are produced in relatively high volumes, are standarized in design, with short manufacturing delivery times. Service life is comparatively short, ranging from 8,000 to 12,000 hours dependent on severity of service and maintenance practices.

OPERATIONAL PRACTICES

- Forces applied during dozing and the profile of the cut made are shown in Figure 2.7.
- It generally requires 30 to 50 ft. to develop a full load on the blade.
- There is a tendency for the operator to try to overcontrol the blade and cut at too high a speed.
- Should cut at low speed; travel with load at higher speeds

Photograph 2.5 LeTourneau D800 wheel dozer working in a coal storage area

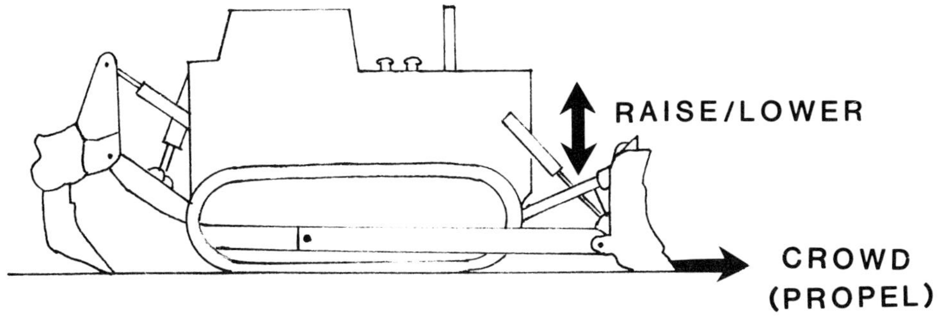

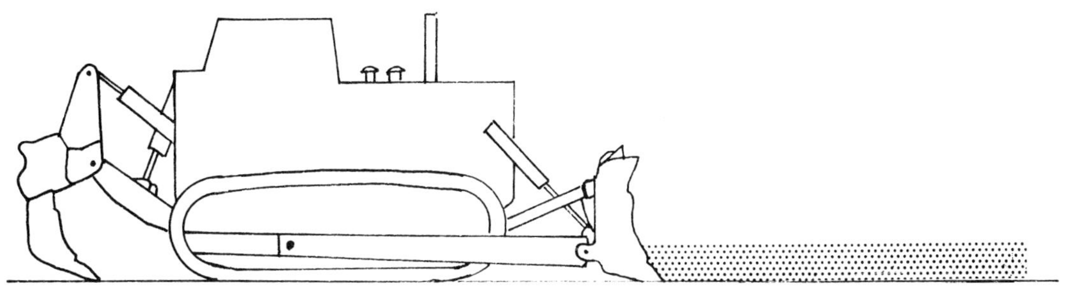

Figure 2.7 Dozer Digging Forces & Digging Profile

- Maximize dozing production (minimize end spillage) by: operating in self-cut slot, running two or more units side by side, using widest blade possible, dozing downhill so blade can carry more materials and use less power.

- On soft materials more gradual turns are desirable to reduce pile up of material on tracks

- When backing away from a load, float blade to reduce front end loading

- Avoid spinning tracks or tires since this accelerates drive component wear and decreases available traction.

- High moisture generally makes dozing more difficult. Freezing makes dozing more difficult, dependent on moisture content.

- The common technique for push loading scrapers is shown in Figure 2.5 (referenced earlier).

- During ripping make slow and even cuts to uniform depth, most ripping is done at 1 to 1.5 mph.

- During ripping it is better to add more or deeper shanks than to work in high gear ranges.

- If some material cannot be ripped by conventional approaches, the area could be given a light blasting charge to improve its rippability. If the ripper has a pusher plate, another dozer could also be used to push the first.

DESIGN FEATURES

Figure 2.8 provides the basic nomenclature for a dozer equipped with a ripper. Most crawler dozer manufacturers make their own engines so that this is generally not an option for the purchaser (See Appendix B for further discussion of engines). Since tractive effort is quite closely tied to machine weight, the engine power varies directly with dozer weight with relatively little difference between crawler and rubber tired units. (See Graph 2.1 given earlier in this chapter). If horsepower is used as a dozer size criteria and operating weight per horsepower is plotted on this basis it shows a greater spread between different units. (See Graph 2.2). The significance of this variance could be that higher weight to horsepower ratio reflects a generally heavier and more rugged machine.

Dozer performance is tied to the effectiveness of the propel system and the prime difference between dozers is in the propel power train drive. The units vary significantly in design, durability

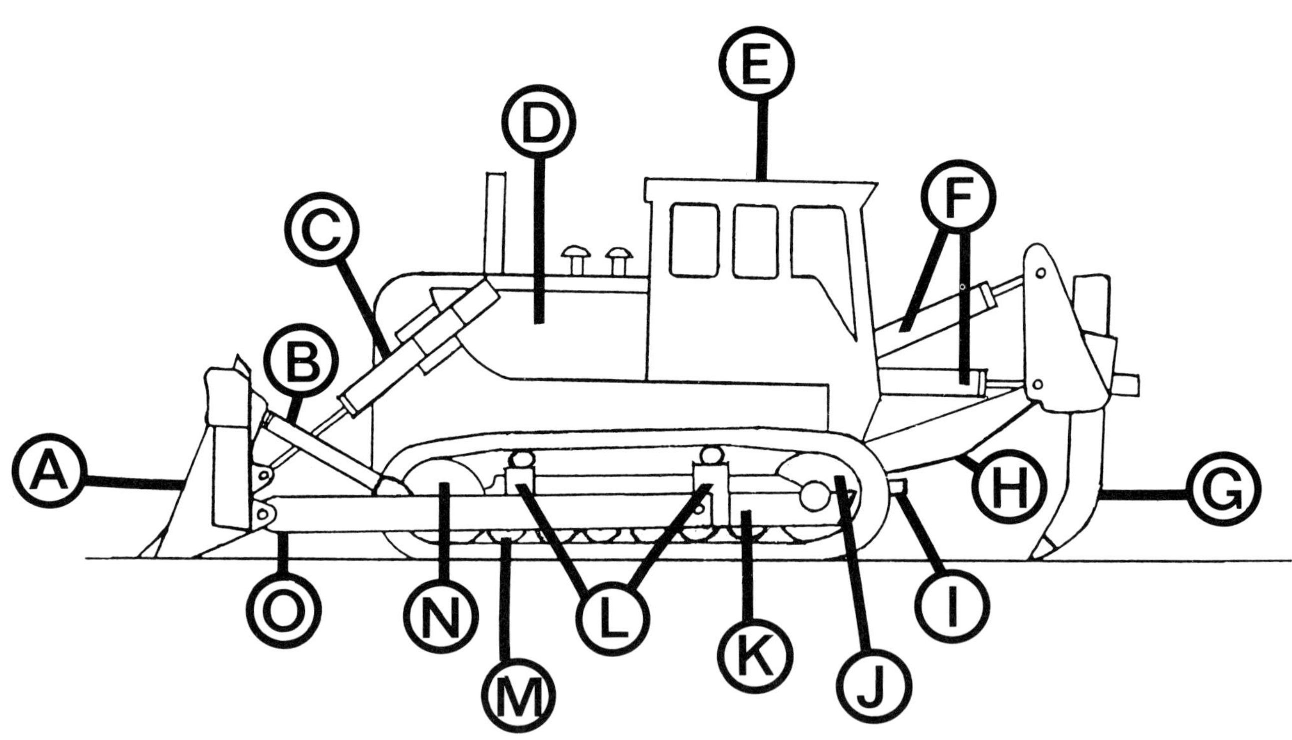

A	Blade	I	Ripper Beam
B	Pitch Strut	J	Sprocket
C	Hoist Cylinder	K	Crawler Frame
D	Engine	L	Track Carrier Roller
E	ROPS–Cab	M	Track Roller
F	Ripper Cylinders	N	Track Idler
G	Drawbar	O	Push Arm
H	Ripper Shank		

Figure 2.8 Dozer Nomenclature

and control response. Unfortunately the differences are quite sophisticated technically so that any detailed evaluation is beyond both the scope of this book and the interest of most users. The most useful yardstick is actual comparative tests supported by observation of production capabilities and operator comments. These will reveal operational response; power matching to demand; and operator effort. But, unfortunately, they do not provide a meaningful measure of component life, ease of maintenance, and operating costs. Available drawbar pull, or rimpull in the case of wheel dozers, is primarily a function of total horsepower (see Graph 2.3 given earlier in the text), machine weight, and traction conditions. It provides limited insight into production capabilities.

Figure 2.9 illustrates what can be termed the proven conventional power train components. Because of the severe duty cycle, it is imperative that all these components, together with the shafts and couplings, be of a very rugged design. The torque converter assures a smooth power flow, provides a torque multiplication and has load matching characteristics. The final drive in virtually all cases is a planetary gear reduction in the hub of the drive sprocket. In this configuration steering is accomplished by disconnecting and/or braking the inside track.

Efforts to improve these drives and to provide steering with power in both tracks have led to a variety of approaches in the newer models. Figure 2.10 attempts schematically to show some of the variations starting with the conventional Type A. The hydrostatic drive of Type C and the twin drive of Type D provide independent reversible power to each track and therefore permit pivot (spin) turns by propelling one track forward and the other in reverse. The improved steering and power utilization of the independent drive must, of course, be weighed against the added complexity of the system and its impact on machine cost and reliability.

The different types of blades (see Figure 2.11) applicable to various operational requirements are discussed later under selection considerations, but some observations are pertinent. Dozing requires both a blade penetration action to break loose the material and a containment capability to carry the load during the transport phase. The cutting action is dependent on the nature of the formation, cutting edge orientation and applied forces. For the standard blades, most dozers based on past experience, develop approximately the same force per unit of blade length, increasing with size (see Graph 2.4 referenced earlier) and have approximately the same orientation. Blades have similar vertical curves to create a rolling action as the load forms in front of the blade; the blade ends are pulled forward (U shape) and the height increased to a maximum if "carry" capability is stressed. The use of longer blades for increased carrying capacity must be traded off against the need for high penetrating forces. Cutting edge segments are replaceable-reversible,

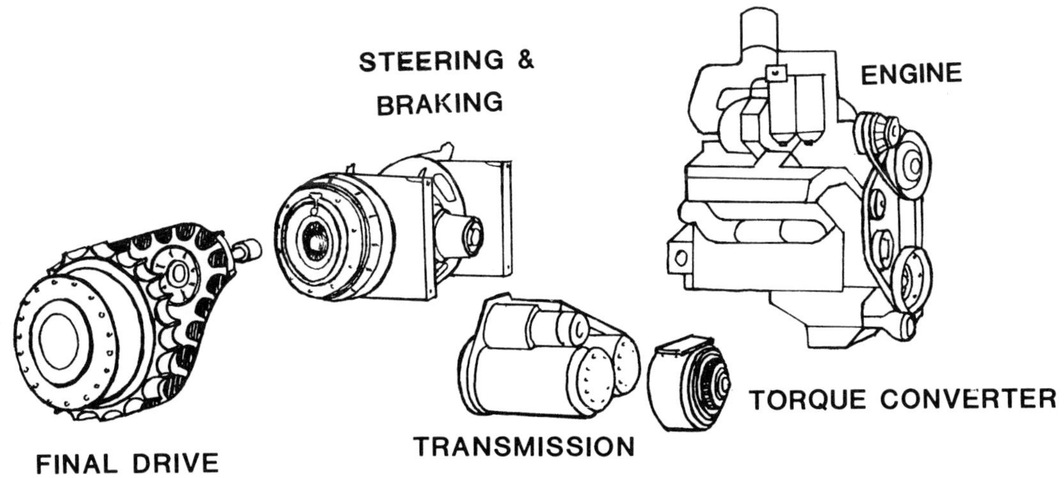

Figure 2.9 Power Train Components

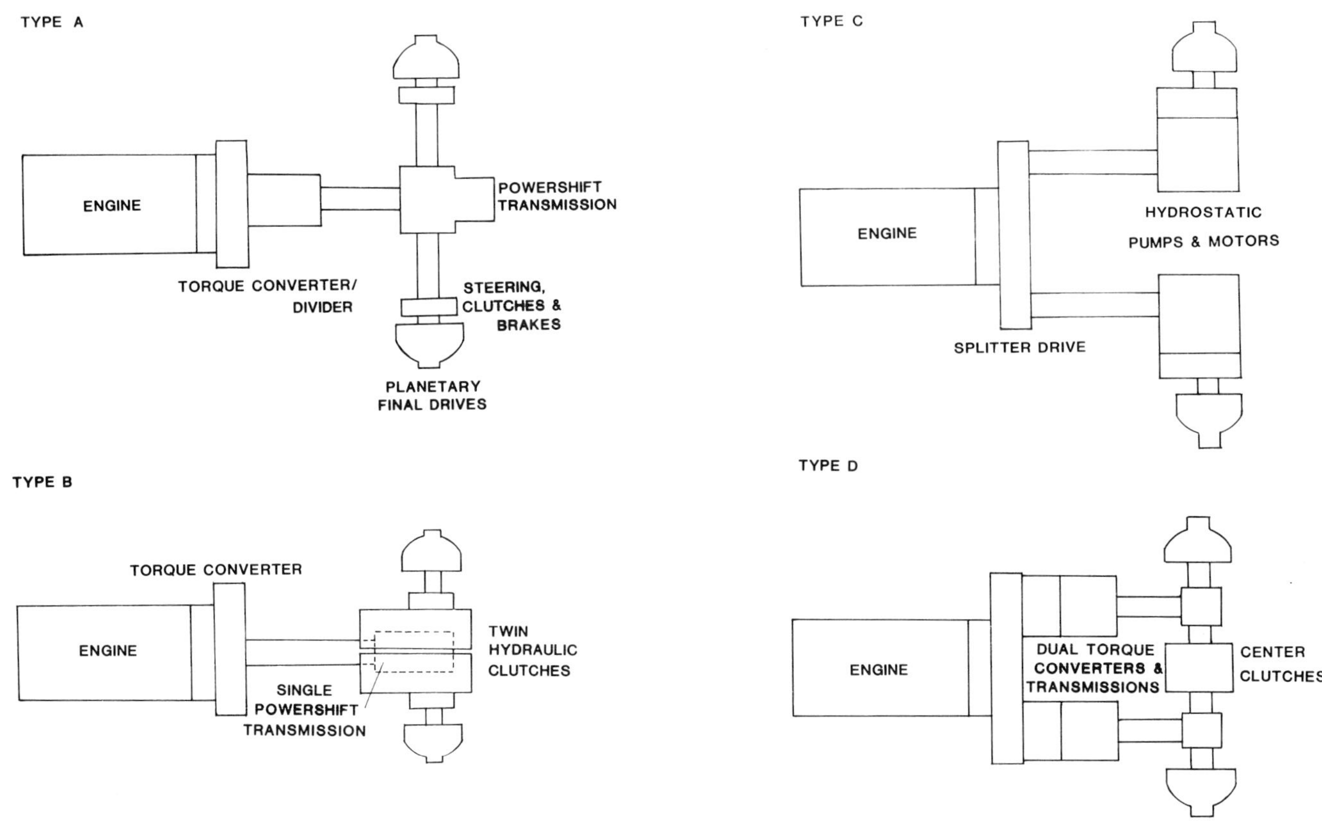

Figure 2.10 Drive Train Options

recognizing the severe abuse and wear to which this portion of the blade is subjected.

The dozer's hydraulic system is a relatively simple design employing cylinders for all the attachment powered functions. (See Figure 2.6 referenced earlier). These systems are well proven by years of field experience and now have a high reliability. Gear type pump equipment is often used because of its low cost and ability to work at higher levels of oil contamination. (See Figure 2.12) Medium pressure levels of 2,000 to 3,000 psi assure long life. The duty cycle consists principally of periodic blade or ripper tooth positioning and does not necessitate complex valve, tank or fitting designs.

Track designs up until the last few years have been highly standardized and stressed a rugged, rigid design which would take the high wear and service abuse of dozer operation. Refinements have focused on extending wear life and reducing the lubrication frequency for rollers and pins. Reduced wear decreases the frequency of track adjustment and noise levels.

Traversing of obstructions with a rigid mounting places a high point loading on individual track links and rollers. Caterpillar and Komatsu have introduced a flexible track on their latest new models. (See Figure 2.13) These designs, paralleling, to some extent, military tank configurations, introduce boggie mounted rollers which can deflect against rubber discs permitting the track to partially conform to the surface. The result is increased ground contact and distribution of the weight over multiple rollers. The improved load distribution means that the peak loads on track rollers and pins are reduced, permitting smaller components and some reduction in cost to compensate for the added complexity. The ground adherence characteristics improve traction and ride quality.

A variety of track grouser designs are available to meet special traction conditions such as ice.

Because of the variety of work assignments given the dozer, however, the tendency has been in most cases to use the standard grouser.

Tire selection options and characteristics are summarized in Appendix A. Severe floor conditions, which reduce tire life, generally mean that crawler machines are employed. It is not common to use tire protection methods such as chains or to ballast the tires. Ground pressures are appreciably higher which must be recognized in the application analysis. The tires deflect under load which does make grading a little more complicated.

The main frames of the dozers are of a box type, all welded, with some casting incorporated in critical load transfer areas. The frame structure provides a base for the engine, cylinder mounting supports and incorporates the final drive gear boxes and pivot shaft brackets. These designs have been continuously refined and upgraded with improved steel and welding techniques.

Adding a ripper to the rear of the dozer increases the overall machine versatility as well as adding additional weight. The weight increases the machine's traction capabilities as well as counteracting the blade weight on the front to improve the balance. Ripping places severe loading on the main frame and the means of attachment should be studied in cases where extended operations of this type are anticipated. The three basic ripper configurations are radial, parallelogram and adjustable. The last two are the most popular for the larger machines and are illustrated in Figure 2.14. The differences relate to the degree the operator can exercise control of the angle at which the tooth attacks the formation. The various arrangements allow for side by side mounting at up to five teeth and pinning of the teeth in a number of vertical positions. Tips and shank guards are selected to meet site requirements.

As machine size increases the operator's view of the blade decreases. To compensate for this loss of visibility in the larger machine sizes, the operator's cab is either offset or has an angular orientation, or both. Rollover protective devices are necessary on all units. Noise level regulations have resulted in sealed and isolated cabs which provide a closed operator environment which can be modified with a range of heat and/or air conditioning accessories.

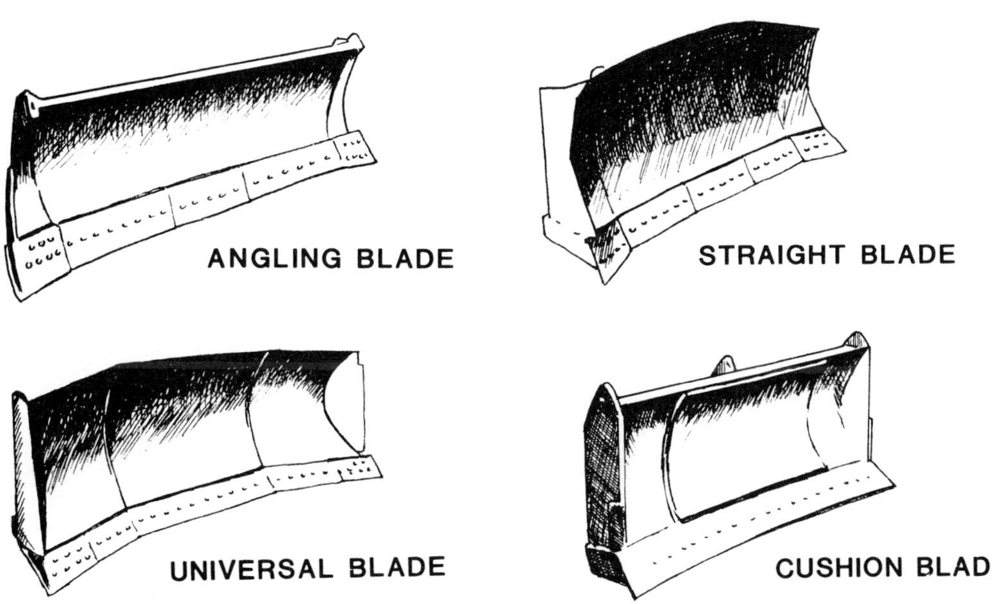

Figure 2.11 Blade Types

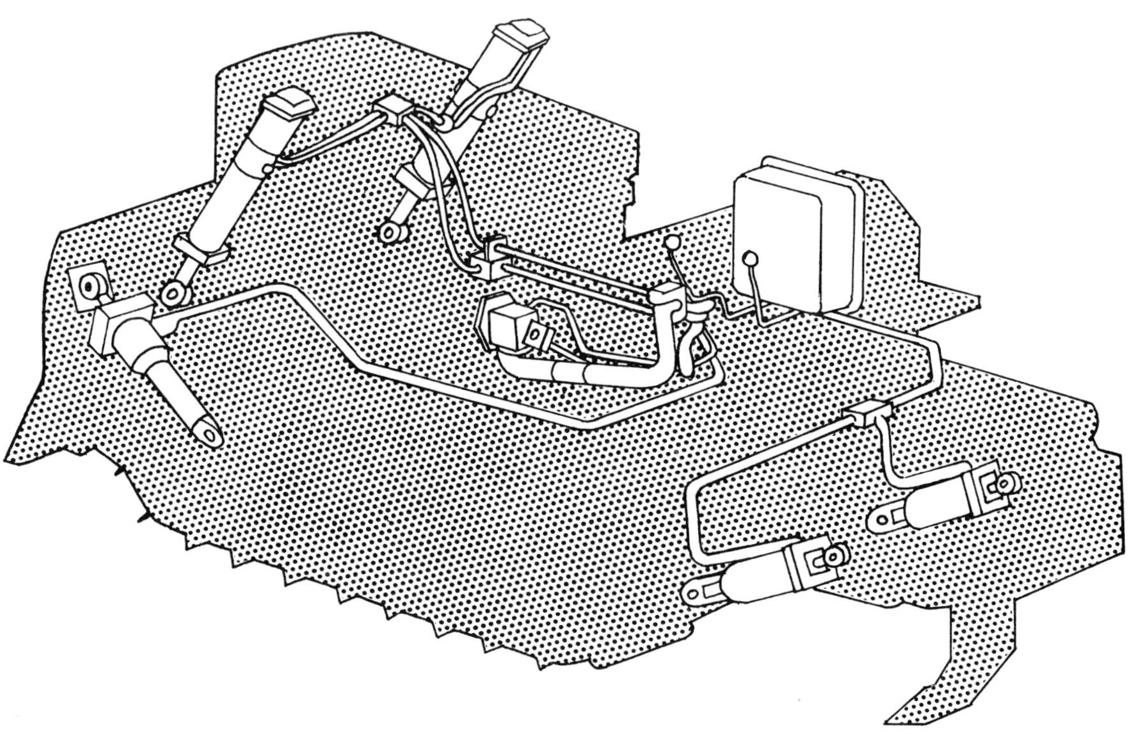

Figure 2.12 Typical Hydraulic System

Photograph 2.6 Paired caterpillar dozers in tandem with single reclamation blade

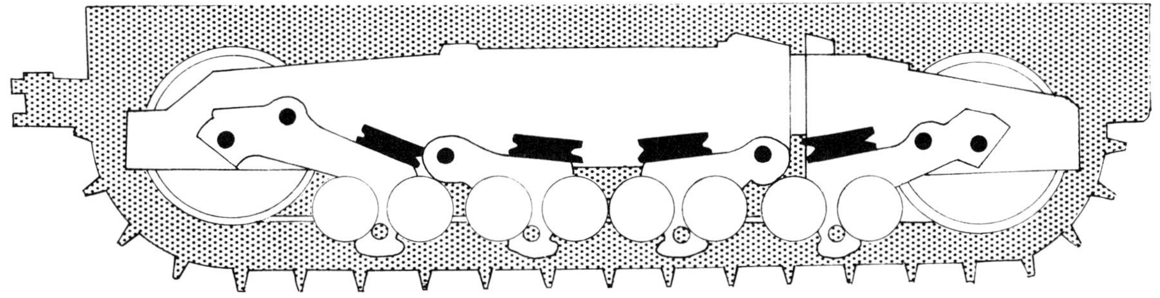

Figure 2.13 Flexible Track

Considerable emphasis has been placed on service accessibility and maintainability. The following characterize and illustrate the direction of these efforts.

- Service indicators
- Sight gauges to simplify checking fluid levels
- Disposable spin-on oil and fuel filters
- Extensive use of filters in hydraulic circuits
- Test panels
- High speed oil change systems
- Dry type air cleaners
- Permanently lubricated and sealed track componentry
- Segmented sprockets
- Reversible cutting edges on blade to increase edge life
- Hydraulic track adjusters

These are reinforced by a modular design approach which facilitates unit replacement and/or interchange. The service support by the manufacturer's dealers organization is extensive with parts exchange and rebuild programs. Oil sampling and analysis are employed extensively to monitor oil and main drive assembly conditions.

SELECTION CONSIDERATIONS

Track vs. wheel selection dependent on:

- Grades and side slopes; crawler dozers have greater gradeability and stability
- Travel distances and frequency of machine relocation; wheel dozers can travel 2 to 4 times faster than crawler dozers
- Ground flotation characteristics; crawler dozers, especially with wide tracks, have less ground pressure
- Weather conditions; crawler units are somewhat less affected by adverse weather
- Traction; crawler units generally have a higher, coefficient of traction, under similar ground conditions, than a wheel unit

Blade selection is dependent on:

- Straight blade is fixed at right angles to travel. It can be raised/lowered below grade, and the top edge can be pitched forwards or backwards 5 to 10 degrees. This blade is most efficient pushing material straight ahead with medium length passes. It provides good penetration (pitched backward), and can handle heavy materials (pitched forward).

- Angling blade is a straight blade (long and narrow) set at an angle (25 degrees) to direction of travel; often combined with a tilt feature. It is designed for side casting at higher speeds than for push dozing. It has high capacity for short side displacement.

- U blade (universal) has forward curving side wings, and like the straight blade is fixed at right angles to travel. It is longer than the straight blade and can be pitched. It is designed for light, easily dozed material; it can handle large loads for long distances.

- Push (cushion) blade is used to push scrapers (or trucks): it is generally cushioned. It is small to avoid scraper tire damage; it can be used for general dozing jobs but is not designed for production earth moving.

- Special (light materials–reclamation) blades can be designed for specific applications.

Production volume calculation procedures and representative performance estimates are provided in available manufacturers' literature. These, together with production cost estimates, are considered a separate subject too extensive to be comprehensively treated in this book. Estimates must be directly correlated with detailed site conditions and matched to specific machine performance characteristics. All of the equipment manufacturers will provide production and cost estimates for stated conditions prepared by their sales technical staff personnel based on their accumulated experience with similar applications. These frequently are developed by in-house computer programs so that alternative fleets of matched equipment can be readily evaluated to find the optimum combination.

It must be recognized that all of the above estimates are based on average data extrapolated to meet anticipated average site conditions over the life of the operations. The spread in these "averages" can be substantial and the correction factors introduced require a considerable amount of judgement. Costs, in particular vary locally with time and economic conditions which make generalization dangerous. Some of the considerations entering into production calculations are:

- Material swell

- Material unit weight

- Formation characteristics

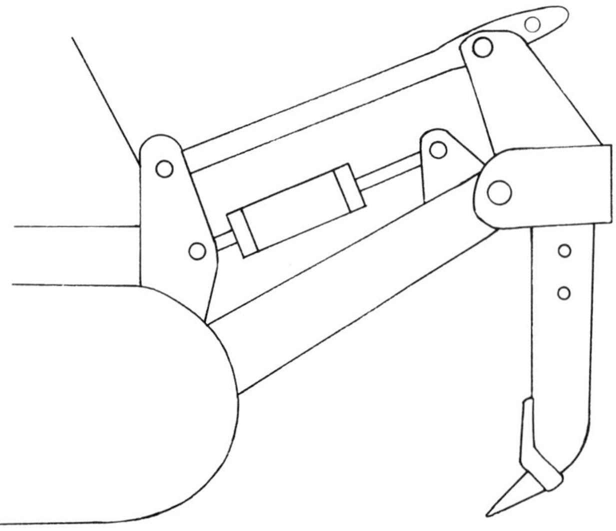

PARALLELOGRAM DESIGN

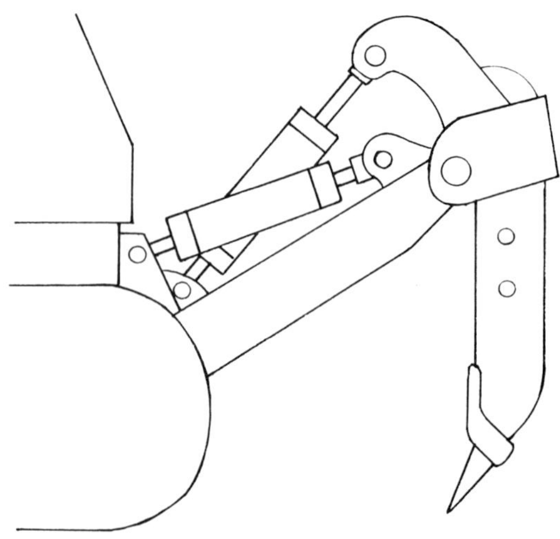

ADJUSTABLE DESIGN

Figure 2.14 Ripper Types

- Job efficiency
- Machine availability
- Grades
- Operator skill
- Type of blade
- Type of operation
- Distances
- Engine power curves

Some of the factors in the cost calculations are:

- Machine prices
- Salvage value
- Machine life
- Interest
- Insurance
- Taxes
- Fuel consumption
- Lube oil, grease, filters, etc.
- Maintenance
- Operator wages
- Special adjustments:
  — Tire cost
  — Undercarriage
  — Cutting edges

If the dozer is to be equipped with a ripper, the following are some of the selection criteria:

- Primary factors in selecting equipment are the required down pressure on tooth, the angle of penetration, the tractor horsepower and tractor weight.
- Consideration must be given to desired degree of fragmentation. It is controlled by the distance between ripper teeth and/or the gap between ripping passes, and depth of cut. Desired fragmentation is dictated by subsequent loading, hauling, and processing equipment characteristics.
- Rippability is dependent on material properties. Favorable properties include weathering effects, fractures, planes of weakness, brittleness, large grain size and low compressive strength.
- Performance can be correlated with seismic wave velocities of the formation (data available in manufacturers' literature).
- Number of teeth, and style of design is dependent on tractor power, depth requirement and need for control of tooth penetration angle.
- Tooth selection is generally by trial and error.
- Ripping is a severe application for tractors; usually larger, track type units are required.

Other production attachments are available, some from small specialty manufacturers.

- Winch (towing)
- Push block
- Pull hook
- Rear cushion pusher
- Compactor
- Side boom
- Railcar coupler for moving railcars
- Root and rock rakes
- Stingers and stumpers for toppling trees
- V-shaped blade for cutting brush
- Logging sweep
- Power angling blades
- Quick disconnect blade couplings

Various types, sizes and/or models are typically available for the following machine components; user has to select one:

- Track shoes for various ground conditions: grouser height and shoe width are common variables
- Transmission; mechanical direct drive or automatic mechanical (for some in the smaller size ranges)
- Blades; straight, semi-U, angle, full-U, or push block

The machine comes equipped with the following components; however, optional sizes and/or types are often available:

- High amp alternator
- Reversible blade fan
- Swinging drawbar
- Long drawbar
- Tilt cylinder for the blade
- Self-cleaning sprocket (Terex) on track unit
- Perforated engine enclosures to improve air flow
- High altitude specifications

In addition to the components supplied with the basic machine, the following optional equipment is typically available:

- ROPS cab
- Sound suppression for cab
- Man lift designed to lift operator from ground to cab on larger machines
- Low temperature starting system
- Track roller guards
- Fan blast deflector
- Fire extinguisher/system

Photograph 2.7 Caterpillar D-10 with ripper working in abrasive floor conditions

- Extra guards
- Engine coolant heater
- Auxiliary/night lighting system
- Vandalism protection
- Tinted glass
- Windshield washers and wipers
- Fast-fill fuel system
- Seat belt
- Reverse alarm (usually standard)
- Mirrors (inside and outside, flat or convex)
- Engine coolant filter
- Special engine enclosures
- Horn
- Automatic lubrication system
- Tool kit
- Operator comfort/convenience items such as heater, air conditioner, radio, etc.
- Direct drive decelerator
- Electric hour meter
- Prescreener
- Quick service oil change
- Radiator core protector grid
- Hydraulic ripper pin puller (single shank)
- Counterweight
- Starting receptacle
- PTO shaft
- Audible low flow and high coolant temperature alarm

NEW DEVELOPMENTS & TRENDS

At the component level, efforts continue to provide added refinements:

- Common lube oil for all cases
- Hydrostatic fan drives so that fan speed can be adjusted to cooling requirement.
- Simplified maintenance
- Reduced grease fittings
- Improved visibility
- Improved operator comfort and simplification of controls
- Reduced engine noise levels (fan, waterpump, mufflers, etc.)
- Modular construction

Several new dozers have been introduced in recent years that are significantly larger. Maximum size appears to be still undetermined; new applications such as expanded ripping, extensive reclamation leveling and the overall larger size of new mining operations are providing new product development opportunities. Increasing size has introduced new problems which require new design concepts.

- Minimizing maneuvering space requirements
- Operator visibility
- Transport between locations

Large rubber tired units have created considerable interest for reclamation work. Some very large units have been tried and, as of now, set aside for future re-evaluation. The Melroe 870 illustrates a rather unique design presently available. (See Photograph 2.8).

The U.S. Bureau of Mines has sponsored extensive studies of various blades and dozer combinations aimed at improving the productivity of reclamation dozer leveling operations. Single and

Photograph 2.8 Melroe M870 multiple wheel dozer in a spreading operation

two machine combinations were worked with super wide (up to 65 feet) straight and angle blades. Other less conventional approaches were also studied and tested:

- Grading bars
- Polymer injection into blade surface to reduce friction
- Combustion assisted side casting
- Winch-dozer system

These concepts are still under evaluation with limited field experience to-date.

Two government studies in recent years have attacked the problem of improving the dozer operator productivity. One involved an automatic surveying system which would control the blade positioning for final grading. The second, a dozer draft power sensor device, provides an audio feedback to the operator of the traction power being utilized as an indication of machine operating efficiency. Fuel consumption and the hours required for dozing per acre, during tests, were reduced significantly.

Periodic efforts continue to improve blade cutting action and ripper penetration by superimposing a vibratory motion to the cutting unit. Rippers of this type are available for the smaller size machines.

MACHINE SPECIFICATIONS

See Figure 2.15—Dozer Dimensions
See Table 2.2—Crawler Dozer Specifications
See Table 2.3—Wheel Dozer Specifications

Crawler Dozers

Crawler dozers are listed alphabetically by manufacturer, and then in ascending order according to the manufacturer's listed flywheel horsepower. Published manufacturer data does not include engine options. Transmission options are available on some machine models. In general, the standard transmission is a powershift/torque converter unit, with a powershift/direct drive unit available as an option. The maximum speeds listed correspond to the standard transmission.

Operating weights are approximate. Basic machine weight with fuel and fluids, 175 lb. operator, ROPS cab and either a straight or semi-"U" blade are included in the operating weight. Rippers and other special attachments are not included.

The drawbar pull is also approximate. The numbers were derived by intersecting the manufacturer's published drawbar pull versus speed curves for first gear at 0.5 mph. In specific applications, it will be affected by traction, grade, etc.

When available, data for a machine equipped with a straight blade was used to compile dimension specifications. When that information was not available, data for a machine with semi-"U" blade was used instead. Special application blades will change some dimensions and blade capabilities significantly. Maximum ripping depth listed is based on the machine equipped with a single shank ripper.

Ground pressure is based on the machine model equipped as outlined above; the standard shoe width is always used. Ground pressure will change when optional, wider or special application shoes are used.

Wheel Dozers

Wheel dozers are listed alphabetically by manufacturer, and then in ascending order according to the manufacturer's listed flywheel horsepower. Optional engines are available in some cases. All the machine models listed, except LeTourneau, utilize a powershift/torque converter transmission. LeTourneau machines utilize an electric drive system, with individual wheel motors.

Blade dimensions are based on the standard blade available with the machine. In most cases this is a semi-"U" configuration. Those machine models marked with an asterisk are special reclamation models, with correspondingly different blade sizes and configurations.

Hydraulic capacities do not include the steering circuits.

Steering angle is the angle of articulation at the hinge-point from straight ahead to maximum turn in one direction. Turning diameter is defined as blade clearance circle.

All specifications, capacities, capabilities and dimensions are based on published manufacturer data. Although the information is believed to be current and the interpretation to be consistent and correct, it is possible that omissions and inaccuracies have occurred and that out-of-date information may have been included. These specification charts are not meant to be used either as an in-depth analysis, or as a head-to-head comparison of the available equipment, but rather as a general overview of the equipment, available sizes, and approximate operating data. Specific questions relating to performance or purchase should be directed to the manufacturer or authorized distributer/dealer in the user's specific area.

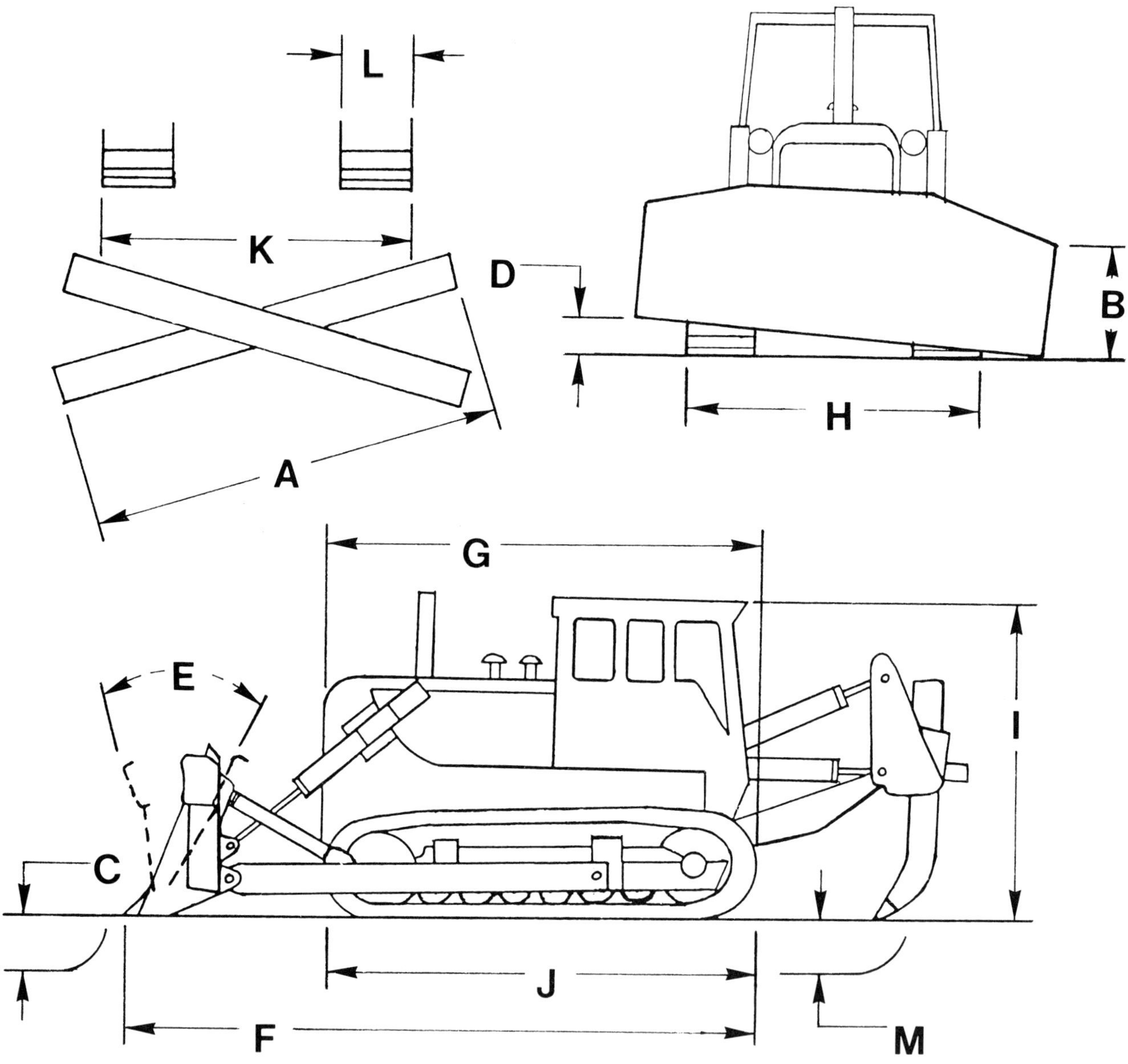

A	Blade Width (ft)	H	Overall Width w/o Blade (ft)
B	Blade Height (ft)	I	Height to Top of Cab (ft)
C	Digging Depth (ft)	J	Length of Crawler (ft)
D	Tilt (in)	K	Width over Crawlers (ft)
E	Pitch (deg)	L	Width of Shoe (in)
F	Overall Length (ft)	M	Ripper Depth (in)
G	Overall Length w/o Blade (ft)		

Figure 2.15 Dozer Dimensions

Table 2.2
CRAWLER DOZER SPECIFICATIONS

MAKE	MODEL	ENGINE HP	WEIGHT (lbs)	RIMPULL (lbs)	FUEL TANK (gal)	COOLANT (gal)	MAX SPEED F/R (mph)	BLADE WIDTH (ft)	BLADE HEIGHT (ft)	DIG DEPTH (in)	BLADE TILT (in)	BLADE PITCH (degrees)	RIPPER DEPTH (in)
Case	1450-B	140	29,890	49,000	65	14	5.4/6.5	10.1	3.7	20	15	10	18
Caterpillar	D6-D	140	31,500	47,500	78	10	6.9/6.0	10.5	3.7	19	32	—	21
Caterpillar	D7-G	200	44,300	70,000	115	12	6.3/5.8	12.0	4.2	18	28	—	29
Caterpillar	D8-K	300	70,500	97,500	170	32	7.0/7.1	13.3	5.0	20	40	—	47
Caterpillar	D9-L	460	111,910	140,000	255	34	7.7/9.6	14.9	6.5	25	46	—	54
Caterpillar	D10	700	190,300	242,000	382	52	7.2/8.6	18.0	7.1	27	32	—	70
Deere	850/6545	145	36,785	50,000	82	9	6.5/6.5	10.3	3.7	19	15	—	—
Fiat-Allis	14-C	150	35,110	56,000	93	11	5.9/7.1	10.3	3.9	18	28	12	27
Fiat-Allis	16-B	195	42,900	67,000	117	18	6.2/7.0	11.8	4.3	17	16	10	27
Fiat-Allis	FD-20	223	54,013	78,000	124	17	5.9/6.5	12.0	4.4	20	30	10	28
Fiat-Allis	FD-30	300	—	—	—	—	—	13.0	—	—	—	—	34
Fiat-Allis	31	425	93,185	133,000	235	35	6.7/7.5	15.8	6.2	22	44	10	40
Fiat-Allis	41-B	524	113,210	—	300	38	6.5/7.4	17.0	7.1	24	22	10	42
International	TD-15C	140	31,520	57,000	80	11	6.0/7.0	10.3	3.8	16	26	10	20
International	TD-20E	210	47,525	86,500	111	18	6.4/7.5	11.5	4.3	18	23	9	24
International	TD-25E	310	71,700	151,000	170	27	6.4/7.7	13.2	5.0	20	35	10	30
Komatsu	D65E-6	155	38,220	52,000	74	15	6.6/8.5	11.2	3.8	23	34	—	22
Komatsu	D85E-18	220	57,180	75,000	119	21	6.9/8.2	12.7	4.4	19	30	—	26
Komatsu	D155A-1	320	81,020	101,000	159	44	7.3/8.5	13.9	5.2	22	40	—	49
Komatsu	D355A-3	410	108,500	150,000	198	48	7.9/7.8	15.5	6.2	28	39	—	55
Komatsu	D455A-1	620	167,435	192,000	338	53	9.1/9.0	16.5	7.5	32	45	—	71
Terex	D600-D	144	34,595	—	63	13	7.0/7.0	—	—	—	—	—	42
Terex	D700-D	201	43,512	—	106	17	7.3/7.3	—	—	—	—	—	31
Terex	82-20B	205	49,303	80,000	110	30	7.0/7.9	11.5	4.3	18	20	14	—
Terex	82-30B	260	61,077	98,000	125	36	7.0/8.4	12.3	4.5	20	19	15	—
Terex	82-50	370	81,515	120,000	195	49	7.0/8.4	13.3	5.5	20	20	15	—

Table 2.2 (Continued)

CRAWLER DOZER SPECIFICATIONS

MAKE	MODEL	ENGINE MAKE	ENGINE MODEL	TRANSMISSION TYPE	NO. OF GEARS F/R	BRAKES	HYDRAULIC SYSTEM DATA: PUMP FLOW (gpm)	HYDRAULIC SYSTEM DATA: RELIEF VALVE PRESSURE (psi)	HYDRAULIC SYSTEM DATA: PUMP TYPE	HYDRAULIC SYSTEM DATA: NO. OF CYLINDERS	HYDRAULIC SYSTEM DATA: NO. OF MOTORS	STEERING TYPE
Case	1450-B	Case	A504-BDT	mech auto	4/4	disc	52	1800	gear	3	0	indep clutch
Caterpillar	D6-D	Caterpillar	3306	mech auto	3/3	band	44	2250	gear	3	0	clutch/brake
				mech direct	5/4							
Caterpillar	D7-G	Caterpillar	3306	mech auto	3/3	band	60	2250	vane	3	0	clutch/brake
				mech direct	6/4							
Caterpillar	D8-K	Caterpillar	D-342	mech auto	3/3	band	78	2400	—	3	0	clutch/brake
				mech direct	6/6							
Caterpillar	D9-L	Caterpillar	3412	mech auto	3/3	disc	103	2400	vane	3	0	clutch/brake
Caterpillar	D10	Caterpillar	D-348	mech auto	3/3	disc	153	2500	gear	3	0	clutch/brake
Deere	850/6545	Deere	—	hydrostatic	n.a.	hydrostatic	46	2250	vane	3	2	indep motor
Fiat-Allis	14-C	Fiat	8205	mech auto	3/3	band	49	2275	gear	3	0	clutch/brake
Fiat-Allis	16-B	Allis-Chalmers	17000 MKII	mech auto	2/2	band	64	2000	—	3	0	clutch/brake
				mech direct	6/3							
Fiat-Allis	FD-20	Fiat	8215	mech auto	3/3	band	79	2250	gear	3	0	clutch/brake
Fiat-Allis	FD-30	Fiat	8285-T	mech auto	3/3	band	—	—	gear	3	0	clutch/brake
Fiat-Allis	31	Cummins	KT1150-C	mech auto	3/3	band	95	2000	—	3	0	clutch/brake
Fiat-Allis	41-B	Cummins	VT1710-C	mech auto	3/3	band	95	2300	—	3	0	clutch/brake
International	TD-15C	International	DT-466B	mech auto	2/2	disc	60	1850	gear	3	0	clutch/brake
International	TD-20E	International	DVT-800	mech auto	3/3	disc	70	1900	gear	3	0	clutch/brake
International	TD-25E	International	DTI-817C	mech auto	6/6	disc	76	2250	gear	3	0	clutch/brake
Komatsu	D65E-6	Cummins	NH220-CI	mech auto	3/3	band	66	2000	gear	3	0	clutch/brake
Komatsu	D85E-18	Cummins	NT855-C	mech auto	3/3	band	68	2000	gear	3	0	clutch/brake
Komatsu	D155A-1	Komatsu	S6D155-4	mech auto	3/3	band	94	2000	gear	3	0	clutch/brake
Komatsu	D355A-3	Komatsu	SA6D155-4A	mech auto	4/4	band	105	2000	gear	3	0	clutch/brake
Komatsu	D455A-1	Cummins	VTA1710-C800	mech auto	4/4	disc	124	2490	gear	3	0	indep trans
Terex	D600-D	Hanomag	D962-K	mech auto	3/3	disc	48	1990	gear	3	0	clutch/brake
Terex	D700-D	Hanomag	D963-A1	mech auto	3/3	disc	61	2135	gear	3	0	clutch/brake
Terex	82-20B	Detroit	6V-71T	mech auto	3/3	disc	63	2000	gear	2	0	clutch/brake
Terex	82-30B	Detroit	8V-71T	mech auto	3/3	disc	72	1750	gear	2	0	clutch/brake
Terex	82-50	Detroit	12V-71T	mech auto	3/3	disc	72	2250	gear	2	0	clutch/brake

Table 2.2 (Continued)

CRAWLER DOZER SPECIFICATIONS

							CRAWLER DATA				
MAKE	MODEL	LENGTH (ft)	LENGTH W/O BLADE (ft)	MAX WIDTH W/O BLADE (ft)	HEIGHT TO TOP OF CAB (ft)	GROUND CLEARANCE (in)	OVERALL LENGTH (ft)	OVERALL WIDTH (ft)	STANDARD SHOE WIDTH (in)	GROUND PRESSURE (psi)	GRADEABILITY (%)
Case	1450-B	15.9	12.3	7.7	10.3	15	—	7.7	18	8.8	—
Caterpillar	D6-D	15.8	12.3	7.8	10.3	12	—	7.8	18	9.4	—
Caterpillar	D7-G	17.3	13.8	8.6	10.7	14	—	8.4	20	10.4	—
Caterpillar	D8-K	21.6	17.3	10.0	11.2	17	—	9.2	22	13.0	—
Caterpillar	D9-L	22.9	17.4	11.9	13.8	24	—	10.2	24	16.7	—
Caterpillar	D10	24.8	19.4	13.8	14.8	28	—	11.8	28	22.1	—
Deere	850/6545	16.0	—	—	10.5	16	—	—	20	9.7	—
Fiat-Allis	14-C	16.1	12.1	8.6	10.2	18	—	7.8	20	9.4	—
Fiat-Allis	16-B	17.7	14.1	—	10.5	13	—	—	22	9.1	—
Fiat-Allis	FD-20	17.3	14.8	9.7	11.5	21	—	—	22	11.1	—
Fiat-Allis	FD-30	—	—	—	—	—	—	—	—	—	—
Fiat-Allis	31	23.5	17.0	—	12.5	22	—	10.8	28	12.6	—
Fiat-Allis	41-B	24.5	—	—	12.6	23	—	—	32	12.3	—
International	TD-15C	16.5	12.7	9.0	9.7	17	—	7.7	18	9.1	—
International	TD-20E	19.4	14.7	9.5	10.6	18	—	8.3	22	10.0	—
International	TD-25E	21.9	17.2	10.6	11.2	20	—	9.0	22	13.1	—
Komatsu	D65E-6	17.7	13.7	—	9.8	16	—	—	22	8.4	—
Komatsu	D85E-18	21.8	—	—	11.1	—	—	8.6	22	9.9	—
Komatsu	D155A-1	23.9	14.6	—	11.9	20	—	9.0	24	13.6	—
Komatsu	D355A-3	25.0	15.3	—	12.6	23	—	9.8	28	14.7	—
Komatsu	D455A-1	29.3	—	—	13.7	21	—	11.4	32	17.0	—
Terex	D600-D	—	—	—	—	—	—	—	22	7.2	—
Terex	D700-D	—	—	—	—	—	—	—	22	8.9	—
Terex	82-20B	17.6	13.0	—	10.9	17	—	8.4	20	11.5	—
Terex	82-30B	20.0	15.3	—	10.8	18	—	8.3	22	11.2	—
Terex	82-50	20.2	16.2	—	12.5	21	—	8.4	24	12.9	—

Table 2.3
WHEEL DOZER SPECIFICATIONS

MAKE	MODEL	ENGINE HP	WEIGHT (lbs)	RIMPULL (lbs)	FUEL TANK (gal)	COOLANT (gal)	MAX SPEED F/R (mph)	STEERING ANGLE (degrees)	BLADE WIDTH (ft)	BLADE HEIGHT (ft)	DIG DEPTH (in)	BLADE TILT (in)	BLADE PITCH (degrees)
Caterpillar	814	170	41,400	39,683	93	13	20/24	44	12.0	3.5	18	25	25
Caterpillar	824-C	310	69,621	76,000	158	23	21/24	42	13.8	4.0	18	—	23
Clark	280	274	69,890	—	145	—	30/30	45	12.7	4.3	28	11	34
Clark	280	274	69,700	—	145	—	30/30	45	15.9	5.9	23	13	34
Clark	380-B	473	122,950	—	240	—	18/17	45	15.8	5.4	22	10	37
Clark	380-B	572	128,460	—	240	—	18/19	45	20.0	7.3	17	10	37
International	H-400C	580	140,000	—	250	38	20/20	40	20.0	7.3	30	—	8
LeTourneau	D-800	800	214,400	182,000	500	55	10/10	45	21.3	7.0	22	12	32
LeTourneau	D-800	820	193,000	182,000	500	65	10/10	45	22.4	9.3	22	12	32
Melroe	M-870	670	175,000	—	700	—	14/14	0 (skid)	22.0	7.0	—	—	—
Melroe	M-880	750	180,000	—	700	—	14/14	0 (skid)	22.0	7.0	—	—	—
Terex	WD-3000	280	48,500	—	—	air	40/40	43	11.5	4.4	16	—	5

Table 2.3 (Continued)

WHEEL DOZER SPECIFICATIONS

MAKE	MODEL	ENGINE MAKE	ENGINE MODEL	TRANSMISSION TYPE	NO. OF GEARS F/R	BRAKES	HYDRAULIC SYSTEM DATA: PUMP FLOW (gpm)	RELIEF VALVE PRESSURE (psi)	PUMP TYPE	NO. OF CYLINDERS
Caterpillar	814	Caterpillar	3306	mech auto	4/4	shoe	33	2250	—	4
Caterpillar	824-C	Caterpillar	3406	mech auto	4/4	disc	68	2250	—	3
Clark	280	GM	8V-71N	mech auto	8/4	shoe	61	1750	gear	5
		Cummins	NT-855-C334							
Clark	280	GM	8V-71N	mech auto	8/4	shoe	61	1750	gear	5
		Cummins	NT-855-C335							
Clark	380-B	Cummins	KTA-1150-C525	mech auto	6/3	shoe	80	1850	gear	5
		GM	16V-71N55							
Clark	380-B	Cummins	VT-1710-C635	mech auto	6/3	shoe	80	1850	gear	5
International	H-400C	Cummins	VT-1710-C635	mech auto	2/2	shoe	90	2500	gear	3
LeTourneau	D-800	GM	16V-92T	electric	n.a.	disc	107	2050	—	6
		Cummins	KT-2300-C860							
LeTourneau	D-800	Cummins	KT-2300-C900	electric	n.a.	disc	107	2050	—	6
		GM	16V-92T							
Melroe	M-870	Cummins	NT-855-P355	mech auto	4/4	—	112	2000	—	6
		GM	8V-71T							
		Caterpillar	3406-DIT							
Melroe	M-880	Caterpillar	3406-PCTA	mech auto	4/4	—	112	2000	—	6
Melroe		GM	8V-92T							
Terex	WD-3000	Deutz	V-10	mech auto	4/4	disc	—	—	—	—

Table 2.3 (Continued)
WHEEL DOZER SPECIFICATIONS

MAKE	MODEL	LENGTH (ft)	LENGTH W/O BLADE (ft)	MAX WIDTH W/O BLADE (ft)	HEIGHT TO TOP OF CAB (ft)	GROUND CLEARANCE (in)	CLEARANCE CIRCLE (ft)	WHEELBASE (ft)	TREAD F/R (ft)	STANDARD TIRES
Caterpillar	814	21.3	17.8	9.1	10.6	14	39.3	10.2	7.1/7.1	23.5-25 12PR
Caterpillar	824-C	25.2	22.0	10.1	13.0	19	—	11.6	7.8/7.8	29.5-25 16PR
Clark	280	27.3	23.0	11.0	12.6	16	47.0	12.5	7.8/8.5	29.5-29 22PR
Clark	280	31.3	23.0	11.0	12.6	16	50.0	12.5	7.8/8.5	29.5-29 22PR
Clark	380-B	33.0	27.8	11.8	14.5	20	56.3	15.3	9.6/9.6	33.25-35 20PR
Clark	380-B	35.0	28.0	12.5	14.7	22	60.3	15.3	9.2/9.2	37.25-35 30PR
International	H-400C	34.3	—	—	15.0	22	63.7	15.0	—	6540-39 30PR
LeTourneau	D-800	42.4	33.2	15.1	18.2	31	78.8	18.0	12.0/12.0	37.5-39 36PR
LeTourneau	D-800	43.4	33.2	15.1	18.2	31	81.2	18.0	12.0/12.0	37.5-39 36PR
Melroe	M-870	38.9	30.5	15.0	16.0	—	—	21.5	—	29.5-29 28PR
Melroe	M-880	38.9	30.5	15.0	16.0	—	—	21.5	—	29.5-29 28PR
Terex	WD-3000	26.8	—	9.0	9.8	18	—	11.3	6.5/6.5	26.5-25

Chapter 3

SCRAPERS

Figure 3.1 Typical Standard Tandem Scraper

The scraper is a rather unique machine because of its ability to excavate material in thin horizontal layers, transport the material considerable distances and then discharge it in a spreading action.

Scrapers have been around for over fifty years. Modern units are rubber tired, self-propelled, high powered machines — some capable of carrying up to 72 cu. yd. They are designed for off-highway use with unusually high transport speeds and good maneuverability. This discussion will be limited to the larger-sized, two axle scrapers used in the mining industry. The scrapers self-load under favorable conditions, and provide both intermediate distance transport and controlled spreading capabilities.

TYPICAL UNITS

All units are two axle, articulated body designs with diesel engines and high maximum travel speeds. There are four basic configurations available; they are illustrated in figure 3.2. The types differ in the number of engines and whether or not they have an elevator to assist in material loading. The primary operating difference between these types is available power as it relates to excavating, loading, and traveling.

Standard

- Single, front mounted engine on tractor
- Front axle drive (two wheel)
- Capacities from 24 to 64 tons (20 to 54 cubic yards)
- Horsepower from 325 to 615

Tandem

- Front mounted engine on tractor and rear engine on scraper body
- Two axle drive (four wheel)
- Capacities from 24 to 52 tons (20 to 50 cubic yards)
- Horsepower from 288 to 950

Elevating

- Single, front mounted engine on tractor
- Front axle drive (two wheel)
- Elevating conveyor forward and above cutting edge to assist loading
- Capacities from 13 to 38 tons (11 to 34 cubic yards)
- Horsepower from 150 to 550

Tandem elevating

- Front mounted engine on tractor and rear engine on scraper body
- Two axle drive (four wheel)
- Capacities from 19 to 38 tons (16 to 34 cubic yards)
- Horsepower from 280 to 700

Manufacturers

Domestic manufacturers dominate the U.S. market; they are listed in Table 3.1. Many are also suppliers of dozers and front-end loaders; sales and service are conducted through a dealer network. Until recently, WABCO also made and marketed scrapers.

BASIC MACHINE OPERATION

The basic operation of the scraper is excavating material and transporting it over medium distances.

- The scraper is loaded by propelling forward with the cutting blade (mounted on the forward edge of the bowl) lowered below the level of the front wheels so as to take a thin horizontal cut of material. The slice of material is pushed by the machine's forward motion into a body cavity (bowl) for storage during transport. The depth of the cut is controlled by raising or lowering the bowl with respect to the original ground surface, normally 3 to 15 inches. Cut width is normally 10 to 12 feet.
- When the bowl has been filled, it is raised clear of the ground and an "apron", which acts similar to a vertical gate, is lowered, blocking the front of the bowl so that the material is contained.
- The machine then propels at relatively high speeds (normally 15 to 25 mph) along haul roads to the material disposal area.

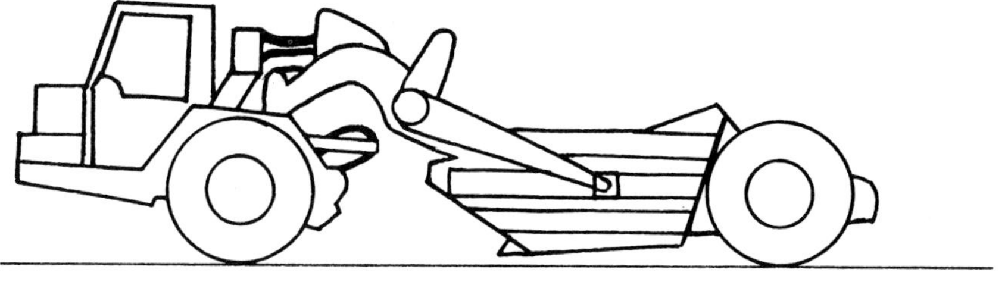

STANDARD (SINGLE ENGINE)

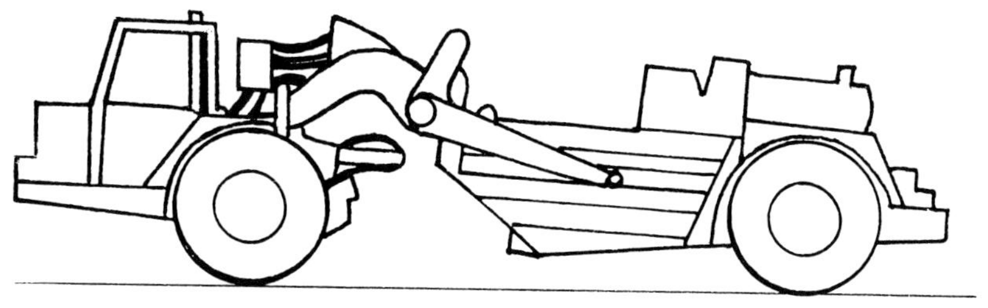

TANDEM (DUAL ENGINE)

STANDARD ELEVATING

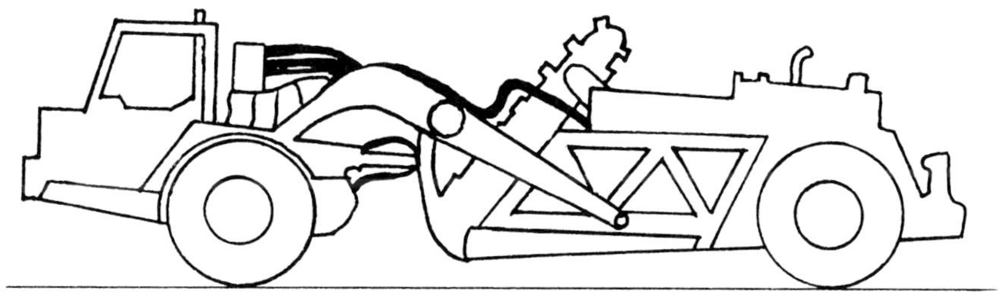

TANDEM ELEVATING

Figure 3.2 Scraper Types

Table 3.1

SCRAPER MANUFACTURERS

Big Bud Tractors, Inc. P.O. Box 1111 Havre, Montana 59501	Standard 470 FWHP 25 cu. yd.
Caterpillar Tractor Co. 100 N.E. Adams Street Peoria, Illinois 61629	Standard 330–550 FWHP 20–54 cu. yd. Standard elevating 150–450 FWHP 11 –34 cu. yd. Tandem-powered 450–950 FWHP 20–44 cu. yd. Tandem-powered elevating 700 FWHP 34 cu. yd.
Clark Equipment Company Construction Machinery Div. Pipestone Rd., P.O. Box 547 Benton Harbor, MI 49022	Standard elevating 175–325 FWHP 11–23 cu. yd. Tandem-powered elevating 300 FWHP 16 cu. yd.
Deere & Company John Deere Road Moline, Illinois 61265	Standard elevating 175–250 FWHP 11–16 cu. yd.
Fiat Allis Construction Machinery, Inc. Box F, 106 Wilmot Road Deerfield, Illinois 60015	Standard 325 FWHP 21 cu. yd. Standard elevating 325 FWHP 23 cu. yd. Tandem-powered 500 FWHP 21 cu. yd. Tandem-powered elevating 500 FWHP 23 cu. yd.
International Harvester Construction Equipment Group 600 Woodfield Avenue Schaumburg, Illinois 60196	Standard 330 FWHP 21 cu. yd. Standard elevating 170–330 FWHP 11–22 cu. yd. Tandem-powered 510 FWHP 21 cu. yd. Tandem-powered elevating 510 FWHP 22 cu. yd.

Table 3.1 (Continued)

SCRAPER MANUFACTURERS

Komatsu Limited Komatsu Building 2-3-6 Akasaka Minato-Ku Tokyo, Japan	Standard 425 FWHP 30 cu. yd.
M-R-S Manufacturing Co. P.O. Box 199 Flora, Mississippi 39071	Standard 420–620 FWHP 30–39 cu. yd. Standard elevating 420–550 FWHP 27–32 cu. yd.
Terex, Corp. IBH Group Hudson, Ohio 44236	Standard 475 FWHP 34 cu. yd. Standard elevating 310 FWHP 23 cu. yd. Tandem-powered 290–720 FWHP 20–50 cu. yd.

Photograph 3.1 Terex S-24B tandem scraper self loading in overburden

Photograph 3.2 Terex TS-24B tandem scraper dumping

- The load is dumped by rotating the bowl back plate forward, or using a ram to move the back of the bowl forward pushing the load out the front with the apron raised. The machine continues to move forward while dumping so that the load is spread uniformly: the depth is controlled by the positioning of the height of the bowl above the ground. Alternately the scraper can be discharged into a drive-over type of grizzly.

- After completion of the dumping, the bowl is raised to a travel position and the machine returns at high speed (normally 25 to 35 mph) on the haul roads to the loading site.

- The overall operating cycle is illustrated in Figure 3.3. Several techniques will be discussed later for providing additional power, if required, during the excavating portion of the cycle.

APPLICATIONS

Since the scraper both digs and hauls the material, like the dozer, it is a very versatile piece of earthmoving equipment. It is found at many mine sites performing a variety of primary mining tasks as well as general utility work.

- Topsoil removal: The scraper is broadly used in those activities which involve selective removal of horizontal horizons and transport to storage.

- General reclamation: The scraper is applied in the rough leveling and contouring phase and for replacement of the upper horizons prior to revegetation.

- Overburden removal (with or without prior ripping): These can be either initial cuts or prebenching operations for other excavating

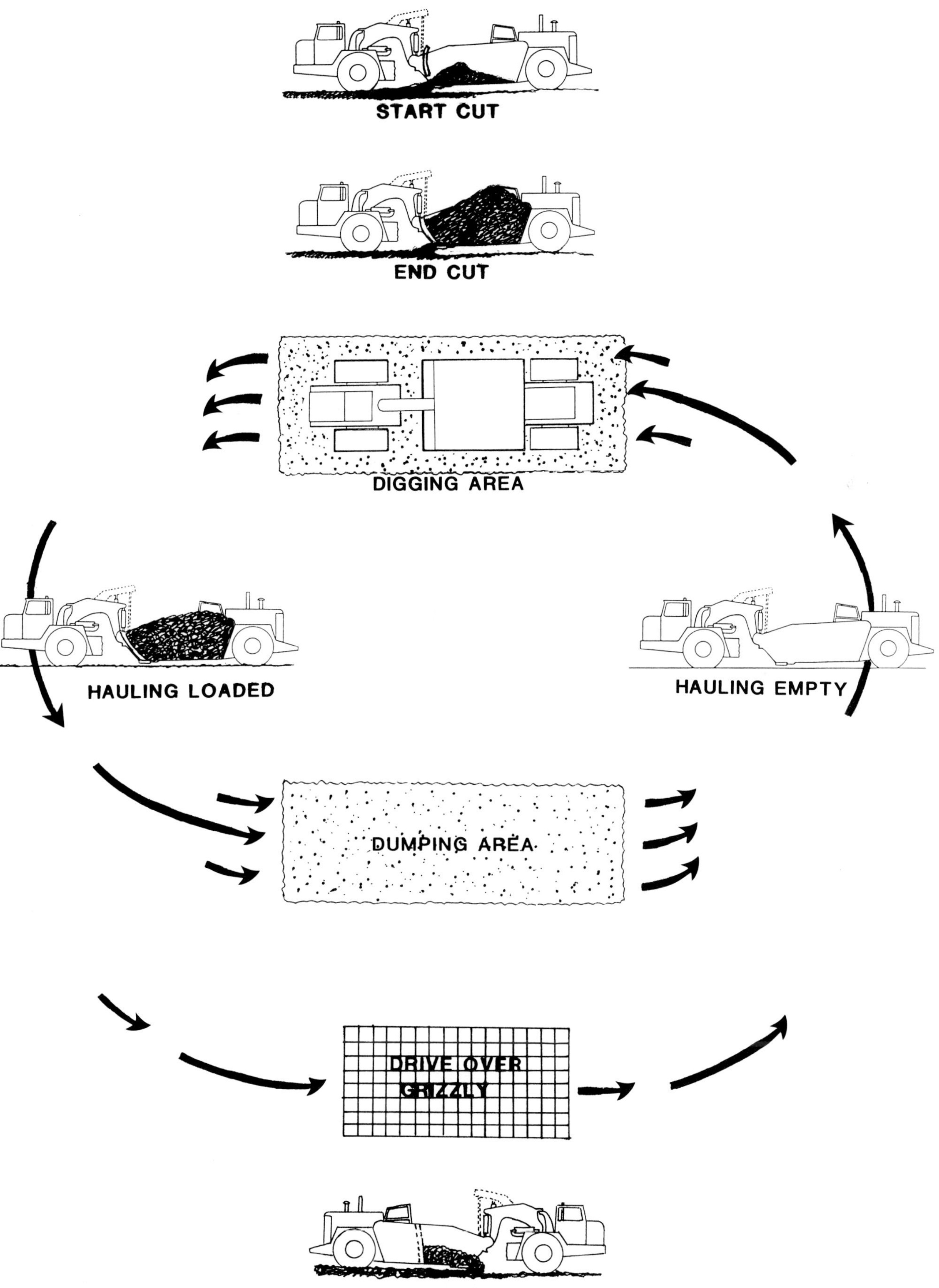

Figure 3.3 Typical Scraper Production Cycle

equipment, or complete overburden removal. The latter case requires a well planned circular operational layout to minimize travel distances and utlilize downgrade loading and dumping. Typically, operations of this type use dozers for preshaping, supplementary material transport and push-pull scraper loading techniques.

- Ore/Coal removal (with or without ripping): Scrapers are employed in cases where the seams are thin and other types of excavating equipment are inefficient.

- Parting removal: similar to above.

- General utility work:
  - — Stockpiling and reclaiming
  - — Site preparation
  - — Road construction
  - — Ponds and ditches

GENERAL CHARACTERISTICS

The scraper, with its combination of excavating and transporting capabilities, is a more complex design than the dozer. The powered functions include the following. (See Figure 3.4):

- Two or four wheel propel drive; torque converter

- Articulated steering, hydraulic cylinders

- Bowl raise/lower, hydraulic cylinders

- Apron raise/lower, hydraulic cylinders

- Dump action, hydraulic cylinders

- Elevator (some models), hydraulic motors

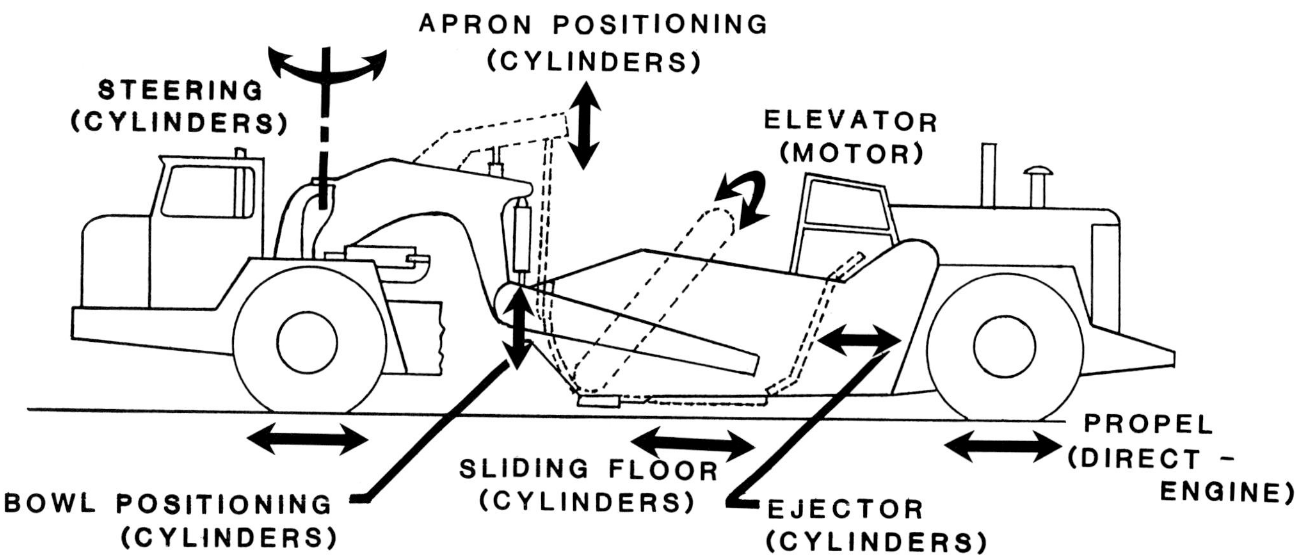

Figure 3.4 Scraper Powered Functions

- Push/pull hook (some models), hydraulic cylinders

While there are seven powered functions, it should be noted that other than the main propel drive these are all relatively simple mechanisms.

In comparison with other types of excavating equipment the following general features of scrapers should be recognized.

- Readily available, proven design
- Excellent mobility with high propel speeds (to 35 mph)
- Good stability
- Low operator skill required
- One operator per unit
- Maintenance can be performed in a shop
- Low initial capital investment
- Relatively short service life (8,000 to 12,000 hours)

From an operational viewpoint the scraper has the following characteristics.

- Has the ability to remove and deposit materials in thin layers
- Can excavate unconsolidated materials
- Requires advanced preparation, such as ripping, in consolidated materials
- Competitive haulage distances ranging from 400 to 5000 feet
- Moderate gradeability
- Moderate turning radius (30 to 60 ft.)
- High ground pressures

The scrapers leave a clean floor and can blend materials over the pit area. Overall performance can be reduced and/or completely curtailed in severe weather, or where bad traction conditions or visibility exist. Auxiliary equipment (dozers) may be required to assist in loading and in some cases to grade hard roads and dump areas to reduce rolling resistance.

Power demand for efficient scraper loading often exceeds that available on the standard single engine machines. Poor traction conditions will similarly restrict loading operations. Four approaches to solving this problem are available. (See Figure 3.5):

- Push loading with a dozer
- Additional machine power and four wheel drive available with tandem designs employing two engines
- Adding elevating devices to assist in moving the material from the cutting edge back into the bowl.
- Push-pull operation of a pair of scrapers to utilize their combined power

Combinations of the above are also employed. However, the elevating device imposes several restraints, specifically, elevating units should not be push loaded or used in a push-pull configuration. The maximum boulder size that can be tolerated is also reduced.

Graphs 3.1, 3.2, and 3.3 show the relationships between weight and horsepower for the four types of scrapers. Empty (net) machine weight increases in the tandem design with the addition of the second engine. Total horsepower for any size (capacity) unit is increased by 30 to 70%. The second rear mounted engine is smaller than the front unit. It is important to recognize that the standard single engine provides a single axle, two wheel drive, while the tandem has a two axle four wheel drive. The four wheel drive has superior performance under bad traction conditions. Gross (loaded) machine weight per horsepower is as expected, considerably lower for tandem scrapers, but is increased with the addition of the elevating device. There is a considerable increase in this ratio for standard scrapers as the capacity increases.

STANDARD SCRAPER, PUSHED BY DOZER

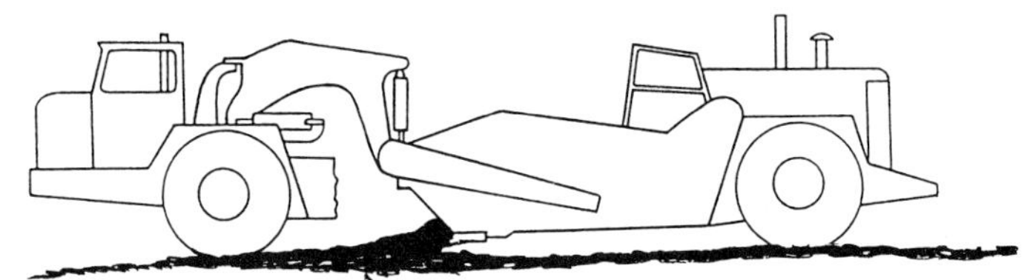

TANDEM SCRAPER, SELF-LOADING

ELEVATING SCRAPER, SELF-LOADING

PUSH-PULL SCRAPERS, ALTERNATE LOADING

Figure 3.5 Scraper Loading Configurations

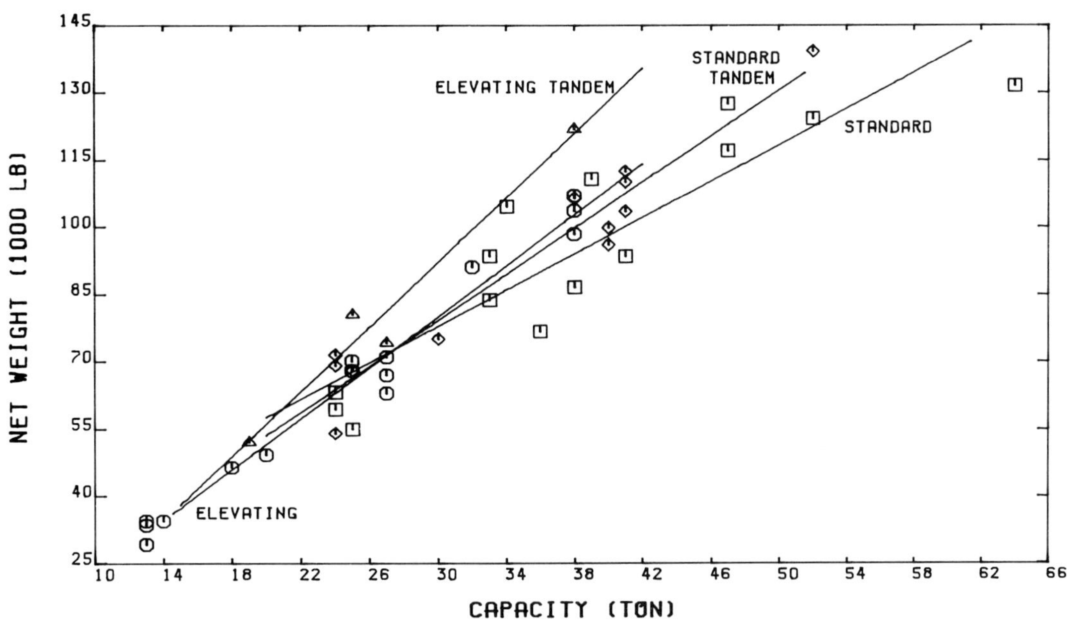

Graph 3.1 Net weight/ton of capacity

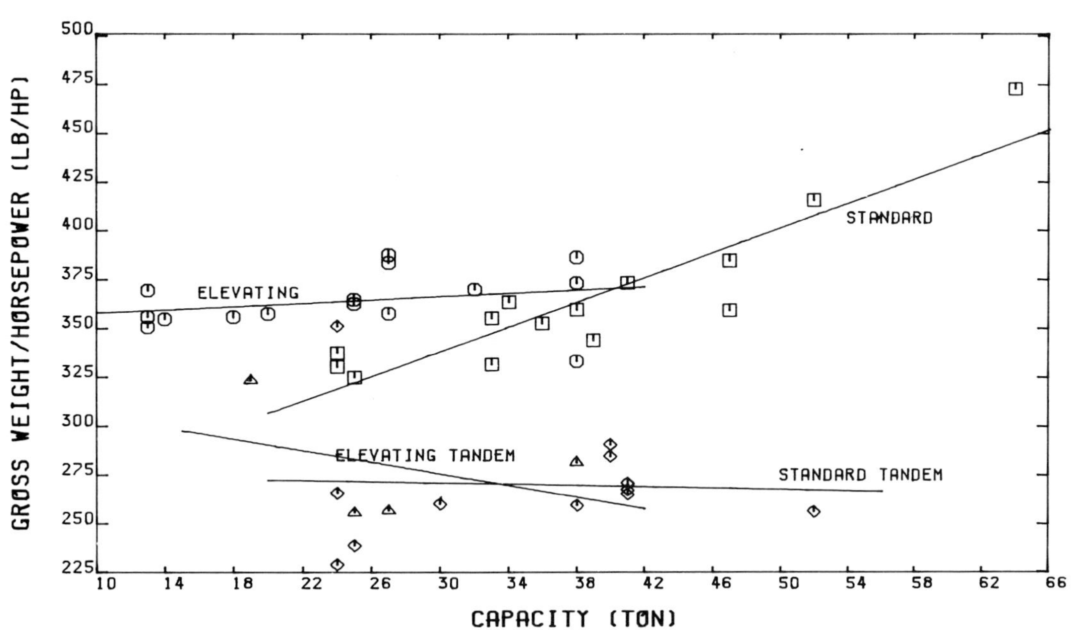

Graph 3.2 Gross vehicle weight per HP/ton of capacity

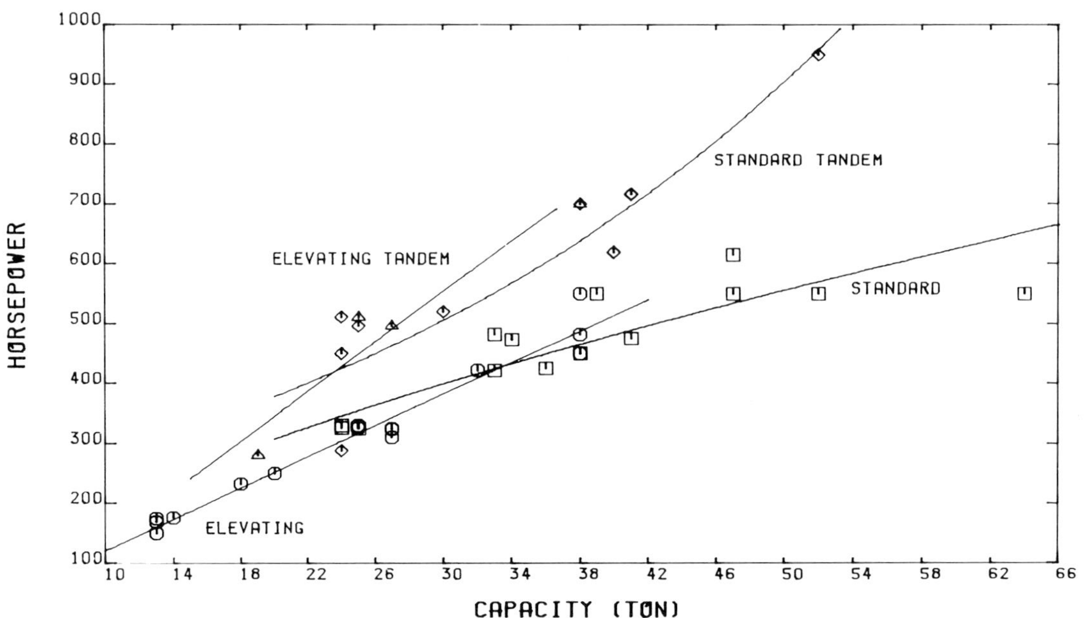

Graph 3.3 HP/ton of capacity

OPERATIONAL PRACTICES

- Forces applied during loading and the profile of the cut are shown in Figure 3.6.
- A typical work cycle consists of:
  — Loading with 3 to 18 inch cuts
  — Hauling at 15 to 25 mph
  — Spreading (dumping) to 24 inches deep
  — Returning at 20 to 35 mph
- Advanced preparation (ripping/blasting) is required for consolidated materials.
- Some characteristics of suitable material are unconsolidated, friable shale and coals, minimal rock and large boulders, and minimal large lumps.
- Cutting depth depends on: machine geometry, available power, traction conditions and job layout
- Scraper cutting edge designs are matched to materials being loaded. Teeth are added for breaking up soft shale and coal.
- Cohesive clays require long, thin cuts
- Loose material is pump loaded with short deep cuts
- Slot loading can reduce fill time
- Downhill loading reduces cutting time because gravity increases the digging force.
- Ejector back is retracted rearward while loading to improve fill boiling action (increase density)
- The loading rate is not uniform over the cutting time. As time increases the filling rate decreases. There is, therefore, a trade-off between loading time and payload which should be considered.
- Loading can be assisted by push loading with dozers, the additional power of tandem units, elevating devices, or push-pull combinations. (See Figure 3.5 referenced earlier) Push-pull scrapers are specially equipped machines which load in pairs. Two machines are physically hooked-up to each other with front and rear end attachments. The rear machine

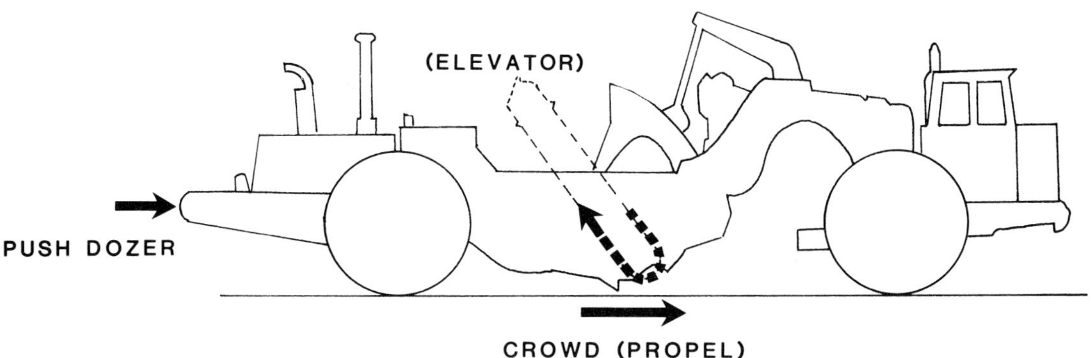

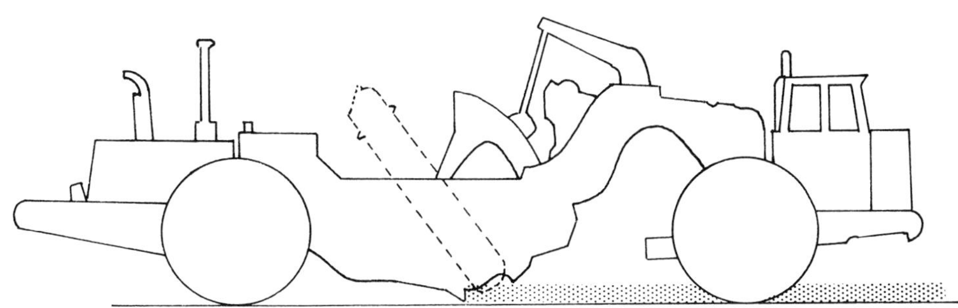

Figure 3.6 Scraper Digging Forces & Digging Profile

push-loads the front machine, and the front machine pull-loads the rear machine; each machine hauls and dumps independently. Digging cuts are longer (approximately 200 ft.). Units cannot negotiate turns while hooked together. The loaded weight of the forward machine during its pull loading phase significantly improves its traction.

- When the dozer assisted or push-pull techniques are employed in fleet operations, machine matching and scheduling becomes critical. Excessive delays can develop in the loading area if the scrapers must wait for either the push unit or a second scraper. Additional supervision may be required and care exercised in the matching up to avoid machine damage. The self loading elevating scrapers avoid these operating inefficiencies.

- Standard scrapers can be top loaded by other loading equipment, such as front-end loaders or shovels.

- Haul distances (400 to 5,000 ft) should be minimized, with well planned and maintained haul roads, and one-way traffic if practical.

- Hauling and returning speeds are limited by: road conditions, traction, safe speeds, traffic, and grades.

- Maximum grades for standard scrapers are 12% empty, 6% loaded; for tandem scrapers 25% empty, 12% loaded.

- Tire wear can be severe if operating on ripped or blasted rock.

Photograph 3.3 Caterpillar tandem scraper loading with dozer assist

Photograph 3.4 Caterpillar 637 tandem scrapers in a push-pull overburden loading operation

Photograph 3.5 Fiat-Allis 263B tandem elevating scraper self loading

- Three axle units (2 axle tractor) while not commonly employed currently in mining applications, for special situations, provide lower ground pressures, improved traction capabilities and have better long haul characteristics.

- Wet weather restricts operations and becomes particularily troublesome if it occurs during non-operating periods when the scraper is not continually redoing the surface in the load and dump areas. Scheduling must recognize lost days because of weather conditions.

DESIGN FEATURES

Figure 3.7 identifies basic scraper nomenclature.

Most of the scraper manufacturers use their own engines. General engine characteristics and specifications are summarized in Appendix B. As noted earlier the second engine in the tandem machines is a smaller unit to reduce the potential for jack-knifing under bad traction conditions. The second engine and drive train module are virtually totally independent of the forward unit with essentially only electrical start and shifting synchronizing control interconnections. (See Figure 3.8).

The typical drive train includes:

- 6 to 8 forward speed ranges, one reverse range

- Semi-automatic power shift

- Combination converter or direct drive operation

- Transmission — torque converter

- Downspeed inhibitor to prevent engine overspeed

- Locking differential

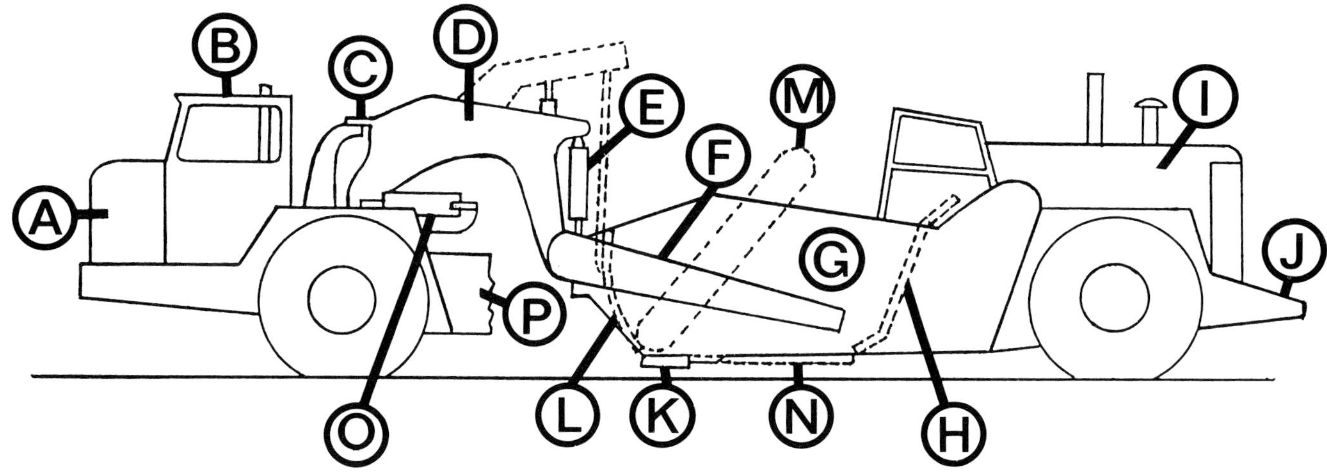

A	Tractor Engine	I	Scraper Engine
B	Cab	J	Push Block
C	Hitch	K	Cutting Edge
D	Gooseneck	L	Apron
E	Bowl Cylinder	M	Elevator
F	Draft Arm	N	Sliding Floor
G	Bowl	O	Steering Cylinders
H	Ejector	P	Transmission

Figure 3.7 Scraper Nomenclature

The torque converter with its load sensing capabilities is used at lower speed and locks up to provide a direct drive at higher speeds with accompanying lower fuel consumption. It is of interest to note that John Deere has, on its machines, introduced a microprocessor transmission control for load sensing shifting in an effort to improve loading efficiency and reduce fuel consumption.

Because of the machine's weight and the typically high propel speeds, the brakes are an important consideration. Shoe type air brakes on each wheel are common, with those on the rear axle being applied first. Emergency systems are provided with automatic application in the event of low air pressure in the system. For applications requiring long downhill runs while loaded, retarders should be added to absorb some of the energy and reduce the demands on the brakes.

The apron contains the load in the scraper bowl and also provides a throttling means to control the discharge. It is a simple, pivoted plate curved to match the shape of the bowl, and actuated by a single or pair of hydraulic cylinders mounted on the gooseneck. (See Figure 3.9).

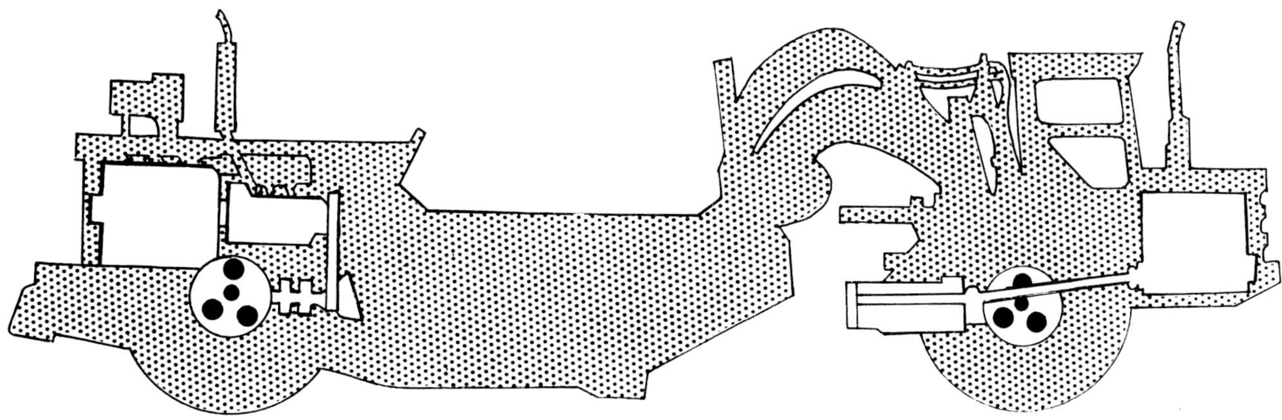

Figure 3.8 Power Train–Tandem Scraper

The bowl is similarly raised and lowered by a pair of cylinders which can be mounted either on the gooseneck or from the draft arms. These cylinders must be capable of transmitting high vertical forces to the cutting edge as required for penetration during excavating in hard formations. There are significant differences in design with respect to the exposure of these positioning cylinders and in particular the rods and seals to the material passing in and out of the bowl.

Caterpillar includes a cushioning device in the gooseneck between the tractor and scraper body. It is a nitrogen accumulator arrangement which acts as a shock absorber and reduces machine bouncing and loping at higher haul road speeds. The unit can be locked out for maximum down pressure or load lift requirements.

There are a number of designs for providing a positive ejection of the load from the bowl. Two of these are illustrated in Figure 3.10. The material can be discharged by pivoting the floor (and bowl) to dump the material forward, or alternatively, a dozer type push plate serving as the back of the bowl can produce the same action. Material cannot be pushed through the elevator from the rear so that machines so equipped must permit discharge behind the cutting blade and elevating mechanism. If a push plate is used in this case it must be coupled with a sliding floor that is automatically retracted rearward for load discharge. Both arrangements are commonly actuated by hydraulic cylinders mounted behind the bowl. An automatic kickout is provided at the end of the cylinder ejection stroke. There are differences of opinion as to which configurations are

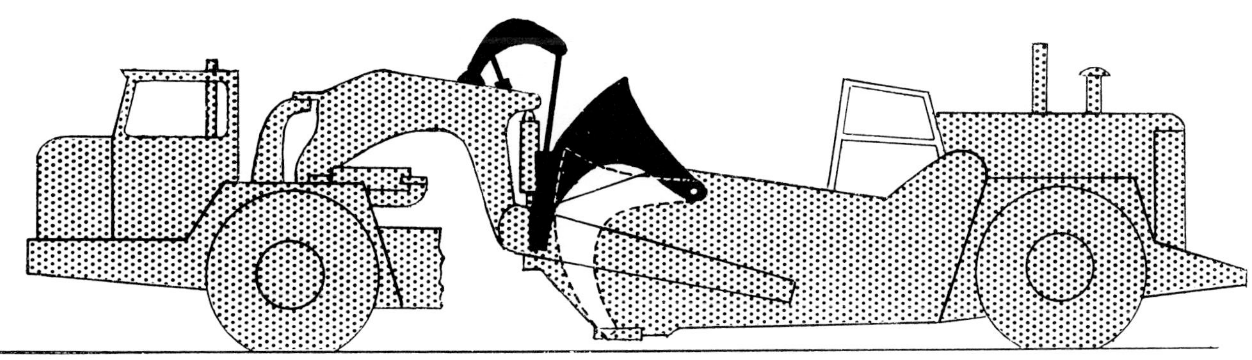

Figure 3.9 Apron Operation

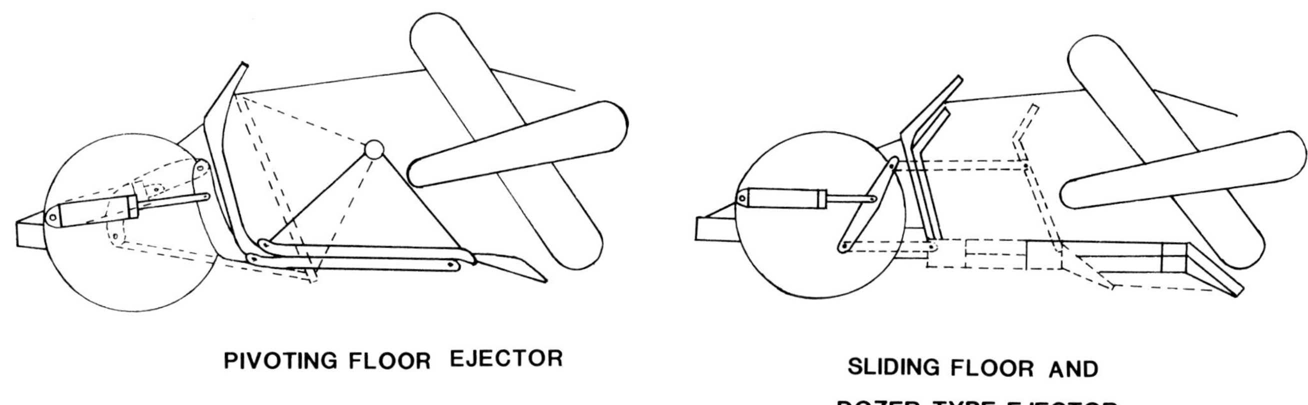

Figure 3.10 Bowl Ejectors

most effective when sticky materials must be handled. Either the cutting blade or an additional strike-off plate act to grade the material during discharge.

The latest Fiat-Allis machines have a liftable elevator which is automatically raised during discharge. This is combined with a variable speed control of the ejection.

The scraper bowl can be subjected to high loads as the material is sheared and forced back into the bowl under the propel power. This, together with the abrasive action, necessitates high strength steels and a double bottom (honey-comb) construction. The side profile for an elevating unit is generally higher because of the higher stacking of the material. Non-elevating versions can have low or high sides dependent on the anticipated material flow characteristics during loading.

Elevator design involves two primary considerations; one, the portion of available power diverted to the elevator drive and two, the incorporation of some type of flexible mounting to reduce impact loads and permit the mechanism to deflect to accomodate oversize material. (See Figure 3.11). Power is generally provided by hydraulic motors through a planetary gear reduction. Sprockets drive twin high strength chains which carry the flights (paddles) which help pull the material up and into the bowl. The main propel drive must not rob power from the elevator drive and thereby interfere with its ability to function under severe conditions. Elevator speed ranges from 100 to 300 fpm, in either direction. The direction is reversed during discharge. Some adjustment is provided in the elevator mounting to permit changing the spacing with respect to the cutting edge to match site conditions.

For operation in pairs to utilize the combined propel power during excavating, the units can be physically locked together in a push-pull configuration. The scrapers must be especially equipped for this matching with a hook added to the rear of each tractor and a positionable bail on the front. (See Figure 3.12). This permits the scraper to operate as either the lead or trailing unit. The bail arrangement includes a push block which is cushioned to ease initial contact impact and also to take up some of the slack during operation. Hydraulic cylinders lift the bail for engagement and then lower it into position over the hook. Any significant turns while hooked-up causes severe loading on the components.

The hydraulic system for the hydraulic actuators are low to medium pressure, closed and filtered. The steering system is commonly independent as a safety precaution. The circuit providing power for the hydraulic motor on the elevating device is more complex but generally only one or two spees are provided with gear pump and motors so that the full sophistication of the variable volume systems is avoided.

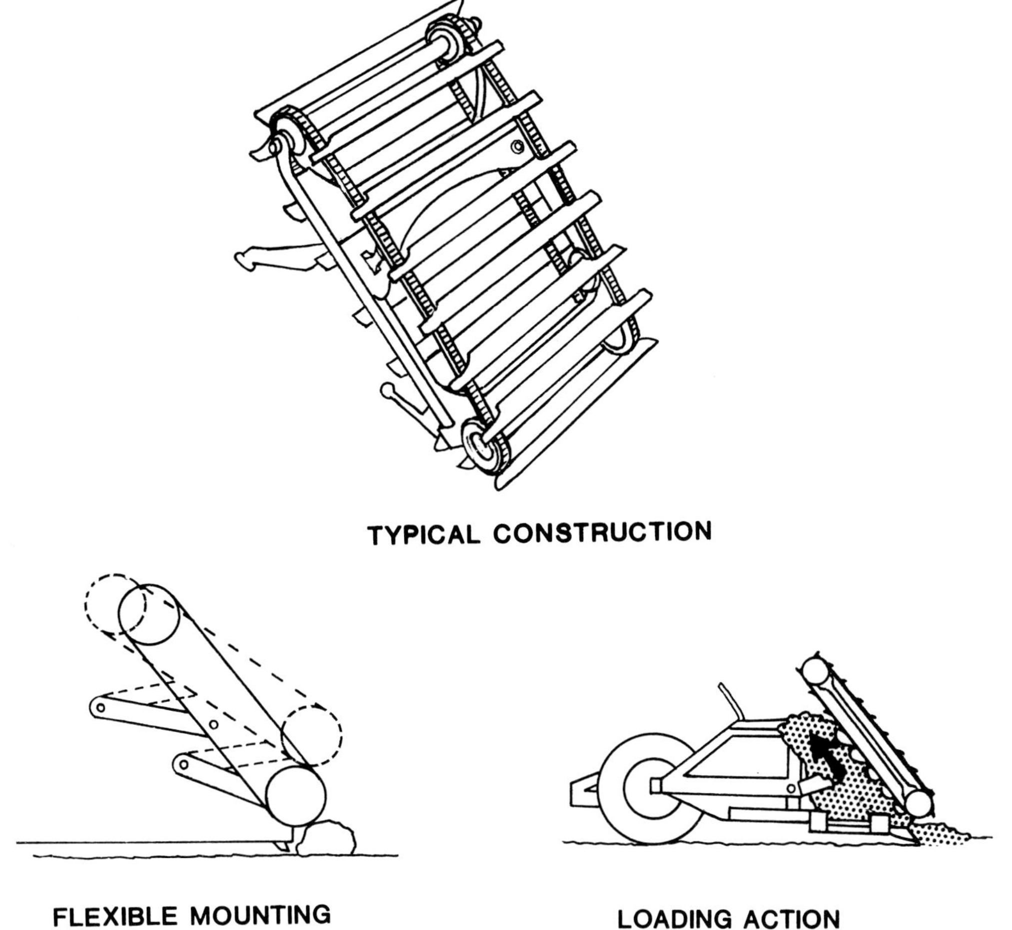

Figure 3.11 Elevating Mechanism

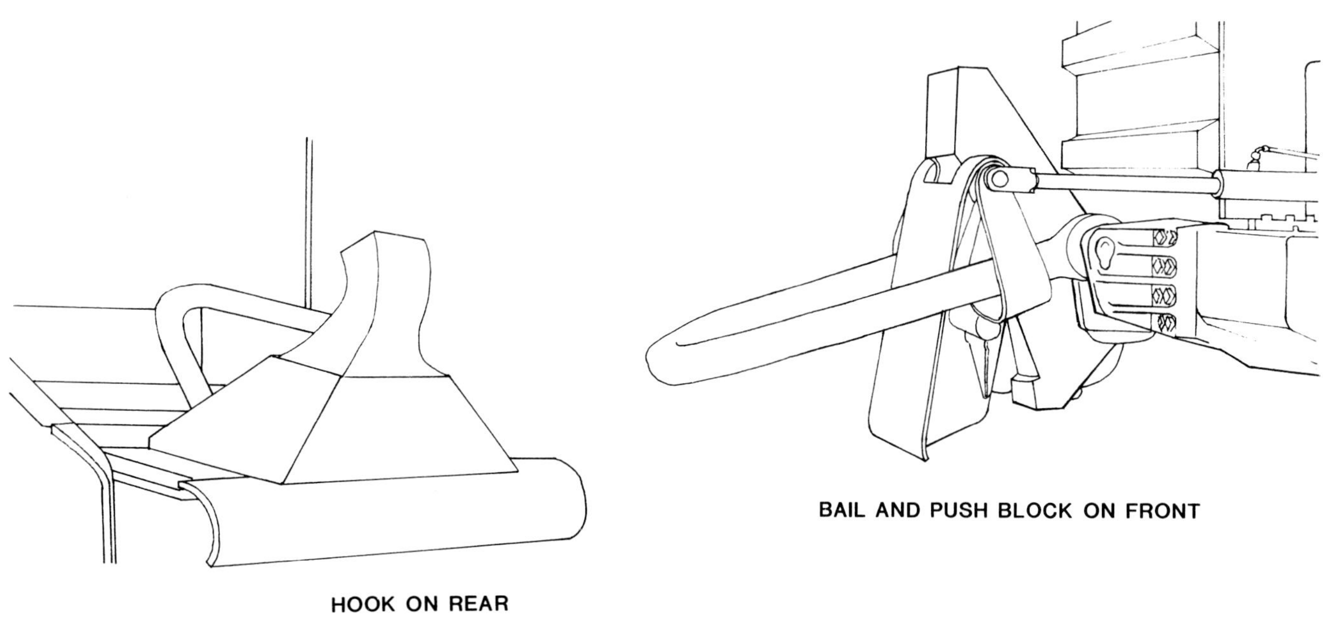

Figure 3.12 Push-Pull Devices

The scraper is a simple, proven design concept, fairly standardized with good accessibility for service and maintenance. The primary maintenance considerations are tied to wear and abuse in the following areas:

- Tire replacement
- Engine servicing
- Hydraulic system contamination
- Wear and damage
  - Cutting edges and teeth
  - Bowl bottom
  - Ejector
  - Elevator flights

SELECTION CONSIDERATIONS

The choice of the type of scraper to apply is generally dictated by the following considerations based on experience.

Standard scraper

- Generally requires push loading
- Total grade and rolling resistance of 3% to 10%
- Will handle clay, silt, sand, gravel, coal, well broken rock-dirt, ripped rock
- Travel distances from 500 to 5000 feet
- Most versatile
- Lowest owning and operating cost

Tandem Powered

- Best under bad traction conditions
- Reduced loading times and improved machine acceleration potential
- Dependent on formation, may require push loading
- Total grade and rolling resistance over 10%
- Will handle clay, silt, sand, gravel, coal, well broken rock-dirt
- Travel distances from 300 to 3500 feet
- Intermediate owning and operating costs

Single Engine-Elevating

- Self loading, restricted by traction conditions
- Total grade and rolling resistance of 3% to 7%
- Will handle clay, silt, sand, some gravels, coal (limited tolerance) for large boulders, blocky material
- Travel distances from 500 to 2500 feet
- High owning and operating cost

Tandem Elevating

- Self loading, less restricted by traction conditions
- Total grade and rolling resistance over 10%
- Will handle clay, silt, sand, some gravels, coal (limited tolerance for large boulders and blocky materials)
- Travel distances from 500 to 2000 feet
- Highest owning and operating cost

The following graphs provide insight into the relative capabilities of the various types of scrapers available. The data is quite scattered but the average horsepower per ton of capacity is about 13 for single engine units and 18 for two tandem machines. (See Graph 3.4). The relative weight efficiency of the design in terms of carrying capacity per pound of machine weight is shown in Graph 3.5. The comparative cutting power available per unit of cutting edge width is shown in Graph 3.6. The last illustrates the influence of machine size.

Graph 3.7 gives a simple conversion for ton of capacity into either struck or heaped cubic yards for the typical scraper. Little difference is noted for the four types of scrapers.

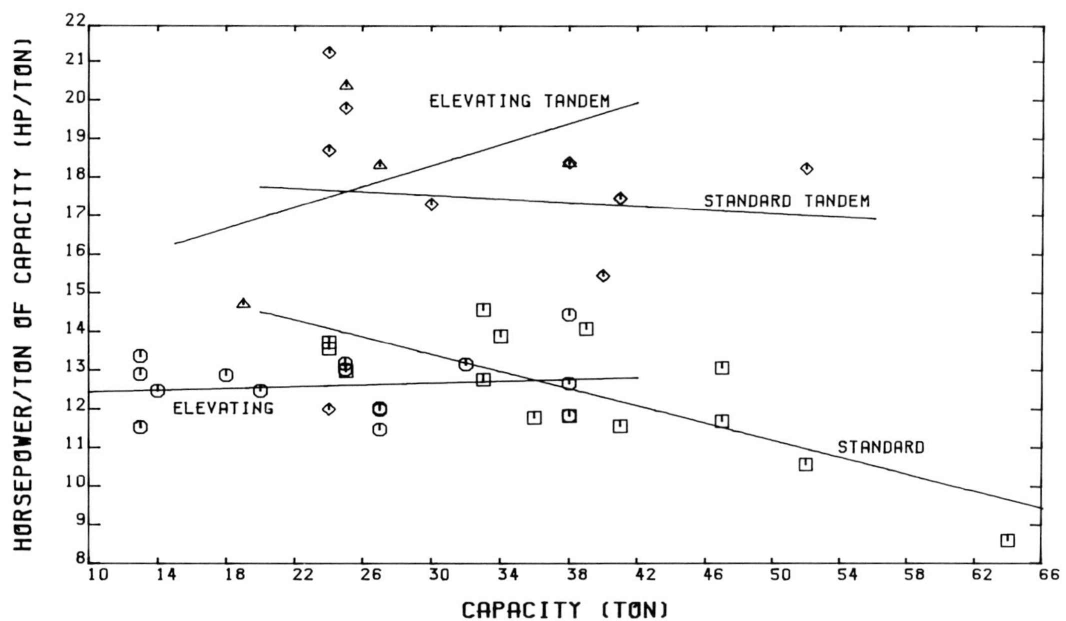

Graph 3.4 HP per ton of capacity/ton of capacity

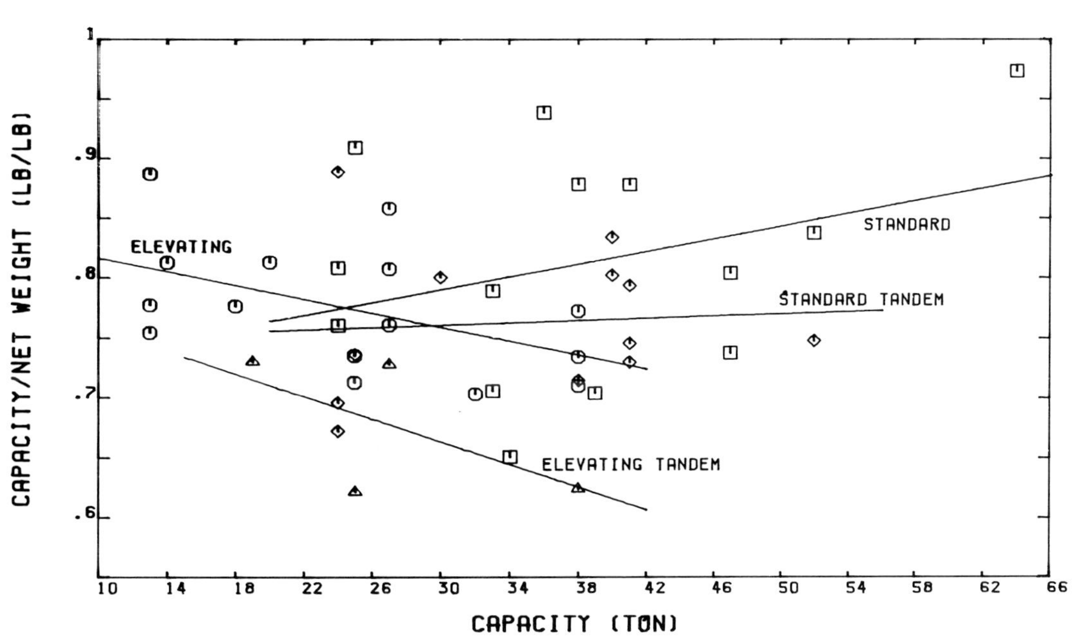

Graph 3.5 Payload per net vehicle weight/ton of capacity

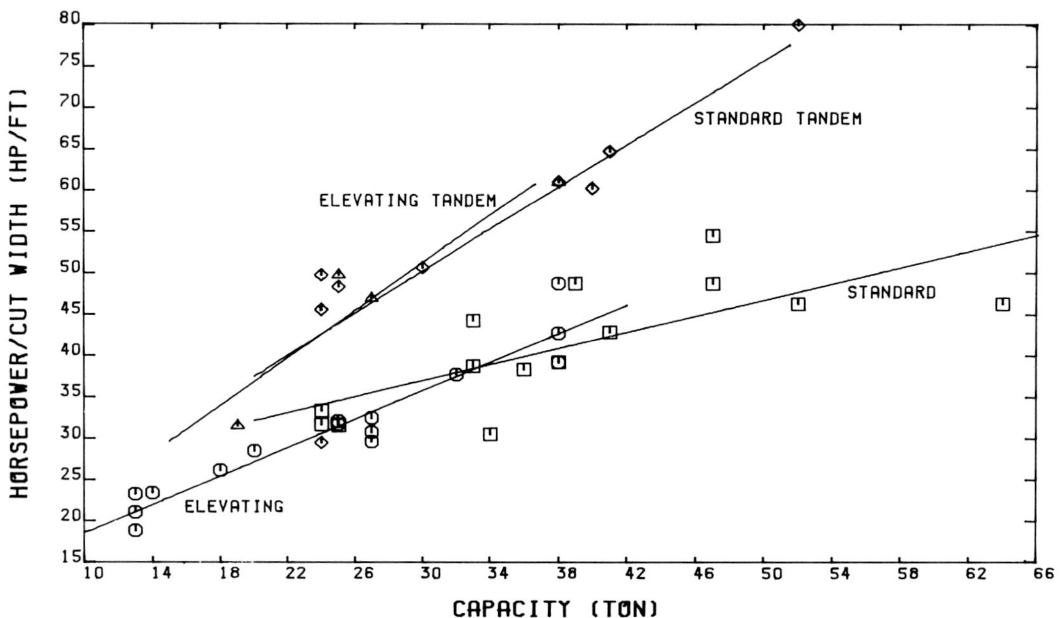

Graph 3.6 HP per cutting width/ton of capacity

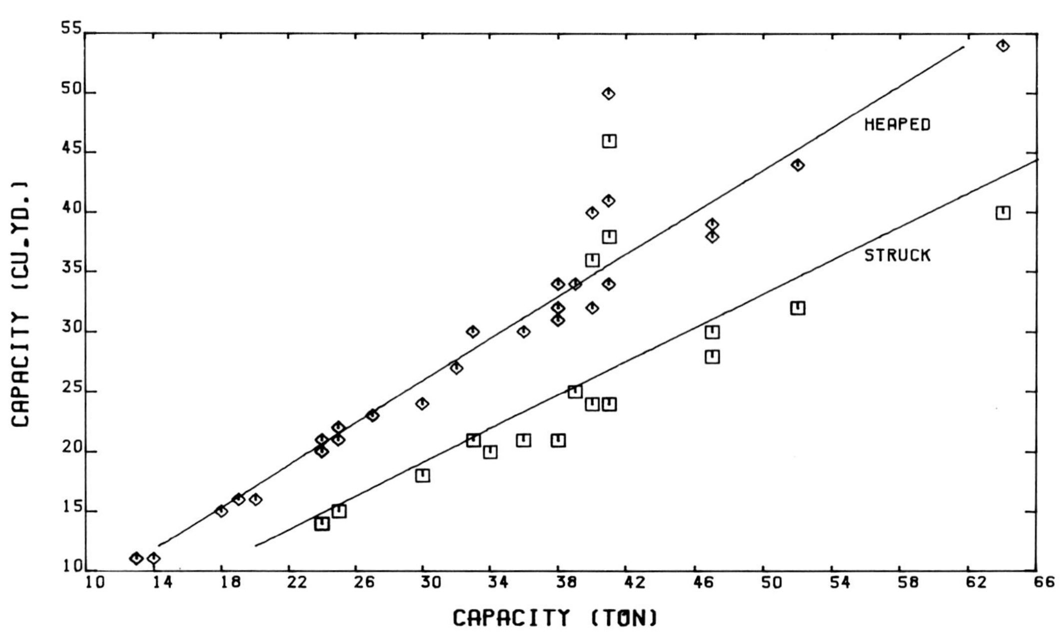

Graph 3.7 Cubic yards of capacity/ton of capacity

Production calculation procedures and representative performance estimates are provided in available manufacturers' literature. These, together with production cost estimates, are considered a separate subject too extensive to be comprehensively treated in this book. Estimates must be directly correlated with detailed site conditions and matched to specific machine performance characteristics. All of the equipment manufacturers will provide production and cost estimates for stated conditions prepared by their sales technical staff personnel based on their accumulated experience on similar applications. These frequently are developed by in-house computer programs so that alternative fleets of matched equipment can be readily evaluated to find the optimum combination.

It must be recognized that all of the above estimates are usually based on average data extrapolated to meet anticipated average site conditions over the life of the operations. The spread in these "averages" can be substantial and the correction factors introduced require a considerable amount of judgement. Costs, in particular, vary locally with time and economic conditions which make generalization dangerous. Some of the production considerations entering into these calculations are:

- Material swell
- Material unit weight
- Grades
- Rolling resistance
- Travel distance
- Maximum speeds
- Job efficiency
- Machine availability
- Engine power
- Type of scraper

Some of the cost considerations entering into these calculations are:

- Machine price
- Salvage value
- Machine life
- Interest
- Insurance
- Taxes
- Fuel consumption
- Lube, oils, greases, filter, etc.
- Operators' wages
- Tire cost
- Cutting edges
- Maintenance

The options available vary with the manufacturer. Alternative sizes and types are generally offered for such items as tires. Additional equipment options can be broken into categories as follows (note that some of these may be included as standard on some manufacturer's models):

Special Application

- Auxiliary lighting systems
- Locking differential
- Extended side cutters and cutting edges
- Ejector gate spill screen
- Cushion hitch
- Push-pull devices
- Retarders
- Counterweight and ballasting
- Special paint

Severe Duty Service

- Transmission guards

- Crankcase guards
- Bowl reinforcements
- Brake shields
- Mud flaps or fenders
- Radiator core protector grid

Performance Extras

- Reversible fan
- Engine coolant filters
- Supplemental steering
- Reverse inhibitor
- Downshift inhibitor
- High amperage alternator

Cold weather aids

- Starting receptacle
- Radiator shutters
- Engine coolant heater
- Ether starting aid
- Air line dryer

Servicing

- Automatic lubrication
- Tire pressure maintenance system
- Fast fill fuel system
- Fast oil change system
- Hydraulic elevator chain adjuster

Safety

- Vandalism protection system
- Fire protection system
- ROPS cab
- Back-up alarm

Operator Station

- Air conditioning
- Heater
- Throttle lock
- Windshield washer-wiper
- Mirrors
- Electric hour meter
- Tachometer

NEW DEVELOPMENTS & TRENDS

Scrapers have not changed greatly in overall design or maximum size in recent years. The development focus has been on detail refinements for improved reliability, reduced noise levels, and reduced fuel consumption. Greater efficiency in servicing activities has also been emphasized.

Earlier reference was made to the Fiat Allis liftable elevator mechanism and its variable speed drive which represents an effort to improve its effectiveness.

Caterpillar has tested and demonstrated for several years a vertical auger design as a replacement for the flight conveyor arrangement currently employed on elevator. (See Photograph 3.6). This approach appears to have the following advantages:

- Reduced dust
- Reduced shock loading
- Less operator skill is required
- Improved capability in sandy soils

Photograph 3.6 Caterpillar twin vertical auger scraper

More power can be put in the auger and as this is done the load time and tire wear can be reduced. Conventional ejection techniques are applied with the load passing through the auger. The latest configuration uses twin vertical augers.

MACHINE SPECIFICATIONS

See Figure 3.13– Scraper Dimensions
See Table 3.2 – Standard Scraper Specifications
See Table 3.3 – Standard Tandem Scraper Specifications
See Table 3.4 – Elevating Scraper Specifications
See Table 3.5 – Elevating Tandem Scraper Specifications

Scrapers are listed alphabetically by manufacturer, and then in ascending order by payload capacity in tons. Net weight is the combined weight of tractor unit and scraper body; including operator, fuel, fluids and ROPS cab. Gross weight is the net weight plus the maximum payload in pounds.

Steering angle is the angle of swing of the tractor from straight ahead to full turn in one direction. In some instances, the angle of swing in one direction is constrained by the operator's cab location. Turning diameter is defined as the vehicle clearance circle.

Hydraulic capacities include the bowl, apron and ejector circuits, but not the steering circuits. The hydraulic specifications for elevator scrapers also include all the elevator circuits.

Ground clearance, as listed, is the maximum clearance under the scraper bowl with a full load. Height is the maximum overall height of the machine, at its highest point.

All specifications, capacities, capabilities and dimensions are based on published manufacturer data. Although the information is believed to be current and the interpretation to be consistent and correct, it is possible that omissions and inaccuracies have occurred and that out-of-date information may have been included. These specifications charts are not meant to be used either as an in-depth analysis, or as a head-to-head comparison of the available equipment, but rather as a general overview of the equipment, available sizes, and approximate operating data. Specific questions relating to performance or purchase should be directed to the manufacturer or authorized distributor/dealer in the user's specific area.

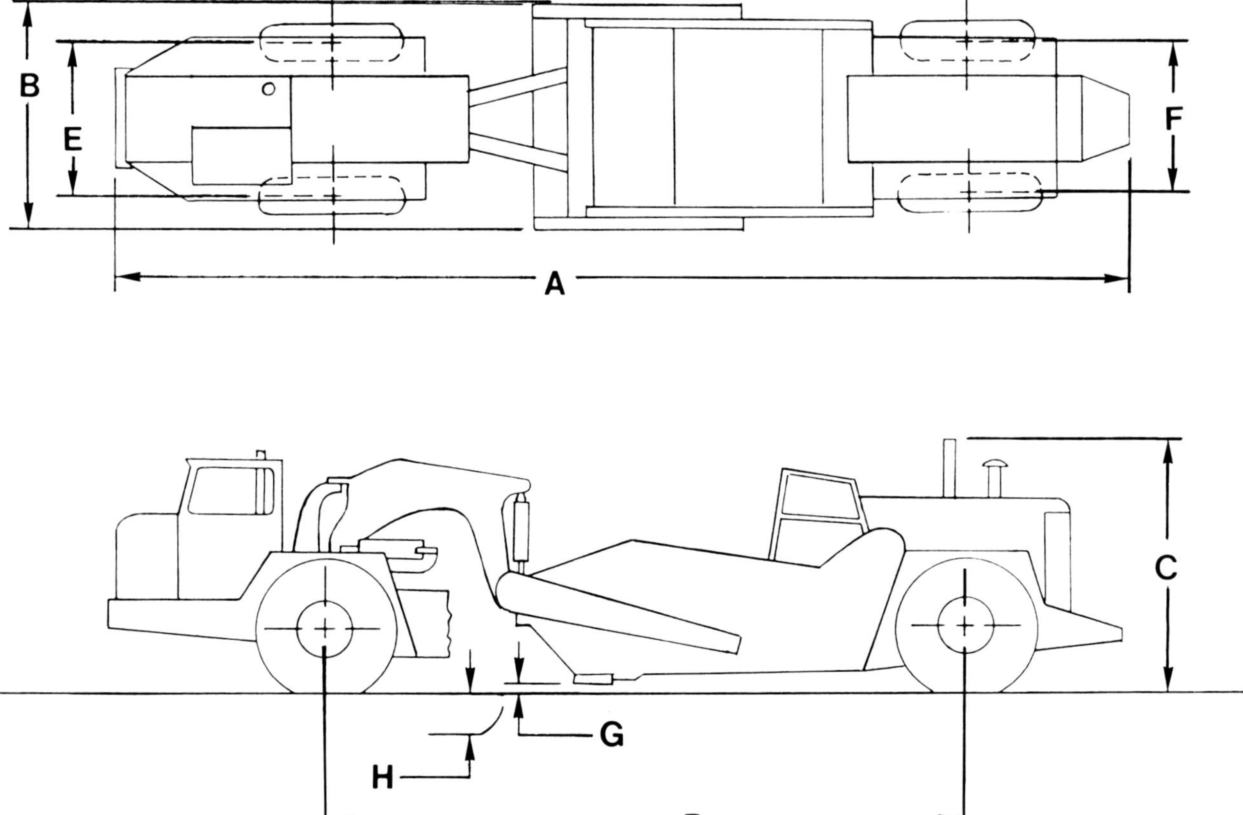

A	Overall Length (ft)	E	Tread/Tractor (ft)
B	Overall Width (ft)	F	Tread/Scraper (ft)
C	Overall Height (ft)	G	Ground Clearance/Scraper (in)
D	Wheelbase (ft)	H	Digging Depth (in)

Figure 3.13 Scraper Dimensions

Table 3.2
STANDARD SCRAPER SPECIFICATIONS

MAKE	MODEL	LOAD RATING (tons)	STRUCK CAPACITY (cu yd)	HEAPED CAPACITY (cu yd)	ENGINE HP	NET WEIGHT (lbs)	GROSS WEIGHT (lbs)	STEERING ANGLE (degrees)	FUEL TANK (gal)	COOLANT (gal)	MAX SPEED F/R (mph)
Big Bud	525/84	34	20	—	473	104,500	171,500	40	450	—	23/17
Caterpillar	621-B	24	14	20	330	63,130	111,130	90	135	20	31/–
Caterpillar	631-D	38	21	31	450	86,500	161,500	90	250	37	30/–
Caterpillar	641-B	47	28	38	550	117,000	211,000	—	280	41	34/–
Caterpillar	651-B	52	32	44	550	124,200	228,200	—	280	41	34/–
Caterpillar	660-B	64	40	54	550	131,500	259,500	—	—	—	31/–
Fiat-Allis	260-B	25	15	21	325	55,000	105,400	90	143	24	34/–
International	431-B	24	14	21	326	59,420	107,420	90	156	15	—/–
Komatsu	WS-23S	36	21	30	425	76,720	149,470	90	225	37	35/6
MRS	I-105S	33	21	30	422	83,660	149,660	—	240	26	—/–
MRS	I-110S	33	21	30	482	93,500	159,500	—	300	26	—/–
MRS	I-115S	39	25	34	550	110,800	188,800	—	300	30	—/–
MRS	I-120S	47	30	39	615	127,500	220,500	—	350	45	—/–
Terex	S-24B	41	24	34	475	93,375	176,875	90	240	37	32/–

Table 3.2 (Continued)

STANDARD SCRAPER SPECIFICATIONS

MAKE	MODEL	ENGINE MAKE	ENGINE MODEL	TRANSMISSION TYPE	NO. OF GEARS F/R	BRAKES	HYDRAULIC SYSTEM DATA				
							PUMP FLOW (gpm)	RELIEF VALVE PRESSURE (psi)	PUMP TYPE	NO. OF CYLINDERS	NO. OF MOTORS
Big Bud	525/84	Cummins	KTA-1150-C525	mech auto	8/4	shoe	60	1850	piston	3	0
Caterpillar	621-B	Caterpillar	3406	mech auto	8/1	shoe	74	2000	vane	4	0
Caterpillar	631-D	Caterpillar	3408	mech auto	8/1	shoe	91	2000	—	4	0
Caterpillar	641-B	Caterpillar	D-346	mech auto	8/1	shoe	153	2000	—	4	0
Caterpillar	651-B	Caterpillar	D-346	mech auto	8/1	shoe	153	2000	—	4	0
Caterpillar	660-B	Caterpillar	D-346	mech auto	8/1	-	—	—	—	—	0
Fiat-Allis	260-B	Allis-Chalmers	25000 MKII	mech auto	11/1	shoe	68	2000	gear	4	0
International	431-B	International	DVT-800	mech auto	9/1	shoe	74	1650	—	5	0
Komatsu	WS-23S	Cummins	KT-1150-C	mech auto	8/1	disc	—	2200	gear	4	0
MRS	I-105S	GM	12V-71N	mech auto	6/1	disc	—	—	—	5	0
MRS	I-110S	GM	12V-71T	mech auto	6/1	disc	—	—	—	6	0
MRS	I-115S	Cummins	KTA-1150-C600	mech auto	6/1	disc	—	—	—	5	0
MRS	I-120S	Cummins	VTA-1710-C700	mech auto	6/1	disc	—	—	—	6	0
Terex	S-24B	Detroit	12V-71T	mech auto	6/1	shoe	80	2250	gear	5	0

Table 3.2 (Continued)
STANDARD SCRAPER SPECIFICATIONS

MAKE	MODEL	LENGTH (ft)	MAX WIDTH (ft)	MAX HEIGHT (ft)	WHEELBASE (ft)	TRACTOR TREAD (ft)	SCRAPER TREAD (ft)	GROUND CLEARANCE (in)	CLEARANCE CIRCLE (ft)	STANDARD TIRES
Big Bud	525/84	—	16.7	—	—	—	—	—	50.5	30.5-32
Caterpillar	621-B	41.6	11.3	11.9	24.3	7.3	7.2	21	36.5	29.5-29 28PR
Caterpillar	631-D	46.8	13.0	13.7	28.7	8.1	8.1	19	40.1	33.25-35 38PR
Caterpillar	641-B	49.1	13.3	13.9	31.0	8.2	8.3	—	42.8	37.5-39 36PR
Caterpillar	651-B	50.3	14.2	14.1	31.9	8.5	8.9	—	44.2	37.5-39 36PR
Caterpillar	660-B	56.7	14.2	14.3	32.4	—	—	—	46.0	37.5-51 44PR
Fiat-Allis	260-B	38.7	11.9	11.4	23.5	—	7.5	21	33.5	29.5-29 22PR
International	431-B	38.5	11.8	11.8	23.9	7.6	7.6	19	34.0	29.5-29 28PR
Komatsu	WS-23S	44.2	12.1	11.6	27.5	7.9	8.0	20	39.4	33.5-33 32PR
MRS	I-105S	48.2	12.0	12.4	26.3	8.7	7.9	—	54.0	33.25-35 26PR
MRS	I-110S	48.2	12.4	13.3	26.3	9.4	7.9	—	56.0	37.5-33 24PR
MRS	I-115S	51.1	13.0	13.3	28.7	9.4	8.5	—	50.0	37.5-39 28PR
MRS	I-120S	52.1	13.2	13.6	29.2	10.0	8.5	—	50.0	37.5-39 28PR
Terex	S-24B	47.6	12.5	14.0	30.6	8.1	7.7	20	45.0	37.5-33 30PR

Table 3.2 (Continued)

STANDARD SCRAPER SPECIFICATIONS

MAKE	MODEL	WIDTH OF CUT (ft)	DEPTH OF CUT (in)	CUT FORCE (lbs)	DEPTH OF SPREAD (in)	APRON OPENING (ft)	APRON CLOSING FORCE (lbs)
Big Bud	525/84	15.5	15	—	—	—	—
Caterpillar	621-B	9.9	13	33,800	18	5.8	24,000
Caterpillar	631-D	11.5	19	48,000	17	6.6	38,000
Caterpillar	641-B	11.3	16	66,100	20	7.6	39,000
Caterpillar	651-B	11.9	16	77,000	20	7.3	36,000
Caterpillar	660-B	11.9	19	94,000	24	9.2	37,000
Fiat-Allis	260-B	10.3	14	—	18	6.7	—
International	431-B	10.3	13	35,600	—	5.8	26,100
Komatsu	WS-23S	11.1	35	—	20	7.1	—
MRS	I-105S	10.9	9	—	18	7.0	—
MRS	I-110S	10.9	9	—	18	7.0	—
MRS	I-115S	11.3	20	—	20	8.0	—
MRS	I-120S	11.3	20	—	20	8.0	—
Terex	S-24B	11.1	19	75,000	16	7.3	27,500

Table 3.3

STANDARD TANDEM SCRAPER SPECIFICATIONS

MAKE	MODEL	LOAD RATING (tons)	STRUCK CAPACITY (cu yd)	HEAPED CAPACITY (cu yd)	ENGINE HP	NET WEIGHT (lbs)	GROSS WEIGHT (lbs)	STEERING ANGLE (degrees)	FUEL TANK (gal)	COOLANT (gal)	MAX SPEED F/R (mph)
Caterpillar	627-B	24	14	20	225/225	71,390	119,390	90	265	37	34/-
Caterpillar	637-D	38	21	31	450/250	106,325	181,325	90	420	57	30/-
Caterpillar	657-B	52	32	44	550/400	139,100	243,100	-	480	72	33/-
Fiat-Allis	262-B	25	15	21	325/171	67,900	118,300	90	238	37	—/-
International	433-B	24	14	21	326/185	68,980	116,980	90	256	28	—/-
Terex	TS-14B	24	14	20	144/144	54,000	101,000	90	175	20	23/-
Terex	TS-18	30	18	24	295/225	74,950	134,950	90	290	41	31/-
Terex	TS-24	40	24	32	394/225	95,935	175,935	90	340	41	30/-
Terex	TS-36	40	36	40	394/225	99,735	179,735	90	340	41	30/-
Terex	TS-24B	41	24	34	475/242	103,315	189,815	90	360	63	32/-
Terex	TS-38B	41	38	41	475/242	110,000	191,500	90	360	63	32/-
Terex	TS-46B	41	46	50	475/242	112,355	193,855	90	360	63	32/-

							HYDRAULIC SYSTEM DATA				
MAKE	MODEL	ENGINE MAKE	ENGINE MODEL	TRANSMISSION TYPE	NO. OF GEARS F/R	BRAKES	PUMP FLOW (gpm)	RELIEF VALVE PRESSURE (psi)	PUMP TYPE	NO. OF CYLINDERS	NO. OF MOTORS
Caterpillar	627-B	Caterpillar	3306/3306	mech auto	8/1	shoe	71	2000	vane	4	0
Caterpillar	637-D	Caterpillar	3408/3306	mech auto	8/1	shoe	91	2000	—	4	0
Caterpillar	657-B	Caterpillar	D-346/D-343	mech auto	8/1	shoe	153	2000	—	5	0
Fiat-Allis	262-B	Allis-Chalmers	25000/11000	mech auto	11/1	shoe	68	2000	gear	5	0
International	433-B	International	DVT-800/DT-466B	mech auto	9/1	shoe	74	1650	—	5	0
Terex	TS-14B	Detroit	4-71N/4-71N	mech auto	6/1	shoe	52	1500	gear	4	0
Terex	TS-18	Detroit	8V-71N/6V-71N	mech auto	5/1	shoe	52	1850	gear	5	0
Terex	TS-24	Detroit	12V-71N/6V-71N	mech auto	6/1	shoe	80	1750	gear	5	0
Terex	TS-36	Detroit	12V-71N/6V-71N	mech auto	6/1	shoe	80	1750	gear	5	0
Terex	TS-24B	Detroit	12V-71T/6V-71T	mech auto	6/1	shoe	80	2250	gear	5	0
Terex	TS-38B	Detroit	12V-71T/6V-71T	mech auto	6/1	shoe	80	2250	gear	5	0
Terex	TS-46B	Detroit	12V-71T/6V-71T	mech auto	6/1	shoe	80	2250	gear	5	0

Table 3.3 (Continued)

STANDARD TANDEM SCRAPER SPECIFICATIONS

MAKE	MODEL	LENGTH (ft)	MAX WIDTH (ft)	MAX HEIGHT (ft)	WHEELBASE (ft)	TRACTOR TREAD (ft)	SCRAPER TREAD (ft)	GROUND CLEARANCE (in)	CLEARANCE CIRCLE (ft)	STANDARD TIRES
Caterpillar	627-B	43.8	11.3	11.9	25.3	7.3	7.2	18	36.5	29.5-29 28PR
Caterpillar	637-D	48.7	13.0	13.7	28.7	8.1	8.1	16	40.1	33.25-35 38PR
Caterpillar	657-B	51.7	14.2	14.5	32.9	8.5	8.8	-	45.1	37.5-39 44PR
Fiat-Allis	262-B	41.5	11.9	11.4	25.0	-	7.5	-	35.0	29.5-29 28PR
International	433-B	41.4	11.8	12.2	23.9	7.6	7.6	19	34.0	29.5-29 28PR
Terex	TS-14B	39.6	11.3	11.2	23.2	7.5	7.5	-	33.0	29.5-25 22PR
Terex	TS-18	41.9	11.8	12.6	25.5	7.5	7.5	24	37.7	29.5-29 34PR
Terex	TS-24	45.6	11.9	12.6	28.0	7.6	76	24	39.1	33.5-33 38PR
Terex	TS-36	47.4	11.9	12.6	29.8	7.6	7.6	24	40.9	33.5-33 38PR
Terex	TS-24B	47.6	12.5	12.9	30.6	8.1	7.7	20	45.0	37.5-33 30PR
Terex	TS-38B	47.6	12.5	12.9	30.6	8.1	7.7	20	45.0	37.5-33 30PR
Terex	TS-46B	49.9	12.5	12.9	32.9	8.1	7.7	20	47.3	37.5-33 36PR

MAKE	MODEL	WIDTH OF CUT (ft)	DEPTH OF CUT (in)	CUT FORCE (lbs)	DEPTH OF SPREAD (in)	APRON OPENING (ft)	APRON CLOSING FORCE (lbs)
Caterpillar	627-B	9.9	13	48,300	18	5.8	24,000
Caterpillar	637-D	11.5	19	70,000	17	6.6	38,000
Caterpillar	657-B	11.9	16	268,800	20	7.7	34,800
Fiat-Allis	262-B	10.3	13	—	18	6.7	-
International	433-B	10.3	13	50,600	19	5.8	26,100
Terex	TS-14B	9.8	14	—	28	6.9	-
Terex	TS-18	10.3	12	56,000	23	6.6	26,000
Terex	TS-24	10.3	12	—	24	7.3	23,000
Terex	TS-36	10.3	12	—	24	7.3	23,000
Terex	TS-24B	11.1	19	75,000	16	7.3	27,500
Terex	TS-38B	11.1	19	75,000	16	7.3	27,500
Terex	TS-46B	11.1	19	75,000	16	7.3	27,500

Table 3.4
ELEVATING SCRAPER SPECIFICATIONS

MAKE	MODEL	LOAD RATING (tons)	STRUCK CAPACITY (cu yd)	HEAPED CAPACITY (cu yd)	ENGINE HP	NET WEIGHT (lbs)	GROSS WEIGHT (lbs)	STEERING ANGLE (degrees)	FUEL TANK (gal)	COOLANT (gal)	MAX SPEED F/R (mph)
Caterpillar	613-B	13	—	11	150	29,300	55,300	—	65	10	26/–
Caterpillar	623-B	25	—	22	330	70,090	120,090	90	135	20	31/–
Caterpillar	633-D	38	—	34	450	98,400	173,400	90	250	37	30/–
Clark	110-11B	13	—	11	174	34,470	60,870	90	77	11	33/5
Clark	110-15	18	—	15	232	46,360	82,360	90	158	18	34/7
Clark	210-HB	27	—	23	324	70,970	123,970	90	180	22	33/–
Deere	JD-762	14	—	11	175	34,450	61,950	90	72	9	30/–
Deere	JD-862	20	—	16	250	49,189	89,189	90	110	15	31/–
Fiat-Allis	261-B	27	—	23	325	62,900	116,000	90	143	21	34/–
International	412-B	13	—	11	168	33,450	59,650	90	70	12	—/–
International	442-B	25	—	22	326	68,000	118,000	90	156	15	—/–
MRS	I-105ES	32	—	27	422	90,990	155,790	—	240	26	32/–
MRS	I-110ES	38	—	32	482	103,500	179,500	—	240	26	—/–
MRS	I-115ES	38	—	32	550	107,000	183,000	—	300	30	—/–
Terex	S-23E	27	—	23	310	66,880	119,880	90	150	23	29/–

Table 3.4 (Continued)
ELEVATING SCRAPER SPECIFICATIONS

							HYDRAULIC SYSTEM DATA				
MAKE	MODEL	ENGINE MAKE	ENGINE MODEL	TRANSMISSION TYPE	NO. OF GEARS F/R	BRAKES	PUMP FLOW (gpm)	RELIEF VALVE PRESSURES (psi)	PUMP TYPE	NO. OF CYLINDERS	NO. OF MOTORS
Caterpillar	613-B	Caterpillar	3208	manual	4/2	disc	85	—	vane	—	1
Caterpillar	623-B	Caterpillar	3406	mech auto	8/1	shoe	150	2500	vane	4	1
Caterpillar	633-D	Caterpillar	3408	mech auto	8/1	shoe	276	2500	vane	4	1
Clark	110-11B	Cummins	V504-C	mech auto	5/1	disc	64	5500	piston	4	1
Clark	110-15	GM	6V-71T	mech auto	8/2	disc	83	5500	piston	4	1
Clark	210-HB	Cummins	NTA-855-C	mech auto	9/2	shoe	192	2350	gear	4	1
Deere	JD-762	Deere	—	mech auto	5/1	disc/shoe	84	2250	—	5	1
Deere	JD-862	Deere	—	mech auto	6/1	disc/shoe	137	5000	—	5	1
Fiat-Allis	261-B	Allis-Chalmers	25000 MKII	mech auto	11/1	shoe	201	2250	gear	3	1
International	412-B	International	DT-466B	mech auto	4/1	shoe	77	2100	—	3	1
International	442-B	International	DVT-800	mech auto	9/2	shoe	155	1900	—	4	1
MRS	I-105ES	GM	12V-71N	mech auto	6/1	disc	—	—	—	8	1
MRS	I-110ES	GM	12V-71T	mech auto	6/1	disc	—	—	—	7	1
MRS	I-115ES	Cummins	KTA-1150-C600	mech auto	6/1	disc	—	—	—	7	1
Terex	S-23E	Detroit	8V-71T	mech auto	6/1	shoe	184	2000	gear	4	1

Table 3.4 (Continued)

ELEVATING SCRAPER SPECIFICATIONS

MAKE	MODEL	LENGTH (ft)	MAX WIDTH (ft)	MAX HEIGHT (ft)	WHEELBASE (ft)	TRACTOR TREAD (ft)	SCRAPER TREAD (ft)	GROUND CLEARANCE (in)	CLEARANCE CIRCLE (ft)	STANDARD TIRES
Caterpillar	613-B	32.1	8.0	9.7	20.8	5.9	5.9	14	29.3	18.00-25 12PR
Caterpillar	623-B	41.1	11.7	12.5	26.2	7.3	7.2	15	37.3	29.5-29 28PR
Caterpillar	633-D	47.3	13.0	13.9	29.2	8.1	8.1	24	40.6	33.25-35 38PR
Clark	110-11B	31.4	8.3	9.9	18.7	6.2	6.2	19	29.2	23.5-25 20PR
Clark	110-15	34.0	9.5	10.3	21.6	6.8	6.8	18	31.7	26.5-25 24PR
Clark	210-HB	41.1	11.6	12.4	25.5	7.7	8.1	21	36.0	29.5-29 34PR
Deere	JD-762	33.0	8.0	9.8	21.0	6.0	6.0	12	30.0	23.5-25 16PR
Deere	JD-862	36.2	9.5	10.4	22.8	6.4	7.0	7	32.8	26.5-25 24PR
Fiat-Allis	261-B	39.1	12.0	11.7	24.4	—	7.5	19	34.7	29.5-29 28PR
International	412-B	32.5	8.0	11.0	19.8	6.0	6.0	19	27.8	23.5-25 16PR
International	442-B	41.3	11.8	11.8	26.7	7.6	7.6	14	36.7	29.5-29 28PR
MRS	I-105ES	52.5	11.9	12.7	27.0	8.7	8.0	19	50.0	33.25-35 26PR
MRS	I-110ES	54.3	13.0	13.1	28.8	9.0	8.3	23	56.0	37.5-39 28PR
MRS	I-115ES	54.3	13.0	13.3	29.8	9.4	8.3	23	56.0	37.5-39 28PR
Terex	S-23E	40.8	10.7	12.5	25.7	7.5	7.7	22	36.8	29.5-29 28PR

Table 3.4 (Continued)
ELEVATING SCRAPER SPECIFICATIONS

MAKE	MODEL	WIDTH OF CUT (ft)	DEPTH OF CUT (in)	DEPTH OF SPREAD (in)	WIDTH OF ELEVATOR (ft)	LENGTH OF ELEVATOR (ft)	NO. OF ELEVATOR FLIGHTS	MAX ELEVATOR SPEED F/R (fpm)
Caterpillar	613-B	8.0	7	—	5.4	8.1	16	225/116
Caterpillar	623-B	10.3	13	—	7.4	12.3	15	243/109
Caterpillar	633-D	11.5	15	—	9.3	10.3	13	300/200
Clark	110-11B	7.5	8	—	6.4	7.8	16	280/280
Clark	110-15	8.9	7	—	7.6	9.7	18	276/276
Clark	210-HB	10.0	14	—	6.8	13.5	15	262/262
Deere	JD-762	7.5	—	—	5.6	9.5	18	236/236
Deere	JD-862	8.8	—	—	6.5	12.0	23	240/240
Fiat-Allis	261-B	10.6	10	12	6.8	12.2	18	250/205
International	412-B	8.0	9	—	5.1	—	16	248/248
International	442-B	10.3	10	—	6.8	—	18	278/278
MRS	I-105ES	11.2	11	11	—	11.7	21	260/260
MRS	I-110ES	11.3	11	11	7.8	12.9	22	260/260
MRS	I-115ES	11.3	11	11	7.8	12.9	22	260/260
Terex	S-23E	10.5	12	10	6.9	—	15	240/183

Table 3.5

TANDEM ELEVATING SCRAPER SPECIFICATIONS

MAKE	MODEL	LOAD RATING (tons)	STRUCK CAPACITY (cu yd)	HEAPED CAPACITY (cu yd)	ENGINE HP	NET WEIGHT (lbs)	GROSS WEIGHT (lbs)	STEERING ANGLE (degrees)	FUEL TANK (gal)	COOLANT (gal)	MAX SPEED F/R (mph)
Caterpillar	639-D	38	—	34	450/250	121,825	196,825	90	420	57	30/–
Clark	110H-TB	19	—	16	140/140	52,062	90,462	90	154	14	30/8
Fiat-Allis	263-B	27	—	23	325/171	74,200	127,200	90	238	37	—/–
International	444-B	25	—	22	326/185	80,510	130,510	90	256	28	—/–

MAKE	MODEL	ENGINE MAKE	ENGINE MODEL	TRANSMISSION TYPE	NO. OF GEARS F/R	BRAKES	HYDRAULIC SYSTEM DATA: PUMP FLOW (gpm)	RELIEF VALVE PRESSURE (psi)	PUMP TYPE	NO. OF CYLINDERS	NO. OF MOTORS
Caterpillar	639-D	Caterpillar	3408/3306	mech auto	8/1	shoe	276	2500	—	4	1
Clark	110H-TB	GM	4-71N/4-71N	mech auto	5/2	shoe	123	2250	gear	4	1
Fiat-Allis	263-B	Allis-Chalmers	25000/11000	mech auto	11/1	shoe	201	2250	gear	5	1
International	444-B	International	DVT-800/DT-466B	mech auto	9/2	shoe	220	—	—	4	1

Table 3.5 (Continued)
TANDEM ELEVATING SCRAPER SPECIFICATIONS

MAKE	MODEL	LENGTH (ft)	MAX WIDTH (ft)	MAX HEIGHT (ft)	WHEELBASE (ft)	TRACTOR TREAD (ft)	SCRAPER TREAD (ft)	GROUND CLEARANCE (in)	CLEARANCE CIRCLE (ft)	STANDARD TIRES
Caterpillar	639-D	47.7	13.0	14.3	29.5	8.1	8.1	21	40.6	37.25-35 36PR
Clark	110H-TB	37.3	9.3	10.7	22.3	6.8	6.8	18	31.5	26.5-29 26PR
Fiat-Allis	263-B	42.3	12.0	11.8	25.8	-	7.5	19	35.8	29.5-29 34PR
International	444-B	44.2	11.8	12.2	26.7	7.6	7.6	15	36.7	29.5-29 34PR

MAKE	MODEL	WIDTH OF CUT (ft)	DEPTH OF CUT (in)	DEPTH OF SPREAD (in)	WIDTH OF ELEVATOR (ft)	LENGTH OF ELEVATOR (ft)	NO. OF ELEVATOR FLIGHTS	MAX ELEVATOR SPEED F/R (fpm)
Caterpillar	639-D	11.5	15	—	9.3	10.4	20	300/200
Clark	110H-TB	8.9	8	—	6.5	8.6	16	247/247
Fiat-Allis	263-B	10.6	10	12	6.8	12.2	18	234/180
International	444-B	10.3	10	—	6.8	—	16	298/218

Chapter 4

TRUCKS

Figure 4.1 Typical Rear Dump Truck

A truck is simply a mobile piece of equipment for hauling material. It is often an integral part of material handling activities in the mine either for transport of ore from the face to processing or stockpile or for transport of overburden to spoil. With the need for flexible mining plans and new reclamation regulations, there has been increased application of shovel/truck systems for overburden handling. Expanding use with broader applications and the growth of surface mining have led to the development of off-highway units of very high power and capacity. This discussion will be limited to the large off-highway trucks utilized in mining. The very special side and end dump units will not be considered.

Figure 4.2 Typical Bottom Dump Truck

TYPICAL UNITS

There are two basic truck designs utilized in surface mining — the rear dump and the bottom dump. As their names imply, they differ in their dumping method. The rear dump discharges material by tilting the body and spilling the contents out the rear by gravity. The bottom dump discharges material by opening the bottom of the body. As illustrated in Figure 4.3, there are a number of variations of these basic designs. The primary operating differences between these two designs centers around the choice of dumping method, inherent ruggedness and the gradeability of each design.

Rear Dumps

- Generally 2 axle units (one steering, one driving), 3 axle on very large sizes or on smaller articulated units (one steering, two driving).
- Single axle converter or two wheel electric drive systems
- Rigid frame with heavy duty body
- Front axle steering
- Single wheels on front axle, dual wheels on rear axle
- Capacities from 50 to 350 tons
- Horsepowers from 405 to 3000
- Vehicle weights from 58,000 to 520,000 lb.

Bottom Dumps

- Tractor-trailer 3 axle units (one steering, one driving, one trailing) or rigid frame 2 axle units (one steering, one driving) with various combinations of single or dual wheels.
- Single axle torque converter or 2 wheel electric drive systems
- Capacities from 70 to 180 tons
- Horsepowers from 420 to 1450
- Vehicle weights from 91,000 to 233,000 lb.

Manufacturers

Trucks used in the larger domestic mines are almost exclusively from U.S. manufacturers. Eight manufacturers offer a range of bottom and rear dump models. An additional four others offer one type, either rear or bottom dump trucks. Four overseas manufacturers market trucks generally in the 20 to 85 ton capacity range. The majority of

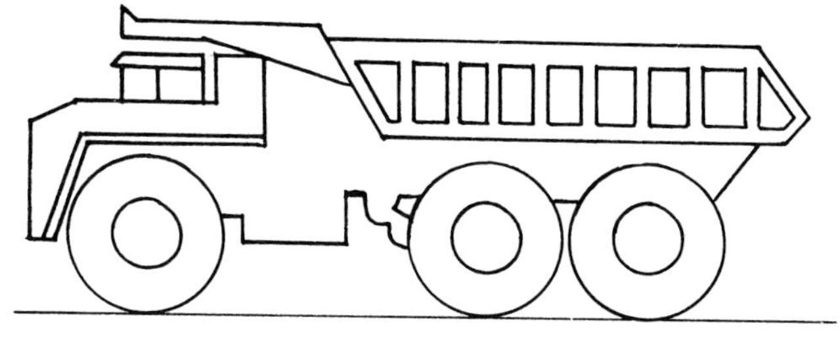

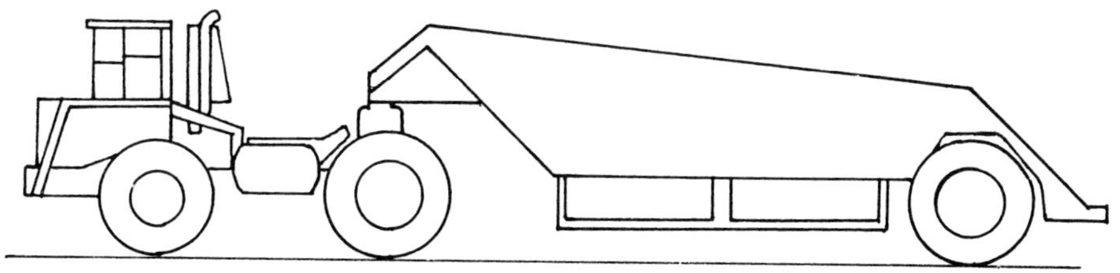

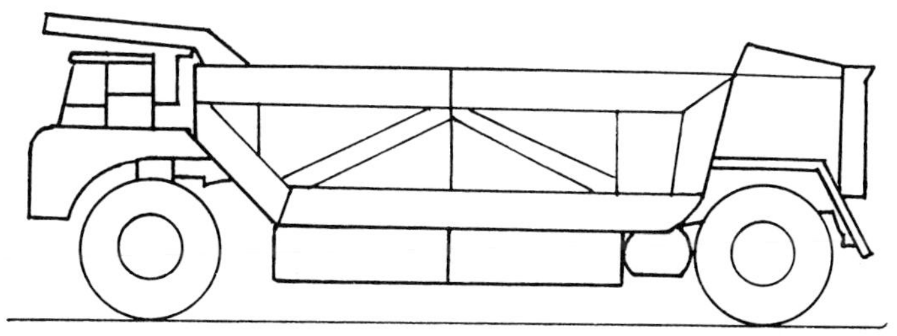

Figure 4.3 Truck Types

units presently in service are 170 ton or less in capacity. However, three U.S. manufacturers are currently offering larger models. Terex, division of General Motors, used to market the world's largest rear dump truck of 350 tons. Although General Motors has divested itself of its Terex Division, it retained the 350 ton, model number 33-19. The truck is not currently being marketed.

Table 4.1

OFF-HIGHWAY-TRUCK MANUFACTURERS

Atlas Hoist & Body, Inc. 7500 Cote de Liesse Road Montreal, Quebec Canada H4T 1E8	Bottom dump 450–870 FWHP 70–150 ton cap.
Caterpillar Tractor Co. 100 N.E. Adams Street Peoria, Illinois 61629	Rear dump 450–870 FWHP 35–85 ton cap.
Dart Truck Company 1301 Chouteau Tr. P.O. Box 321 Kansas City, Missouri 64141	Rear dump 640 -1200 FWHP 65–120 ton cap. Bottom dump 635–1100 FWHP 100–160 ton cap.
DJB Engineering Ltd. Peterlee, Co. Durham England, SR8 2HX	Rear dump 235–450 FWHP 27.5–55 ton cap.
DJB Sales, Inc. 8280 Patuxent Range Road Jessup, Maryland 20794	
Euclid, Inc. 22221 St. Clair Avenue Cleveland, Ohio 44117	Rear dump 228–1520 FWHP 25–170 ton cap. Bottom dump 300–1050 FWHP 30–150 ton cap.
General Motors, Diesel Div. General Motors of Canada, Ltd. P.O. Box 5160 London, Ontario N6A 4N5	Rear dump 1445–3000 FWHP 170–350 ton cap.
Goodbary Engineering Co. P.O. Box 100 Cardin, Oklahoma 74335	Bottom dump 1000–1325 FWHP 100–170 ton cap.
International Harvester Construction Equipment Group 600 Woodfield Avenue Schaumburg, Illinois 60196	Rear dump 395–600 FWHP 36–50 ton cap.
Kenworth Truck Co. P.O.Box 1000 Kirkland, Washington 98033	Rear dump 405 FWHP 50 ton cap.

Table 4.1 (Continued)

OFF-HIGHWAY-TRUCK MANUFACTURERS

Komatsu Limited Komatsu Building 2-3-6 Akasaka Minato-Ku Tokyo, Japan	Rear dump 230–1600 FWHP 20–176 ton cap.
Kress Corporation P.O. Box 368 Brimfield, Illinois 61517	Bottom dump 800–1200 FWHP 150–160 ton cap.
Rimpull Corporation Box 748 U.S. 169 South Olathe, Kansas 66061	Rear dump 600–1050 FWHP 65–120 ton cap. Bottom dump 635–1050 FWHP 100–150 ton cap.
T & J Industries, Inc. 13850 Wyandotte Street P.O. Box 8620 Kansas City, Missouri 64114	Rear dump 160–575 FWHP 13–65 ton cap.
Terex, Corporation IBH Group Hudson, Ohio 44236	Rear dump 225–1200 FWHP 17–130 ton cap. Bottom dump 840 FWHP 150 ton cap.
Unit Rig & Equipment Co. P.O. Box 3107 Tulsa, Oklahoma 74101	Rear dump 1000–2250 FWHP 85–200 ton cap. Bottom dump 1050–1600 FWHP 145–180 ton cap.
WABCO Construction & Mining Equip. Div. of American Standard, Inc. 2300 N.E. Adams Street Peoria, Illinois 61639	Rear dump 420–2250 FWHP 35–250 ton cap. Bottom dump 1000–1450 FWHP 150–170 ton cap.

BASIC MACHINE OPERATIONS

Mine site operation of trucks for haulage of material introduces some unique operational requirements.

- Truck bodies are filled by front-end loaders, hydraulic excavators, electric shovels and/or bucket wheel excavators, while the truck is parked (positioned) so as to optimize the operational cycle of the specific excavating equipment. If the units are properly matched, cyclic loading machines normally require 3 to 5 passes to fill the truck.

- The haul cycle includes periods of acceleration and deceleration, operation on grades with speeds dictated by road conditions, traffic, truck rimpull capabilities and haul distances.

- Dependent on body design, the dumping phase consists of discharging the load over the highwall, into a hopper or ground level grizzly. The amount of maneuvering required and the time required varies with the situation.

- Return to the loading area, empty, is at higher speed and includes some time for maneuvering into position loading. (See Figure 4.4)

APPLICATIONS

These trucks are used exclusively for material transport. The material can be just about anything but, in mining, the broad classifications are:

- Overburden

- Ore/Coal

When trucks are used to haul overburden, the mine normally has an open pit or area mine plan with dumping off of spoil benches. Trucks can be used to haul ore/coal to a hopper or stockpile, in virtually any surface mine plan. Dumping to stockpile is generally done in shallow lifts. Bottom dump units, driving over a grizzly, are used to feed a hopper. A back-in hopper station is utilized with rear dumps.

In some cases the trucks carrying coal directly to a nearby power plant will on the return trip transport ash back into the pit for burial.

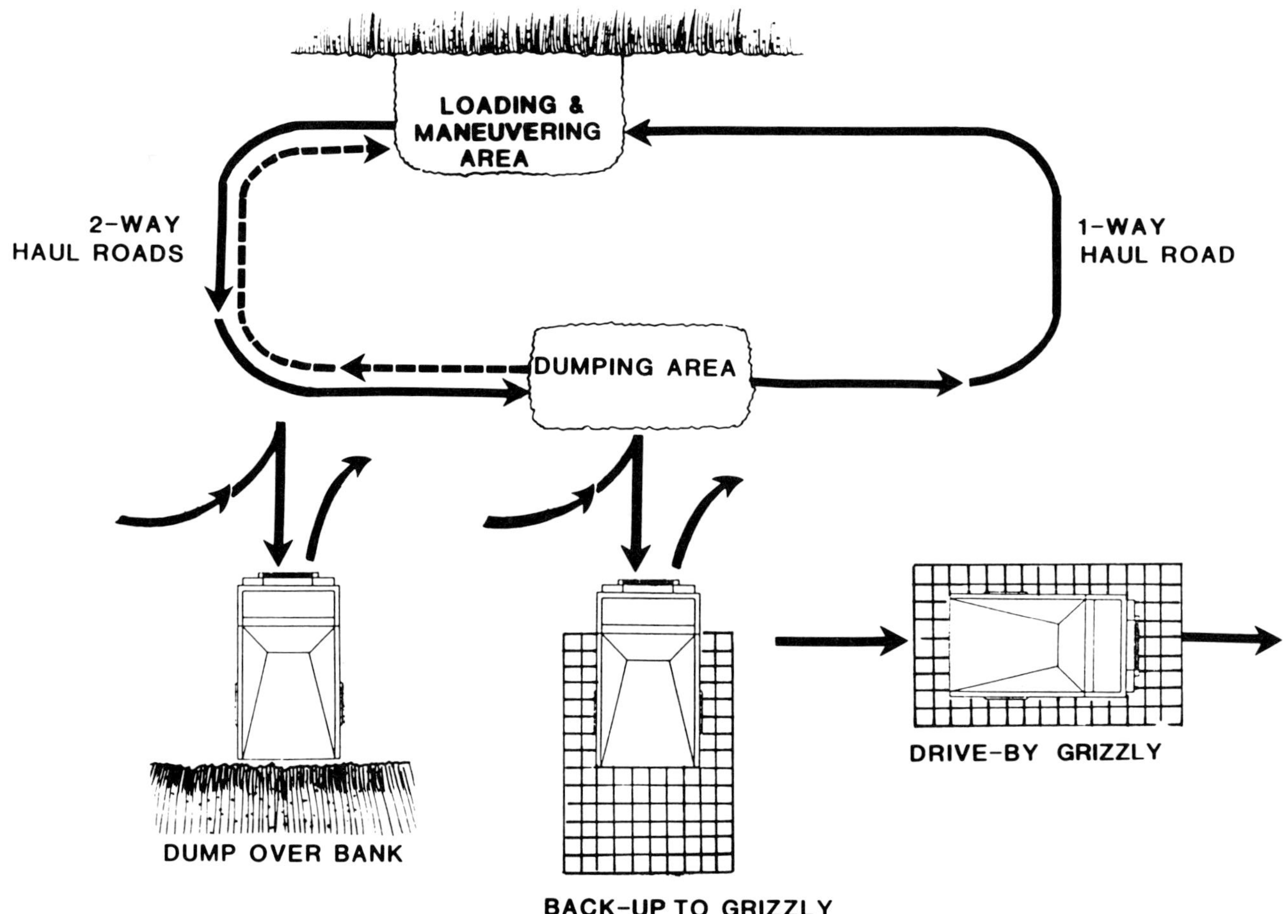

Figure 4.4 Typical Truck Production Cycle

GENERAL CHARACTERISTICS

Since the truck is a very specialized, single purpose piece of equipment, it has very few functions to operate and control. (See Figure 4.5)

- Propel, direct mechanical-torque converter drive, or individual electric wheel motors
- Steering, hydraulic cylinders
- Dumping, hydraulic or pneumatic cylinders

In general, trucks have:

- Diesel Power
- Good maneuverability
- Good traction with maximum weight applied to the drive axle
- Ground clearances of 20 to 40 inches
- Top speed of 30 to 45 mph
- Service life for units up to 100 ton capacity is approximately 20,000 hours, larger sizes up to 30,000 hours

The rear dump units are commonly employed to haul heavy, abrasive material such as overburden, while the bottom dumps carry the free flowing materials such as coal. The two designs have unique features aimed to optimize their performance for their intended application.

Rear dumps (two and three axle)

- Average payload to net vehicle weight ratio of approximately 1.45
- Average gross vehicle weight to horsepower ratio 350.

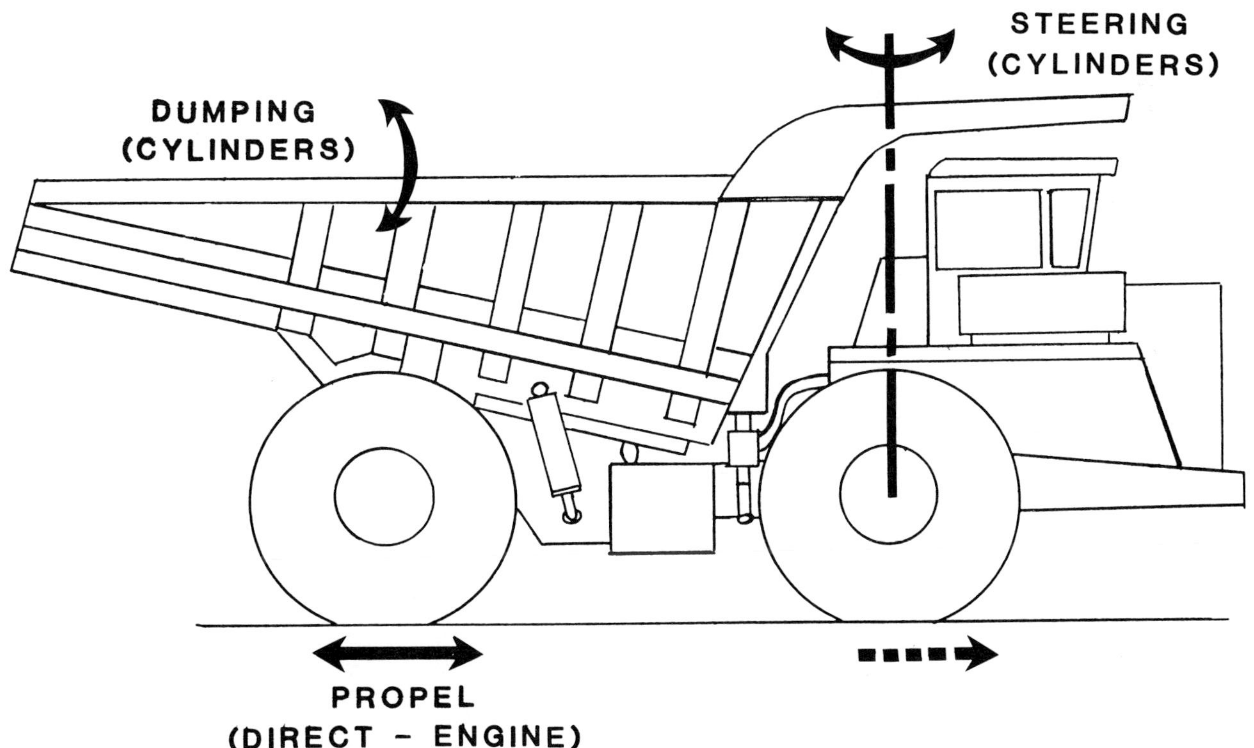

Figure 4.5 Truck Powered Functions

- Average horsepower per ton of capacity ratio of about 9.5.
- High power and gradeability (loaded).
- A turning radius about equal to 1.1 to 1.2 times overall length.
- A rugged body designed to withstand high loading impact.
- Forward mounted engine
- Mechanical drive system to 130 tons
- Electrical wheel drive systems from about 85 tons to 200 tons.
- Electrical axle drive systems from 235 to 350 tons
- Unit body construction
- Rear axle-dual wheel drives
- For 2 axle units, the front axle will carry approximately 47% of the net weight of the load and 32% of the gross weight. The rear axle will carry 53% of the net weight and 68% of the gross weight.
- For 3 axle units, the front axle will carry approximately 36% of the net weight of the load and 20% of the gross weight. The rear axles will carry 64% of the net weight and 80% of the gross weight.

Bottom Dumps (two axle)

- Average payload to net vehicle weight ratio of 1.7
- Average vehicle weight to horsepower ratio of 150
- Average horsepower per ton of capacity ratio of 7.6
- Medium gradeability (loaded)

Photograph 4.1 Euclid R-170, 170 ton, two axle rear dump hauling ore.

Photograph 4.2 Wabco 3200B, 250 ton, three axle rear dump

- A turning radius to overall length ratio of .63

- Clam-shell bottom discharge doors are designed for free flowing material and cannot withstand high impact loading

- Unit body construction

- Average width 19.9 feet

- Average loading height of 15 feet

- The front axle will carry approximately 42% of the net weight of the load and 47% of the gross weight. The rear axle will carry approximately 58% of the net weight and 53% of the gross weight.

- Drive systems, wheel combinations and steering do not follow consistent patterns. Goodbary has a front engine, rear drive, electric 2 axle, 4 wheel, and 62-64 degree steering. Kress has a rear engine, rear drive, mechanical 2 axle, 8 wheel, and 90 degree steering. Unit Rig has a rear engine, rear drive, electric 2 axle, 8 wheel and 90 degree steering. Wabco has a rear engine, rear drive, electric 2 axle, 6 wheel and 90 degree steering.

Bottom dumps (three axle)

- Average payload to net vehicle weight ratio of 1.6

- Average gross vehicle weight to horsepower ratio of 500

- Average horsepower per ton of capacity ratio of 6.5

- Limited gradeability (long grades less than 6%) and acceleration capability

- A turning radius of 1.8 times overall length

- Clam-shell bottom discharge doors are designed for free flowing material and cannot withstand high impact loading

- Low gradeability

- Lower tire loading

- Forward mounted engine
- Articulated tractor-trailer design with dual wheel drive on rear tractor axle
- Mechanical drive system to 160 tons
- Electrical drive system from 130 tons to 180 tons
- Average width 16.4 feet
- Average loading height about 13 feet
- The front axle will carry approximately 29% of the net weight of the load and 15% of the gross weight. The rear axle will carry approximately 37% of the net weight and 40% of the gross weight. The trailer axle will carry approximately 34% of the net weight and 45% of the gross weight.

The actual net weight for trucks of comparable capacity in the two basic configurations does not differ greatly, as shown in Graph 4.1. Rated capacity in terms of a ratio with net vehicle weight, however, varies substantially as shown in Graph 4.2. Because of the difference in materials expected to be handled, rear dump trucks are of more rugged construction thus weigh more relative to payload. Graphs 4.3, 4.4 and 4.5 provide greater perspective to the other average numbers referenced earlier and, in particular, emphasize the overall trend with respect to increasing size. It is also obvious that there is in reality a wide range in the values masked by the use of an average. These horsepower comparisons again illustrate that rear dump trucks generally have more power than comparable size bottom dump trucks. Gross vehicle weight to gross horsepower is often considered a fundamental indicator of vehicle performance level.

For many years the rear dump trucks were quite standardized as two axle units with dual wheels on a rear drive axle. There were three axle units only in the smaller sizes to reduce axle loads for highway transport. The recent supersized machines were forced to return to the three axle design to support the required loads on available tires. The multiple axle configuration reduces maneuverability.

International Harvester manufactures a unique two axle, 50 ton rear dump, with both axles powered and dual wheels front and rear.

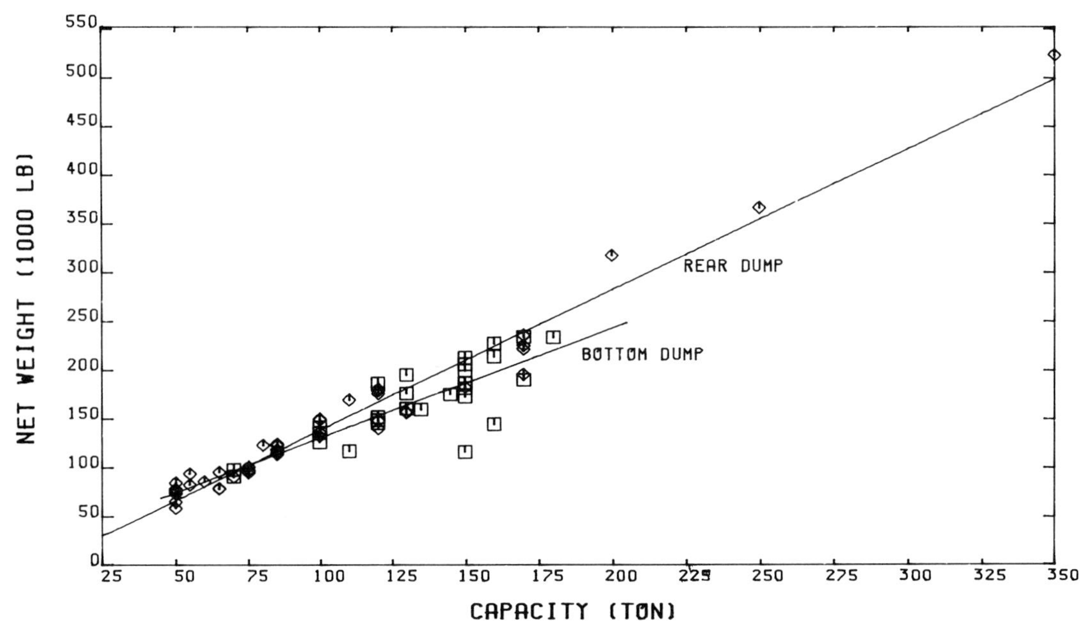

Graph 4.1 Net weight/ton of capacity

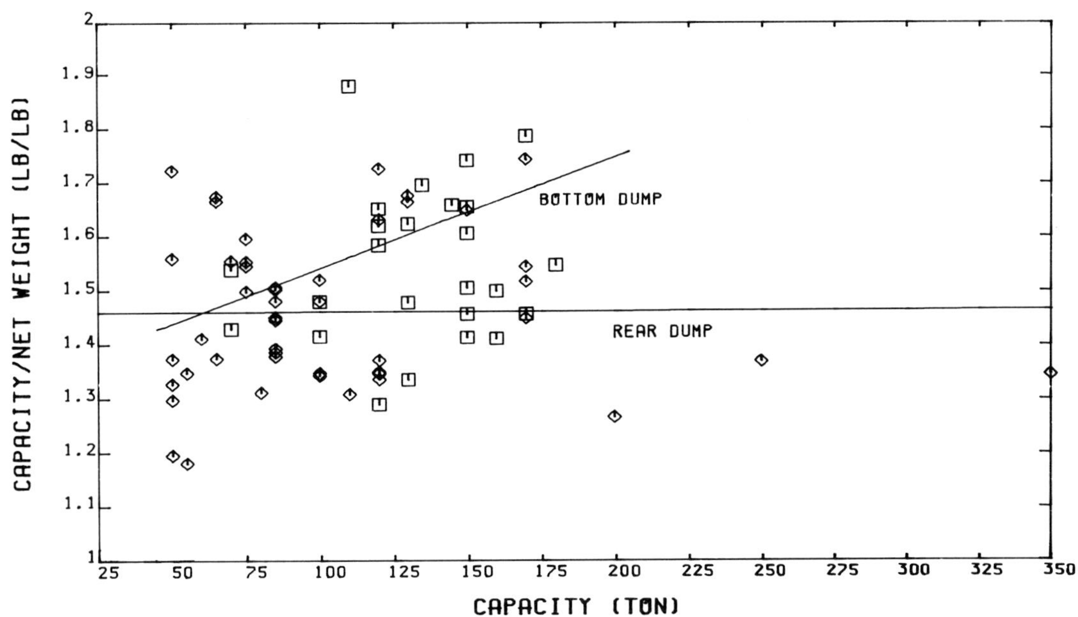

Graph 4.2 Payload per net vehicle weight/ton of capacity

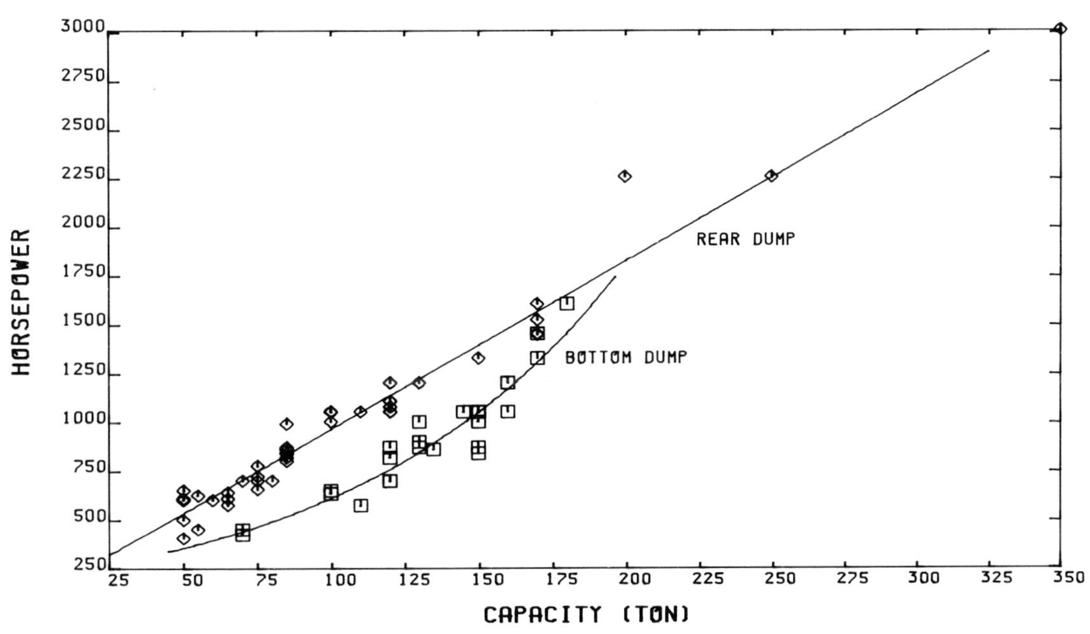

Graph 4.3 HP/ton of capacity

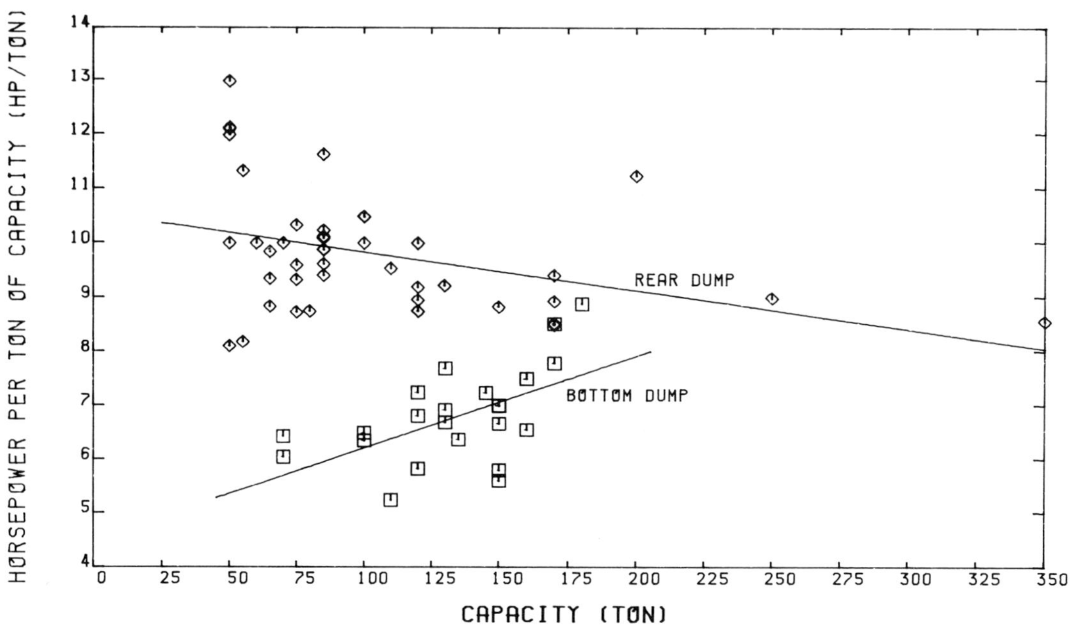

Graph 4.4 HP per ton of capacity/ton of capacity

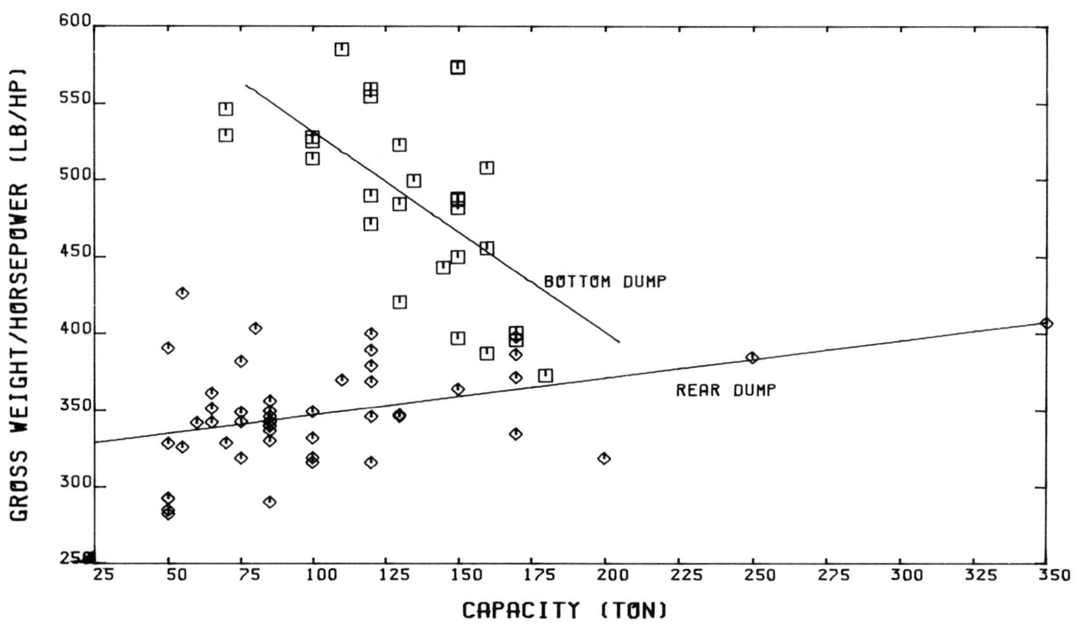

Graph 4.5 Gross weight per horsepower/ton of capacity

Photograph 4.3 Rimpull three axle bottom dump coal hauler

In the rear dump design the wheels are under the body so that with increasing wheel size the height of the body increases. This means that care must be exercised to be sure that any equipment used to load these trucks has adequate dump height.

The conventional three axle bottom dump truck is being challenged by the newer two axle designs. Machine differences are significant, hence the separation in the earlier summary listing. The two axle designs have improved maneuverability with their shorter turning radius, but the body height is also raised and it is wider. The more recent designs, dependent on the manufacturer, also bring in electric drives, and rear mounted engines. Graphs 4.6 and 4.7 show that, while there is considerable variation, the two axle bottom dump trucks tend to be lighter and more powerful than three axle units. Design considerations related to these options will be discussed in a subsequent section.

Tires are rated for E3 or E4 service. (See Appendix A) Until very recently most driving axles used a dual wheel arrangement to support the load and minimize the diameter. The duals increase the exposure to damage by rocks forced into the gap between tires, reduce the heat dissipation, make replacement procedures more complex, and may increase machine width. They increase the risk of one tire being partially deflated and not carrying its share of the load. However, on the positive side, they can reduce the consequence of a tire blowout. The Goodbury Company's latest two-axle bottom dump truck has gone to single larger tires on the driving axle. Tires are occasionally oversized to provide added dampening under shock loading. The same size tires are applied on all wheels to permit interchanging.

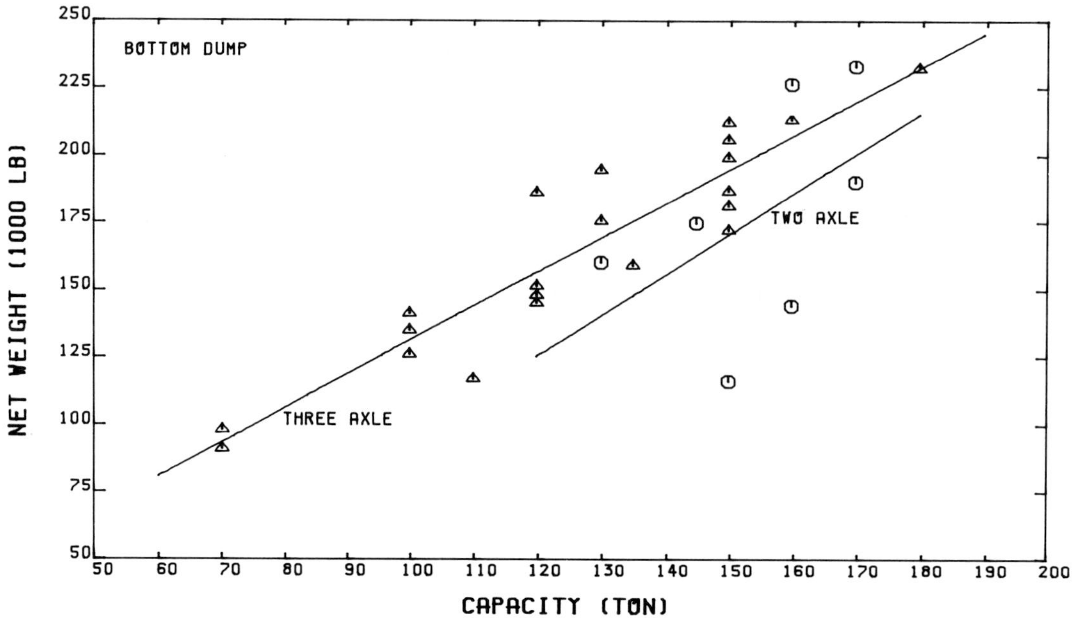

Graph 4.6 Net weight/ton of capacity (axle no., bottom dump)

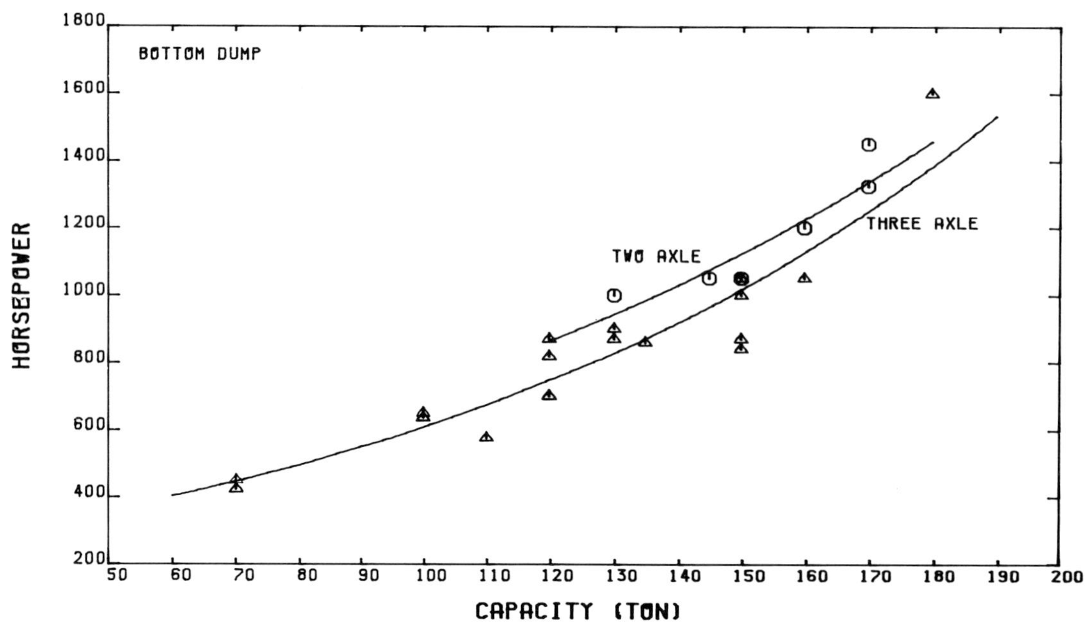

Graph 4.7 HP/ton of capacity (axle no., bottom dump)

Photograph 4.4 Wabco 170 two axle bottom dump truck

OPERATIONAL PRACTICES

Truck performance and economy are tied closely to haul road quality factors such as:

- Grades
- Curve radii
- Width
- Surface condition
- Number of crossings

These are functions of mine planning and operational practices. Top transport speeds are generally limited by safety, traffic, visibility, weather conditions, and tire heating. Many operations enforce speed limits which may be reduced for night work. Care must be exercised in operating on slippery down grades to guard against jack-knifing of the tractor-trailer type bottom dump units. The trucks have a relatively high center of gravity and, therefore, comparatively low tipping speeds, making elevated curves desirable.

Gradeability is a function of the combined rimpull required to overcome the grade and the rolling resistance of the tires. Rolling resistances increase with tire penetration and the general quality of the road surface. Engine power must be matched to the site requirements. Turbocharging of the engine is required at higher altitudes. (See Appendix B) Axle ratio reductions can be determined (in conjunction with the manufacturer) to optimize performance for specific haul route conditions.

Tires are a major operating cost item so that their selection, care, and maintenance is a prime consideration. Experience has shown that tire life can differ substantially between operations. A trial and error approach may be necessary to optimize the tire configuration. High design loads,

Photograph 4.5 Terex 32-19, 350 ton, three axle rear dump

high operating speeds and abuse in the loading or dumping areas or on the haul roads from bad surface conditions, will significantly reduce tire life. Wear is greater on the drive axles and may be increased on the rear tires because of the load transfer that occurs on grades.

Sideboards are frequently added to the truck body to increase the volume carrying capacity when hauling lighter materials than anticipated at the time of purchase. Liners, wear bars and plates are sometimes added under very adverse conditions to extend the body life and reduce maintenance. To extend the wear life, the use of rubber type body liners is growing. These materials are relatively light weight and have the ability to absorb some of the shock loading.

Braking capacity can be a problem in long steep downhill runs. Special attention, in these cases, must be given to brake maintenance and consideration given to adding retarding capacity.

The excavator must be matched to the truck with respect to size and capability to deposit the load in the center of the truck body. Proper size correlation, is generally considered to be from 3 to 5 passes of the excavator to fill the truck. Adequate dump height and reach are essential to prevent damage to the truck and avoid lost excavator cycle time because of excessive maneuvering and positioning in their dumping phase. Excavators differ not only in the size of the lumps or boulders they can discharge, but also in the severity of impact of the discharge. With bottom dump trucks, both of these criteria should be evaluated when matching units.

The typical truck maneuvering sequences at the loading site are illustrated in Figure 4.6. The drive-by method can have both excavator and truck on the same bench level or, alternatively, the excavator can be either on the bench above or below. Hydraulic hoes are commonly used in those cases where the excavator is positioned above the truck level. Long range electric shovels are used when the excavator is sitting on the lower bench. The single or double back-up approach, as the title suggests, can be applied with trucks backed into position on one or both sides of the excavator. The two truck method, if properly scheduled, permits sequential loading with minimum lost excavating time. It should be noted that not all mine operational plans permit the double back-up method.

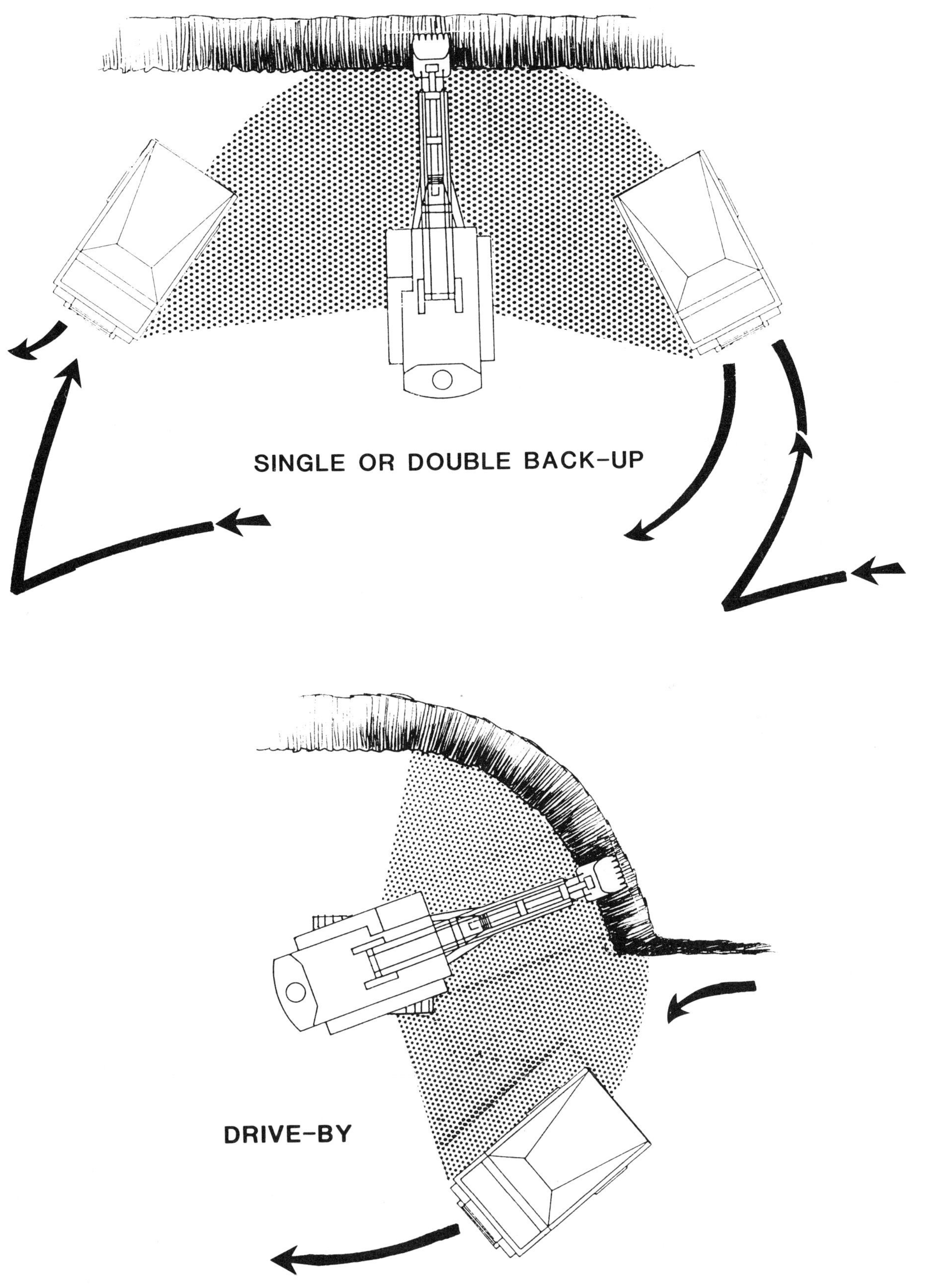

Figure 4.6 Truck Load And Spotting Configurations

Similar to other fleet operations, good scheduling and dispatching are essential to maximize output. Any bunch-up in the loading or discharge areas means idle equipment and reduced haul tonnages. Shifting of trucks between loading units, if any breakdowns or delays occur, optimizes overall fleet performance.

Carrying excessive loads can result in both reduced tire life and long term damage to the truck. Conversely, undersize loads reduce system capacity and increase transport costs. On-board weightometers are available to indicate status of the load and to record production. They are particularly valuable for applications which involve high capacity trucks on long hauls.

Under some conditions tire ground bearing pressures and/or flotation is critical because of traction conditions and the fact that resistance increases with greater tire penetration. Large, low pressure high flotation tires are available for smaller size trucks which may be desirable on some sites. The DJB company is currently offering a series of articulated rear dump trucks with high flotation tires which are claimed to offer improved performance on soft floor conditions.

Studies continue on automatic guidance systems for trucks such as those which will follow a buried cable between load and dump sites. Full scale operational tests have been made by the Unit Rig Company. Safety, efficiency, operational techniques and cost considerations are still being evaluted.

High fuel costs are forcing a re-evaluation of mine plans to minimize transport distances and operations on grades. Fuel consumption varies directly with engine horsepower and duty cycle requirements, with a 100% increase possible from light to severe duty. As a spin-off of this thinking, there has been considerable work done by General Electric and Unit Rig in the development of trolley assist systems for electric drive trucks. On steep grades, the truck with a trolley mounted on the front (see Photograph 4.6) contacts an overhead power line which supplies additional power for this segment of the haul route. Systems of this type have been in use as far back as 1956.

Photograph 4.6 Unit rig rear dump with trolley assist equipment

Probably the most extensive application was at the Quebec Cartier Mining Company, LacJeanine Mine, from 1971 to 1977. This system apparently improved productivity about 23% and reduced fuel consumption 87%. Extensive installations are currently in progress in South Africa. In addition to the benefits noted above, it is anticipated that engine maintenance costs would also be reduced. A major capital investment is required for such installations and special haul road layouts and maintenance requirements must be considered. Future applications may occur at mines with steep grades, as well as in areas associated with high diesel costs or limited availability and where cheap electric power is available.

Operator visibility, particularly to the right side and rear, is limited on the larger trucks. Care must be exercised to restrict the movement of people and other vehicles in the truck maneuvering areas. Special Fresnel lens in mirror arrangements have been studied under a U.S. Bureau of Mines contract to increase the field of vision. These units will see around a 70 degree corner but currently need a fairly high contrast for good visibility. In some cases, closed circuit cameras are being mounted to allow the operator sight access to blind spots.

Oil sampling programs are common to monitor the condition of the engine and transmission units.

DESIGN FEATURES

Refer to Figure 4.7 for rear dump truck nomenclature.

Optional engines and horsepower ratings are available on most truck models. (See Appendix B) Unique to trucks of more than 200 ton capacity, is the application of General Motors locomotives engines. These are well proven two cycle versions but differ from the other engines in their heavy low speed design.

Front mounting provides a cleaner environment for the engine resulting in less filter maintenance; it also provides a good cooling system air flow and a less complicated control system interconnection to the operator station. Rear mounting provides added protection to the driver from fire hazards, and a better weight distribution. Radiator shutters are used in both cases to control engine temperatures.

The general arrangement of the mechanical drive system is illustrated in Figure 4.8. Typically these consist of the following (service life estimates are based on field experience and are hours before any basic overhaul):

- Fully automatic powershift transmission designed to minimize drive line shocks thus increasing operator comfort and reducing component stresses.

- Three to six gear ratios forward and one reverse.

- Transmissions manufactured by the Allison Division of General Motors, Clark Equipment, Caterpillar, Fuller, and International Harvester (5,000 to 8,000 hours of service before rebuilding).

- Torque converters capable of providing high starting torque multiplications with a hydraulic clutch for automatic lock-up at road speeds. Often a single stage, three element configuration.

- Hydraulic retarding capability included to reduce braking demand.

- High capacity final drive shaft (15,000 to 18,000 hours of average service life).

- Limited slip or non-slip differential with a gear reduction (10,000 to 12,000 hours of service life).

- Free floating axles which can be disassembled without removal of final gear reduction.

- Final planetary gear reduction in the wheel hubs. This arrangement reduces the torsional stress in the axles and differential (18,000 to 20,000 hours of service life).

- Total drive reduction ranges between 17:1 and 28:1.

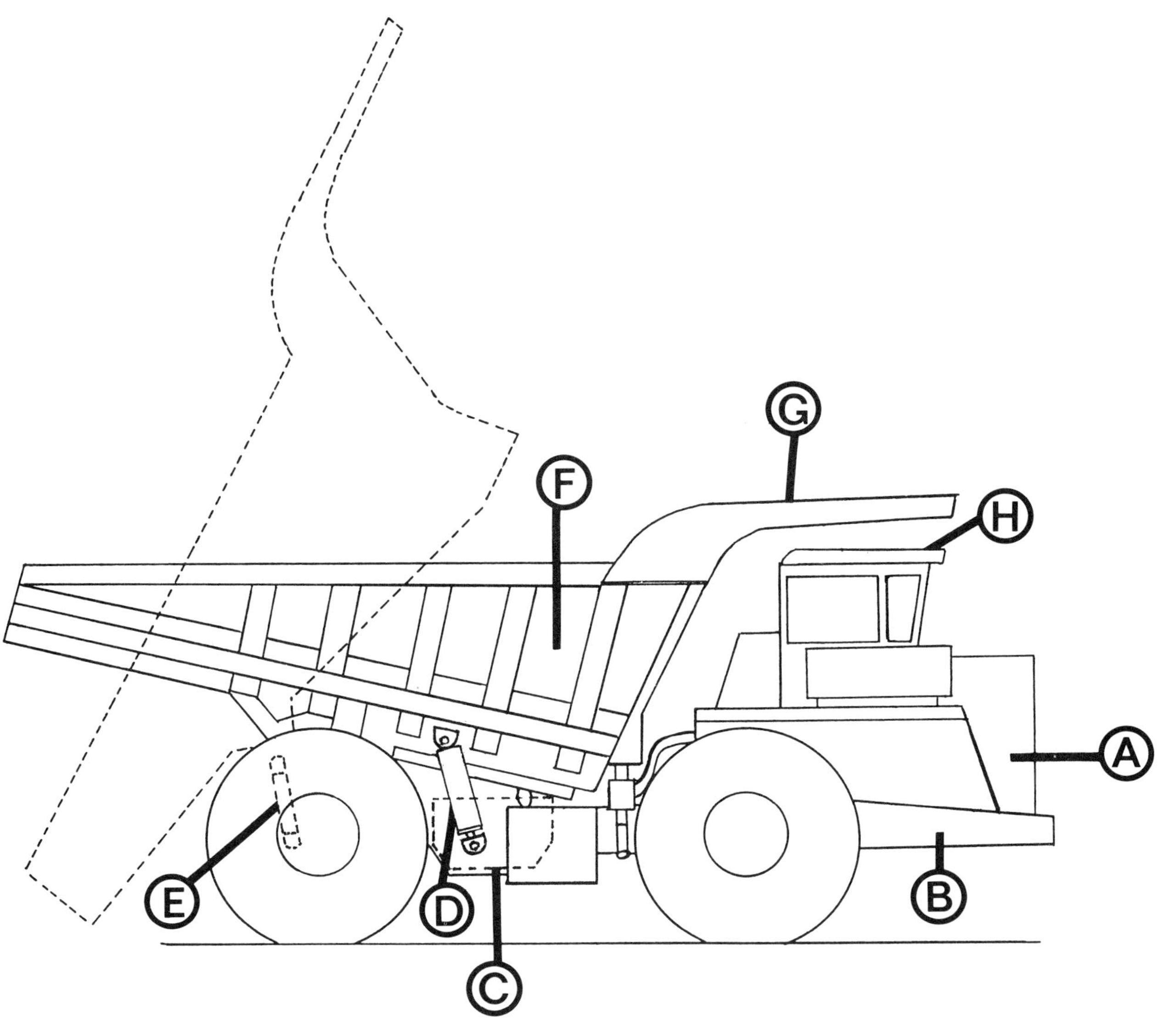

A	Engine	E	Suspension Cylinders
B	Main Frame	F	Body
C	Transmission	G	Canopy
D	Body Lift Cylinder	H	Cab

Figure 4.7 Truck Nomenclature

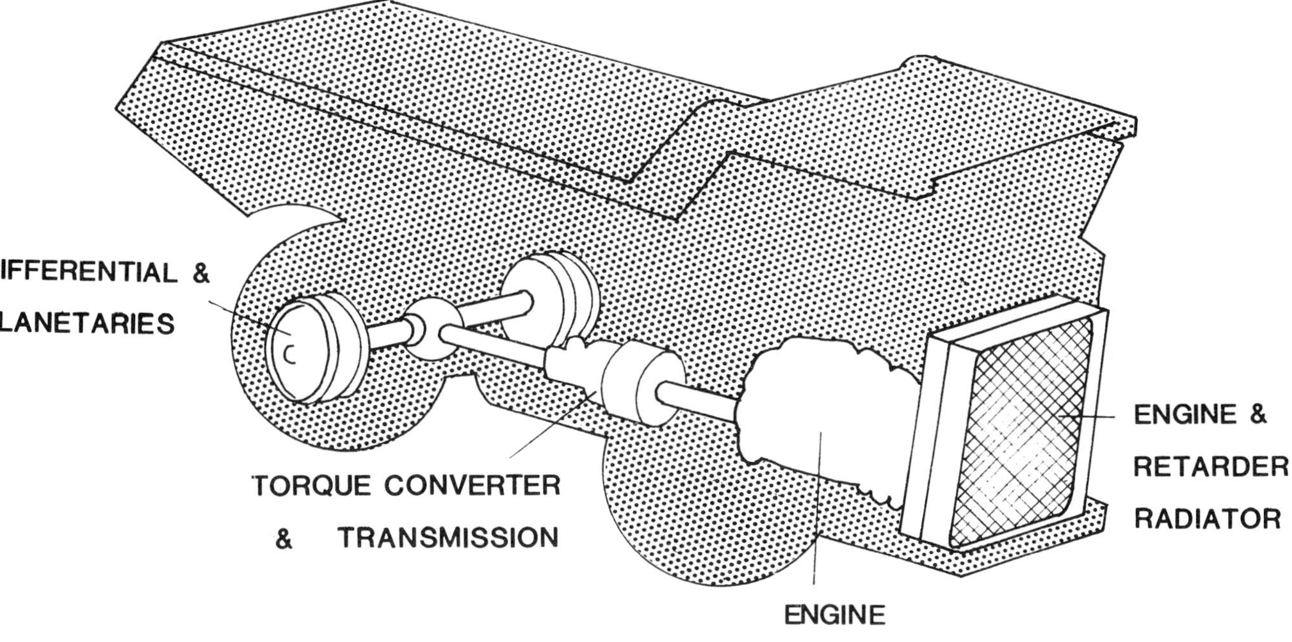

Figure 4.8 Mechanical Drive Train

The electric drive train is illustrated in Figure 4.9. Typically the components consist of the following (service life estimates are based on field experience and are hours before any basic overhaul):

- Direct coupled D.C. generator for unit up to 1000 H.P. (12,000 to 20,000 hours service life) and an alternator with rectifier for larger units. Units are manufactured by General Electric, Westinghouse and Reliance Electric.

- Solid state control system.

- Grid box for dissipation of heat generated during retarding conditions

- D.C. motors in the hub of each drive wheel or each driving axle in the largest sizes. Motors are series wound for high torque/low speed or low torque/high speed operations (15,000 to 18,000 hours service life). Matched units are supplied by the manufacturers of the generators noted above. (Unit Rig offers their own generator/wheel motor sets as options for most of their product line.)

- Motor module includes gear reduction and brake

- Blowers for generators and motors

Very large trucks utilize the slow speed General Motors diesel and the locomotive traction motors. These units mount the drive motors in the rear above the axles and drive through a gear reduction to the axle which has conventional differential and planetary gear reduction and brakes in the wheel hub.

Due to the component availability, electric drives are used on larger trucks; there is, however,

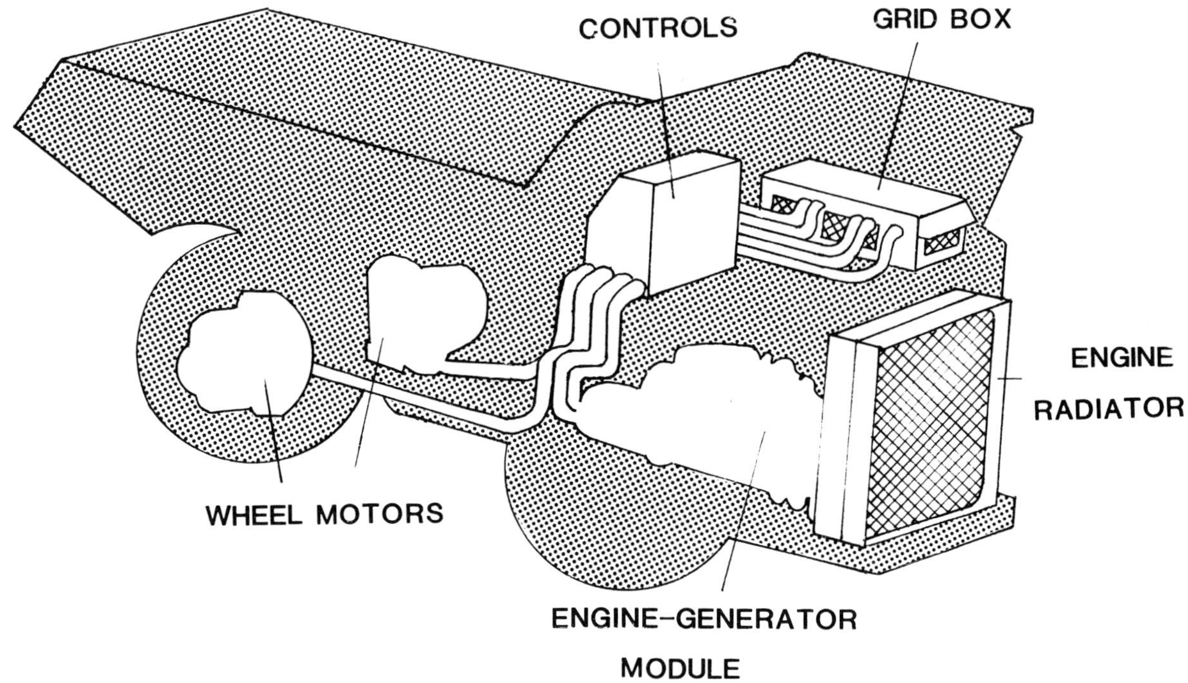

Figure 4.9 Electrical Drive Train

considerable overlap in the sizes. (See Graph 4.8) The pros and cons of mechanical and electrical drive systems are complex and difficult to assess in total because of the offsetting features.

- Overall efficiency is approximately 65% to 75%. Mechanical drive is superior at low power demands. Electric drive is superior at high power demands.

- Slightly higher horsepower diverted in electric drive to power auxiliary equipment.

- Fuel consumption approximately 0.50 to 0.60 lbs/rim hp/hour. The number is slightly higher for electric drives at speeds above 5 mph.

- Retarding capability is roughly comparable for both types of drives at speeds above 10 mph. Below 10 mph the mechanical drives can be substantially better. Electrical options are available to provide superior performance above 10 mph.

- Engines for electric drives are operated at rated maximum rpm to extend engine life.

- There are only minor differences in operating costs.

- The electric drives are heavier. This difference does not appear to have any major impact on total machine weight. (See Graph 4.9)

The service brakes are commonly of either shoe or disc design. (See Figure 4.10) There are two types of disc brakes — multiple disc or a single disc with a caliber actuating device (a small button of friction material applied with a lateral clamping action). The disc-caliber is the most common, with only Caterpillar using a multiple disc oil cooled arrangement. The shoe brakes have internal shoes which expand against the inside of

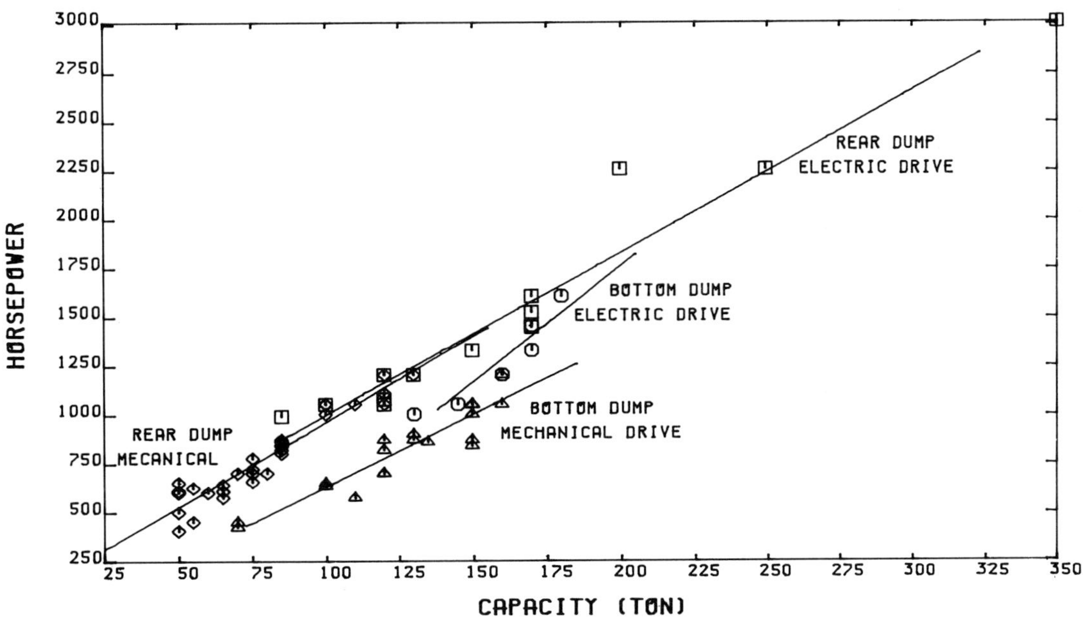

Graph 4.8 HP/ton of capacity (drive type)

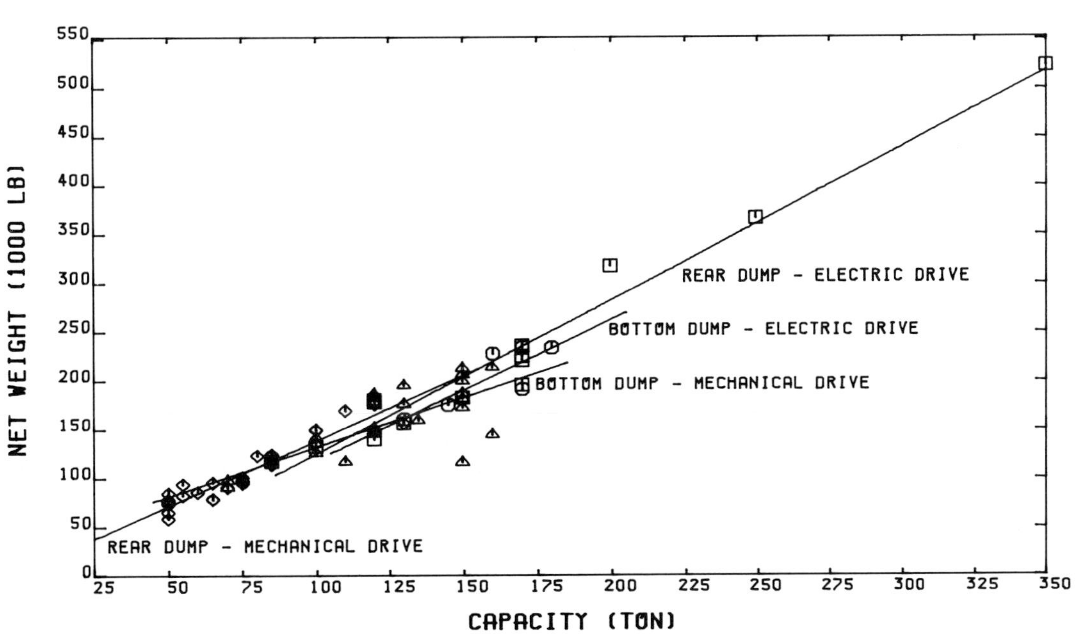

Graph 4.9 Net weight/ton of capacity (drive type)

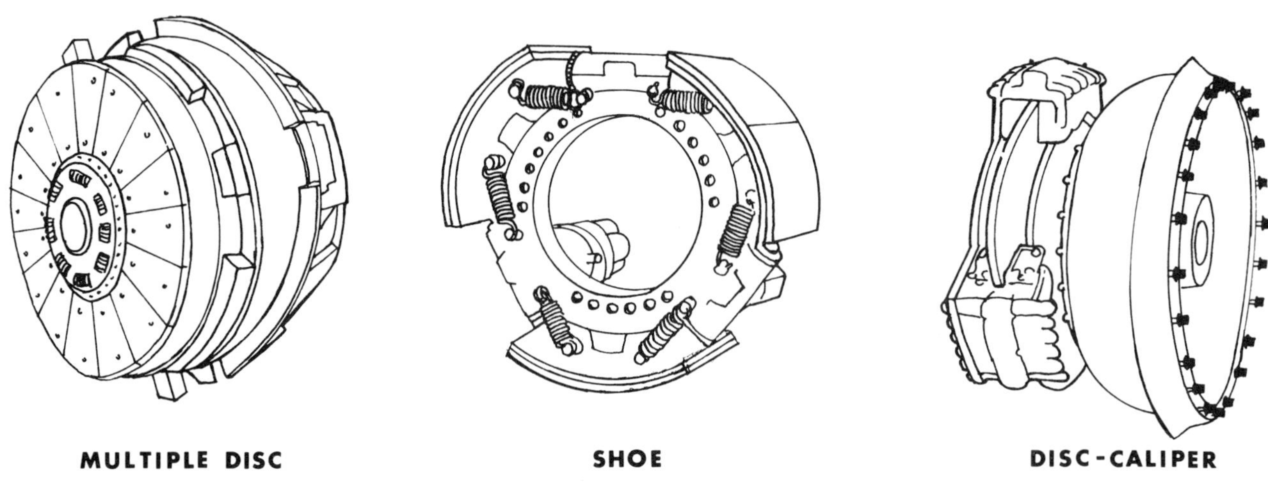

Figure 4.10 Brake Configurations

a drum surface. The shoes are free floating and self adjusting. Atlas uses a pneumatically actuated tube brake in its smaller models. Brake design is dictated primarily by available space and heat dissipation. The disc-caliber design is popular because of its light weight, simplicity and lower cost. Generally they are air or hydraulic actuated with multiple circuits for safety and independent circuits for emergency braking. Parking brakes located on the transmission shaft or the armature of the traction motor are a separate system applied automatically when the air pressure drops.

Mechanical retarders are capable of braking 135% to 140% of engine rating, excluding engine friction. Located between the engine and the final drive, mechanical retarders consist, essentially, of a bladed rotor revolving in oil between fixed stators. They are effective in all speed ranges but retarding action is applied only to driving wheels. In a single operator pedal system, the inital pressure activates the retarder and subsequent pressure activates the service brakes. The two pedal system has one pedal for the service brakes and a second for the retarder so that independent action is possible.

The electric drive incorporates dynamic retarding by converting the wheel motor temporarily into generators and dissipating the power developed as heat in a fan cooled resister bank. Single or two pedal operation is available similar to that for mechanical retarders.

A suspension system mounted between the frame of the truck and the axles insulates the body from road roughness and the impact that occurs during loading. While providing both vehicle stability and ride comfort, a suspension system must also be simple, durable, readily maintainable and have a predictable performance. Dual tires reduce the distance between suspension on either side and increases the design problem. There are two basic types of configurations; mechanical springs or a mechanical linkage with an energy absorbing member. (See Figure 4.11) Mechanical leaf or coil springs (occasionally torsion tubes) are used on smaller trucks below 75 tons in capacity. They have limited deflection and limited self dampening characteristics; they require additional shock absorbers or friction dampers, to reduce bouncing. They are heavy, and failure causes a shut down of the machine. The linkage-energy absorbing types of suspension have two basic designs. Hydro-pneumatic systems displace oil under pressure and produce dampening by a throttling action, with a constant spring rate regardless of temperature. Rubber discs are generally bonded to metal and stacked inside a telescoping strut; these discs are efficient, self-dampening, require little maintenance and have a long service life.

To dump their load, rear dump trucks commonly employ two double acting hydraulic cylinders to rotate the body around a rear hinge. In some cases two or three stage telescoping

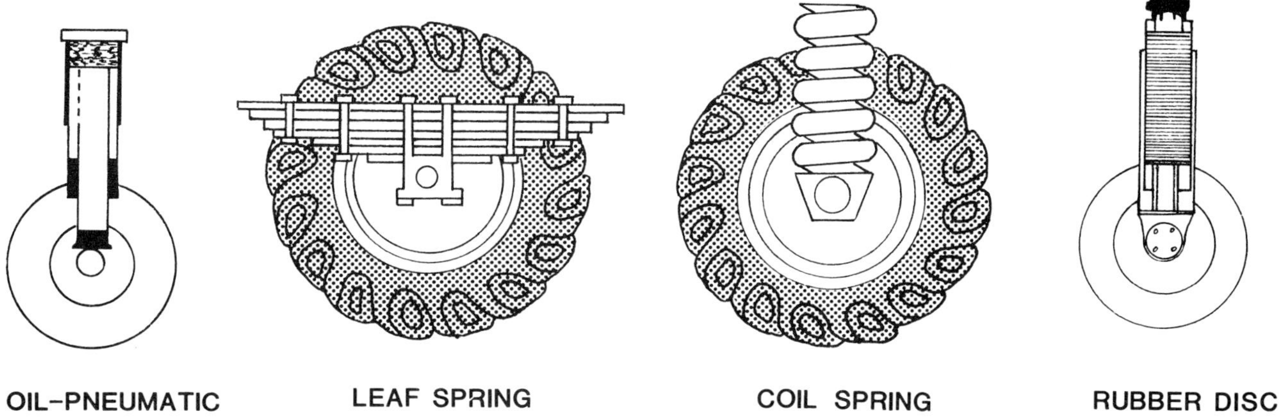

Figure 4.11 Truck Suspension Alternatives

cylinders are required for the long strokes involved. Raise time for the medium size units ranges from 16 to 24 seconds and increases to approximately 30 seconds for the largest units. Care must be taken in locating these cylinders to avoid damage from rocks and debris thrown by the tires.

To dump, bottom dumps utilize twin doors on the bottom of the body either hinged or mounted on a linkage which produces a semi-parallelogram action. Air or low pressure hydraulic cylinders provide the power for opening and closing the doors. These designs are complicated by generally poor accessibility, the need to maximize ruggedness, and lack of ground clearance when the doors are open.

Rear dump bodies have a dual slope floor with a V-bottom. They are generally made from low alloy, high strength steels. Aluminum is occasionally substituted for less severe applications to decrease machine weight and thus increase the payload. Bodies are usually heated by routing the exhaust gas through the structural support; the connection is through a flexible hose with a mechanical control damper. This helps to prevent wet material from freezing to the body in cold temperatures.

Rubber pads are provided at frame contacts to absorb shocks and allow for distortion and misalignment. Body height is normally minimized to allow loading by a range of loading equipment. Width varies for different unit material weights (extra wide for coal). Power tailgates are sometimes added to increase body capacity.

The hydraulic steering on the rear dump and tractor-trailer bottom dump truck is conventional. The newer two-axle bottom trucks manufactured by Kress, Unit Rig, and Wabco have an individual vertical post type mounting for the front wheels which permit steering angles of 90 degrees.

SELECTION CONSIDERATIONS

The type of truck to apply is dependent on the following site conditions and mine requirements:

- Material characteristics (abrasiveness, unit weight, swell, lump or rock size)
- Loading equipment (see Figure 4.5 and 4.6) (dump height, reach, discharge severity). Graph 4.10 plots typical truck loading heights and illustrates the effects truck type and size can have on making the proper loader/truck match.
- Haul route requirements (grades, distances, and curves)

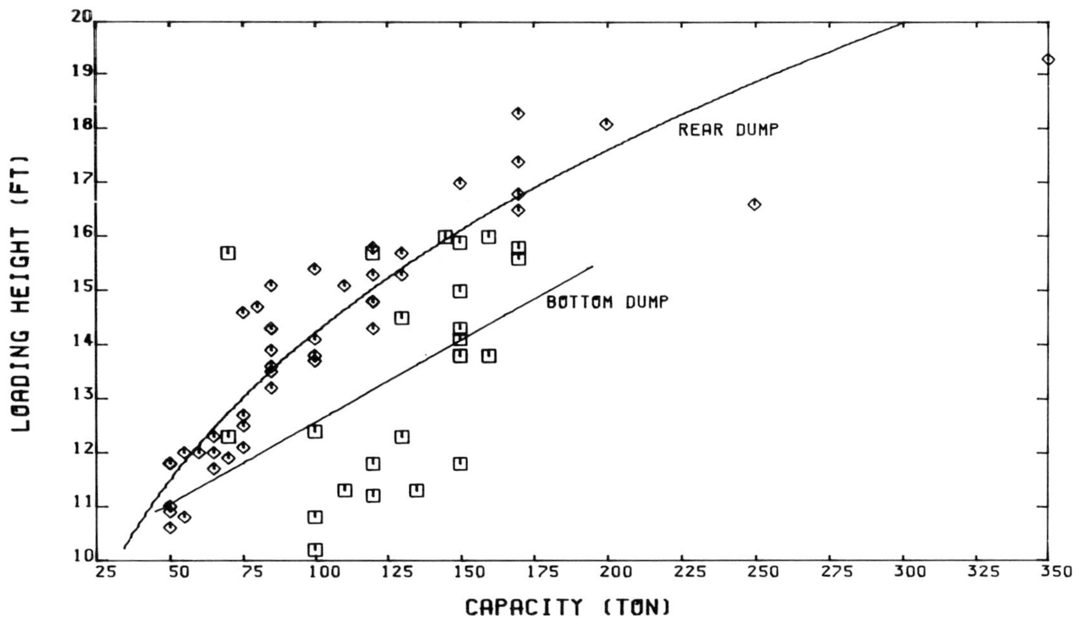

Graph 4.10 Loading height/ton of capacity

- Maneuvering space (turning radius)
- Dumping conditions (spreading, over the bank, hopper, crusher capacity, etc.)

Basic truck specifications must be chosen based on prior experience, job requirements and/or design evaluation.

- Capacity (see Graph 4.11 for typical cubic yard to ton capacity relationships)
- Engine power and manufacturer (altitude limitations)
- Final drive gear ratios for mechanical drives (dictated by anticipated rimpull to meet combined grade and rolling resistance requirements)
- Two axle or three axle configuration
- Mechanical or electrical drive system
- Tires size, tread and ply rating
- Options (listed later)

Truck operational performance is related to the tractive effort available (rimpull) which is a function of net engine power, drive train reduction and efficiency, and tire diameter. This available rimpull varies with vehicle speed dependent on the engine torque-speed characteristics and the transmission speed ranges. Rimpull must then be equated to job requirements in terms of the "effective grade", which is the combined resistance of the grades to be negotiated and the rolling resistance of the operating surfaces. Grade resistance (1% of machine weight per percent of grade) is readily calculated knowing the truck weight. Rolling resistance is established from available tables selecting the anticipated haul road and/or dump and dig area surfaces. Manufacturers' performance charts are available for each model showing:

- Rimpull requirements for various combinations of vehicle weight and effective grade
- Vehicle speed vs. rimpull (vehicle speed varies with the transmission range for mechanical driver)

The vehicle speed can also be established from the chart data under retarding conditions.

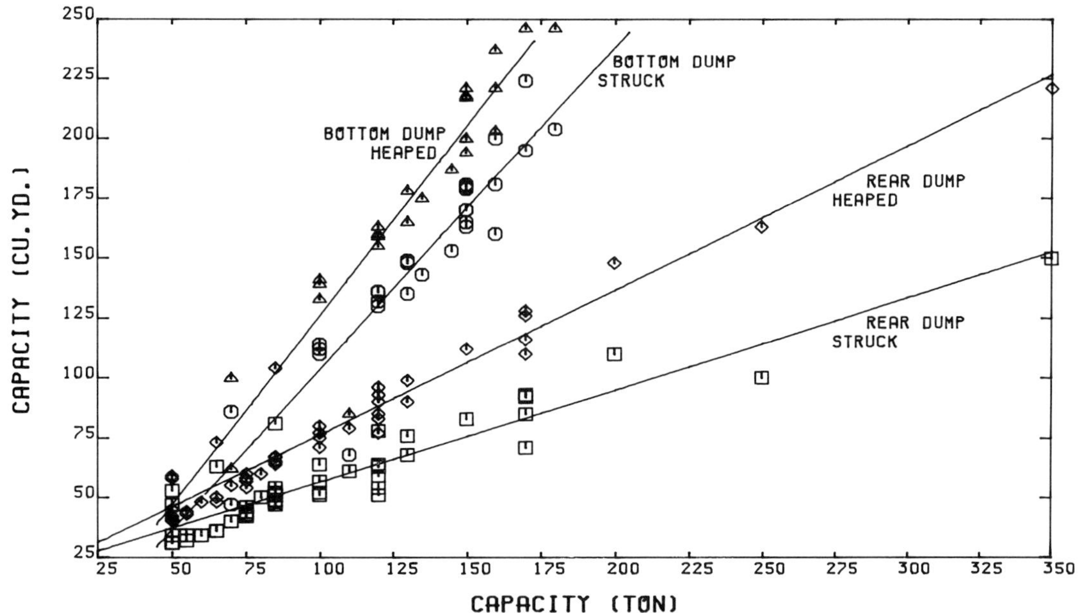

Graph 4.11 Cubic yards of capacity/ton of capacity

Truck production calculation procedures and estimates are provided in available manufacturers' literature. These, together with ownership and operating cost estimates, are considered a separate subject too extensive to be comprehensively treated in this manual. Estimates must be directly correlated with detailed site conditions and matched to specific machine performance characteristics. All of the equipment manufacturers will provide production and cost estimates for stated conditions prepared by their sales technical staff personnel based on their accumulated experience on similar applications. These are usually developed by in-house computer programs so that alternative fleets of matched equipment can be readily evaluated to find the optimum combination.

It must be recognized that these estimates are based on data extrapolated to meet anticipated site conditions over the life of the operations. The spread in these "averages" can be substantial and the factors introduced require a considerable amount of judgement. Costs, in particular, vary locally with time and economic conditions which make generalization dangerous. Haul routes will also usually vary significantly over the life of a mine. Some of the production considerations entering into these calculations are:

- Material swell
- Material unit weight
- Road grades
- Rolling resistance
- Travel distance
- Curves
- Traffic
- Maximum speeds
- Job efficiency
- Mechanical availability
- Engine power
- Transmission characteristics

- Loading system (including delays)
- Dump arrangement

Some of the cost considerations entering into these calculations are:

- Machine price
- Salvage value
- Machine life
- Interest
- Insurance
- Taxes
- Fuel consumption
- Lube, oils, greases, filter, etc.
- Tire cost
- Maintenance
- Operator wages

In general:

- Productivity per hour increases with truck size.
- Productivity in tons/hour on 0% grades is superior for the bottom dump trucks.
- Productivity in tons/hour on 10% grades is approximately equal for bottom dump and rear dump trucks.

The options available vary with the manufacturer but can be summarized as follows:

Performance features

- Retarders
- No-spin differential
- Oversize brakes
- Thermostatic fan
- Side boards
- High capacity fuel tank
- Drive axle oil cooler

Control refinements

- Independent trailer brake control
- Dual steering
- Supplemental electric steering
- Reverse inhibitor
- Hopper doors open indicator
- Body down indicator
- Traction motor cut-off switch

Operator station extras

- Air conditioner
- Windshield washer
- Turn signals
- Speedometer
- Load counter

Safety

- Fire protection system
- Rear window guard (bottom dumps)
- Vandalism protection system

Servicing

- Automatic lubrication
- Tire air pressure maintenance system
- Fast fuel filler system
- External air and electrical connections
- Underhood light

Severe service extras

- Transmission guards
- Body liners
- Rock ejectors
- Mud flaps
- Engine coolant heater
- Air dryer
- Radiator shutters

NEW DEVELOPMENTS & TRENDS

Maximum size appears to have been already reached, possibly exceeded, so that future developments should focus on design refinements and filling in gaps in the size ranges.

Higher capacity mechanical transmissions have recently become available resulting in new units up to 130 ton capacity being marketed with mechanical drives up to 130 ton capacity. Komastsu has test units in service of 160 ton capacity.

General Electric is introducing a new electric wheel motor for trucks up to 220 tons in size, which will encourage development on larger electric wheel trucks.

The two axle bottom dump design continues to dominate the new model introductions for units of this type. Kress is offering a unique 2 axle, 160 ton cross body truck. (See photograph 4.7)

DJB is marketing in the smaller size range, three axle, articulated rear dump trucks. The two rear axles are powered and have a single wheel, low profile, and high flotation tires. (See Photograph 4.8)

The new rubber type composites appear to offer improved wear protection as body liners.

Photograph 4.7 Kress 160 ton two axle, cross body, bottom dump truck

Photograph 4.8 DJB model D550, three axle, rear dump articulated truck

These are lighter in weight and shed better than comparable steel liner materials.

Continued improvements are anticipated in engine load matching, control, and design to reduce fuel consumption per unit of production.

Driverless truck systems are being studied. These devices presently appear to be in the future for any broad application but may have a place in special situations.

Trolley assist systems are proven and operational. Diesel fuel cost and availability will dictate the future application. As mining goes to greater depths, accompanied by steeper haulroad grades and increased length of operation on grades, justification for the use of trolleys will become apparent.

MACHINE SPECIFICATIONS

See Figure 4.12—Truck Dimensions
See Table 4.2—Rear Dump Truck Specifications
See Table 4.3—Bottom Dump Truck Specifications

Trucks are listed alphabetically by manufacturer, and then in ascending order by payload capacity in tons. Payload capacity is based on published manufacturer data. In addition to payload capacity, rated capacity in cubic yards (both struck and 2:1 heaped) is also listed. The cubic yard capacity rating is based on the manufacturer's standard rock body where possible, or on the rock body used in the manufacturer's published data. In some cases, there are machines that have a rated capacity in cubic yards significantly greater than other machines of similar payload. These machines are equipped with a coal or other light material body. Most manufacturers offer, as options, high capacity or special duty bodies for their trucks.

Engine horsepower can be misleading because of various ways engines can be rated. Engine horsepower, as it is given in these charts, refers to SAE net or rated horsepower. When the truck manufacturer did not give a rated or net horsepower figure, data from the engine manufacturer was used. The horsepower listed in the following chart, then, is the approximate net or rated horsepower of the highest rated standard engine listed by the manufacturer.

Steering angle is defined as the arc, in degrees, of the front tires from full lock in one direction to straight ahead. The abbreviation 'art' designates an articulated machine. Max speed is given for forward speeds only. Reverse speed data is not common. Where reverse speeds are listed, the average is less than 10 miles per hour. Turning diameter is the SAE clearance circle, measured to the outside front corner of the machine. Wheelbase on two axle trucks is measured in the conventional manner; wheelbase dimensions for the three-axle trucks are measured to the mid-point between the two rear axles. Tire size, as listed, is the smallest, lowest tread rated tire available as standard equipment.

All specifications, capacities, capabilities and dimensions are based on published manufacturer data. Although the information is believed to be current and the interpretation to be consistent and correct, it is possible that omissions and inaccuracies have occurred and that out-of-date information may have been included. These specifications charts are not meant to be used either as an in-depth analysis or as a head-to-head comparison of the available equipment, but rather as a general overview of the equipment, available sizes, and approximate operating data. Specific questions relating to performance or purchase should be directed to the manufacturer or authorized distributor/dealer in the user's specific area.

Table 4.2
REAR DUMP TRUCK SPECIFICATIONS

MAKE	MODEL	RATED LOAD (tons)	STRUCK CAPACITY (cu yd)	HEAPED CAPACITY (cu yd)	ENGINE HP	STEERING ANGLE (degrees)	FUEL TANK (gal)	COOLANT (gal)	NO. OF AXLES F/R	TOTAL NO. OF WHEELS	MAX SPEED (mph)
Caterpillar	773B	50	34	45	650	31	185	40	1/1	6	34
Caterpillar	777	85	48	67	870	30	250	76	1/1	6	30
Dart	3065	65	36	48	640	35	400	46	1/1	6	—
Dart	3075	75	43	54	720	35	400	47	1/1	6	—
Dart	2085	85	81	104	800	35	250	47	1/2	10	—
Dart	3100	100	51	75	1050	40	—	—	1/1	6	—
Dart	3110	110	61	79	1050	41	—	—	1/1	6	—
Dart	3120	120	59	90	1200	41	—	—	1/1	6	—
DJB	D550	55	32	43	450	art	230	28	1/2	6	30
Euclid	R-50	50	31	41	606	41	185	49	1/1	6	40
Euclid	R-75	75	46	60	655	41	—	—	1/1	6	38
Euclid	R-85	85	52	67	818	41	265	58	1/1	6	38
Euclid	R-100	100	52	77	1000	37	325	95	1/1	6	41
Euclid	R-120E	120	63	93	1050	42	510	95	1/1	6	32
Euclid	R-170E	170	93	128	1519	44	510	115	1/1	6	34
GM	33-15B	170	85	116	1445	40	480	90	1/1	6	31
GM	33-19	350	150	221	3000	28	960	258	1/2	10	—
Isco	250C	50	47	59	500	—	100	24	1/2	10	33
Isco	265-C	65	63	73	575	—	150	55	1/2	10	35
International	350B	50	31	42	607	28	180	49	1/1	8	42

Kenworth	548CH	50	53	58	405	—	—	—	1/2	10	28
Komatsu	HD680-2	75	42	58	775	—	—	—	1/1	6	—
Rimpull	RD-65	65	36	50	608	35	160	53	1/1	6	40
Rimpull	RD-70	70	40	55	700	35	160	55	1/1	6	40
Rimpull	Rd-80	80	50	60	700	41	240	55	1/1	6	40
Rimpull	RD-85	85	49	64	860	41	240	55	1/1	6	40
Rimpull	RD-100	100	57	71	1050	41	360	105	1/1	6	40
Rimpull	RD-120	120	64	85	1050	41	360	105	1/1	6	40
Terex	33-09	55	34	44	624	42	220	51	1/1	6	42
Terex	33-11C	85	51	65	840	38	325	56	1/1	6	37
Terex	33-11D	85	51	65	840	38	325	56	1/1	6	37
Terex	33-14	130	76	99	1200	38	—	—	1/1	6	34
Unit Rig	M-85	85	54	65	990	—	500	—	1/1	6	30
Unit Rig	M-100	100	64	80	1050	—	500	—	1/1	6	31
Unit Rig	M-120-15	120	78	96	1200	—	500	—	1/1	6	31
Unit Rig	Mark 30	130	68	90	1200	—	600	—	1/1	6	34
Unit Rig	Mark 33	150	83	112	1325	—	500	—	1/1	6	33
Unit Rig	Mark 36	170	92	126	1600	—	700	—	1/1	6	33
Unit Rig	M-200	200	110	148	2250	—	800	—	1/1	6	24
Wabco	50B	50	31	40	600	—	154	42	1/1	6	40
Wabco	60B	60	34	48	600	—	240	42	1/1	6	47
Wabco	75C	75	44	57	700	—	240	57	1/1	6	44
Wabco	85D	85	47	67	858	—	320	68	1/1	6	40
Wabco	120C	120	54	83	1075	—	500	80	1/1	6	32
Wabco	120CM	120	51	77	1104	—	500	120	1/1	6	34
Wabco	170D	170	71	110	1450	—	500	110	1/1	6	35
Wabco	3200B	250	100	163	2250	—	700	195	1/2	10	25

Table 4.2 (Continued)
REAR DUMP TRUCK SPECIFICATIONS

MAKE	MODEL	ENGINE MAKE	ENGINE MODEL	TRANSMISSION TYPE	NO. OF GEARS F/R	FRONT BRAKES	REAR BRAKES	FRONT SUSPENSION	REAR SUSPENSION
Caterpillar	773B	Caterpillar	3412	mech auto	7/1	shoe	multi-disc	hyd/pneumatic	hyd/pneumatic
Caterpillar	777	Caterpillar	D348	mech auto	7/1	shoe	multi-disc	hyd/pneumatic	hyd/pneumatic
Dart	3065	Detroit	16V92-N70	mech auto	6/1	shoe	shoe	leaf springs	leaf springs
		Cummins	VTA-1710-C635						
Dart	3075	Detroit	16V92-N80	mech auto	6/1	shoe	shoe	leaf springs	leaf springs
		Cummins	VTA-1710-C700						
Dart	2085	Detroit	16V92T	mech auto	6/1	shoe	shoe	leaf springs	torque rods
		Cummins	VTA-1710						
Dart	3100	Cummins	KTA2300-C1050	mech auto	6/1	shoe	shoe	rubber struts	rubber struts
		Detroit	12V-149T						
Dart	3110	Cummins	KTA2300-C1050	mech auto	6/1	shoe	shoe	leaf springs	torque rods
Dart	3120	Cummins	KTA2300-C1200	mech auto	6/1	shoe	shoe	rubber struts	rubber struts
		Detroit	12V-149TI						
DJB	D550	Caterpillar	3408TA	mech auto	4/4	tube	tube	hyd/pneumatic	hyd/pneumatic
Euclid	R-50	Cummins	VT-1710-C	mech auto	6/1	shoe	shoe	rubber struts	rubber struts
		Detroit	16V-71N						
Euclid	R-75	Detroit	16V-71T	mech auto	6/1	shoe	shoe	rubber struts	rubber struts
		Cummins	VTA-1710-C						
Euclid	R-85	Detroit	16V-92T	mech auto	6/1	shoe	shoe	rubber struts	rubber struts
		Cummins	VTA-1710-C						
Euclid	R-100	Cummins	KTA2300-C	mech auto	6/1	shoe	shoe	rubber struts	rubber struts
		Detroit	12V-149T						
Euclid	R-120E	Cummins	KTA2300-C	electric	n.a.	disc	disc	rubber struts	rubber struts
		Detroit	12V-149TI						
Euclid	R-170E	Cummins	KTA-3067	electric	n.a.	disc	disc	rubber struts	rubber struts
		Detroit	12V-149TI						
GM	33-15B	Detroit	16V-149TI	electric	n.a.	shoe	shoe	rubber struts	rubber struts
GM	33-19	GM-EMD	16-645E4	electric	n.a.	2-shoe	3-shoe	rubber struts	hyd/pneumatic
Isco	250C	Cummins	KT-1150-C	mech auto	6/1	shoe	shoe	leaf springs	rubber struts
Isco	265-C	Caterpillar	3412 DIT-ST	mech auto	6/1	shoe	shoe	leaf springs	rubber struts
		Caterpillar	3412 DIT-DT						
International	350B	Cummins	VT-1710-C	mech auto	6/1	disc	disc	leaf springs	leaf springs
		Detroit	16V-71N-65						

Kenworth	548CH	Cummins	KT-450	mech auto	5/1	shoe	shoe	leaf springs	leaf springs
Komatsu	HD680-2	Cummins	VTA1710-C800	mech auto	6/1	shoe	disc	hyd/pneumatic	hyd/pneumatic
Rimpull	RD-65	Detroit	16V-71N	mech auto	6/1	shoe	shoe	coil springs	leaf springs
Rimpull	RD-70	Cummins Detroit	VTA1710-C700 16V-71N	mech auto	6/1	shoe	shoe	coil springs	leaf springs
Rimpull	RD-80	Cummins Detroit	VTA1710-C700 16V-71T	mech auto	6/1	shoe	shoe	coil springs	leaf springs
Rimpull	RD-85	Cummins Detroit	VTA1710-C800 16V-92T	mech auto	6/1	shoe	shoe	coil springs	leaf springs
Rimpull	RD-100	Cummins Detroit	KTA2300-C1050 12V-149T	mech auto	6/1	shoe	shoe	coil springs	leaf springs
Rimpull	RD-120	Cummins Detroit	KTA2300-C1050 12V-149T	mech auto	6/1	shoe	shoe	coil springs	leaf springs
Terex	33-09	Detroit	16V-71T	mech auto	6/1	shoe	shoe	hyd/pneumatic	hyd/pneumatic
Terex	33-11C	Detroit	16V-92TA	mech auto	6/1	shoe	shoe	hyd/pneumatic	hyd/pneumatic
Terex	33-11D	Detroit	16V-92TA	mech auto	6/1	shoe	disc	hyd/pneumatic	hyd/pneumatic
Terex	33-14	Detroit	12V-149TI	mech auto	6/1	disc	disc	hyd/pneumatic	hyd/pneumatic
Unit Rig	M-85	optional	optional	electric	n.a.	shoe	disc	rubber struts	rubber struts
Unit Rig	M-100	optional	optional	electric	n.a.	shoe	disc	rubber struts	rubber struts
Unit Rig	M-120-15	optional	optional	electric	n.a.	shoe	disc	rubber struts	rubber struts
Unit Rig	Mark 30	optional	optional	electric	n.a.	shoe	disc	rubber struts	rubber struts
Unit Rig	Mark 33	optional	optional	electric	n.a.	shoe	disc	rubber struts	rubber struts
Unit Rig	Mark 36	optional	optional	electric	n.a.	shoe	disc	rubber struts	rubber struts
Unit Rig	M-200	optional	optional	electric	n.a.	shoe	disc	rubber struts	rubber struts
Wabco	50B	Cummins Detroit	VT-1710-C 16V-71T	mech auto	6/1	shoe	shoe	hyd/pneumatic	hyd/pneumatic
Wabco	60B	Cummins Detroit	VT-1710-C 16V-71T	mech auto	6/1	shoe	shoe	hyd/pneumatic	hyd/pneumatic
Wabco	75C	Cummins Detroit	VTA-1710-C 16V-71T	mech auto	6/1	shoe	shoe	hyd/pneumatic	hyd/pneumatic
Wabco	85D	Cummins Detroit	KT-2300-C 16V-92T	mech auto	6/1	shoe	shoe	hyd/pneumatic	hyd/pneumatic
Wabco	120C	Cummins	KTA-2300-C	electric	n.a.	shoe	disc	hyd/pneumatic	hyd/pneumatic
Wabco	120CM	Cummins Detroit	KTA-2300-C 12V-149TI	mech auto	8/1	shoe	shoe	hyd/pneumatic	hyd/pneumatic
Wabco	170D	Cummins Detroit	KTA-3067-C 16V-149TI	electric	n.a.	disc	disc	hyd/pneumatic	hyd/pneymatic
Wabco	3200B	GM-EMD	12-645E4	electric	n.a.	shoe	shoe	hyd/pneumatic	hyd/pneumatic

Table 4.2 (Continued)
REAR DUMP TRUCK SPECIFICATIONS

MAKE	MODEL	LENGTH (ft)	WIDTH (ft)	HEIGHT W/BOX DOWN (ft)	WHEELBASE (ft)	FRONT TREAD (ft)	REAR TREAD (ft)	GROUND CLEARANCE (in)	CLEARANCE CIRCLE (ft)	LOADING HEIGHT (ft)	MAX HEIGHT W/BOX RAISED (ft)	DUMP ANGLE (degrees)	DUMP CLEARANCE (in)
Caterpillar	773B	29.9	13.3	10.9	13.8	10.4	9.0	25	77.0	11.8	30.3	55	27
Caterpillar	777	32.1	16.0	16.1	15.0	13.0	11.1	28	88.0	13.6	30.6	52	29
Dart	3065	30.8	16.3	14.9	15.0	11.2	10.8	25	—	11.7	27.7	60	33
Dart	3075	30.8	16.3	14.9	15.0	11.2	10.8	25	—	12.5	27.7	60	33
Dart	2085	40.6	13.7	14.9	18.8	10.4	9.0	22	90.0	13.2	34.8	—	13
Dart	3100	36.0	18.5	17.4	15.8	15.3	12.6	26	81.0	14.1	33.0	59	34
Dart	3110	35.7	20.2	17.8	15.8	16.2	13.1	30	83.0	15.1	33.1	59	40
Dart	3120	38.8	19.8	17.7	17.5	16.2	13.1	30	87.0	14.3	34.7	60	30
DJB	D550	37.3	12.0	14.3	22.0	—	—	23	62.0	10.8	26.8	75	24
Euclid	R-50	30.3	14.0	14.7	13.8	10.5	9.6	25	—	11.8	28.3	60	31
Euclid	R-75	30.4	16.2	15.4	14.5	—	—	25	—	12.7	—	58	21
Euclid	R-85	30.4	16.2	17.1	14.5	13.8	11.3	25	—	14.3	29.7	58	30
Euclid	R-100	34.3	19.0	17.7	16.0	14.8	12.0	28	—	13.8	31.9	60	58
Euclid	R-120E	37.0	19.0	17.8	17.4	15.8	12.5	22	—	15.8	34.7	60	44
Euclid	R-170E	39.0	22.6	18.7	18.5	17.6	13.7	34	—	17.4	36.6	58	51
GM	33-15B	41.4	20.6	19.4	18.0	16.2	13.0	26	92.8	16.8	39.0	52	63
GM	33-19	65.9	25.7	22.6	29.9	21.5	17.2	29	—	19.3	56.0	55	51
Isco	250C	34.1	12.0	11.5	20.8	8.7	8.8	16	—	10.6	25.9	45	30
Isco	265-C	35.0	13.0	12.7	18.4	9.4	8.8	16	—	12.3	22.3	45	21
International	350B	30.3	13.3	13.4	11.7	9.6	9.6	18	72.8	10.9	28.0	70	23

Kenworth	548CH	33.3	12.8	11.0	18.2	—	—	14	82.5	11.0	—	—	—
Komatsu	HD680-2	32.1	15.3	14.1	15.6	12.6	12.6	—	—	12.1	28.6	65	21
Rimpull	RD-65	32.0	15.5	14.3	14.0	11.4	10.9	36	—	12.0	27.6	55	30
Rimpull	RD-70	33.0	15.5	14.4	15.0	11.4	10.9	—	—	11.9	28.5	55	36
Rimpull	RD-80	33.2	16.3	16.5	15.0	13.9	11.1	—	—	14.7	30.2	55	45
Rimpull	RD-85	35.6	16.5	16.5	17.7	13.9	11.1	—	—	13.5	32.3	55	48
Rimpull	RD-100	37.3	16.5	16.5	17.7	13.7	11.1	45	—	13.7	32.6	55	45
Rimpull	RD-120	35.7	19.3	17.8	17.7	13.3	13.3	48	—	14.8	33.3	55	54
Terex	33-09	31.7	14.8	14.8	14.0	11.8	9.8	25	70.8	12.0	26.7	58	22
Terex	33-11C	34.3	15.5	15.8	15.0	12.3	10.7	28	78.7	14.3	28.8	57	25
Terex	33-11D	34.3	15.8	15.7	15.0	12.3	10.7	27	78.7	14.3	28.7	57	24
Terex	33-14	37.4	19.1	17.8	—	—	—	25	—	15.7	—	55	—
Unit Rig	M-85	32.8	17.7	17.3	15.0	14.0	14.0	26	—	15.1	29.6	—	25
Unit Rig	M-100	32.8	17.7	17.4	15.0	14.3	14.3	28	—	15.4	29.8	—	27
Unit Rig	M-120-15	32.8	18.8	17.4	15.0	14.3	14.3	28	—	15.3	29.8	—	27
Unit Rig	Mark 30	34.8	19.3	17.8	17.0	14.8	14.8	—	—	15.3	31.8	50	30
Unit Rig	Mark 33	38.8	21.1	18.8	17.5	15.6	15.6	31	82.0	17.0	34.1	50	24
Unit Rig	Mark 36	38.8	21.1	19.0	17.5	15.8	15.8	34	83.0	18.3	34.3	50	27
Unit Rig	M-200	48.0	25.5	21.2	22.0	20.0	20.0	—	108.0	18.1	41.6	50	34
Wabco	50B	28.9	13.4	14.2	12.6	11.3	9.3	20	66.0	11.8	27.3	50	13
Wabco	60B	31.5	15.9	14.4	13.3	12.0	10.8	21	—	12.0	28.5	49	11
Wabco	75C	30.6	15.9	14.6	13.3	12.0	10.8	21	—	14.6	28.0	49	18
Wabco	85D	37.3	16.4	16.3	15.4	13.7	10.8	30	80.0	13.9	32.3	50	18
Wabco	120CM	37.3	20.6	19.0	16.8	15.4	13.7	24	—	14.8	35.5	45	44
Wabco	120C	36.6	19.3	19.0	16.8	15.3	13.7	26	—	14.8	35.5	45	46
Wabco	170D	39.0	22.9	20.0	17.8	17.7	14.5	31	—	16.5	37.7	45	41
Wabco	3200B	54.3	25.2	20.3	33.6	18.4	—	29	—	16.6	—	51	—

Table 4.2 (Continued)

REAR DUMP TRUCK SPECIFICATIONS

MAKE	MODEL	NET WEIGHT FRONT (lbs)	NET WEIGHT REAR (lbs)	NET WEIGHT TOTAL (lbs)	GROSS WEIGHT FRONT (lbs)	GROSS WEIGHT REAR (lbs)	GROSS WEIGHT TOTAL (lbs)	STANDARD TIRES
Caterpillar	773B	39,200	44,445	83,645	61,625	122,020	183,645	21.00-35 32PR
Caterpillar	777	57,000	65,600	122,600	97,600	195,000	292,600	24.00-49 42PR
Dart	3065	45,400	49,150	94,550	75,400	149,150	224,550	24.00-35 36PR
Dart	3075	46,275	50,225	96,500	77,275	169,225	246,500	24.00-35 42PR
Dart	2085	41,078	73,662	114,740	51,078	233,662	284,740	21.00-35 32PR
Dart	3100	71,625	76,750	148,375	114,225	234,150	348,375	27.00-49 48PR
Dart	3110	81,575	86,575	168,150	128,975	259,175	388,150	30.00-51 46PR
Dart	3120	89,600	85,400	175,000	138,300	276,700	415,000	30.00-51 46PR
DJB	D550	45,200	36,400	81,600	62,800	128,800	191,600	33.25-29 R
Euclid	R-50	37,625	39,450	77,075	59,975	117,100	177,075	21.00-35 32PR
Euclid	R-75	48,000	52,000	100,000	83,000	167,000	250,000	24.00-35 42PR
Euclid	R-85	52,100	60,700	112,800	92,100	190,700	282,800	24.00-49 42PR
Euclid	R-100	71,100	77,200	148,900	117,700	231,200	348,900	27.00-49 48PR
Euclid	R-120E	85,240	94,360	179,600	133,650	285,950	419,600	30.00-51 52PR
Euclid	R-170E	110,450	113,450	223,900	187,950	375,950	563,900	36.00-51 50PR
GM	33-15B	111,200	123,200	234,400	191,500	382,900	574,400	36.00-51 50PR
GM	33-19	180,580	339,920	520,400	232,240	988,160	1,220,400	40.00-57 68PR
Isco	250C	22,790	41,290	64,080	28,295	135,785	164,080	16.00-25 XRB
Isco	265-C	27,600	50,000	77,600	39,700	167,800	207,500	18.00-25 XRB
International	350B	51,400	21,400	72,800	86,500	86,300	172,800	18.00-25 32PR

Kenworth	548CH	19,000	39,000	58,000	28,000	130,000	158,000	16.00-25	24PR
Komatsu	HD680-2	—	—	97,000	—	—	247,000	24.00-35	42PR
Rimpull	RD-65	37,000	41,000	78,000	58,900	149,100	208,000	24.00-35	36PR
Rimpull	RD-70	45,000	45,000	90,000	64,200	165,800	230,000	24.00-35	48PR
Rimpull	RD-80	60,000	62,000	122,000	94,000	188,000	282,000	24.00-49	42PR
Rimpull	RD-85	60,000	62,000	122,000	97,000	195,000	292,000	24.00-49	48PR
Rimpull	RD-100	65,000	70,000	135,000	109,000	226,000	335,000	27.00-49	42PR
Rimpull	RD-120	70,000	77,000	147,000	119,500	267,500	387,000	30.00-51	46PR
Terex	33-09	45,200	48,000	93,200	67,800	135,400	203,200	24.00-35	36PR
Terex	33-11C	55,900	61,600	117,500	95,400	192,100	287,500	24.00-49	42PR
Terex	33-11D	57,200	66,100	123,300	96,500	196,800	293,300	24.00-49	42PR
Terex	33-14	74,400	80,600	155,000	137,000	278,100	415,100	30.00-51	46PR
Unit Rig	M-85	52,878	64,272	117,150	94,710	192,440	287,150	24.00-49	
Unit Rig	M-100	60,750	70,750	131,500	109,493	222,007	331,500	27.00-49	
Unit Rig	M-120-15	63,768	75,160	138,928	119,741	259,187	378,928	27.00-49	42PR
Unit Rig	Mark 30	71,788	84,273	156,061	128,563	287,498	416,061	30.00-51	46PR
Unit Rig	Mark 33	89,741	92,381	181,852	160,457	321,395	481,852	33.00-51	46PR
Unit Rig	Mark 36	85,918	108,907	194,825	178,097	356,728	534,825	36.00-51	50PR
Unit Rig	M-200	158,743	157,857	316,000	235,045	481,555	716,600	40.00-57	60PR
Wabco	50B	36,166	39,174	75,340	58,440	116,900	175,340	21.00-35	32PR
Wabco	60B	41,576	43,424	85,000	67,184	137,816	205,000	24.00-35	36PR
Wabco	75C	44,540	49,360	93,900	81,100	162,800	243,900	24.00-35	42PR
Wabco	85D	58,800	54,300	113,100	94,150	188,950	283,100	24.00-49	42PR
Wabco	120CM	85,360	92,470	177,830	139,280	278,550	417,830	30.00-51	46PR
Wabco	120C	77,895	100,250	178,235	139,662	278,573	418,235	30.00-51	46PR
Wabco	170D	101,156	118,806	219,962	185,053	374,909	559,962	36.00-51	50PR
Wabco	3200B	125,600	239,400	365,000	173,000	692,000	865,000	36.00-51	50PR

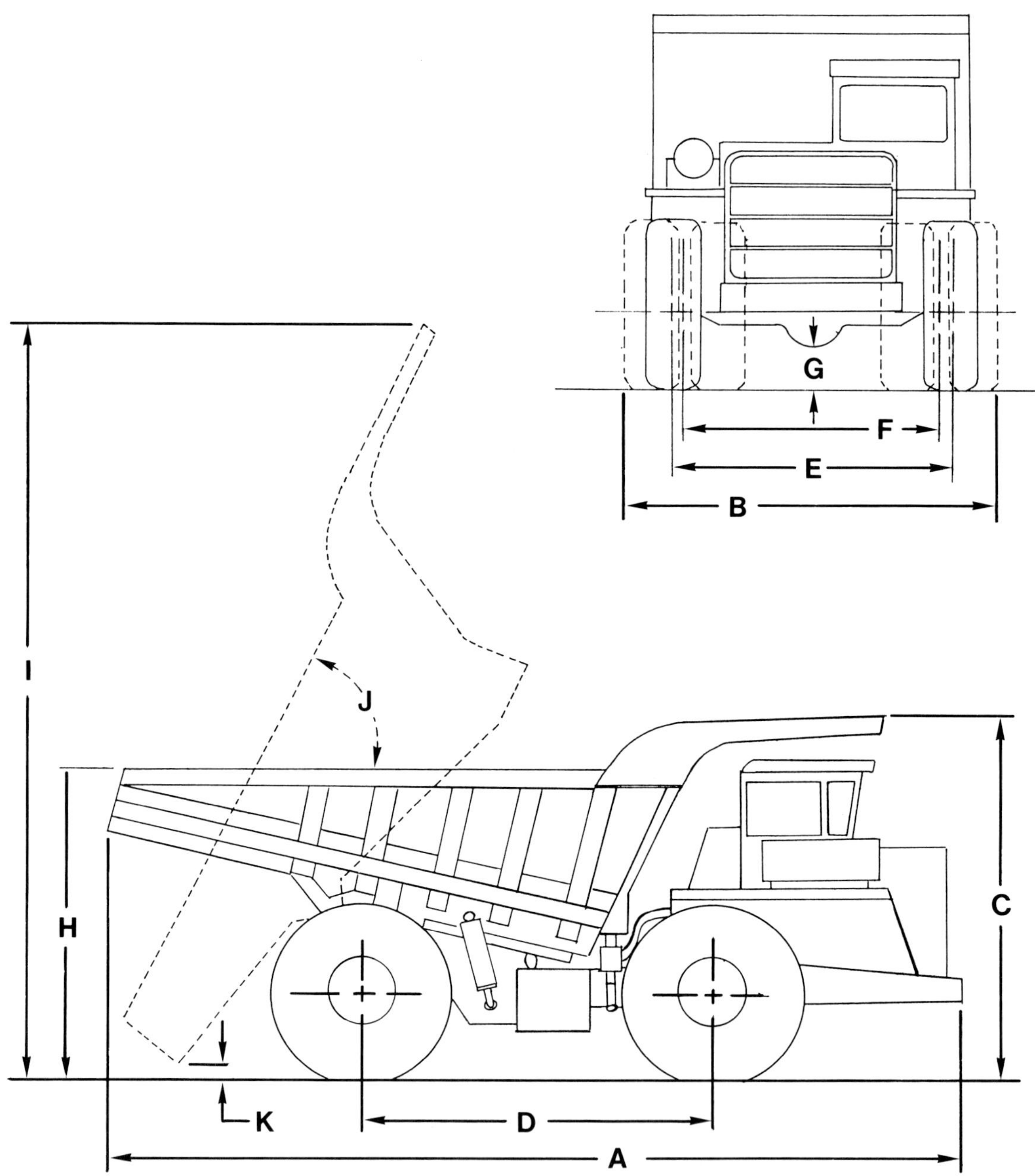

A	Overall Length (ft)	G	Ground Clearance (in)
B	Overall Width (ft)	H	Loading Height (ft)
C	Overall Height/Body Down (ft)	I	Overall Height/Body Raised (ft)
D	Wheelbase (ft)	J	Dump Angle (deg)
E	Tread/Front (ft)	K	Dump Clearance (in)
F	Tread/Rear (ft)		

Figure 4.12 Truck Dimensions

Table 4.3
BOTTOM DUMP TRUCK SPECIFICATIONS

MAKE	MODEL	RATED LOAD (tons)	STRUCK CAPACITY (cu yd)	HEAPED CAPACITY (cu yd)	ENGINE HP	STEERING ANGLE (degrees)	FUEL TANK (gal)	COOLANT (gal)	NO. OF AXLES F/R	TOTAL NO. OF WHEELS	MAX SPEED (mph)
Atlas	CH-70E	70	86	100	450	39	140	30	2/1	10	—
Atlas	CH-100E	100	114	133	650	31	185	40	2/1	10	—
Atlas	CH-120E	120	136	155	870	30	250	76	2/1	10	—
Atlas	CH-130D	130	148	178	870	30	250	76	2/1	10	—
Atlas	CH-150D	150	170	200	870	30	250	76	2/1	10	39
Dart	4100	100	110	141	635	32	400	—	2/1	10	—
Dart	4120	120	132	163	700	38	400	—	2/1	10	—
Dart	4130	130	135	165	900	32	—	—	2/1	10	—
Dart	4150	150	181	221	1050	40	450	—	2/1	10	—
Dart	4160	160	181	221	1050	40	450	—	2/1	10	—
Euclid	B-70	70	47	62	423	40	175	31	2/1	10	37
Euclid	B-110	110	68	85	576	40	250	42	2/1	10	47
Euclid	CH-120	120	136	159	818	41	265	58	2/1	10	38
Euclid	CH-150	150	179	218	1050	37	325	95	2/1	10	41
Goodbary	2400/36	130	149	—	1000	64	500	75	1/1	4	—
Goodbary	2400/40	170	224	—	1325	62	500	75	1/1	4	—
Kress	CH150	150	163	—	1050	90	—	—	1/1	8	50
Kress	CH160	160	200	237	1200	90	300	—	1/1	8	52
Rimpull	CW-100	100	112	139	635	35	240	55	2/1	10	40
Rimpull	CW-120	120	130	160	700	35	240	55	2/1	10	40
Rimpull	CW-135	135	143	175	860	35	240	65	2/1	10	40
Rimpull	CW-150	150	165	200	1050	41	240	105	2/1	10	40
Terex	34-11C	150	180	217	840	35	325	56	2/1	10	39
Unit Rig	BD-145	145	153	187	1050	90	600	—	1/1	8	33
Unit Rig	BD-30	160	160	203	1200	90	500	—	1/1	8	34
Unit Rig	BD-180	180	204	246	1600	—	600	—	2/1	10	34
Wabco	150-CT	150	170	194	1000	—	400	75	2/1	10	41
Wabco	170-CP	170	195	246	1450	90	500	130	1/1	6	35

Table 4.3 (Continued)

BOTTOM DUMP TRUCK SPECIFICATIONS

MAKE	MODEL	ENGINE MAKE	ENGINE MODEL	TRANSMISSION TYPE	NO. OF GEARS F/R	FRONT BRAKES	REAR BRAKES	FRONT SUSPENSION	REAR SUSPENSION
Atlas	CH-70E	Caterpillar	3408	mech auto	7/1	tube	shoe	hyd/pneumatic	hyd/pneumatic
Atlas	CH-100E	Caterpillar	3412	mech auto	7/1	tube	shoe	hyd/pneumatic	hyd/pneumatic
Atlas	CH-120E	Caterpillar	D-348	mech auto	7/1	shoe	shoe	hyd/pneumatic	hyd/pneumatic
Atlas	CH-130D	Caterpillar	D-348	mech auto	7/1	shoe	disc	hyd/pneumatic	hyd/pneumatic
Atlas	CH-150D	Caterpillar	D-348	mech auto	7/1	shoe	disc	hyd/pneumatic	hyd/pneumatic
Dart	4100	Cummins	VTA1710-C635	mech auto	6/1	shoe	shoe	leaf springs	leaf springs
		Detroit	16V-92N						
		Cummins	VTA1710-C700						
Dart	4120	Cummins	VTA1710-C700	mech auto	6/1	shoe	shoe	leaf springs	leaf springs
		Detroit	16V-92N						
Dart	4130	Cummins	KT2300-C900	mech auto	6/1	shoe	shoe	leaf springs	leaf springs
		Detroit	16V-92T						
Dart	4150	Cummins	KTA2300-C1050	mech auto	6/1	shoe	shoe	rubber struts	rubber struts
		Detroit	12V-149T						
Dart	4160	Cummins	KTA2300-C1050	mech auto	6/1	shoe	shoe	rubber struts	rubber struts
		Detroit	12V-149T						
Euclid	B-70	Detroit	12V-71N	mech auto	6/1	shoe	shoe	leaf springs	rigid
Euclid	B-110	Detroit	16V-71N	mech auto	6/1	shoe	shoe	leaf springs	rigid
Euclid	CH-120	Cummins	VTA1710-C700	mech auto	6/1	shoe	shoe	rubber struts	rubber struts
		Detroit	16V-71T						
		Cummins	VTA1710-C800						
		Detroit	16V-92T						
Euclid	CH-150	Cummins	KTA2300-C1050	mech auto	6/1	shoe	shoe	rubber struts	rubber struts
		Detroit	12V-149T						

Goodbary	2400/36	Detroit Cummins Caterpillar Cummins Detroit	12V-149T KT2300-C1000 D348 KTA2300-C1200 12V-149TI	electric	n.a.	disc	disc	rubber struts	rubber struts
Goodbary	2400/40	Detroit Cummins Caterpillar Detroit	12V-149TI KTA2300-C1200 D349 16V-149TI	electric	n.a.	disc	disc	rubber struts	rubber struts
Kress	CH150	Detroit Cummins	12V-149T KTA2300-C1050	mech auto	6/1	shoe	shoe	hyd/pneumatic	hyd/pneumatic
Kress	CH160	Detroit	12V-149TI	mech auto	6/1	shoe	shoe	rubber struts	rubber struts
Rimpull	CW-100	Cummins Detroit	VTA1710-C635 16V-71T	mech auto	6/1	shoe	shoe	coil springs	rigid
Rimpull	CW-120	Cummins Detroit	VTA1710-C700 16V-71T	mech auto	6/1	shoe	shoe	coil springs	rigid
Rimpull	CW-135	Detroit Cummins	16V-92T VTA1710-C800	mech auto	6/1	shoe	shoe	coil springs	rigid
Rimpull	CW-150	Cummins Detroit	KTA2300-C1050 12V-149T	mech auto	6/1	shoe	shoe	coil springs	rigid
Terex	34-11C	Detroit	16V-92TA	mech auto	6/1	shoe	shoe	hyd/pneumatic	hyd/pneumatic
Unit Rig	BD-145	Detroit Caterpillar Cummins	12V-149TI D348SCAC KTA2300-C1050	electric	n.a.	shoe	disc	rubber struts	rubber struts
Unit Rig	BD-30	Detroit Cummins	16V-149T KTA2300-C1200	electric	n.a.	disc	disc	rubber struts	rubber struts
Unit Rig	BD-180	Detroit Cummins	16V-149TI KTA-3067	electric	n.a.	shoe	disc	rubber struts	rubber struts
Wabco	150-CT	Cummins Detroit	KTA2300-C1050 12V-149T	mech auto	6/1	shoe	shoe	hyd/pneumatic	hyd/pneumatic
Wabco	170-CP	Detroit Cummins	16V-149TI KTA-3067-C	electric	n.a.	disc	disc	hyd/pneumatic	hyd/pneumatic

Table 4.3 (Continued)

BOTTOM DUMP TRUCK SPECIFICATIONS

MAKE	MODEL	LENGTH (ft)	WIDTH (ft)	MAX HEIGHT (ft)	WHEELBASE (ft)	FRONT TREAD (ft)	DRIVE TREAD (ft)	TRAILER TREAD (ft)	GROUND CLEARANCE (in)	CLEARANCE CIRCLE (ft)	LOADING HEIGHT (ft)	DUMP CLEARANCE (in)
Atlas	CH-70E	58.4	12.5	—	33.8	10.2	8.1	—	21	—	12.3	—
Atlas	CH-100E	65.6	15.0	13.0	38.5	10.7	9.6	—	23	77.0	12.4	24
Atlas	CH-120E	75.0	15.0	—	45.7	13.3	11.1	—	25	—	—	25
Atlas	CH-130D	72.5	18.0	—	41.7	13.3	11.1	—	25	—	—	25
Atlas	CH-150D	76.5	18.0	14.9	45.7	13.3	11.1	—	25	88.4	14.1	25
Dart	4100	66.8	15.9	13.7	51.8	11.4	10.8	10.8	24	83.0	10.2	25
Dart	4120	66.8	15.8	15.2	51.8	12.3	10.6	10.6	30	75.0	11.8	32
Dart	4130	71.3	17.0	15.2	56.3	11.2	11.6	11.6	26	85.0	12.3	35
Dart	4150	78.1	18.7	15.9	61.1	15.3	12.6	12.6	26	81.0	13.8	29
Dart	4160	78.1	18.7	15.9	61.1	15.3	12.6	12.6	26	81.0	13.8	29
Euclid	B-70	57.5	12.8	11.0	45.5	7.7	8.1	9.1	32	—	15.7	32
Euclid	B-110	63.8	15.3	12.4	52.3	8.9	8.7	11.0	19	—	11.3	28
Euclid	CH-120	68.5	16.2	15.7	52.1	10.4	11.3	11.2	32	—	15.7	32
Euclid	CH-150	79.2	17.4	15.3	62.2	14.8	12.0	12.0	—	—	15.0	—
Goodbary	2400/36	48.3	19.3	18.6	28.0	16.0	16.0	n.a.	33	—	14.5	24
Goodbary	2400/40	48.8	21.0	19.7	32.0	17.3	17.3	n.a.	40	—	15.6	31
Kress	CH150	54.8	17.2	14.0	—	—	—	n.a.	—	60.0	13.8	20
Kress	CH160	60.2	17.5	14.7	38.4	—	—	n.a.	34	62.6	13.8	20
Rimpull	CW-100	66.3	15.5	13.8	49.5	11.4	10.9	10.9	—	—	10.8	25
Rimpull	CW-120	70.4	15.8	14.6	52.8	11.4	11.1	11.1	—	—	11.2	25
Rimpull	CW-135	71.6	16.3	14.8	53.9	13.7	11.1	11.1	—	—	11.3	26
Rimpull	CW-150	78.6	16.3	15.9	60.2	13.7	11.1	11.1	—	—	11.8	29
Terex	34-11C	76.4	16.5	15.9	58.7	12.3	10.8	11.1	29	82.2	15.9	29
Unit Rig	BD-145	53.3	20.9	17.8	35.0	12.8	14.4	n.a.	26	65.3	16.0	26
Unit Rig	BD-30	54.0	20.9	18.8	35.0	—	—	n.a.	26	85.0	16.0	26
Unit Rig	BD-180	78.0	20.5	—	45.0	14.8	—	—	—	—	—	31
Wabco	150-CT	74.6	16.7	16.0	55.3	13.5	11.3	11.3	—	—	14.3	—
Wabco	170-CP	56.5	22.8	18.0	34.0	—	—	n.a.	—	80.0	15.8	31

TABLE 4.3 (Continued)
BOTTOM DUMP TRUCK SPECIFICATIONS

		NET WEIGHT				GROSS WEIGHT					
MAKE	MODEL	FRONT (lbs)	DRIVE (lbs)	TRAILER (lbs)	TOTAL (lbs)	FRONT (lbs)	DRIVE (lbs)	TRAILER (lbs)	TOTAL (lbs)	STANDARD FRONT TIRES	STANDARD REAR TIRES
Atlas	CH-70E	29,440	35,355	33,075	97,870	37,710	92,325	107,835	237,870	18.00-33 24PR	37.50-39 44PR
Atlas	CH-100E	37,770	53,840	49,585	141,195	49,970	136,140	155,085	341,195	24.00-35 36PR	24.00-49 36PR
Atlas	CH-120E	57,600	70,700	57,700	186,000	74,780	156,600	194,420	425,800	27.00-49 36PR	24.00-49 42PR
Atlas	CH-130D	58,060	73,040	63,400	194,500	76,410	164,790	213,300	454,500	27.00-49 36PR	27.00-49 42PR
Atlas	CH-150D	58,380	74,620	66,000	199,000	79,850	181,980	237,160	499,000	27.00-49 36PR	30.00-51 40PR
Dart	4100	42,982	49,192	42,852	135,026	53,632	134,442	146,952	335,026	21.00-35 32PR	24.00-35 42PR
Dart	4120	46,428	56,052	48,850	151,330	59,228	158,052	174,050	391,330	24.00-49 36PR	24.00-49 36PR
Dart	4130	47,607	70,569	57,408	175,584	60,381	179,146	196,058	435,584	24.00-35 36PR	27.00-49 36PR
Dart	4150	64,240	78,360	69,500	212,100	80,265	214,135	217,700	512,100	27.00-49 42PR	27.00-49 42PR
Dart	4160	64,400	78,800	69,900	213,100	81,600	224,800	226,700	533,100	27.00-49	27.00-49
Euclid	B-70	17,030	38,840	35,000	90,870	22,430	94,440	114,000	230,870	16.00-25 24PR	18.00-49 28PR
Euclid	B-110	22,300	50,800	43,800	116,900	31,700	139,300	165,900	336,900	18.00-25 20PR	21.00-49 40PR
Euclid	CH-120	36,760	55,340	53,000	145,100	56,290	150,930	177,880	385,100	24.00-35 36PR	24.00-49 36PR
Euclid	CH-150	54,695	74,485	75,420	205,600	75,090	194,950	235,560	505,600	27.00-49 42PR	27.00-49 42PR
Goodbary	2400/36	79,520	80,480	n.a.	160,000	208,800	211,200	n.a.	420,000	36.00-51 58PR	36.00-51 58PR
Goodbary	2400/40	94,430	95,570	n.a.	190,000	263,530	266,470	n.a.	530,000	40.00-57 60PR	40.00-57 60PR
Kress	CH150	50,000	66,000	n.a.	116,000	208,000	208,000	n.a.	416,000	27.00-49 42PR	27.00-49 42PR
Kress	CH160	56,590	87,560	n.a.	144,150	219,790	244,360	n.a.	464,150	30.00-57 48PR	30.00-57 48PR
Rimpull	CW-100	36,000	48,000	42,000	126,000	41,700	137.700	146,600	326,000	24.00-35 36PR	24.00-35 36PR
Rimpull	CW-120	40,000	56,000	52,000	148,000	46,300	162,300	179,400	388,000	24.00-35 30PR	24.00-49 36PR
Rimpull	CW-135	45,000	60,000	54,000	159,000	51,700	174,300	203,000	429,000	24.00-35 36PR	24.00-49 42PR
Rimpull	CW-150	52,000	60,000	60,000	172,000	60,500	201,500	210,000	472,000	27.00-49 42PR	27.00-49 42PR
Terex	34-11C	54,000	65,100	62,000	181,000	67,700	203,300	210,100	481,100	27.00-49 42PR	27.00-49 42PR
Unit Rig	BD-145	68,506	106,094	n.a.	174,600	230,398	234,202	n.a.	464,600	30.00-51 40PR	30.00-51 40PR
Unit Rig	BD-30	104,300	122,200	n.a.	226,500	272,200	274,300	n.a.	546,500	33.00-60 46PR	33.00-60 46PR
Unit Rig	BD-180	67,660	90,440	74,200	232,300	92,560	246,940	256,700	596,200	30.00-51	30.00-51
Wabco	150-CT	64,700	63,100	58,800	186,600	82,720	192,200	211,680	486,600	27.00-49 36PR	27.00-49 36PR
Wabco	170-CP	67,500	165,500	n.a.	233,000	200,600	372,400	n.a.	573,000	36.00-51 50PR	36.00-51 50PR

Chapter 5

FRONT-END LOADERS

Figure 5.1 Front-End Loader, Carrying A Load

The front-end loader is a wheel or crawler mounted tractor with a front mounted bucket: it is utilized in excavating, loading, and transporting material. Because of its versatility, the front-end loader is found in a wide variety of mining applications.

Front-end loader type linkages started as attachments for crawler mounted dozers, providing a means for bulldozing, truck or hopper loading, and limited material transport. Current crawler mounted loaders have a heaped bucket capacity of 5 cubic yards or less. Articulated wheel mounted loaders have also been continually developed, now ranging in capacity from 0.5 to 24 cubic yards. The design of these units has been refined and is now relatively standardized in these basic configurations. For mining applications, the trend has been to wheel mounted units in sizes above 5 cubic yards. The larger units, for the most part, are relatively new to production applications in surface mining. They are, however, finding uses as primary production machines, excavating in medium tough conditions, and loading up to 170 ton trucks.

TYPICAL UNITS

This manual will consider only wheel mounted front-end loaders of the type used in surface mining. These loaders are all quite similar in basic design:

- Articulated frame (70 to 90 degree steering)
- Four wheel drive
- 1 or 2 diesel engines
- Mechanical or electric drive train
- Capacities from 3.5 to 30 cubic yards
- Horsepower from 180 to 1,270

While crawler mounted loaders are available, they are small, usually less than 5 cubic yards, and infrequently found in surface mines. Crawler loaders are similar in design to crawler mounted dozers except for the front end linkage used to raise, lower and dump the bucket.

Figure 5.2 Front-End Loader, Dumping Into Hopper

Photograph 5.1 Dart 600 FEL loading coal into a Euclid 120 ton bottom dump truck

Manufacturers

While loaders are available from a number of Japanese and European manufacturers, the U.S. manufacturers are essentially the exclusive suppliers of the large units to the mining industry. The U. S. manufacturers (see Table 5.1) are all large, well-established companies with a number of products aimed at this market. Sales and service are normally conducted through dealer networks.

BASIC MACHINE OPERATIONS

The articulated frame and rubber tire mountings give the machine excellent maneuverability and mobility, which may, however, be limited by operating surface conditions. Truck loading normally involves minimal transport distances, while direct spoiling or loading central hoppers can require significant load-and-carry distances.

- The machine digs by filling its bucket through a combination of crowding action produced by propelling, bucket orientation by a wristing action, and a hoisting motion. The crowd results from the rimpull/traction force developed by the drive, plus the inertia as the unit is propelled into the digging face. The wristing and hoisting actions permit positioning of the cutting edge and teeth to penetrate and take a horizontal or upward slice of the face until the bucket is full. To retain the load in the bucket it is fully rolled back as it leaves the face.

- The unit is backed away from the face making a "Y" turn (to minimize turning space) and then accelerated forward in the direction of the truck or hopper to be loaded. If the transport distance is short, the bucket is raised to its maximum raise position in preparation for dumping. For longer transport distances, the bucket is lowered to just clear the ground, achieving maximum machine stability while traveling. The bucket is raised as indicated earlier as the dump point is approached.

Table 5.1

FRONT-END LOADER MANUFACTURERS

J.I. Case Company Construction Equipment Div. 700 State Street Racine, Wisconsin 53404	Wheel mounted 32–185 FWHP .5– 4 cu. yd.
Caterpillar Tractor Co. 100 NE Adams Street Peoria, Illinois 61629	Wheel mounted 65–690 FWHP 1–13.5 cu. yd.
Clark Equipment Company Construction Machinery Div. Pipestone Road, P.O. Box 547 Benton Harbor, MI 49022	Wheel mounted 80–1270 FWHP 1.5–36 cu. yd.
Dart Truck Company 1301 Chouteau Trafficway P.O. Box 321 Kansas City, MO 64141	Wheel mounted 700–818 FWHP 7–30 cu. yd.
Deere & Company John Deere Road Moline, Illinois 61265	Wheel mounted 71–260 FWHP 1.25–7 cu. yd.
Fiat-Allis Construction Machinery, Inc. Box F, 106 Wilmot Road Deerfield, IL 60015	Wheel mounted 80–335 FWHP 1.5–6.5 cu. yd.
International Harvester, Inc. Construction Equipment Group 600 Woodfield Avenue Schaumburg, Illinois 60196	Wheel mounted 51–1075 FWHP .59–24 cu. yd.
Kawasaki Heavy Industries Ltd. 375 Park Avenue Seagram Building Room 3309, 33rd Floor New York, New York 10022	Wheel mounted 67–305 FWHP 1.5–6 cu. yd.
Marathon LeTourneau Co. Longview Division P.O. Box 2307 Longview, Texas 75606	Wheel mounted 525–1075 FWHP 8–30 cu. yd.
Terex Corporation IBH Group Hudson, Ohio 44236	Wheel mounted 40–434 FWHP .8 –9 cu. yd.
Trojan Industries, Inc. Trojan Circle Batavia, New York 14020	Wheel mounted 82–400 FWHP 1.5 –8.5 cu. yd.

- Dumping when the machine is positioned and stopped consists of wristing the bucket so that it rolls forward, permitting the load to gravity fall out over the cutting edge.

- The return generally consists of a sharp "Y" turn and a higher transport speed with the empty bucket carried close to the ground ready to start the dig cycle at the base of the digging face. (See Figure 5.3)

APPLICATIONS

The wheel loader is a competitive excavator, loader and transporter. It competes with shovels, dozers and, over short transport distances, with scrapers and trucks. Being quite fast, mobile, and versatile, it can be used in a number of mine applications.

Because the FEL has generally not been considered to have the digging ability of a shovel in consolidated digging faces, it finds many of its applications in softer formations, coal/ore and stockpile work. The larger sizes are more rugged and powerful, and are proving themselves in difficult digging. The primary mine applications are the following:

- Loading and/or transporting topsoil

- Loading and/or transporting coal/ore from the digging face

- Loading and/or transporting coal/ore from stockpile

- Loading and/or transporting overburden and waste

In all of the above loading can be into trucks, hoppers, railroad cars, or belt loaders. Transport

Photograph 5.2 Terex 72-81 FEL loading rock into a Terex 55 ton rear dump truck

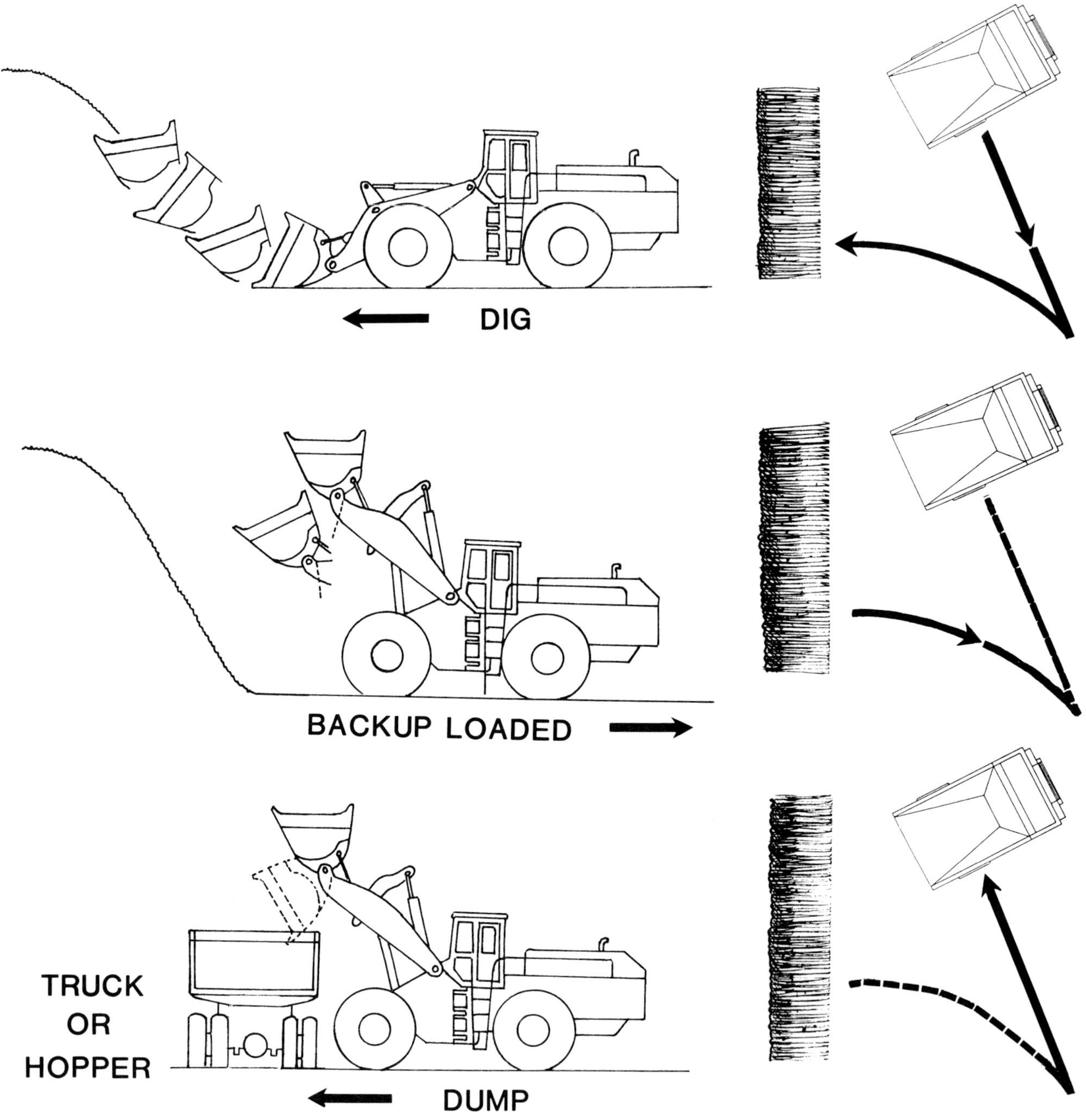

Figure 5.3 Typical FEL Truck Loading Production Cycle

can be for distances up to 1,000 feet on the level or on grades of up to 12%.

Other applications are:

- Pit clean-up
- Stump and boulder removal
- Road maintenance and snow removal
- Logging (with special attachments)
- General utility

GENERAL CHARACTERISTICS

Despite its combined excavating and transport capabilities, the machine is relatively simple in design. The powered functions are (see Figure 5.4):

- 4 wheel drive, direct torque converter, or electric wheel motors
- Articulated steering, hydraulic cylinders
- Bucket raise/lower, hydraulic cylinders
- Bucket wristing, hydraulic cylinders

General machine characteristics can be briefly summarized as follows:

- Excellent mobility and high propel speeds (12 to 24 mph) because of its tire mounting
- Dump height from 10 to 18 feet
- Maximum reach from 4 to 8 feet
- Lift capacities from 10,000 to 72,000 lbs.
- Tipping loads of 60% to 80% of machine weight
- Machine weight/cubic yards of bucket capacity 10,000 to 16,500
- Pounds/hp from 200 to 300
- Good gradeability
- Good turning radius, articulated frame is hinged at midpoint so that rear wheels track front wheels

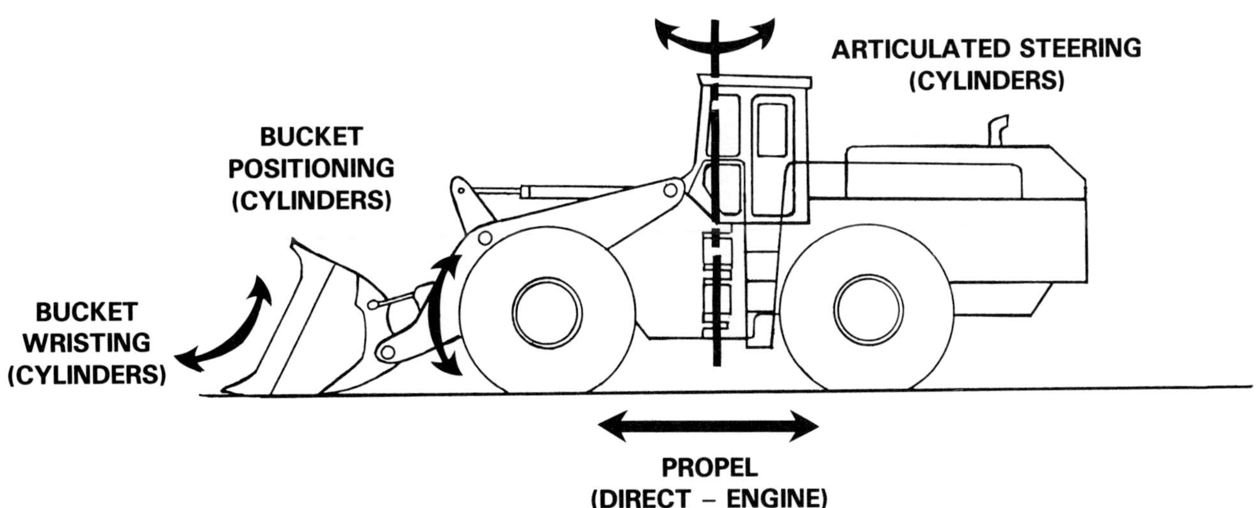

Figure 5.4 FEL Powered Functions

- High ground pressures compared to track units
- Single operator required
- Moderate operator skill required
- Moderate reliability
- Maintenance performed in the shop
- Low service life of 8,000 to 15,000 hours
- Low capital costs relative to other loading equipment
- Short delivery time (very large machines may be an exception)

Some of the positive machine performance features are:

- Wide bucket permits handling large pieces
- Can dig material selectively with a minimum of waste
- High break-out forces while digging (80 to 150% of static tipping load)
- Good blending capabilities at working face
- Leaves a clean floor
- No auxiliary equipment required
- Moderate dust generator
- Dumping action easy on trucks/hoppers (wide bucket can result in spillage)
- Good cutting efficiencies in medium bench heights (6–20 feet)
- Down pressure can be applied for penetration into the floor or digging face

Operational limitations must be recognized.

- Prior preparation (ripping or blasting) required for consolidated materials, may be unsuitable for hard dense rock which fragments badly

Photograph 5.3 Trojan 7500 dumping into a hopper/crusher in quarry

Photograph 5.4 Caterpillar 988B carrying a load

- Considerable maneuvering space required
- Reduced performance with bad traction conditions (wet surfaces, soft ground, clay surfaces)
- High tire wear with sharp abrasive work floor
- Short coupled front-end (minimal reach) limits ability to deposit material deep into truck or hopper
- Reduced stability in load and carry position
- Relatively poor visibility rearward while maneuvering: this together with fast cycles can make it dangerous to work around
- Operator is in a relatively exposed position when working high faces
- Relatively high operator fatique

The digging and dumping heights of a loader can be increased to meet site requirements, by using a special long boom (bucket support arms). Since the lift capacity of the machine is limited by machine stability, this change reduces the maximum bucket load (size) that can be handled.

Graph 5.1 shows the relationship between machine weight and nominal bucket size. The front-end loader has the highest bucket capacity (in tons) to machine weight ratio of any of the cyclic excavators. As illustrated in Graph 5.2, the horsepower increases, as you would expect, consistently, with machine size (nominal bucket size). Graph 5.3 shows that the machine weight per horsepower increases significantly with size. The changes that occur with increasing size are of particular interest in the case of front-end loaders. These units, when initially marketed in the smaller sizes, were designed primarily for miscellaneous utility type work. When these activities expanded into production loading they were not

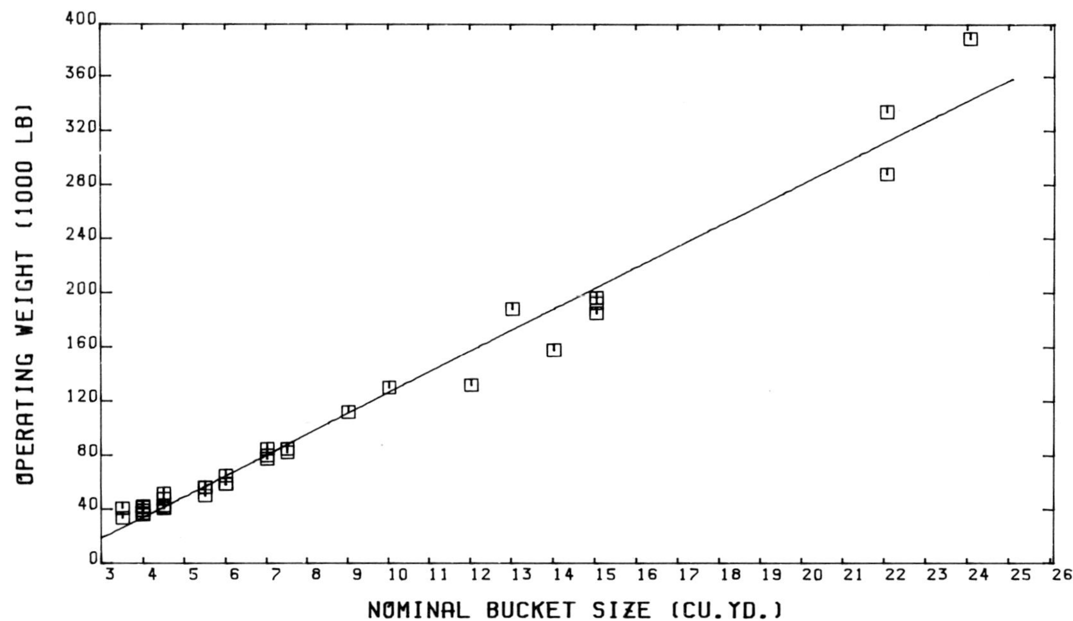

Graph 5.1 Operating weight/nominal bucket size

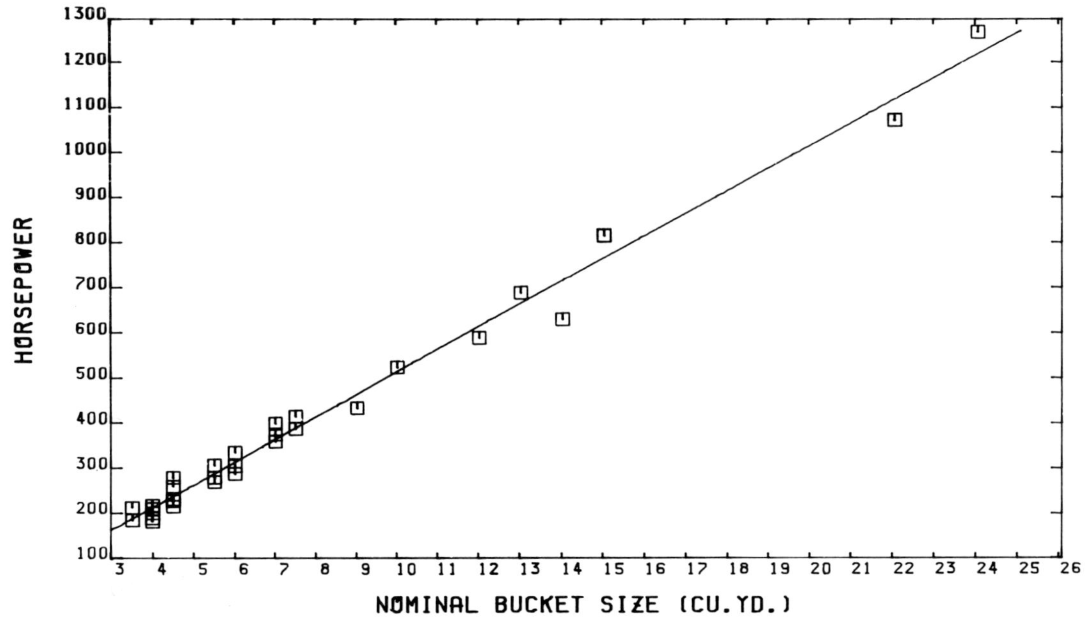

Graph 5.2 Operating weight per HP/nominal bucket size

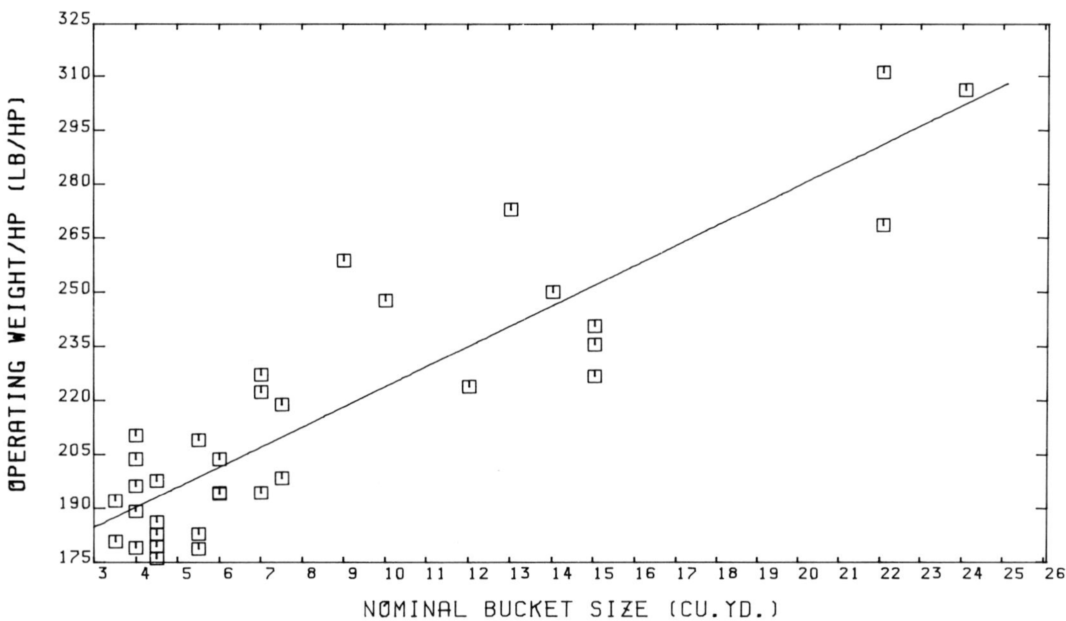

Graph 5.3 HP/nominal bucket size

rugged enough for anything but stockpiling or easy digging conditions. This early experience has, to a degree, hindered the acceptance, for harder digging application, of the new larger models now in service. These latter units have been designed for production operations on large construction projects and surface mining.

Increased machine size has resulted in the following:

- Hydraulic cycle time increases with size.
- Maximum breakout/inch of bucket width increases substantially with size.
- Maximum tractive force/inch of bucket width increases substantially with size.
- Dumping height roughly doubles with a four fold increase in size.
- Machine reach doubles with a four fold increase in size.
- Carrying capacity remains relatively constant at about 20% of machine weight for all sizes.
- Propel speeds change little with size.
- Wheel base does not increase in proportion to machine size.

Some general observations from the above are possible:

- The larger machines appear to be more rugged with increased effective power and, hence, should have improved performance in hard digging conditions.
- The increase in physical size of the larger machines does permit working on higher faces

Photograph 5.5 Caterpillar 988 FEL unit dumping into a belt loader

and dumping into higher and deeper hoppers and trucks. This increased range, however, is not in direct proportion to the increased size.

- Relative carrying capacity, wheel base and travel speeds do not improve with size, indicating that there is no added advantage from the standpoint of productivity in a load-and-carry service.

The more rugged machines are proportionally heavier, have more inertia and require more power with increased fuel consumption. Available power and structural strength restrict maximum total breakout force. To provide close quarter manueverability, these machines have shorter wheel bases and reduced for and aft stability.

Rimpull is the term used to describe the tractive force between the loader's tires and the ground. Maximum rimpull is a function of the engine's power, transmission characteristics, and the gear ratios of the drive train. Usable rimpull can be limited by the traction between the wheels and ground. Rimpull curves showing the relationship between rimpull force and vehicle speed for a specific front-end loader are typically not available from the manufacturer.

Initial bucket penetration of the bank is dependent, to a large degree, on the face conditions. Bank penetration will also vary with the force per unit of bucket width generated by the loader, as modified by the use of teeth, and subject to the sharpness of the teeth and bucket lip. Since bucket width approximates tread width, penetration effectiveness reflects the forward crowd action developed by the loader. (See Figure 5.5) Crowd force is a combination of rimpull (limited by traction) and machine inertia. Inertia is a function of speed and is essentially controlled by the operator and limited by potential abuse to the bucket and machine. The larger machines generally approach the face slower but their increased weight maintains the high inertia forces.

Once the initial inertia forces have been dissipated, forward horizontal crowd forces are essentially dictated by traction conditions. All of

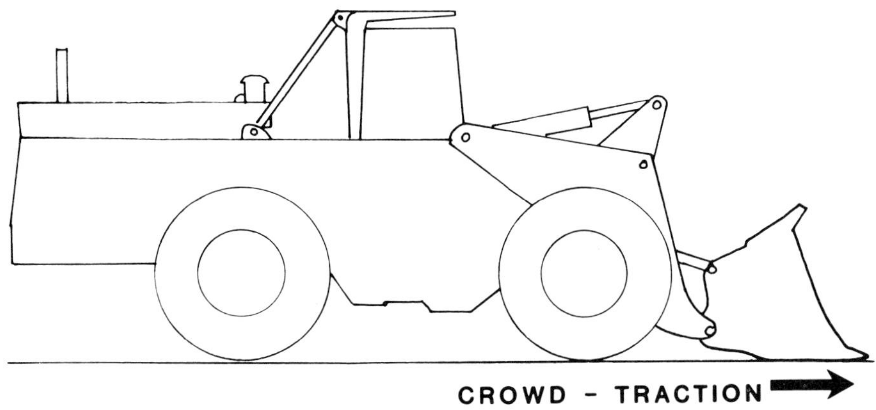

DIGGING FORCES

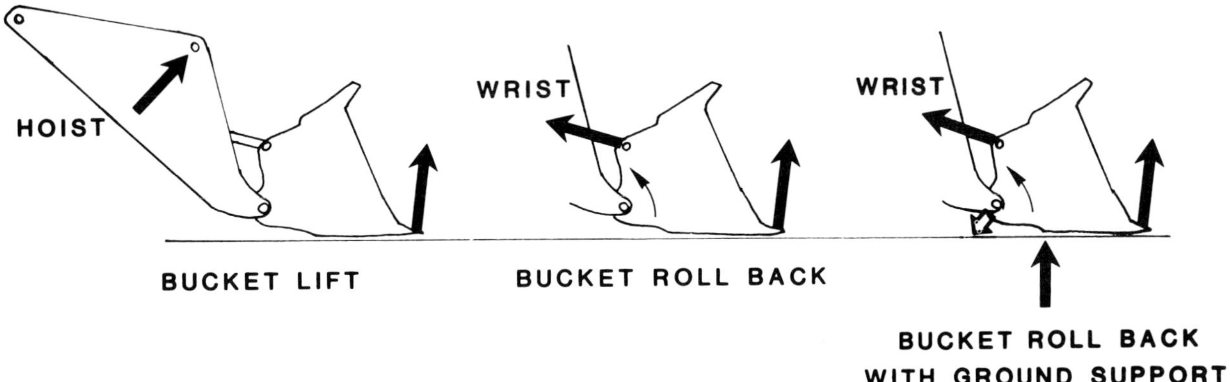

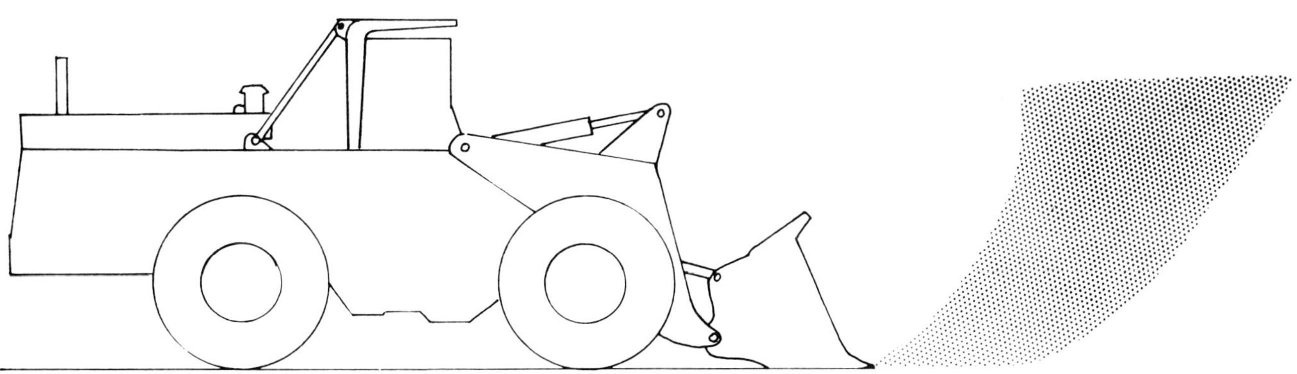

DIGGING PROFILE

Figure 5.5 FEL Digging Forces & Digging Profile

the larger loaders have sufficient rimpull at digging speeds to spin the wheels under most traction conditions.

Efficient bucket filling also requires the ability to pry out boulders and consolidated materials which cannot be effectively penetrated. Further, the bucket must be progressively reoriented to expedite filling and retention of the accumulated load. These actions must occur before the tires penetrate too deeply into the toe of the digging face; this means a relatively short forward motion.

The rollback (rotating) capabilities of the bucket, generally indicated as the break-out force, provides the essential "prying" action. These forces are generated by the hydraulic wrist cylinders rotating the bucket around its hinge pin. Breakout force (see Graph 5.4) is not limited by machine stability if the bucket is supported partially by the material under the bucket. Maximum forces are dictated by design relationships which are limited by bucket strength. The utilization of these wristing forces is dependent on the severity of the digging conditions.

Vertical motion (actually a forward arc) of the bucket penetrating the face and raising the loaded bucket out of the bank is achieved by the lift cylinders rotating the bucket arms about the machine hinge point. Forces are a function of the lever arms, cylinder size and hydraulic pressures. The maximum force is limited by the machine forward "tipping load" which is specified by the manufacturer. Most loaders are designed such that the lift forces, which can be generated, will reach the maximum tipping load.

Crowd, breakout, and tipping load relationships vary with the bucket position in its travel arc; this makes overall digging performance comparisons difficult. Similarly, the relative speeds of the various motions and smoothness of the controls affects the ability of the operator to optimize performance. There are also differences in how the total available engine power is distributed between the propel and hydraulic functions during digging. If possible, comparative operating trials with an experienced operator may add insight into the design balance of the digging forces.

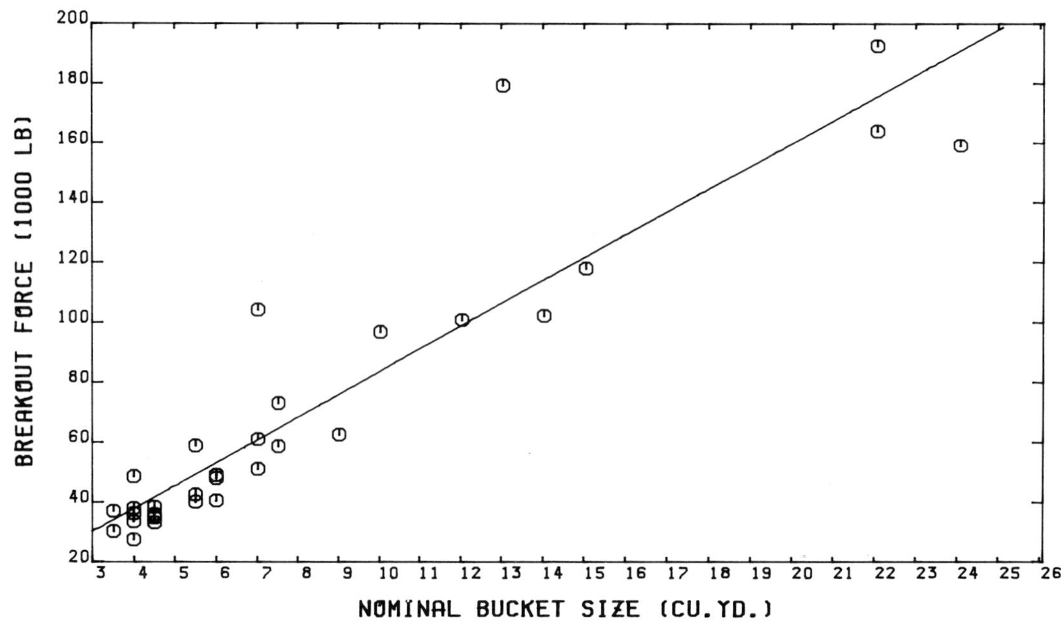

Graph 5.4 Breakout force/nominal bucket size

OPERATIONAL PRACTICES

The loader can be employed in two production operations; direct loading and load-and-carry. Figure 5.3, referenced earlier, illustrated the basic digging, dumping and maneuvering pattern when the unit is working with a single truck. Similar operations would apply in loading a mobile hopper positioned close to the digging face. Figure 5.6 expands on this operation to show that loaders can be paired to load a truck and that with good coordination the overall loading time can be reduced if one of the loads is placed in the truck while it is moving into positioning. The chain loading method, also shown, suggests that if the number of loaders is matched to the bucket loads required to fill the truck, a drive-by technique can be introduced which greatly simplifies the loader maneuvering requirements and reduces the accompanying cycle time.

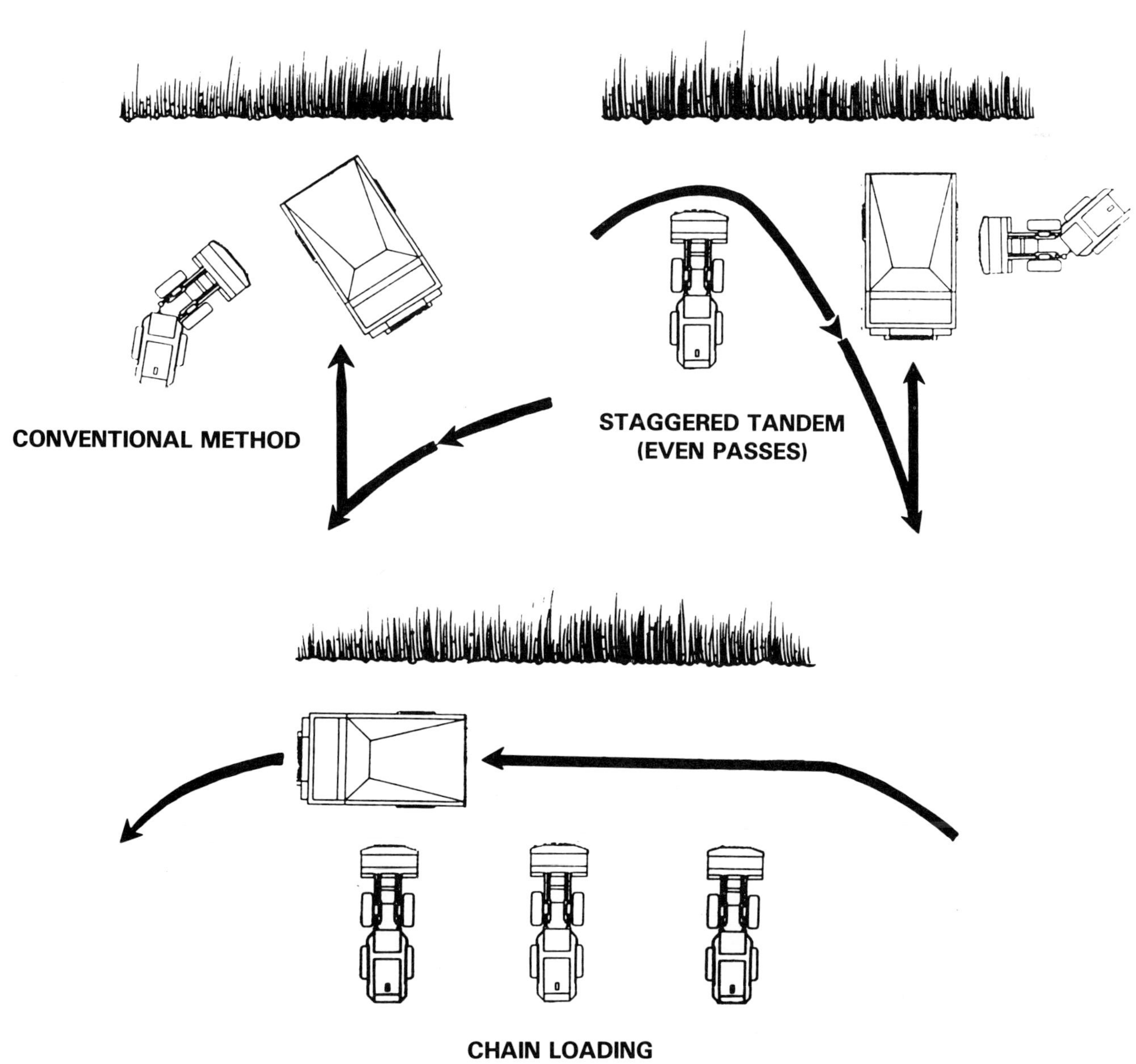

Figure 5.6 FEL Truck Loading Options

The loader provides a unique short haul capability which can be applied to digging and hauling material to convenient hopper locations. Figure 5.7 illustrates a typical operational sequence. The haul route is dependent on site requirements but economics suggest that the distances should not exceed about 900 feet. Ramps are commonly provided for dumping into portable or mobile hoppers; they serve to reduce the dumping height and provide a deceleration zone for the loader.

Machine performance can vary substantially dependent on; weight and swell of the material being handled, hardness/consolidation of the material in the digging face, traction conditions, proper matching of loader to the truck or hopper, haul distances, and operator skill. The dig and maneuver cycles require alertness on the part of the operator and the constant accelerating and decelerating of the machine imposes severe loading on the machine.

Additional weight is often added to the machine to increase stability, breakout force, and lift capacity. The two common methods of adding weight are to counterweight the rear of the machine and/or to ballast the tires. Increasing machine weight adversely effects fuel consumption and adds to the wear and tear on the machine.

Key points to remember in digging are:

- Best bank height (non-caving material) is between the height of the push arm hinge and the maximum lift position of the cutting edge.
- As the toe of the digging face is removed, care must be exercised to avoid any major caving of the entire face supported by this material.
- Crowd force is produced by machine traction and inertia; wheel spinning reduces traction.
- Climbing the face, which places the tires on rough broken material, abuses tires and can significantly reduce their life.
- Hard digging conditions require rock buckets, (spade nose with teeth) heavy duty tires and possibly added machine weight.
- Face should be penetrated smoothly and firmly to minimize machine abuse.
- Bucket should be progressively orientated to take advantage of failure planes.
- Digging floor should be kept clean of obstructions, generally flat but with a slight drainage slope to avoid water pockets which adversely affect tire life and traction.
- Propel inertia can be effectively utilized to assist in bank penetration but if the face is approached too fast it leads to tire spinning.
- Highest digging forces are possible at low bucket positions.
- Excessive bucket loads result in spillage during the maneuvering cycle and tire abuse.
- Constant maneuvering can tear-up the floor and dilute coal/ore.
- Wider buckets on larger machines, while improving the ability to handle larger chunks, make it more difficult to attack obstructions in the face.
- Raising bucket to dumping height just prior to reaching the hopper or truck improves visibility for machine positioning.
- Bucket roll speed during dumping should be reduced when discharging large rocks to minimize damage.
- Sticky material is dislodged by oscillating bucket a small amount to repeatedly strike the dump stops.

Overall productivity can be increased significantly if a dozer with a ripper is employed to assist the loader. The dozer (can be multiple units) pushes the mining face down to the loader, reducing its cutting height; and at the same time the dozer can be ripping to break-up the material. This "trapping" operation is illustrated in Figure 2.4 in the chapter on dozers.

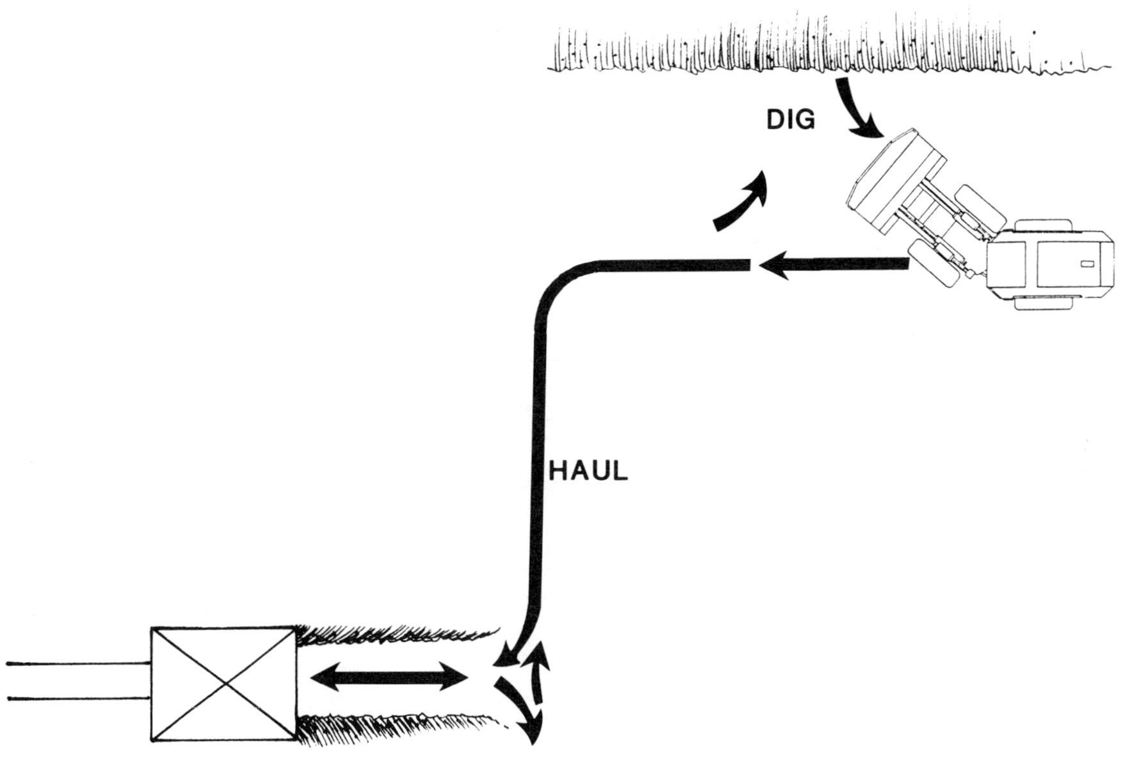

Figure 5.7 Typical FEL Load-And-Carry Production Cycle

Photograph 5.6 Clark 475B loading oversize shot rock

In a load-and-carry operation which involves some haul road travel, there are additional factors which should be considered:

- Greater care should be taken to fill the bucket for the haul; load must be centered to equalize the load distribution on the front tires.
- Bucket must be fully rolled back during carry to minimize spillage onto the haul road.
- Bucket must be carried as low as possible to keep the machine center of gravity low and maximize visibility, but still high enough to prevent ground contact from machine pitching on rough roads.
- Haul roads must be maintained adequately and propel speeds adjusted to control machine pitching and load spillage.
- High travel speeds and accompanying bouncing cause excessive tire sidewall deflection and tire deterioration. Increasd tire air pressure may be desirable to reduce machine side sway.
- Higher travel speeds increase productivity but must be balanced against increased fuel consumption, machine abuse and safety considerations.

Operating costs and the overall acceptance of the loader is critically tied to tire life/cost. Many of the operating factors noted above are aimed at optimizing their performance. Tires are discussed in Appendix A but a few brief points are in order:

- Rear tires normally have substantially more life than front tires. A planned rotation program is desirable.
- Tire abuse is severe if bucket width does not equal or exceed overall width across tires.
- Half tread tires (half smooth) have been helpful when tire cutting from sharp rocks is a problem.
- Smooth tires work well with tire chains.
- Track type tires, available for some loaders, are effective for bad floor conditions at the digging face and on haul roads.

- Foam filled tires provide a means of reducing the risk of tire blowouts but do aggrevate heat build-up.

- Switching to lower service code tires for load-and-carry service reduces heat build-up.

- Tire costs are minimized by regular tire inspection, maintaining proper air pressure, protecting casing for as many retreads as possible, observing wear indicators and avoiding uneven wear.

DESIGN FEATURES

Figure 5.8 shows the basic front-end loader (FEL) nomenclature.

In the sizes under consideration all front-end loaders utilize diesel engines as the primary power sources and those are located in the rear of the machines so that their weight can be used to counteract the bucket loads. The Clark model 675C has twin diesels for power. Basic engine characteristics and specifications are included in Appendix B.

The demanding duty cycle of loader service requires a fast drive system response and a balanced power distribution between the propel and the hydraulic system to optimize performance. These control characteristics vary between units and are achieved by special circuits and feed back systems whose design is beyond the scope of this manual. Trial operation of a machine, if practical, to evaluate these features is recommended.

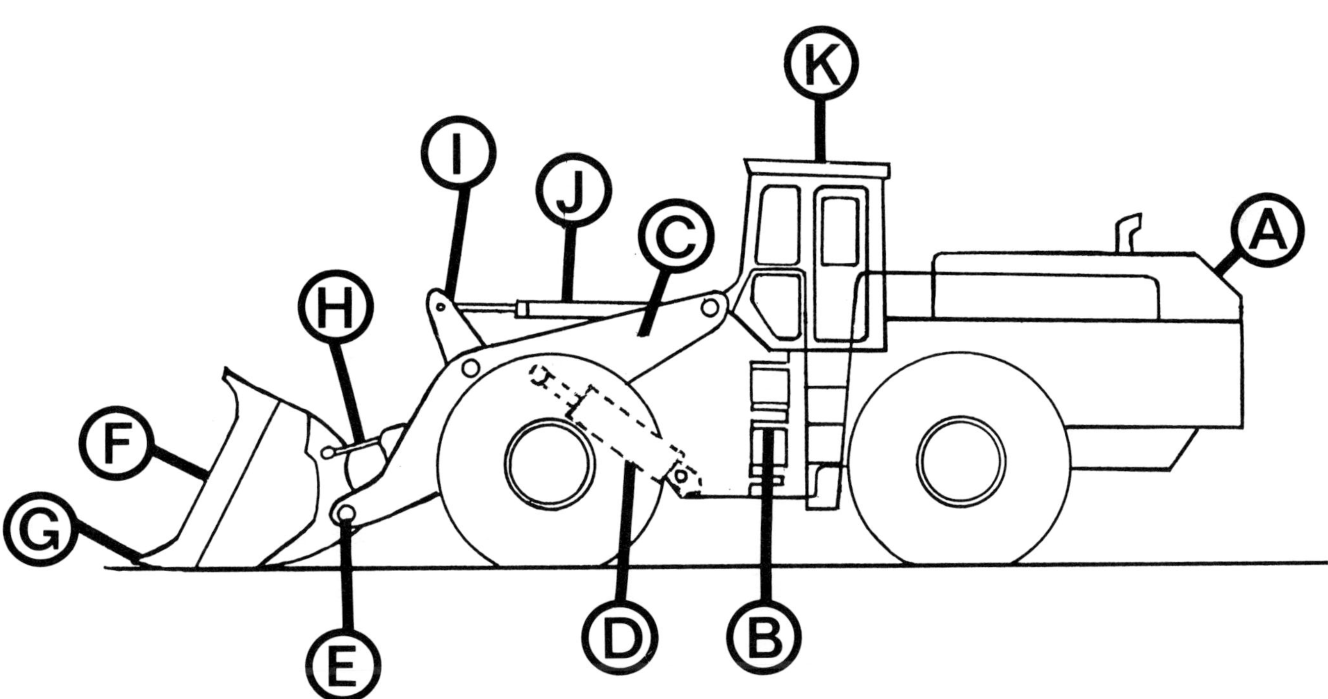

A Engine
B Articulating Hinge Point
C Lift Arm
D Lift Cylinders
E Bucket Hinge Pin
F Bucket
G Bucket Teeth (cutting edge)
H Bucket Link
I Bellcrank
J Bucket Cylinder
K ROPS

Figure 5.8 FEL Nomenclature

Currently, loaders utilize one of four different power train systems. Three of these systems are diesel/mechanical, one is diesel/electric. The three mechanical types can be described:

- Throttle controlled engine rpm. Engine drives through a conventional torque converter/ transmission unit.
- Throttle controlled power modulator located between the engine and a conventional torque converter/transmission unit. Engine rpm and hydraulic pump rpm always remain constant.
- Throttle controlled variable vane transmission with integral, multiple torque converters. Using a work range, torque converters (both forward and reverse), transmission vane speed regulates ground speed, while engine rpm remains constant. Use of travel range torque converter allows for throttle control of engine rpm.

The typical mechanical drive components are (see Figure 5.9):

- Modulated clutch to control power distribution between hydraulics and propel drive.
- Torque converter with stall ratio of 3.0 to 4.8.
- Power shift transmission with single lever control providing multiple forward and reverse speeds (soft shift).
- High capacity drive shafts, U-joints and torque proportioning differential (transfers power to wheel with best traction).
- Planetary gear reductions in wheel hubs.
- Air over hydraulic wheel brakes.

There are three basic types of differentials: torque proportioning, limited slip and no-spin. Each has its own unique operating features and application depends upon traction requirements. The torque proportioning is most efficient on high traction surfaces, diminishes in effectiveness as the traction surface or coefficient of friction decreases. A no-spin differential is a cam operated differential which directs the total axle torque to the slower moving wheel of the axle. A limited-slip differential is a clutch type differential which provides improved traction while maintaining torque to both wheels of the axle.

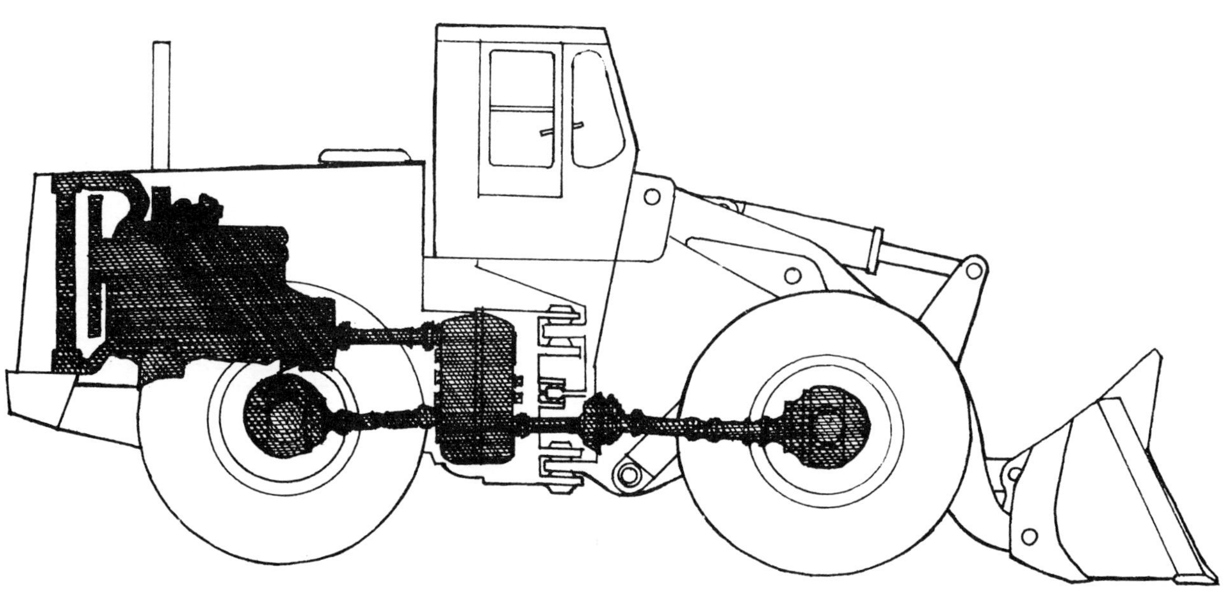

Figure 5.9 FEL Mechanical Drive Train

Only one manufacturer has electric drive models: the Marathon LeTourneau system is of their own design and consists of throttle controlled power resistors that regulate level of dynamic braking and direction of power flow. The engine drives an A.C. generator. Power is then converted to D.C. and delivered to individually adjustable D.C. wheel motors. The D.C. wheel motor powers a planetary drive in each wheel. The throttle, in full down position, allows for full current flow. As the throttle is returned to an upright position, dynamic braking from the wheel motors takes effect. In full up position, power flow reverses to stop the machine. Engine rpm and hydraulic pump rpm remain constant. Simply stated, its major components are (see Figure 5.10):

- Engine alternator (AC-three phase)
- Silicon controlled rectifier to supply D.C. power to individual wheel motors
- D.C. motors for each wheel (2 sizes: 100 and 150 hp)
- Forced air cooling (filtered) for wheel motors, generator, grid box and control box
- Planetary gear reduction
- Air over hydraulic wheel disc brakes

This system includes solid state electronic circuitry, plug-in printed circuit cards to control converters, electronic monitoring of critical parameters, traction control of wheel torque to minimize wheel spin and dynamic retarding.

The brake types commonly used on front-end loaders are expanding shoe, dry disc and liquid cooled disc. All three types can be actuated by either air or hydraulic pressure. The expanding shoe brake is more subject to fade and generally requires the removal of the wheel hub for lining replacement. Dry disc brakes offer easier servicing because linings can be replaced without axle disassembly. Dry disc brakes are not as subject to fade, and lining life is sometimes longer. Liquid cooled disc brakes are enclosed and free from contamination and the life of the brake components is

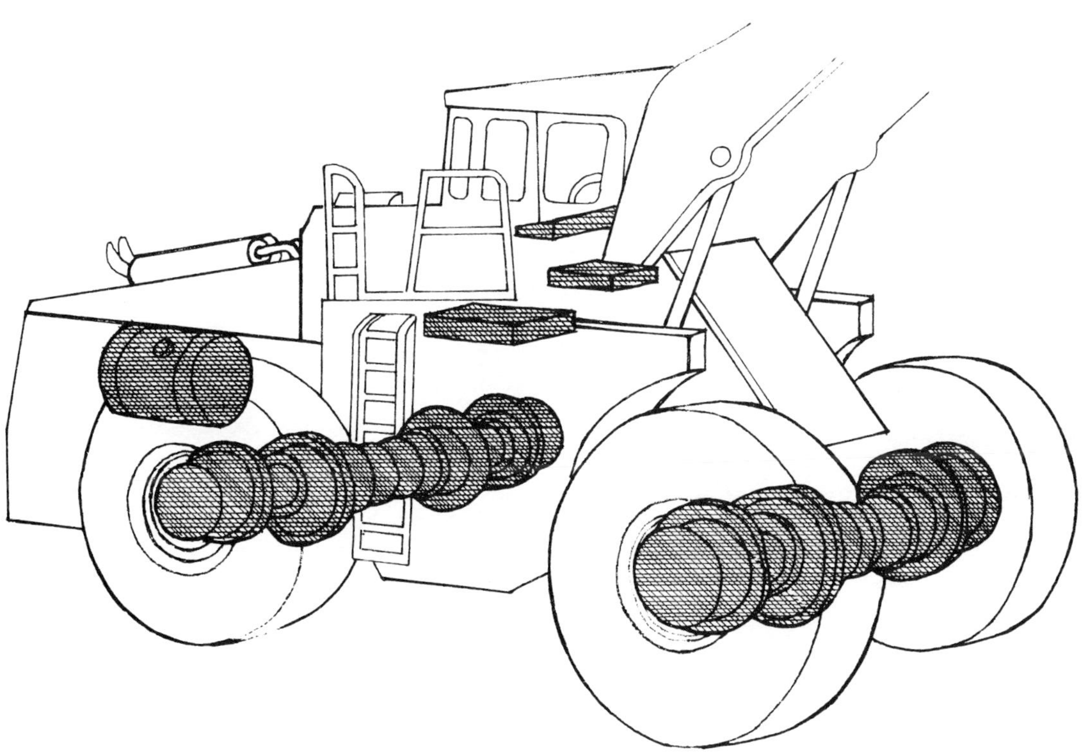

Figure 5.10 LeTourneau Type Electric Drive Train

generally expected to equal or exceed the time between complete axle overhauls. In abrasive or other harsh environments, the higher initial cost of these brakes may be offset by savings in brake maintenance costs.

Buckets must extend across the full width of the machine to protect the front tires as the loader penetrates the bank in digging. This considerable width means the bucket is structurally weaker than that of other excavators with their narrow widths. Minimizing the weight to maximize the payload adds to the design constraints. To some extent, this limits the maximum breakout forces which can be designed into the machine. The bucket is sized to match lift capability with the loose weight of the material to be handled. Three types of bucket construction are generally available, ranging from light duty to general purpose and the most rugged (and heaviest) is the rock dipper. The first two usually have straight cutting edges without teeth and the last has a spade shape cutting edge equipped with teeth. (See Figure 5.11) Tooth size and number are dependent on digging conditions. Cutting edges are reversible to extend their life. Special side dump and bottom dump buckets are available but rarely used in high production operations because of their added complexity and weight.

The lift arm linkage is commonly of either 2-bar or 4-bar configuration, as shown in Figure 5.12. The linkage design defines the bucket path, amplifies the cylinder forces to provide the lift and breakout capabilities, and actuates the bucket roll for positioning and dumping. Good geometric design improves the bucket fill, minimizes the operating pressure during digging and reduces the hydraulic heat build-up. Basic requirements are:

- Adequate structural strength
- Sealed in-line linkage

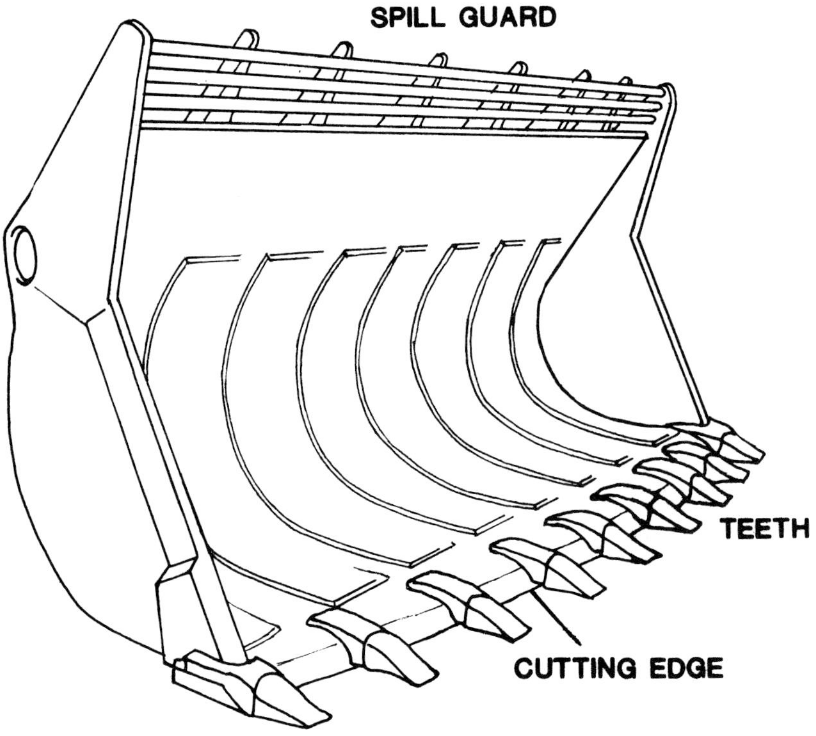

Figure 5.11 FEL Spade Nose Dipper

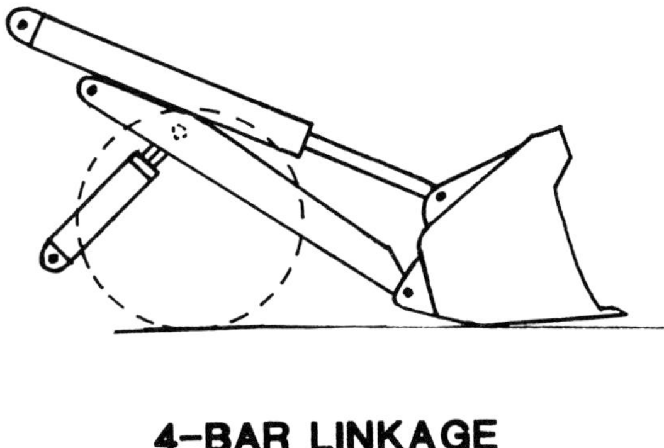

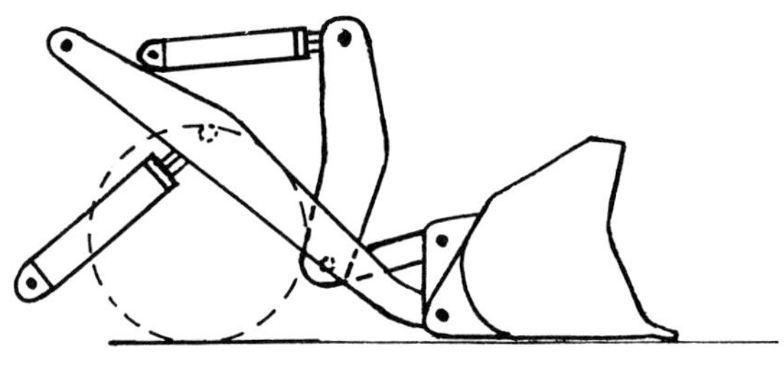

2-BAR LINKAGE

Figure 5.12 FEL Bucket Linkage Configurations

- Minimum parts and joints
- Low overall weight
- Simple and protected hydraulic plumbing
- High bucket rollback to minimize spillage while maneuvering and traveling
- High forward roll for full discharge of material
- Built-in bucket deceleration characteristics to minimize shocks at the end of the dump cycle
- Hold the bucket orientation as it is raised or lowered

After filling the bucket, the loader operator must concentrate on the machine maneuvering, while positioning the machine for dumping. Similarily, after dumping he must maneuver into position for the next cut. The bucket motions can be automated so that during these periods the bucket will be repositioned for the next phase. This is achieved by linkage design and sequencing of the hydraulic actions to, in essence, complete an action initiated by the operator. These automatic cycles are namely; "lift kick-out" (adjustable), which stops the raise when the bucket reaches dump height, "dump kick-out" which stops the action when the bucket reaches full forward roll, and "return to dig" which returns the bucket to a level position as it is lowered to ground level. Such features effectively reduce cycle time, reduce operator fatique and machine abuse. (See Figure 5.13)

The Dart loaders incorporate a counter balanced linkage which compensates for the weight of the linkage and the empty bucket to reduce the power demand and increase lift speed. The counter balancing action is provided by two nitrogen filled cylinders that are compressed and thus store energy as the bucket is lowered to the ground.

Since the hydraulics involve only pumps, valves and cylinders (no hydraulic motors) for power transmission, the systems are relatively simple. Two separate systems are generally provided: one for the steering and one for bucket motions. They are both closed, pressurized and filtered systems operating at medium pressure levels (2500 to 3500 psi). Large reservoirs are common, which aid in hydraulic fluid temperature control. Valves are electric or pneumatic assisted to permit handling large flow volumes with minimum manual effort. Pumps are frequently of the gear type, which are the most tolerant of contamination.

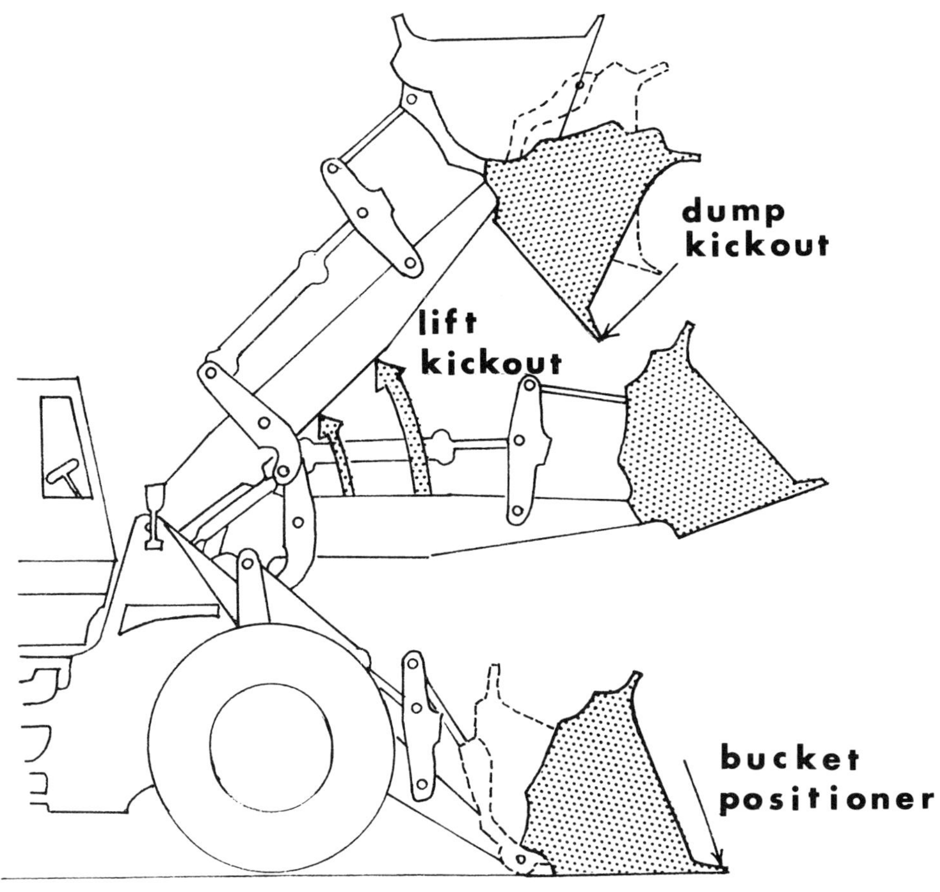

Figure 5.13 FEL Kickout And Positioners

While the systems are well proven, they can be a source of problems due to detail design deficiencies, abusive operation and/or poor maintenance practices. The primary problem areas are:

- Oil contamintion
- Failure of joints, hoses and support brackets
- Seal failures
- Excessive heat build-up

Hoses and fittings should comply with SAE (Society of Automotive Engineers) recommended standards.

Hydraulic cycle times provided by the manufacturers are of interest only for the "dump" and "lower" portions of the cycle, since the actual "hoist" (dig) time is tied to bank conditions. These times, however, represent a very small portion of the load-and-carry cycle and do not differ substantially between machines of the same size.

Some loaders have the operator's cab located forward of the articulation hinge while others locate it behind. There are differences of opinions as to the optimum position. The forward location improves the visibility of the bucket and places the operator further from the noise, heat and vibration of the engine. However, the operator's rearward perspective while maneuvering is constantly changing. Most models currently have rear mounted cabs.

From a service and maintenance standpoint, the machine design should meet the following objectives:

- Good accessibility to all elements requiring servicing or maintenance on a regular basis
- Permit simple and fast inspection and replacement of all filters
- Group lubrication points and filters requiring servicing
- Service points located to permit work to be done while standing on the ground, if possible
- Modular units wherever possible to permit rapid interchange/exchange with units from other machines, in stock, with distributor, etc.
- Standarized componentry to minimize parts inventory and increase mechanic's familiarity
- Diagnostic type instruments which provide an indication of deteriorating conditions
- Clean, simple hydraulic plumbing arrangement
- Maximum use of sealed bearings aimed at reducing servicing frequency

SELECTION CONSIDERATIONS

The site conditions and requirements and their relationship to the machine to be selected can be briefly summarized as follows:

- Mine plan (loading vs. load-and-carry)
- Production requirement and scheduling (machine size, cycle time)
- Digging face conditions (machine power, bucket design, breakout forces, digging height and counterweights)
- Material weights (machine lift capacity, tipping loads)
- Climatic conditions and elevation (engine power, traction conditions)
- Maneuvering area and digging floor conditions (tire selection, turning diameter)
- Haul road conditions and effective grades (maximum speeds, wheel base, tire selection)
- Truck or hopper height and opening dimensions (machine dump height, reach, dump characteristics)
- Machine configuration
  - Mechanical or electric
  - Ruggedness and reliability
  - Servicing requirements
  - Accessibility for maintenance

Graph 5.5 provides a general correlation between machine size and the cut height which must be matched to digging face conditions. It also indicates the typical dump height which can be assumed for matching the loader with either trucks or a hopper.

Graph 5.6 provides a measure of the potential digging power (hp/cubic yard) for the various machine sizes. This graph shows the rather broad range that exists between available machines. The spread is greatest in the smaller sizes where there is a greater divergence in the emphasis placed on different applications.

Since the loader is a relatively low investment unit (in the smaller sizes), it generally is selected with excess capacity to assure that it does not limit overall production.

Production calculation procedures and operating cost estimates are provided in available manufacturers' literature. These are considered a separate subject too extensive to be comprehensively treated in this book. Estimates must be directly correlated with detailed site conditions

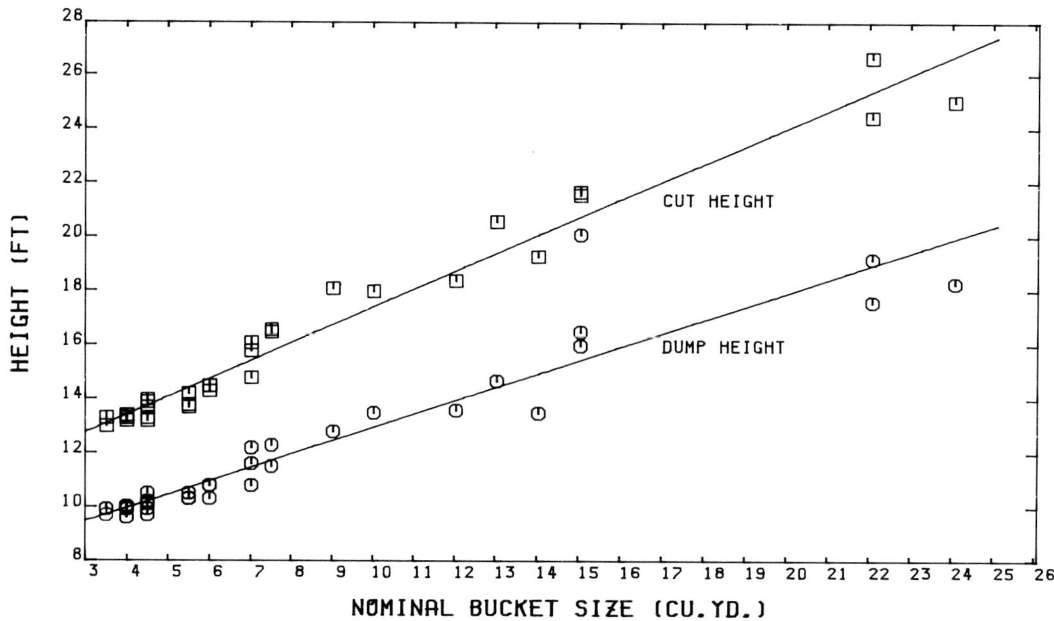

Graph 5.5 Dump height & cut height/nominal bucket size

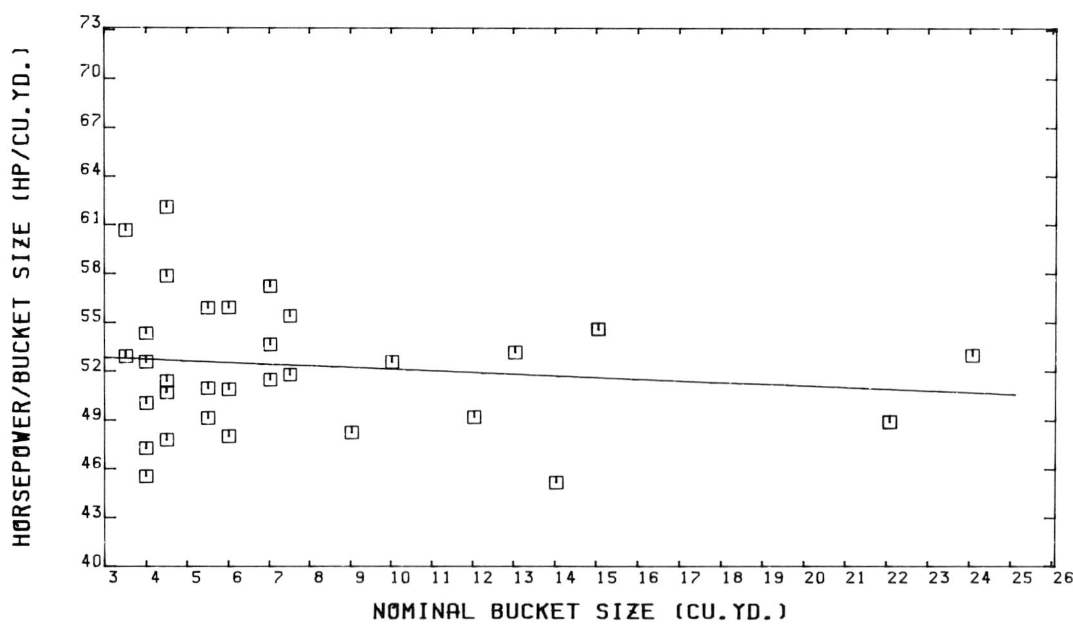

Graph 5.6 HP per nominal bucket size/nominal bucket size

and matched to specific machine performance characteristics. All of the equipment manufacturers will provide production and cost estimates for stated conditions prepared by their sales technical staff personnel based on their accumulated experience on similar applications.

These estimates are generally based on average data and average site conditions over the life of the operations. The spread in these "averages" can be substantial and the correction factors introduced require some judgement. Costs, in particular, vary locally with time and economic conditions making generalizations dangerous. Some of the production considerations entering into these calculations are:

- Material swell
- Material unit weight
- Bucket size
- Machine payload
- Digging conditions
- Loading arrangements (including delays)
- If traveling:
  - — Grades
  - — Rolling resistance
  - — Travel distance
  - — Maximum speeds
- Job efficiency
- Machine availability
- Operating schedule

Some of the cost considerations entering into these calculations are:

- Machine price
- Salvage value
- Machine life
- Interest
- Insurance
- Taxes
- Fuel consumption
- Lube, oils, greases, filters, etc.
- Operator wages
- Tire cost
- Cutting edges
- Maintenance

Decisions are required on a number of performance options to complete the basic machine specifications:

- Tires; size, manufacturer, ply rating, tread, etc.
- Engine; horsepower, turbocharging, after cooling, manufacturer
- Differential; no spin, torque proportioning
- High capacity alternator (electric drives only)
- Bucket; light, general purpose, heavy duty, special purpose
- Bucket accessories; cutting edges, teeth
- Lift arms; standard, high lift, heavy lift
- Counterweights

Special operational conditions may dictate the addition of some of the following:

- Heaters; hydraulic, battery, engine, oil
- Reversible fan
- Supplemental steering system
- Night lighting equipment
- Power train guard
- Starting receptacle
- Radiator shutters

- Air system dryer
- Starting kits
- Front fenders

Special safety servicing and maintenance equipment is available:

- Vandalism protection systems
- Fire suppression system
- Automatic lubrication system
- Fast fuel fill system
- Fast oil fill system
- Tire inflation system

Operator station accessories are available:

- Air conditioning
- Heater/defroster
- Radio
- Mirrors, windshield wipers
- Special instrumentation and warning lights
- Special seat suspension

Photograph 5.7 LeTourneau L-1200, equipped with tire chains

Photograph 5.8 International Harvester 580 alongside a 170 ton rear dump truck

NEW DEVELOPMENTS & TRENDS

The size trend for loaders is complicated by the very early introduction of the Clark 675, 24 cubic yard machine, well over a decade ago. Only within the last two years have units of comparable size, the 22 cubic yard, International 580 (see Photograph 5.7) and the 22 cubic yard Marathon LeTourneau L-1200 (see Photograph 5.8), become available. It is interesting to note that one of these is a mechanical and the other an electric drive train. These sizes can load into most of the trucks presently in use. This, together with current component limitations, suggests that this size level may not increase for a few years. The larger loaders are a relatively new product development so there is considerable work directed towards further improvement in machine effectiveness and component refinement.

The efforts of Goodyear Tire Company and Caterpillar Tractor Company to market the track type tires are aimed at improving machine versatility and broadening the application potential. Their future use depends on improving the economics.

Caterpillar has introduced on its 992C a more sophisticated variable volume, piston pump hydraulic system. The system decreases torque demand at low speeds to unload the engine permitting faster acceleration. Most manufacturers are continuing to refine their matching of the bucket hydraulic system demand with the propel requirements during digging. Such improvements attempt to optimize the power distribution to provide the best performance in terms of both cycle time and digging forces.

The loader, by its basic design, has restricted dump height capability. Bucket discharge, without rolling the bucket too far forward, increases the usable dumping height. Dart's ejector type bucket which utilizes a cylinder operated, hinged, ejector plate to facilitate dumping, is aimed at this problem and does substantially increase the effective dumping height and dumping reach.

Considerable effort has been applied in recent years to improve the operator's environment with special emphasis on noise reduction. This has been achieved by resilient mounting, isolation from the ROPS structure, and bulkhead seals for all control levers and cables.

Significant advances continue to increase access for servicing and reduce service frequency. Applications of electronic monitoring and diagnostic readouts are expanding rapidly and will contribute to future machine operational availability.

MACHINE SPECIFICATIONS

See Figure 5.14—Front-end Loader Dimensions
See Table 5.2—Front-end Loader Specifications

Front-end loaders are listed alphabetically by manufacturer, and then in ascending order based on nominal bucket size (cubic yards). Nominal bucket size is defined as the manufacturer's listed general purpose bucket rated at, or closest to, 3000 lb./cu. yd. material. All dimensions and operating specifications are based on the manufacturer's listed general purpose bucket as described previously. When more than one bucket fits the nominal bucket criteria, a straight-edged bucket without teeth was used. Digging depth, dump height and reach dimensions will vary depending on the bucket selected, whether the bucket is equipped with teeth, and on the lift arms selected (heavy lift, high lift, etc.). All dimensions in the table are based on the standard nominal bucket and standard lift arms. Where only long or short booms are offered, the specifications are based on a machine equipped with a short boom.

Engine horsepower is the manufacturer's stated flywheel or rated horsepower for the highest rated standard engine. Where this data was not available, data from the engine manufacturer was used to determine approximate flywheel or rated horsepower.

Wherever possible, operating weight and tipping load were based on manufacturer's listed basic machine weight, with general purpose bucket, ROPS cab, standard tires, full fuel tank and operator. In some cases counterweights and hydro-inflation weights were built into the manufacturer's listed weights and operating loads. Where this was the case, and when the effects of counterweight and hydro-inflation could not be segregated out, the lightest available numbers were used. Reach is measured from the leading edge of the front tires to the bucket lip or teeth.

The hydraulic specifications are based on the loader circuit only. Steering and pilot pumps have been omitted. Hydraulic cycle times are matched to the standard engine. In some cases, listed optional engines will result in different hydraulic cycle times. Optional transmissions, transmission ratios and tires will result in different travel speeds. The tire size listed is the smallest, lowest ply rated tire available as standard equipment from the manufacturer. No distinction is made in the specifications between caliper disc and multiplate disc brakes.

All specifications, capacities, capabilities and dimensions are based on published manufacturer data. Although the information is believed to be current and the interpretation to be consistent and correct, it is possible that omissions and inaccurancies have occurred and that out-of-date information may have been included in the specification charts. These specifications charts are not meant to be used as either an in-depth analysis or as a head-to-head comparison of the available equipment, but rather as a general overview of the equipment, available sizes, and approximate operating data. Specific questions relating to performance or purchase should be directed to the manufacturer or authorized distributor/dealer in the user's specific area.

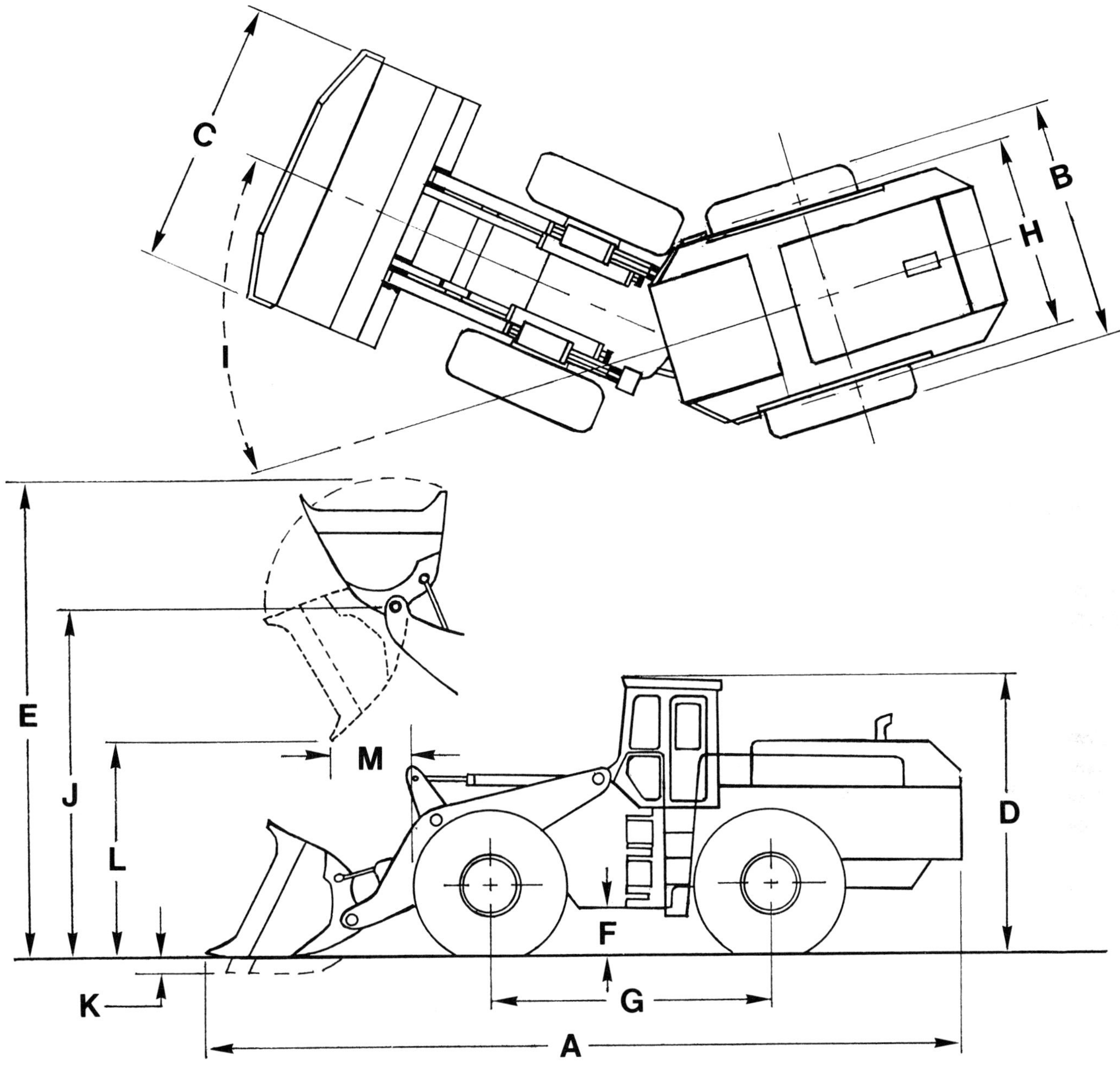

A	Overall Length/Bucket on Ground (ft)	H	Tread/Front & Rear (ft)
B	Overall Width w/o Bucket (ft)	I	Angle of Articulation (deg)
C	Bucket Width (ft)	J	Height to Hinge Pin (ft)
D	Height to Top of Cab (ft)	K	Digging Depth (in)
E	Height to Top Raised Bucket (ft)	L	Max Dump Height (ft) (bucket at 45 degrees)
F	Ground Clearance (in)	M	Reach (ft) (at max lift/bucket at 45 degrees)
G	Wheelbase (ft)		

Figure 5.14 FEL Dimensions

TABLE 5.2
FRONT-END LOADER SPECIFICATIONS

MAKE	MODEL	NOMINAL BUCKET SIZE (cu yd)	AVAILABLE BUCKET SIZES (cu yd)	ENGINE HP	WEIGHT (lbs)	BUCKET PAYLOAD (lbs)	BREAKOUT FORCE (lbs)	STRAIGHT TIPPING LOAD (lbs)	FULL-TURN TIPPING LOAD (lbs)	FUEL TANK (gal)	COOLANT (gal)
Case	W36	3.5	3.5-4	185	33,450	12,000	30,300	25,700	21,975	88	17
Caterpillar	966D	4.0	4-4.5	200	42,097	12,000	48,536	29,461	27,953	81	12
Caterpillar	980C	5.5	5-6	270	56,457	15,400	58,880	39,217	36,473	105	21
Caterpillar	988B	7.0	7-8	375	85,200	21,200	104,100	51,060	46,320	165	28
Caterpillar	992C	13.0	13-13.5	690	188,572	40,500	179,269	106,924	96,020	300	36
Clark	125C	3.5	3.5-6	212	40,740	12,000	37,000	28,870	24,500	75	18
Clark	175C	4.5	4.5-6	279	51,975	15,000	34,700	35,807	31,173	124	20
Clark	275C	7.0	6.5-8	360	80,060	21,000	51,000	55,490	48,960	165	26
Clark	475C	14.0	10-22	632	158,150	42,000	102,200	104,660	97,150	273	68
Clark	675C	24.0	18-36	1270	389,500	72,000	159,250	185,800	165,360	500	122
Dart	600C	15.0	7-30	818	192,750	45,000	—	134,600	118,300	—	—
Dart	600-E	15.0	10-23	818	197,000	45,000	—	134,600	118,300	—	—
Deere	JD844	4.5	4.5-7	260	47,490	13,500	35,070	32,080	27,780	100	20
Fiat-Allis	FR20	4.5	3.5-4.5	215	42,500	13,500	33,150	30,620	25,000	110	20
Fiat-Allis	945-B	6.0	6-6.5	335	65,090	18,000	48,000	46,550	38,642	160	25
International	540	4.0	4-5	189	35,750	13,500	33,540	26,860	24,174	80	16
International	550	5.5	5-6	280	50,050	17,250	40,070	40,770	34,665	135	16
International	560-B	7.5	7.5-15	415	82,384	22,500	73,210	64,145	51,381	166	23
International	570	12.0	12-24	590	132,180	36,000	100,800	97,800	83,130	270	38
International	580	22.0	22	1075	289,100	66,000	192,500	199,100	169,235	525	94
Kawasaki	KSS-80T	4.0	4	217	38,840	10,580	27,500	24,250	22,200	61	17
Kawasaki	95Z-II	6.0	6	305	59,194	17,000	40,565	40,344	36,596	95	23
LeTourneau	L-600	10.0	8-20	525	130,200	30,000	96,636	80,000	72,000	230	35
LeTourneau	L-800	15.0	10-30	818	185,500	45,000	117,972	136,000	116,000	390	55
LeTourneau	L-1200	22.0	22	1075	335,000	66,000	164,000	230,000	197,000	650	90
Terex	66C	4.0	3.5-7	210	41,225	13,225	37,900	28,550	24,525	—	—
Terex	72-51B	4.5	4-5.5	231	40,651	13,500	38,355	34,283	30,186	90	20
Terex	72-61	5.5	5.5-6.5	307	56,140	16,500	42,481	45,182	39,611	125	22
Terex	72-71B	7.5	7.5-8	388	84,970	22,500	58,550	60,577	52,090	180	24
Terex	72-81	9.0	9	434	112,390	27,000	62,500	78,695	69,925	200	26
Trojan	2500	4.0	3.5-5	182	37,090	12,000	36,189	28,651	24,927	76	16
Trojan	3000	4.5	4-5.5	228	40,905	12,750	36,000	29,885	26,000	76	19
Trojan	5500	6.0	5.5-8	288	58,674	18,000	49,050	44,405	40,408	129	28
Trojan	7500	7.0	6.5-8.5	400	77,722	22,500	61,000	57,774	52,574	146	32

Table 5.2 (Continued)

FRONT-END LOADER SPECIFICATIONS

MAKE	MODEL	LENGTH (ft)	MAX WIDTH W/O BUCKET (ft)	BUCKET WIDTH (ft)	HEIGHT TO TOP OF CAB (ft)	HEIGHT TO TOP OF BUCKET (ft)	GROUND CLEARANCE (in)	CLEARANCE CIRCLE (ft)	WHEELBASE (ft)	TREAD (ft)	STANDARD TIRES
Case	W36	24.3	8.9	9.5	10.5	17.3	16	41.8	10.6	7.2	20.5-25 16PR
Caterpillar	966D	25.7	9.4	10.0	11.7	17.8	18	48.0	11.0	7.2	23.5-25 16PR
Caterpillar	980C	28.3	10.2	11.0	12.8	19.0	16	51.3	11.6	7.8	26.5-25 20PR
Caterpillar	988B	32.8	11.5	11.9	13.5	21.3	19	56.0	12.5	8.5	6535-33 24PR
Caterpillar	992C	41.7	14.8	15.6	18.0	28.4	21	71.3	15.8	10.8	6545-45 38PR
Clark	125C	25.2	9.4	10.0	11.6	17.1	17	48.0	10.7	7.3	23.5-25 16PR
Clark	175C	27.5	10.0	10.3	12.2	18.6	21	49.4	11.3	7.4	26.5-25 20PR
Clark	275C	29.8	11.5	11.8	12.5	20.2	19	54.0	12.2	8.8	29.5-29 22PR
Clark	475C	40.2	12.7	13.5	16.2	27.0	22	68.0	15.2	9.5	37.25-35 42PR
Clark	675C	48.9	19.7	20.8	20.5	34.0	40	84.7	18.6	13.8	6751SXT 54PR
Dart	600C	41.3	16.4	15.5	18.5	—	—	66.8	16.5	11.4	37.5-39 36PR
Dart	600-E	—	16.4	—	18.5	—	—	66.8	16.5	11.4	37.5-39 36PR
Deere	JD844	26.1	9.7	10.7	11.7	—	23	45.8	10.5	7.5	26.5-25 20PR
Fiat-Allis	FR20	25.9	9.5	9.8	11.5	17.7	18	40.0	10.8	7.3	23.5-25 16PR
Fiat-Allis	945-B	27.8	10.8	11.3	13.5	18.8	21	45.0	12.3	8.0	29.5-29 22PR
International	540	24.2	—	9.5	11.0	—	15	49.8	9.8	—	—
International	550	26.4	10.2	10.8	12.3	—	14	49.0	10.8	7.8	26.5-25 20PR
International	560-B	30.1	11.1	12.0	14.0	—	20	56.6	12.9	8.5	29.5-29 22PR
International	570	35.5	13.3	14.0	15.0	—	22	60.0	15.0	9.8	6540-39 30PR
International	580	47.8	—	18.0	17.6	—	23	81.2	21.0	—	6550-51 46PR
Kawasaki	KSS-80T	24.1	9.1	9.7	11.3	—	19	—	10.5	7.0	23.5-25 12PR
Kawasaki	95Z-II	27.9	10.2	10.8	11.9	—	18	—	11.5	7.9	26.5-25 24PR
LeTourneau	L-600	37.7	11.8	13.2	14.7	25.4	20	59.5	16.0	9.0	33.25-35 32PR
LeTourneau	L-800	43.8	14.8	16.4	16.0	29.5	19	64.7	18.0	11.6	37.5-39 36PR
LeTourneau	L-1200	53.7	19.0	20.7	20.5	34.0	30	83.0	22.0	14.3	40.00-57 44PR
Terex	66C	24.3	8.9	9.2	10.7	17.5	15	—	9.8	7.0	23.5-25 EM
Terex	72-51B	25.8	9.4	10.0	11.3	18.0	14	44.0	10.0	7.4	23.5-25 16PR
Terex	72-61	27.8	10.6	11.0	11.9	18.8	17	47.0	10.5	8.3	26.5-25 20PR
Terex	72-71B	34.0	11.5	12.1	13.5	22.7	18	54.3	13.3	8.9	29.5-29 22PR
Terex	72-81	35.5	12.0	12.8	13.8	24.0	18	57.3	13.7	9.2	33.25-35 26PR
Trojan	2500	24.7	9.4	9.8	11.6	18.3	16	42.2	10.5	7.3	23.5-25
Trojan	3000	25.4	9.4	9.8	11.6	18.8	16	43.2	10.5	7.3	23.5-25
Trojan	5500	28.8	10.8	11.6	12.1	19.8	17	54.7	12.3	8.3	26.5-25 20PR
Trojan	7500	30.3	11.3	12.0	12.7	22.0	18	56.8	13.0	8.8	29.5-29 22PR

Table 5.2 (Continued)
FRONT-END LOADER SPECIFICATIONS

MAKE	MODEL	ENGINE MAKE	ENGINE MODEL	TRANSMISSION TYPE	NO. OF GEARS F/R	BRAKES	HYDRAULIC SYSTEM DATA: PUMP FLOW (gpm)	RELIEF VALVE PRESSURE (psi)	PUMP TYPE	NO. OF CYLINDERS	CAB POSITION
Case	W36	Case	A504-BDTI	mech auto	4/4	disc	44	2500	gear	4	front
Caterpillar	966D	Caterpillar	3306	mech auto	4/4	shoe	81	2750	vane	3	rear
Caterpillar	980C	Caterpillar	3406	mech auto	4/4	disc	96	3000	gear	4	rear
Caterpillar	988B	Caterpillar	3408	mech auto	4/4	disc	137	3000	gear	4	rear
Caterpillar	992C	Caterpillar	3412	mech auto	3/3	disc	237	3250	piston	4	rear
Clark	125C	Detroit	6V-71N65	mech auto	4/4	disc	87	2250	gear	4	rear
		Cummins	VT-555-C								
Clark	175C	Detroit	8V-71N	mech auto	4/4	disc	125	2200	gear	4	rear
		Cummins	NT-855-C								
Clark	275C	Cummins	KT-1150-C400	mech auto	4/4	shoe	130	2200	gear	4	rear
Clark	475C	Cummins	VTA-1710-C700	mech auto	4/4	shoe	217	2700	gear	4	rear
		Detroit	16V-92N80								
		Cummins	VTA-1710-C725		2/1		282				
Clark	675C	(2)Cummins	VTA-1710-C675	mech auto	4/4	disc	500	2750	gear	6	rear
Dart	600C	Cummins	KT-2300-C860	mech auto	3/3	shoe	279	2000	vane	4	front
Dart	600-E	Cummins	KT-2300-C860	mech auto	3/3	shoe	279	2250	vane	6	front
		Detroit	16V-92T								
		Cummins	VTA-1710-C700								
Deere	JD844	Deere	V-8 Turbo	mech auto	4/3	disc	95	2250	vane	3	rear
Fiat-Allis	FR20	Fiat	8215	mech auto	4/4	disc	91	2000	gear	4	rear
Fiat-Allis	945-B	Allis-Chalmers	25000 MK-II	mech auto	4/4	shoe	135	2250	gear	4	rear

International	540	International	DT-466B	mech auto	3/3	disc	82	2750	gear	3	front
International	550	International	DVT-800	mech auto	3/3	disc	101	3000	gear	3	front
		Cummins	KT-1150-C								
International	560-B	Cummins	KT1150-C450	mech auto	3/3	disc	155	3000	gear	3	front
International	570	Cummins	VT-1710-C635	mech auto	2/2	shoe	214	3000	gear	3	front
International	580	Detroit	12V-149T1	mech auto	3/3	disc	320	3000	gear	3	front
Kawasaki	KSS-80T	Cummins	NTO-6-CI	mech auto	4/2	disc	—	2490	gear	4	rear
		Cummins	NH-220-CI								
Kawasaki	95Z-II	Cummins	NT-855-C335	mech auto	4/2	disc	—	2489	gear	4	rear
LeTourneau	L-600	GM	12V-71T	electric	n.a.	disc	220	2200	-	4	rear
		Cummins	KT-1150-C525								
LeTourneau	L-800	GM	16V-92T	electric	n.a.	disc	350	2000	-	4	rear
		Cummins	KT-2300-C860								
LeTourneau	L-1200	GM	12V-149TI	electric	n.a.	disc	480	2250	-	4	rear
		Cummins	KTA2300-C1200								
Terex	66C	Hanomag	D963-A1	mech auto	4/4	disc	—	-	-	4	front
Terex	72-51B	Detroit	6V-71T	mech auto	2/2	disc	90	2500	gear	4	rear
Terex	72-61	Detroit	8V-71T	mech auto	2/2	shoe	125	2500	gear	4	rear
Terex	72-71B	Detroit	8V-92T	mech auto	3/3	shoe	134	2500	gear	4	rear
Terex	72-81	Detroit	12V-71T	mech auto	3/3	shoe	178	2500	gear	4	rear
Trojan	2500	GM	6V-71N	mech auto	4/4	disc	91	2500	gear	4	rear
Trojan	3000	GM	6V-71N	mech auto	4/4	disc	100	2500	gear	4	rear
		Cummins	V-903-C265								
Trojan	5500	GM	8V-92N70	mech auto	4/4	shoe	120	2500	gear	4	rear
		Cummins	NT-855-C310								
Trojan	7500	GM	8V-92T-N90	mech auto	4/4	shoe	150	2500	gear	4	rear
		Cummins	KT-1150-C450								

TABLE 5.2 (Continued)
FRONT-END LOADER SPECIFICATIONS

MAKE	MODEL	MAX SPEED F/R (mph)	HYDRAULIC CYCLE TIMES: BUCKET RAISE (sec)	BUCKET LOWER (sec)	BUCKET DUMP (sec)	STEERING ANGLE (degrees)	HEIGHT TO HINGE PIN (ft)	DIGGING DEPTH (ft)	DUMP HEIGHT (ft)	REACH AT FULL HEIGHT 45 DEGREES DISCHARGE (ft)
Case	W36	20/20	6.7	5.4	2.0	40	13.0	—	9.7	3.3
Caterpillar	966D	21/24	6.3	3.0	2.0	35	13.4	3	9.9	3.6
Caterpillar	980C	22/25	7.3	3.4	2.0	35	13.7	3	10.3	4.3
Caterpillar	988B	23/26	9.4	4.5	3.0	35	16.1	3	12.2	5.4
Caterpillar	992C	13/14	11.4	3.7	3.4	35	20.6	2	14.7	6.8
Clark	125C	19/19	7.2	3.1	1.1	35	13.3	3	9.9	3.2
Clark	175C	20/20	6.7	4.2	2.0	35	14.0	4	10.2	4.3
Clark	275C	22/22	8.7	6.0	2.1	35	14.8	4	10.8	5.0
Clark	475C	18/18	12.0	5.8	3.4	35	19.3	6	13.5	6.1
Clark	675C	18/18	12.0	6.5	2.5	35	25.1	—	18.3	6.4
Dart	600C	15/15	10.5	5.3	3.8	40	21.6	—	16.0	6.3
Dart	600-E	15/15	10.5	5.3	7.6	40	-	—	20.1	10.8
Deere	JD844	23/15	7.0	4.0	2.0	37	13.9	5	10.5	3.5
Fiat-Allis	FR20	21/23	6.5	4.6	2.6	45	13.3	4	9.9	3.2
Fiat-Allis	945-B	20/22	7.7	4.9	3.6	45	14.5	3	10.8	4.0
International	540	20/21	7.0	5.7	—	35	13.3	3	10.0	3.7
International	550	21/21	7.1	5.6	2.0	35	14.2	6	10.5	3.8
International	560-B	20/20	9.1	6.3	2.8	35	16.6	2	12.3	4.6
International	570	23/23	10.1	5.0	—	40	18.4	12	13.6	6.0
International	580	18/18	17.4	8.0	—	40	24.5	18	17.6	6.4
Kawasaki	KSS-80T	23/10	6.8	3.8	2.0	38	13.2	2	9.6	2.9
Kawasaki	95Z-II	21/9	7.0	4.8	1.3	36	14.3	13	10.3	4.6
LeTourneau	L-600	15/15	11.0	7.5	3.0	45	18.0	5	13.5	4.5
LeTourneau	L-800	15/15	14.0	7.0	3.0	45	21.7	8	16.5	7.0
LeTourneau	L-1200	12/12	13.0	7.0	3.0	42	26.7	5	19.2	9.7
Terex	66C	24/24	5.2	3.5	1.3	40	13.3	5	10.0	4.2
Terex	72-51B	18/19	7.0	4.3	2.5	37	13.2	7	9.7	3.8
Terex	72-61	24/27	6.5	4.2	2.6	37	13.8	4	10.3	4.5
Terex	72-71B	18/17	10.0	4.4	2.4	40	16.5	3	11.5	4.8
Terex	72-81	15/17	11.6	5.5	3.2	40	18.1	5	12.8	5.3
Trojan	2500	21/21	6.3	4.9	2.7	40	13.4	-	9.9	3.3
Trojan	3000	22/22	6.8	4.7	2.7	40	13.7	-	10.1	3.8
Trojan	5500	20/22	8.2	6.9	2.5	35	14.5	-	10.8	4.6
Trojan	7500	20/20	8.8	7.0	2.5	35	15.8	-	11.6	4.8

Chapter 6

HYDRAULIC EXCAVATORS

Figure 6.1 Hydraulic Shovel

Hydraulic hoes and shovels, primarily a European development, have proven themselves on construction projects. They have now reached a level of reliability and have increased in size to the point where units are common in surface mining applications. First development of these machines was aimed at hoes; the shovel front-end on the basic machine is a relatively recent innovation. Shovel configurations initially suffered from limited dump height, but this has been resolved by new bucket designs. These machines are currently going through a very aggressive development period with sizes increasing ten-fold over about a five year period. This discussion will focus on the larger sizes with potential for surface mining applications. For the larger models referenced, there is very limited field operating experience. Both hoe and shovel versions will be considered.

TYPICAL UNITS

All these units are diesel powered with hydraulic powered operating functions. As illustrated in Figure 6.2, the two basic machine configurations are the shovel and the hoe. The design difference between these is in the orientation of the bucket and in front-end geometry. A common basic machine is often offered in both versions. In operation, the two designs differ primarily in digging action and cut profile. Typical units have the following features:

- Hydraulic hoes in sizes 3 to 21 cubic yards
  - Horsepowers from 300 to 2400
  - Machine weights from 110,000 to 930,000 lbs.
- Hydraulic shovels in sizes 3 to 40 cubic yards
  - Horsepowers from 300 to 2400
  - Machine weights 115,000 to 930,000 lbs.
- Crawler mounted
- Full revolving upper works
- All hydraulic drives
- Low profile machinery compartments

Manufacturers

All of the U.S. manufacturers who formerly made small mechanical shovels and hoes, have converted to similar fully hydraulic machines. In addition, a number of U.S. farm and construction equipment manufacturers have entered the market —namely, J.I. Case, Deere, International Harvester and Caterpillar. There are relatively few U.S. machines in sizes above 6 cubic yards. Presently a number of overseas manufacturers have large models available. Originally these companies were slow in penetrating the U.S. market, but are now effective through established U.S. affiliates. Some of these U.S. operations are limited to marketing, servicing and parts warehousing, while others have progressed to final assembly and full manufacture. Sales and service are handled both by dealers and manufacturer's representatives. See Table 6.1.

Photograph 6.1 Koehring 1166E hydraulic hoe digging wet material

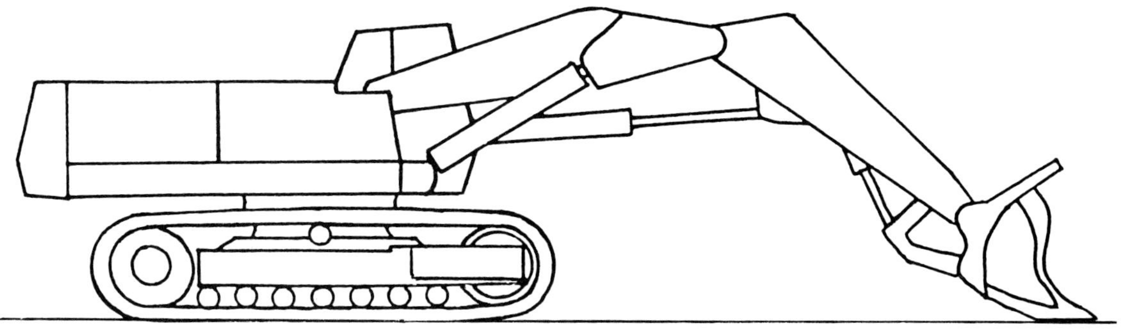

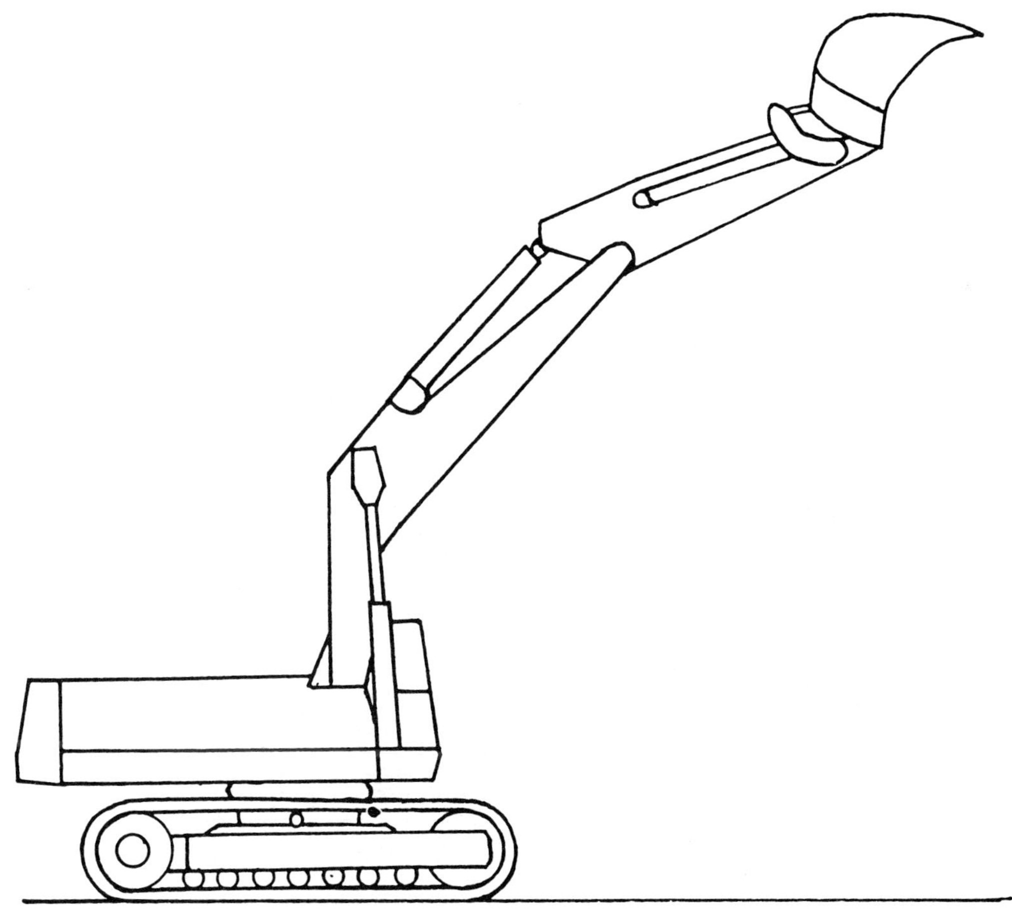

Figure 6.2 Hydraulic Excavator Types

Table 6.1

HYDRAULIC MACHINE MANUFACTURERS

American Hoist & Derrick Co. American Construction Machinery Division 63 S. Robert Street St. Paul, Minnesota 55107	Hoes to 6 cu. yd.
Bucyrus-Erie Company 1100 Milwaukee Avenue South Milwaukee, WI 53172	Hoes to 6 cu. yd.
J.I. Case Drott & Poclain Box 1087 Wausau, WI 54401	Hoes to 17 cu. yd. Shovels to 22 cu. yd.
Caterpillar Tractor Co. 100 N.E. Adams Street Peoria, Illinois 61629	Hoes to 3.25 cu. yd. Shovels to 5 cu. yd.
Demag Corporation 1100 Jorie Blvd. Oakbrook, Illinois 60521	Hoes to 21 cu. yd. Shovels to 28 cu. yd.
FMC Cable Crane & Excavator Div. 1201 6th Street, S.W. Cedar Rapids, Iowa 52406	Hoes to 6 cu. yd.
Harnischfeger Corporation P.O. Box 554 Milwaukee, WI 53201	Hoes to 10 cu. yd. Shovels to 18 cu. yd.
Hein-Werner Corporation Construction Equipment Div. Waukesha, Wisconsin 53186	Hoes to 4 cu. yd.
Hitachi Construction Machinery Co., Ltd. 2-10, 1-Chome Uchikanda, Chiyoda-Ku Tokyo, 101, Japan	Hoes to 15.5 cu. yd. Shovels to 16 cu. yd.
Insley Manufacturing AMCA International Corporation 2118 N. Gale P.O. Box 11308 Indianapolis, Indiana 46201	Hoes to 4.5 cu. yd.
Koehring Crane & Excavator Group P.O. Box 2060 Milwaukee, WI 53202	Hoes to 16 cu. yd. Shovels to 15 cu. yd.

Table 6.1 (Continued)

HYDRAULIC MACHINE MANUFACTURERS

Liebherr-America, Inc. 4100 Chestnut Avenue Newport News, Virginia 23605	Hoes to 7.5 cu. yd. Shovels to 14 cu. yd.
Northwest Engineering Co 201 W. Walnut Street Green Bay, WI 54305	Hoes to 13 cu. yd. Shovels to 7 cu. yd.
O & K (Orenstein & Koppel, Inc.) Box 479 Clifton, New Jersey 07015	Hoes to 21 cu. yd. Shovels to 40 cu. yd.
Warner & Swasey Company Construction Equipment Solon, Ohio 44139	Hoes to 6 cu. yd.

BASIC MACHINE OPERATION

The hydraulic machines are used primarily as excavating and loading machines. The basic operating cycle consists of a cutting pass through the bank, loaded swing to the discharge area, and dump empty swing back to the digging face.

- Crawlers with medium speed capabilities are used to move about the pit and position the machine at the digging face.
- Hoist (vertical) and crowd (horizontal) digging motions are achieved by actuating a pivoted two section boom with hydraulic cylinders. This combination permits variation of the path of the bucket through the bank in either very flat or vertical radial curves.
- As a hoe, the machine digs downward and back towards itself. As a shovel, it digs upward and away from itself.
- The bucket is wristed (rotated) to orientate the teeth for the desired penetration of the face.
- When operating as a hoe, the bucket wristing is also used to carry or to dump the load by rolling the bucket.
- The shovel dipper is most commonly dumped by rotating the forward section of the bucket with respect to the back of the bucket so that the load falls out in an action similar to a clam shell.
- The machine upperworks rotates on the crawler base so that the machine can swing to the right or left from its digging face and discharge the material from the bucket into a hopper, truck or stockpile.

APPLICATIONS

Hydraulic machines are employed in overburden removal, coal/ore loading or, in the smaller sizes, for utility work generally related to mine drainage systems.

The hydraulic shovel is primarily an excavating and loading device. While it can swing and/or propel to transport material short distances, it is used almost exclusively to load trucks or, in some cases, hoppers/crushers.

Hoes have similar uses to shovels. However, their below grade digging capability makes them particularly suited to tasks such as trenching or excavating under water. Hoes are utilized in mining when floor conditions warrant keeping machines off the bottom of the pit.

GENERAL CHARACTERISTICS

These are essentially single purpose machines and, in those sizes used in mining, have been developed within a relatively short span of years. While there are differences in detail design features, the overall configurations are very similar. The powered functions and the actuating means are (see Figure 6.3):

- Crawler propel, hydraulic motors
- Positioning of lower boom section, hydraulic cylinders
- Positioning of upper boom section, hydraulic cylinders
- Positioning of bucket (wristing), hydraulic cylinders

Photograph 6.2 Demag hydraulic hoe dumping overburden into a rear dump truck

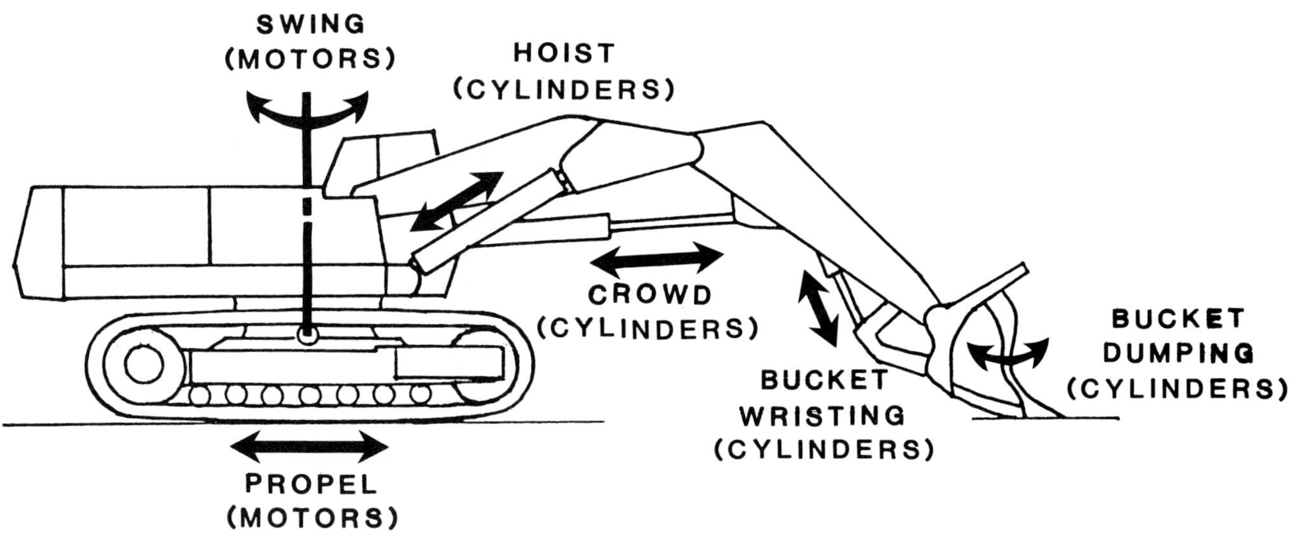

Figure 6.3 Hydraulic Shovel Powered Functions

- Shovel bucket dumping, hydraulic cylinders

- Upper works rotation, hydraulic motors

Note that these are relatively complex machines in comparison with dozers, scrapers and/or trucks. They have an unusual degree of digging flexibility because of the control of the digging path that is possible. Their general characteristics can be summarized as follows:

- Hydraulic hoes
  - Maximum digging heights from 20 to 55 ft.
  - Maximum digging depths from 20 to 55 ft.
  - Maximum dumping heights to 40 ft.
  - Maximum reach at dumping from 20 to 37 ft.

- Hydraulic shovels
  - Maximum digging heights from 29 to 48 ft.
  - Maximum digging depths from 7 to 18 ft.
  - Maximum dumping heights to 37 ft.
  - Maximum reach at dumping from 15 to 32 ft.

- Medium overall mobility with diesel moderate propel speeds (about 1.5 mph)

- Excellent positioning capabilities (spin turns) with independent track drives

- High gradeability

- Moderate ground pressure (13 to 28 psi)

- Good ground clearances (1.7 to 3.3 ft.)

- Good stability

- High swing speeds (2.5 to 5 rpm)

- High breakout forces through wristing

- Digging forces are not entirely limited by machine stability

- Maximum versatility in bucket orientation for face penetration

- The bucket is narrow

- Digging cycle utilizes upper works rotation rather than propel, reducing undercarriage costs and minimizing damage to floor

- Smooth, low shock dumping because dump action can be controlled

- High bucket fill factors
- Compact size and low weight
- Can work in close quarters
- Loads directly into trucks with good placement of load
- Moderate dust generation
- No support equipment is required; can keep a clean floor
- One man operation with low operator fatigue; moderate operator skill required
- Operator position is safe; visibility is good for digging, but may be marginal for loading into larger trucks
- Lower fuel consumption than a comparable front-end loader
- Relatively high hydraulic maintenance
- Typical climatic temperature range of − 30 to + 120 degrees F.
- Medium service life of 20,000 to 30,000 hours
- High initial cost
- Short field assembly cycle

Graphs 6.1 and 6.2 indicate that for the various nominal bucket sizes, the hoes are heavier and have more horsepower. This reflects the practice of using much larger buckets on the shovel versions, with a corresponding reduction in range (reach, etc.) to fit within the machine stability limits. Graphs 6.3 and 6.4 show that there are considerable differences between individual models and that as the size increases the weight per horsepower is very similar for hoes and shovels. Again, the horsepower per cubic yard of bucket capacity, as noted above, is greater for the hoes.

Photograph 6.3 Demag H241 hydraulic shovel dumping with clam shell type action

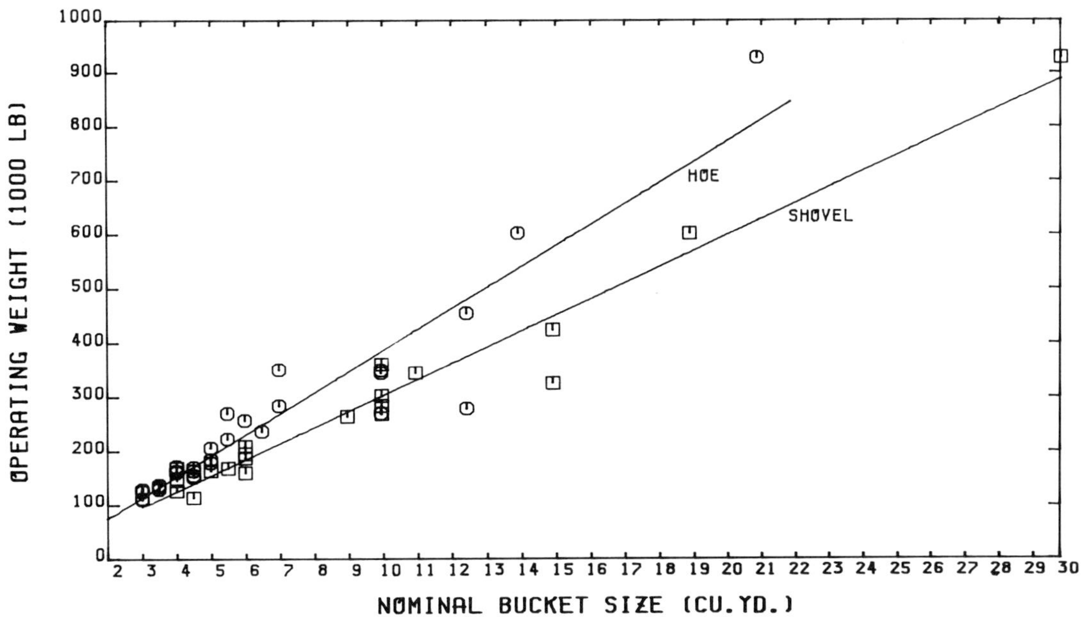

Graph 6.1 Operating weight/nominal bucket size

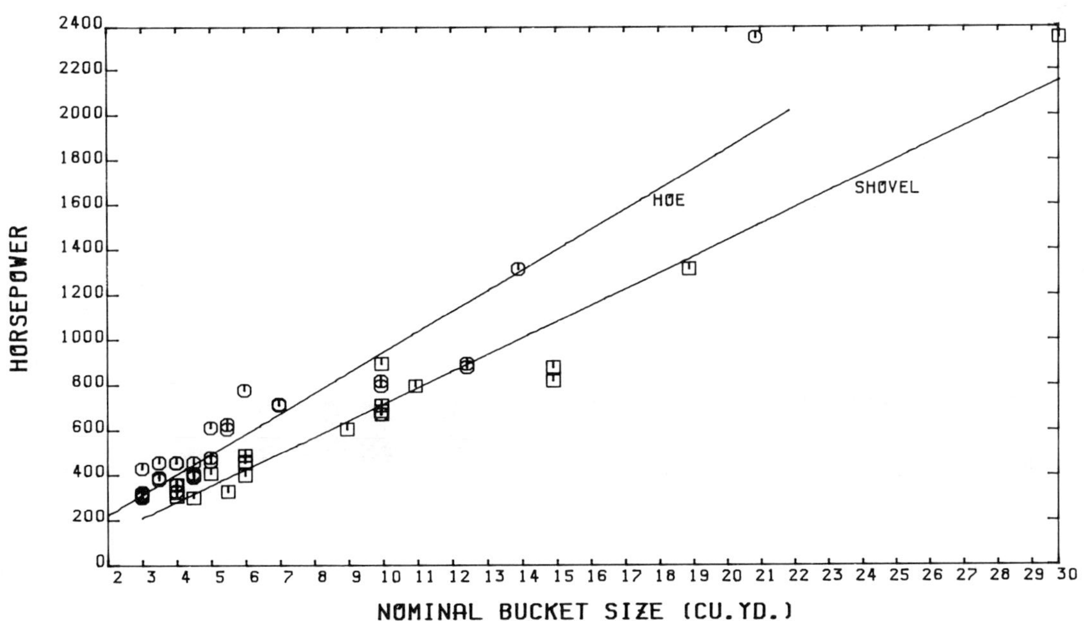

Graph 6.2 HP/nominal bucket size

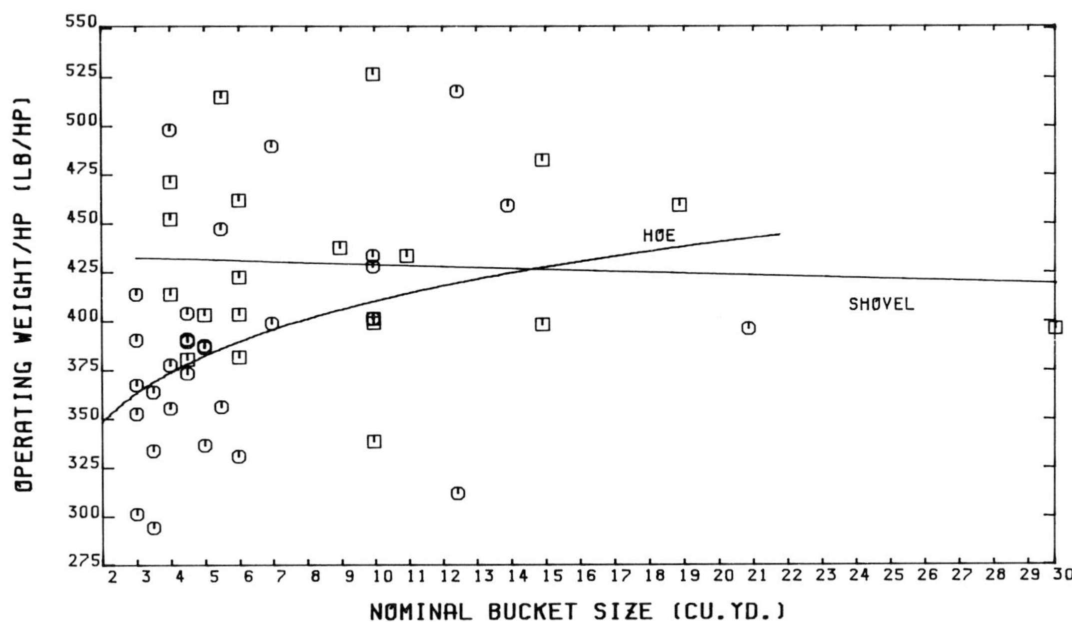

Graph 6.3 Operating weight per HP/nominal bucket size

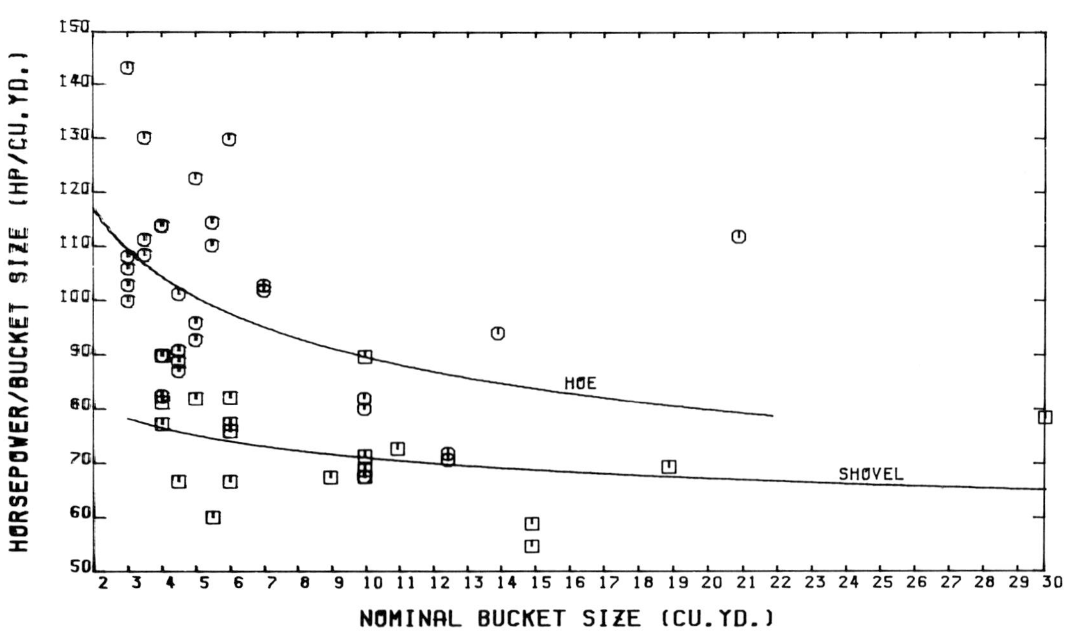

Graph 6.4 HP per nominal bucket size/nominal bucket size

Photograph 6.4 Drott hydraulic shovel digging in rock

The machine statistics just noted are distorted by two new large models: the Demag H241, 14–24 cubic yards, and the O&K RH-300, 23–40 cubic yards. These units might properly be considered as separate size categories. The O&K machine is roughly 50 % heavier and twice as powerful as the Demag, which has a similar relationship to the next largest machine.

Crowd action is produced by hydraulic cylinders pivoting the upper and lower boom sections with respect to the revolving frame; this, in turn, extends the bucket. The hoisting function is similar and, together with the crowd, produces any shape of cutting profile desired. Rotation of the upper works in either direction positions the raised bucket above the truck/hopper for discharge. The bucket is lowered and positioned during the return swing so as to be ready for the next cut.

The hydraulic machines have bucket wristing (rotation) capabilities similar to the front-end loader, provided by cylinders mounted on the upper boom section. The wristing can provide a very high breakout force limited by cylinder size and structural strength of the bucket. If the bucket is supported by the ground, these forces are not limited by machine stability. Therefore, high forces roughly perpendicular to the teeth can be generated with a relatively light machine. (See Figure 6.4)

In the smaller sizes, the basic machine is designed to permit the interchanging of front end equipment to either the hoe or shovel configuration, dependent on the application requirement. With increased size and machine cost, this feature has, to some extent, been de-emphasized to optimize the performance in one or the other of these configurations. Few of the larger machines are ever converted after being placed in service.

The shovel front-end equipment is short coupled, which means that it has less range than a similar hoe front-end. The basic design generally does not provide a machine with adequate dumping height if a single piece bucket is used — one similar to the hoes which must be rotated to dump. Consequently, most units employ a rather shallow bottom dump bucket which is of two piece construction. The back of the bucket and the bottom/sides are

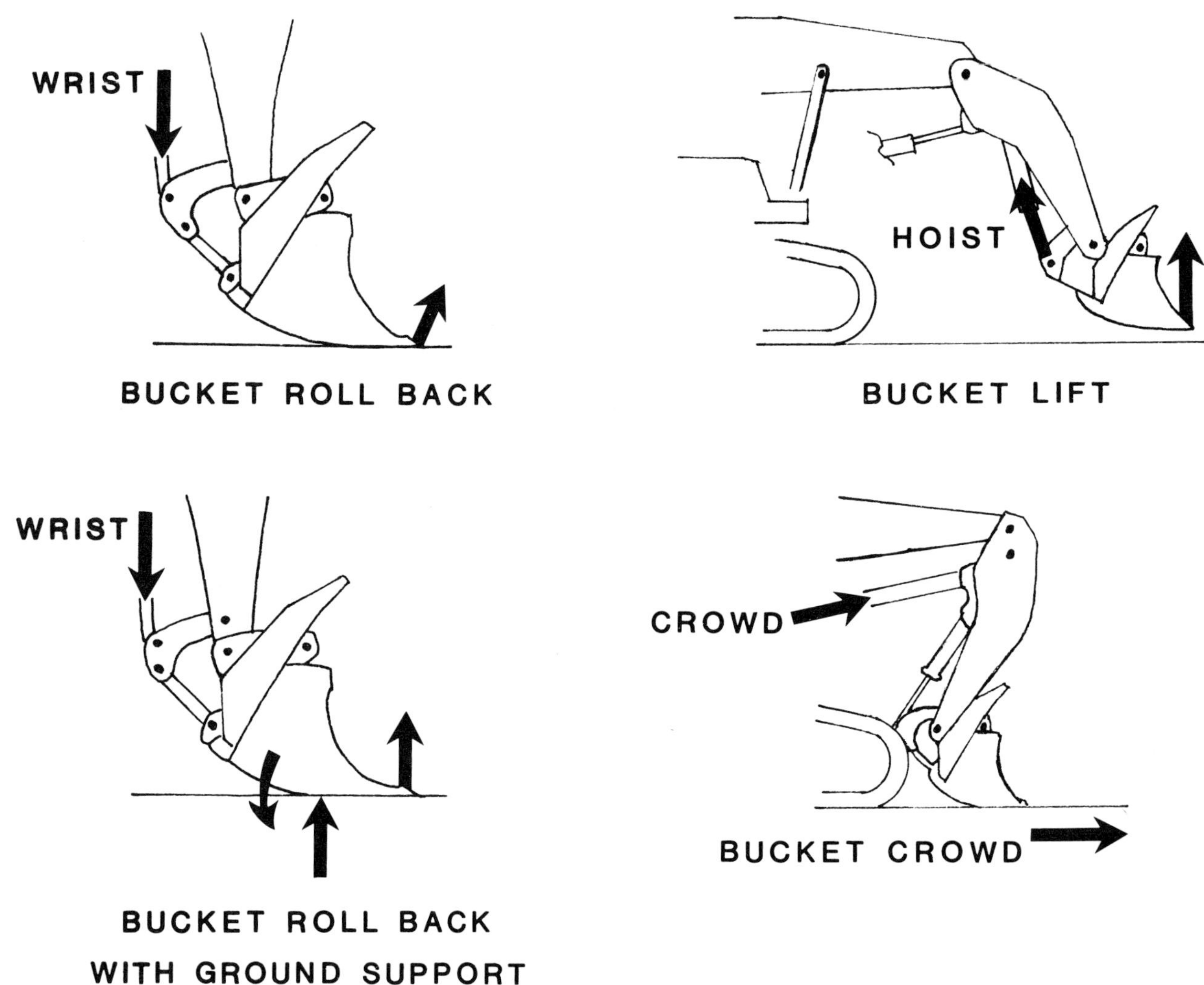

Figure 6.4 Hydraulic Shovel Digging Forces

hinged together. Hydraulic cylinders, located in the back, pivot the front section forward so that the dirt is discharged in a manner similar to a clam shell bucket. By controlling the size of the opening, the rate of discharge and its severity can be regulated. The positive control of the bucket segments is adequate to permit picking up rocks and debris with a clamping action if necessary. These buckets require an additional hydraulic circuit, are heavier, dump cleaner and have a faster dumping cycle, than a one piece shovel bucket.

The hydraulic hoe dumps the material by rotation of the bucket until gravity causes discharge. There is good control of the bucket so that the discharge severity can be regulated to some degree. The bucket is a one piece design which is structurally strong. It tends to be narrow and deep with a high bottom curvature to aid the hydraulic prying action. (See Figure 6.5)

The bucket does not have to be as wide as that on the loader which has to protect the tires. The nar-

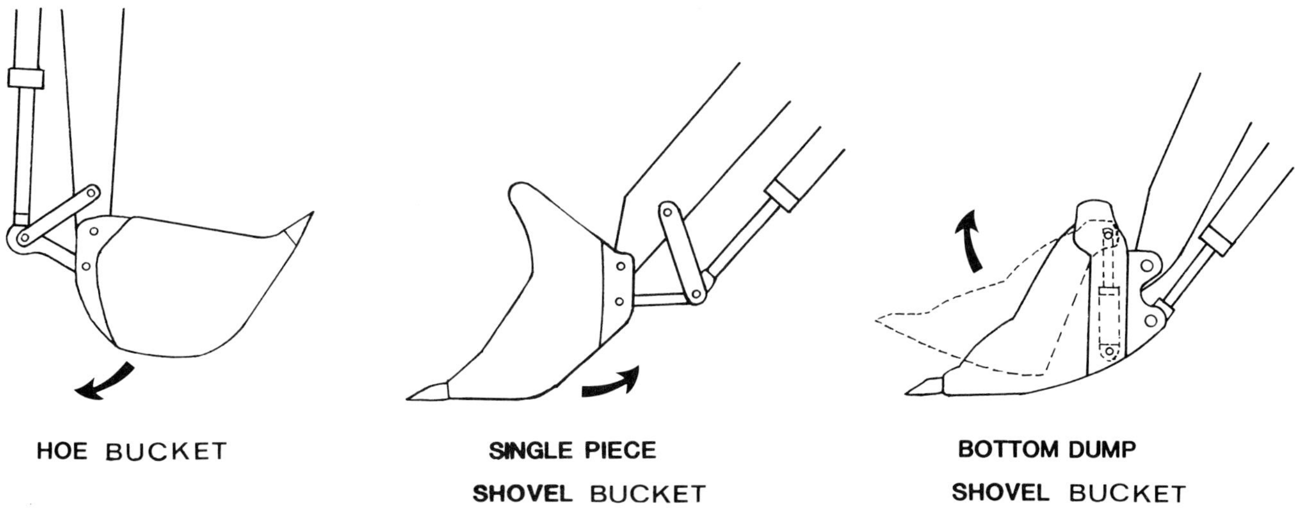

Figure 6.5 Hydraulic Shovel and Hoe Bucket Configurations

rower bucket significantly increases both the structural strength and the applied forces per unit of width. In addition, the bottom of the bucket can be given more curvature to increase the leveraging action associated with wristing, further magnifying the effectiveness of the forces.

As with other excavators, bucket size is varied to match the unit weight of the material to be handled. The front-end design can also be extended to provide greater digging and dumping height (long range front-end) to meet special operational requirements. Stability considerations then necessitate that the total weight in the bucket be reduced, which means a proportional reduction in bucket capacity.

Cylinders can be sized and the geometry of the front-end equipment selected to produce high digging forces. Graphs 6.5 and 6.6 show the maximum crowd and breakout forces for shovels and hoes as a function of nominal bucket size. The shovels have slightly higher maximum forces. Digging forces vary with bucket elevation in the face and, because of the action lines of the forces, tend to increase effective machine weight at the high positions. The machine weight can also be used to force the bucket into the ground by applying down pressure.

The hydraulic machines are intermittently propelled towards the face when the unit is positioned for the dig-dump cycle. The crawler mounting provides a rugged working base impacted relatively little by floor conditions. Maneuvering requirements are minimal so that the floor is not torn up. Track shoe size can be varied to meet ground flotation requirements. There is sufficient machine reach in the case of the shovel to permit keeping the tracks out of the debris at the base of the digging face.

The high power on the machine provided for the digging functions is also available for the intermittent propelling by diverting the oil to hydraulic motors mounted on each track. Sufficient power is available for unusually high tractive forces and gradeability. Crawler design has componentry similar to the dozer but with shorter, multiple grousers. Independent track drives permit counterrotation of opposite tracks and spin turns about the axis of the machine — the optimum in machine steering capability.

Since the machinery mounted on the revolving frame consists only of the diesel engine(s) and totally enclosed hydraulic componentry, the enclosure has a low profile. A separate engine compartment with inspection and service hatches is provided, while fuel and hydraulic tanks, valves and swing drive modules, etc. are conveniently open deck mounted. This design maximizes the operator's rearward visibility.

Most operating experience is with machines 6 cubic yards and smaller; the larger units have not been in operation a sufficient number of years to fully establish their capability and/or limitations. This is particularly true in surface mining applications in the United States, where broad use of this type of excavator has occurred only within approximately the last five years.

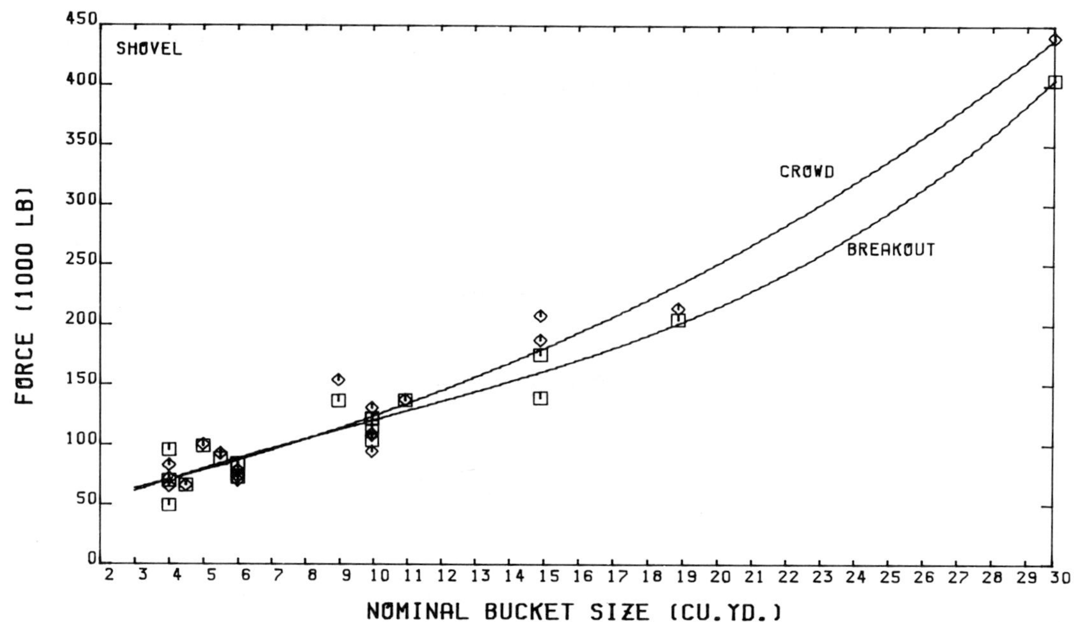

Graph 6.5 Breakout & crowd force/nominal bucket size (shovels)

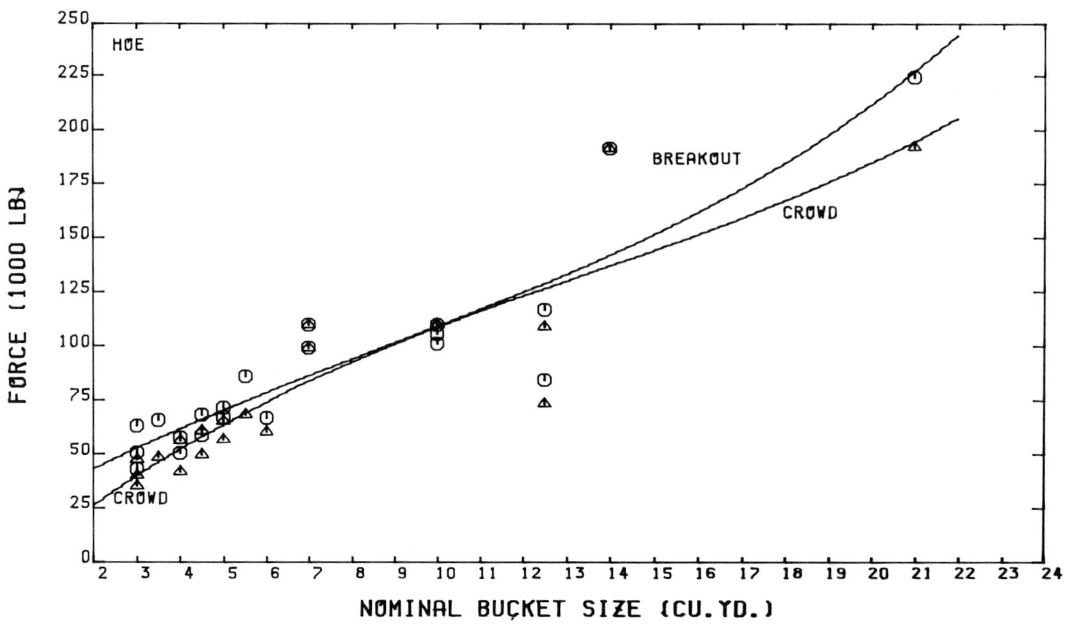

Graph 6.6 Breakout & crowd force/nominal bucket size (hoes)

Photograph 6.5 P&H 1200 hydraulic shovel loading coal into a rear dump truck

OPERATIONAL PRACTICES

Figure 6.6 illustrates the differences in the shovel and hoe digging profiles. While the shovel has the capability to dig below itself, this feature is only used during ramping down to establish a lower operating level. Primary use of the shovel is digging above its floor. Similarly, the hoe, while it can excavate above its floor, is seldom operated in this manner.

Typical maneuvering and digging patterns for a shovel loading a truck are illustrated in Figures 6.7 and 6.8. The best cycle time in each operating plan results when the swing angle can be kept to a minimum. Single side or alternate side truck loading is possible, depending on the mine plan. Loading alternately to trucks on either side will, of course, mean less idle time for the excavator. The machine is repositioned periodically to optimize the digging capability. The unit can effectively control and maintain a clean level floor so that no auxiliary equipment is required.

Hydraulic machines' digging characteristics, with their fine control of bucket position and orientation, are different from cable operated units. Figure 6.9 illustrates some of the unique capabilities of a hydraulic shovel. The wristing of the bucket permits wedging boulders out of the digging face. The bucket can enter the face at any elevation with the teeth properly positioned to take advantage of any horizontal cracks or cleavage planes, and the face can be removed in layers starting from the top and working down. Its level floor crowd facilitates picking up rocks and boulders at the toe of the face. The face profile can be shaped in any manner compatable with the nature of the formation. The fine control facilitates selective mining of materials and thin seam removal.

The down pressure available on the machine can be used to assist the penetration of the teeth into the floor or face. Since the cylinders provide power in

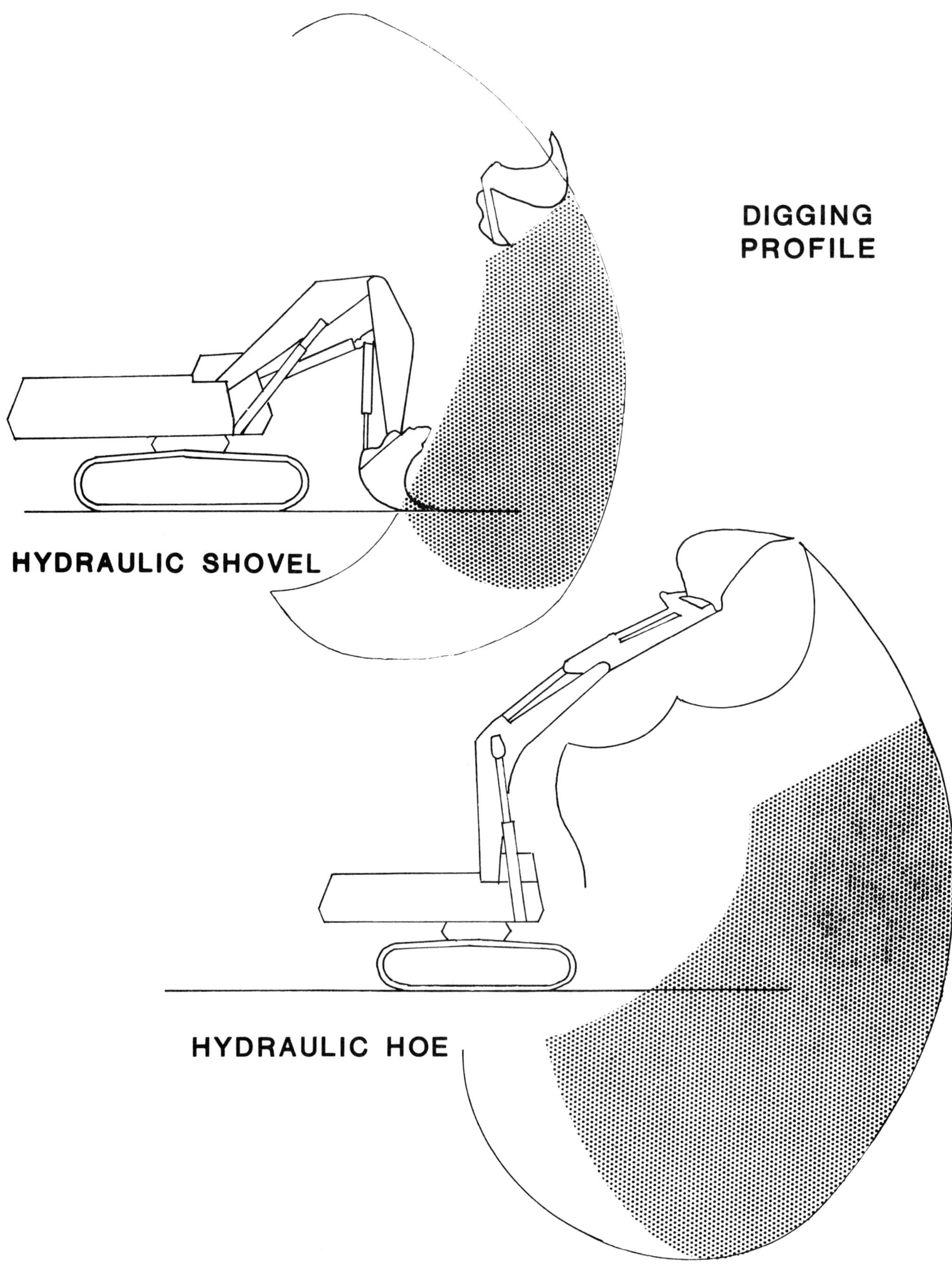

Figure 6.6 Hydraulic Shovel and Hoe Digging Profiles

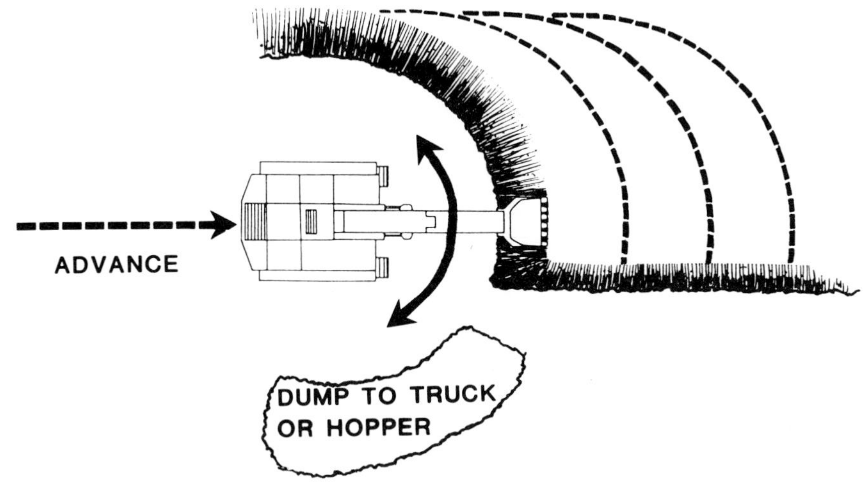

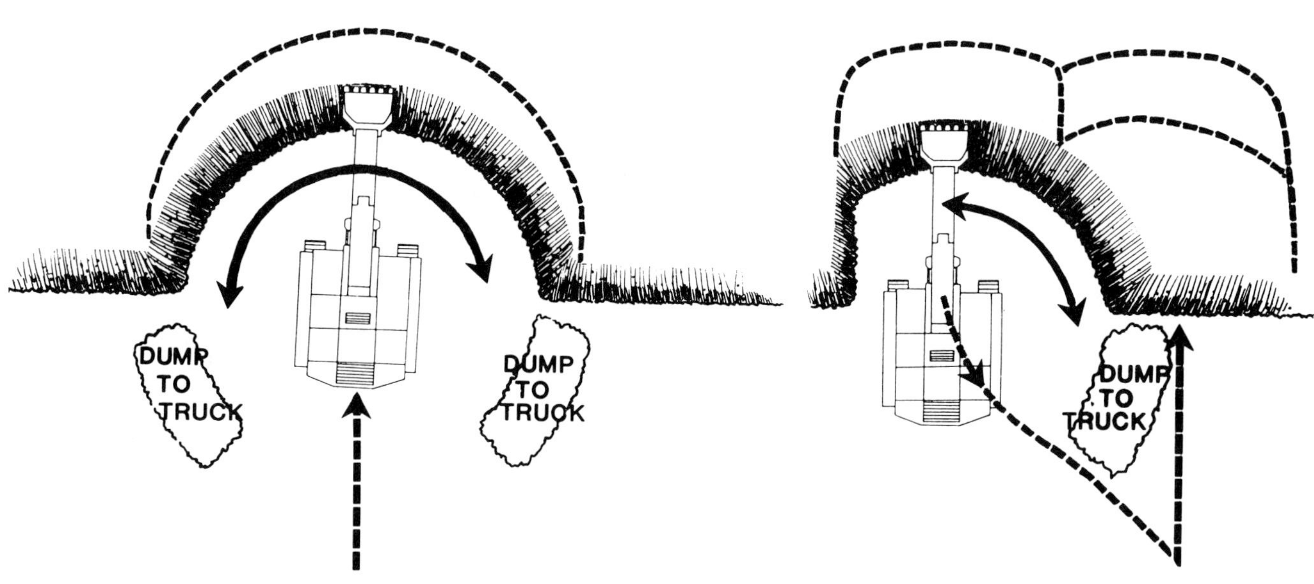

Figure 6.7 Typical Hydraulic Shovel Production Cycle

both directions, the bucket can be dug into the ground and the front end used to pull or push the machine under severe traction conditions.

The reader should also consult the following chapter on electric shovels. Its section on operating practices has a more lengthy discussion on some of the operating procedures for shovels. While electric and hydraulic machines do have design and operating differences, many of these observations will apply to both.

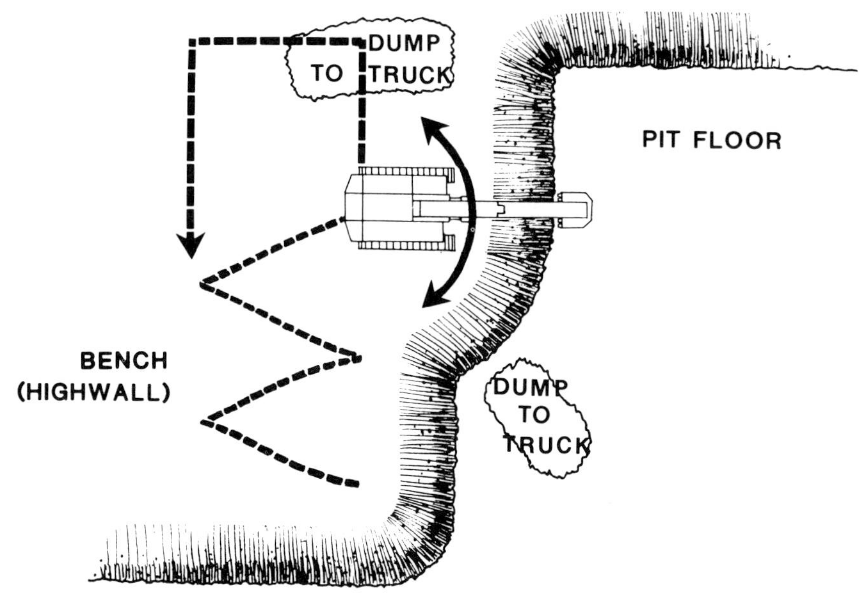

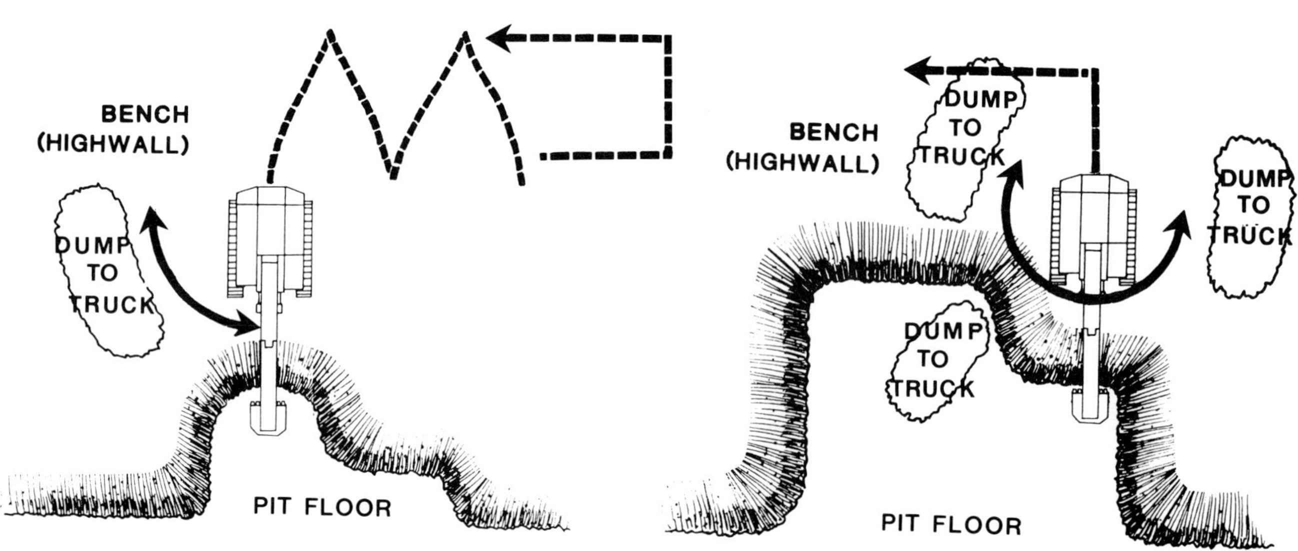

Figure 6.8 Typical Hydraulic Hoe Production Cycle

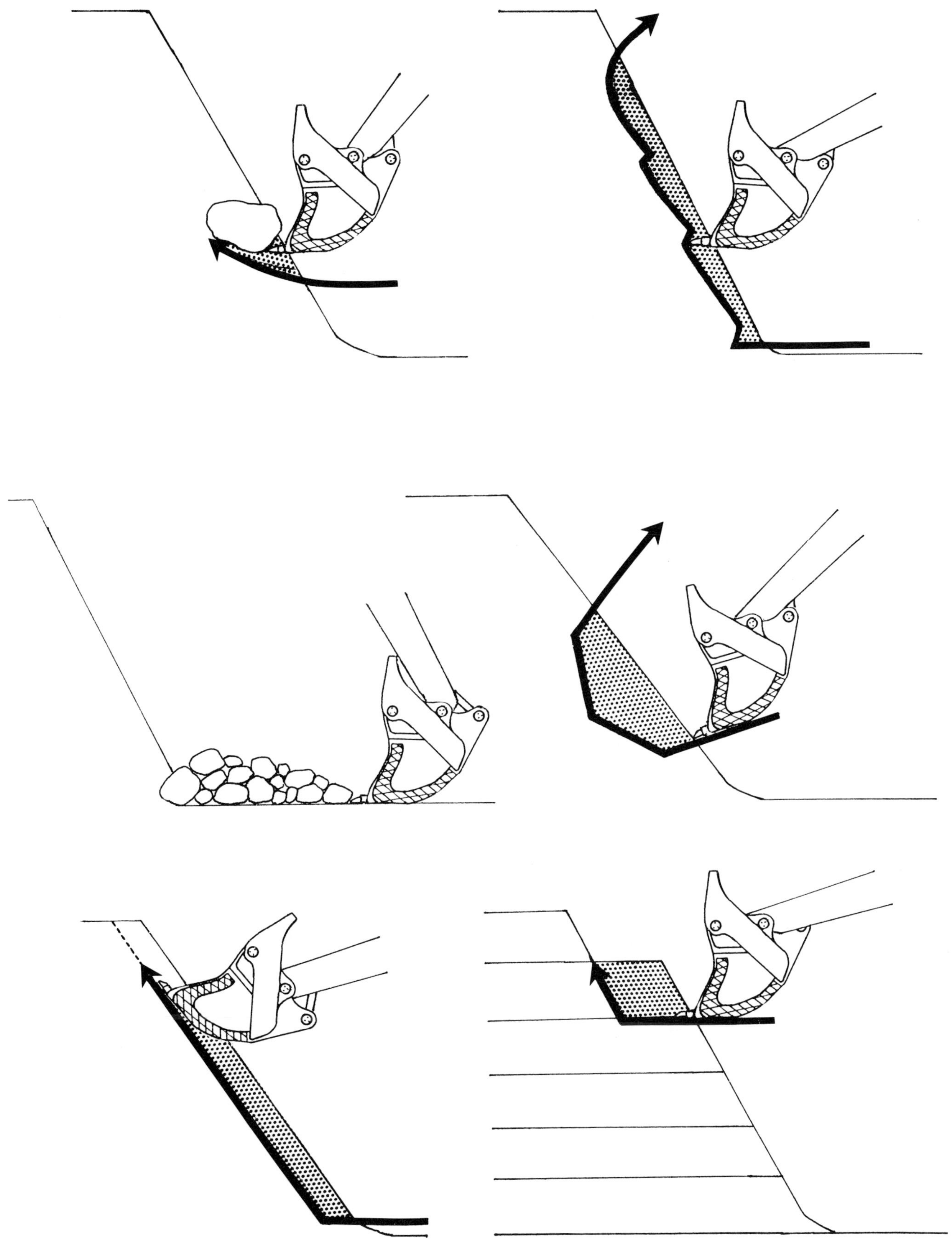

Figure 6.9 Hydraulic Shovel Face Digging Characteristics

DESIGN FEATURES

See Figure 6.10 for basic hydraulic shovel nomenclature.

General boom configurations are quite similar on all machines but cylinder arrangements differ with single or twin units mounted to provide the desired positioning speeds and forces.

Parallel crowd and level digging can be provided by parallelogram linkages, or special linkages and a master cylinder. For smaller models, multiple pin connections may be provided between boom sections to permit rapid changing of front-end geometry.

Single or twin diesel engines are the typical power plants driving multiple hydraulic pumps through a gear box. They are operated in their optimum speed range for high efficiency and low fuel consumption. Low noise level is accomplished through compartment sound proofing, and ready access to the engine compartment is possible from the deck. Reversible fans are normally optional for winter/summer temperature control. Imported air cooled engines had some heating problems caused

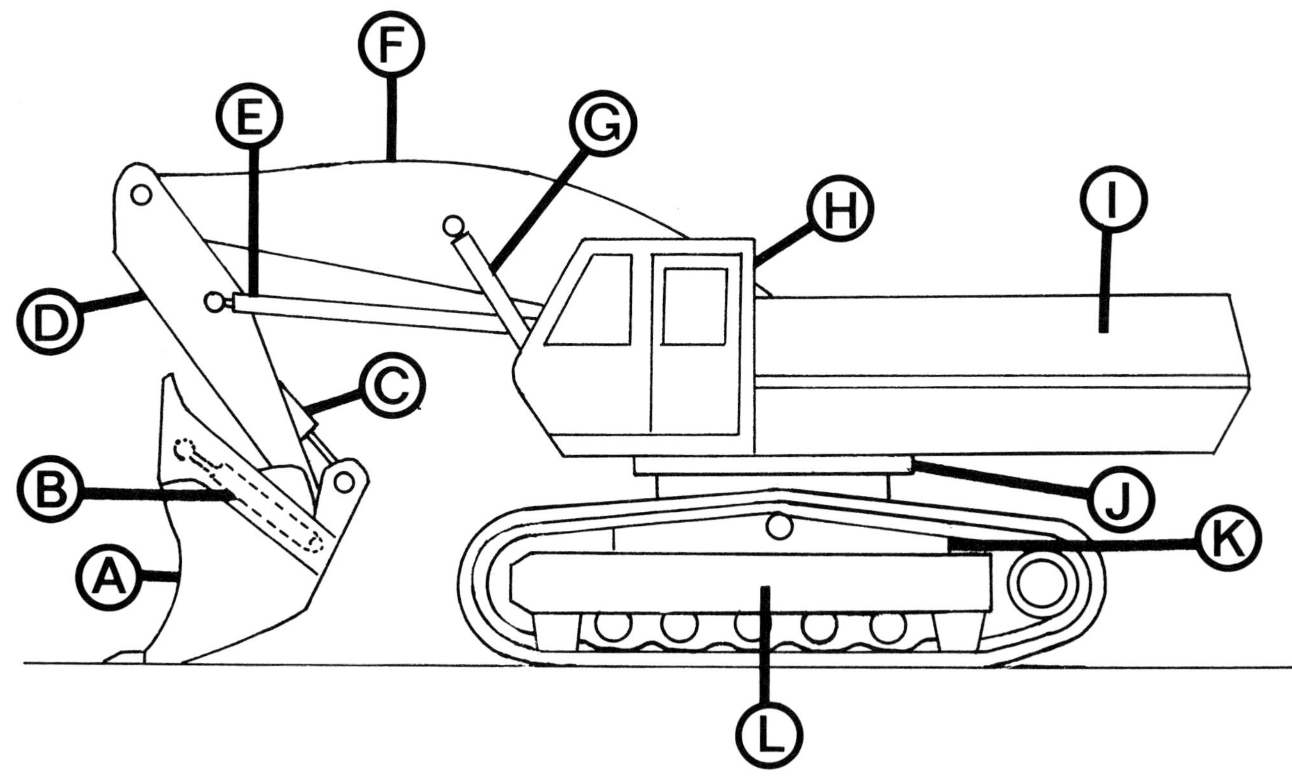

A	Bottom Dump Bucket	G	Boom Cylinder
B	Dump Cylinders	H	Cab
C	Bucket Cylinders	I	Upper Structure
D	Arm (stick)	J	Swing Bearing
E	Arm Cylinder	K	Undercarriage
F	Boom	L	Crawler Frame

Figure 6.10 Hydraulic Shovel Nomenclature

by dust and dirt accumulation so most hydraulic excavators are equipped with water cooled U.S. engines. Single motor A.C. drives are optional on some of the larger machines.

The engines, fuel and hydraulic oil tanks are mounted on the main deck together with the swing drive modules, hydraulic valves and the swivel. The valves and piping take up considerable space and can pose problems with respect to accessibility for maintenance. Multiple oil coolers may be mounted independently on the deck or combined with oversized engine radiators. Figures 6.11 and 6.12 are typical deck plans for single and twin engine installations. Note the engines are located as far to the rear as practical to act as counterweights.

Frames and booms are generally welded box section designs. A counterweight is built in or attached to the rear of the revolving frame to improve machine stability.

Oil for the hydraulic track drive motors is taken down through the center of rotation in a swivel. Because of the number of lines involved, the large oil volumes and the high pressures, the swivel is a relatively large complex unit.

Crawler designs can be summarized as follows: (See Figure 6.13)

- Tractor type roller, link and sprocket design
- Independent motor-planetary gear reductions mounted on the track units (two motors on some sizes)
- Brakes incorporated in drive train
- Track shoe widths matched to flotation requirements; light triple grouser shoes are common.
- Medium speed capability
- Adjustable width track are available on some small machines. Width is set hydraulically; they have removable side frames.

The swing drives are modular in construction, consisting of: a hydraulic motor, brake and gear reduction driving a pinion on the end of a shaft which extends through the deck to contact the swing gear on the lower truck frame. Small and medium size units have a supersized enclosed ball or

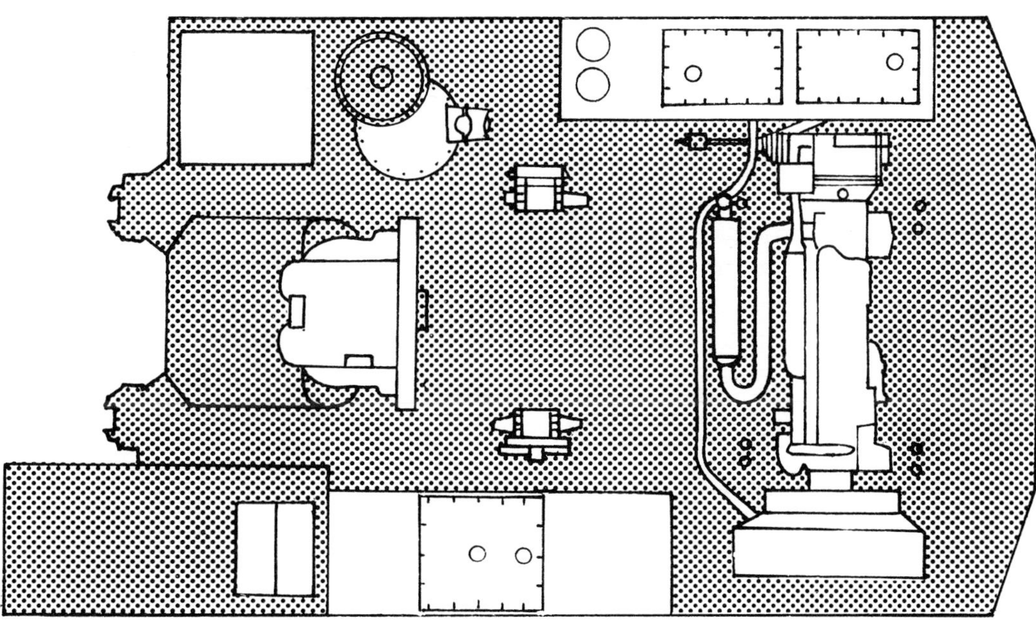

Figure 6.11 Typical Hydraulic Excavator Deck Plan — Single Engine Configuration

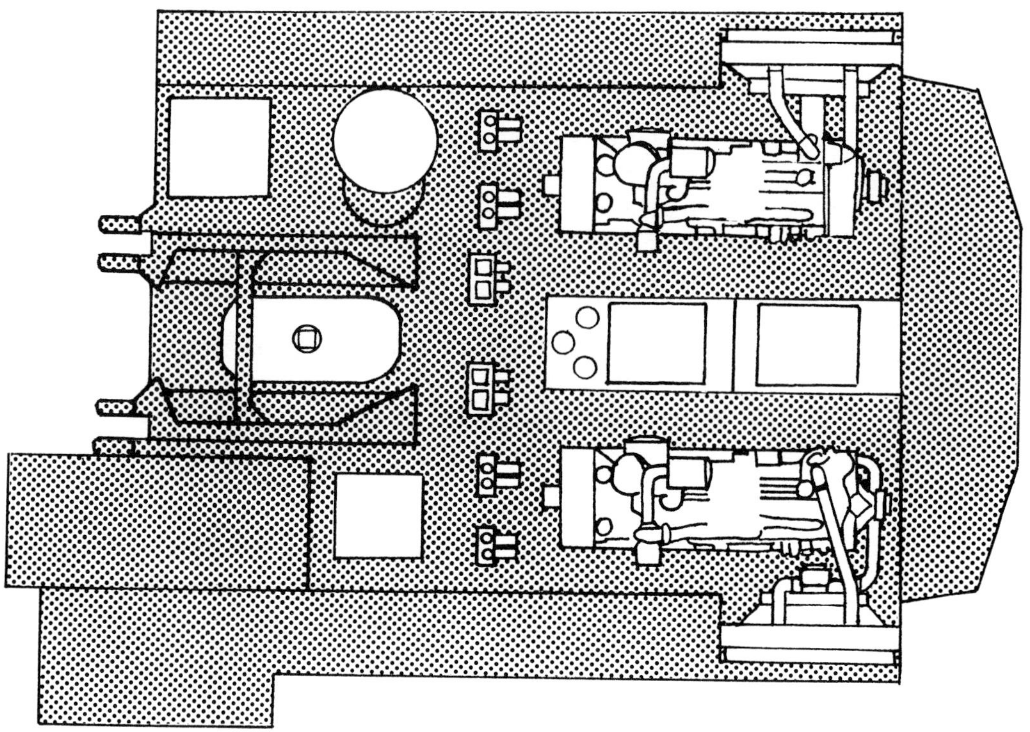

Figure 6.12 Typical Hydraulic Excavator Deck Plan — Twin Engine Configuration

roller swing bearing for upper works rotation. It is a low friction, low maintenance unit which effectively transmits all of the vertical and horizontal forces from the revolving frame to the truck frame and crawlers.

The key to machine operating efficiency is the hydraulic system. These have grown rapidly in sophistication as efforts have been made to make the systems responsive to the operational power demands of each function while utilizing the smallest power plant (and minimizing fuel consumption). Available componentry and system configurations have been improved substantially in recent years. Common features can be summarized as follows:

- Compact, low inertia, high power density systems; gear or vane motors are .4 to .6 lb/HP, piston motors are .2 to .4 lb/HP.

- Basic components (3000 to 5500 psi systems) are pumps, motors, cylinders, valves, tanks, hose and tubes, filters, accumulators, swivels, oil coolers.

- Cylinders for all digging motions. They feature end of stroke cushioning, self aligning bushings with special attention to dirt seals, cylinder thrust angles as close to 90 degrees as possible.

- Pump types are gear (multi-section) with fixed displacement, vane with fixed displacement, axial piston with variable or fixed displacement.

- Motor types (matched to pump) are gear with fixed displacement, axial piston with fixed, or variable displacement, or radial piston with fixed displacement. They typically have 60 to 90% efficiency.

- Filters are 10 to 25 microns; trend is 5 to 10 micron.

- Tanks are pressurized to prevent entrance of contaminants, often incorporate a strainer/filter, and are baffled to minimize oil movement. Tank size is normally three times volume per minute flow (sufficient volume to provide some cooling capability).

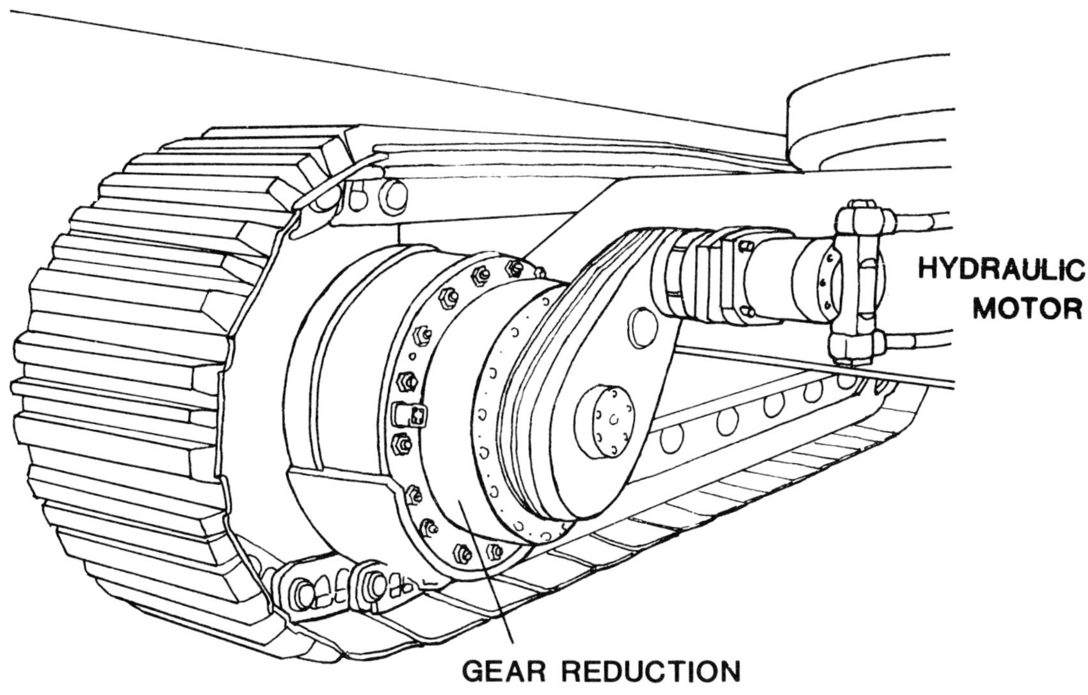

Figure 6.13 Typical Hydraulic Excavator Track Drive

- Hydrostatic transmissions (swing and propel rotary drives) are generally of two types. The variable pump/fixed motor transmission provides constant torque at maximum pressure over full speed range. Speed and direction are controlled by pump; motor speed is limited by maximum pump displacement. The system is capable of full power only at maximum pump displacement. The variable pump/variable motor transmission provides constant torque and increasing power up to maximum pump displacement. It permits an extended range of motor speeds with full power and elevated speeds as motor displacement and torque are reduced.

- Constant horsepower units utilize self regulating variable flow axial piston hydraulic pumps; a pressure sensing piston alters pump output volume. Under heavy loads, the system pressure increases, flow is reduced (speed reduced). Under light loads, the pressure is low and flow is high. Lower total horsepower is required and higher overall machine productivity is achieved.

- Multiple circuits allow the simultaneous operation of all functions; a minimum of two pumps are needed. Pumps are linked to a pressure sensing control piston, all can be utilized for a single powered function. Individual pump regulation and summation responding to pressure changes is possible; when summation is selected, the pumps work at equal presssure and flow. Twin engines introduce a twin self-regulating system; failure of one system still permits operation at half speed; summation doubles operational speeds.

- Swing motors are an independent circuit. Occasionally the system allows free swing in neutral with brakes for deceleration. This reduces oil heating generated by reversing (plugging) with hydraulics.

- Relief valves are included in each individual circuit with the circuits designed to minimize the flow over the valves to reduce the heat generated.

- Power assist controls are generally required; larger valves need high actuating forces.

- Oil temperatures typically run from 120 degrees to 140 degrees F. Oil coolers are used to limit maximum oil temperature to 180 degrees F.

- Hydraulic fluids are petroleum oils with rust and oxidation inhibitors, often fire resistant.

- Higher pressures allow smaller sized components.

- Hoses and piping must be located to minimize flexing and the possibility of external damage.

- Extreme care is required to attach hose and tubing securely to the structure and prevent chafing and rubbing. Bearing blocks are required at every joint or junction.

- Circuitry must minimize high transit pressure peaks in the system.

- System rated pressures are normally below theoretical component ratings.

- Controls are dual joystick for digging functions and separate lever or foot controls for propel. They can be two speed (single or double pumps).

Complex, high pressure hydraulic systems introduce a unique set of service and maintenancee requirements which must be recognized to keep the machine operational. The newness and complexity of the technology has posed problems with finding and/or training qualified service personnel. This has encouraged component exchange programs with machine distributors or specialized service centers.

The number of hoses, tubes and pipes, and the related joints and fittings, presents a special servicing situation. (See Photograph 6.6) Piping on the boom can involve as many as 4 to 10 cylinders and three pivoting joints. Steel piping (tubing) is preferred where possible to reduce cost and improve heat dis-

Photograph 6.6 Hydraulic piping on the boom of a large hydraulic shovel

sipation. However, hose is required for all flexing joints on the boom. Multiple swivels, couplings and connectors are necessary in each circuit and each must be kept tight to prevent leakage and/or entrance of dirt. Tube assembly failures are typically caused by poor forming, defective tubing, forced installation, and inadequate welding. Transient peak pressures, excessive flexing and service damage will take their toll.

Component life is dependent on keeping the system as free of contaminants as possible. Adequate filtering is commonly provided but the system must be kept tight, filters changed regularly, and special care must be exercised not to introduce any dirt during servicing and maintenance work. Oil should be replaced every 5000 hours or after any system failure; there will be some oil deterioration with high operating temperatures.

Special efforts have been made to minimize potential maintenance downtime and spare parts inventory through component part standardization. With proper design, hose sizes and lengths, seals, fittings, filters, etc., can be made of a limited number of common sizes. In some cases, cylinders, pumps and motors have been partially standardized.

SELECTION CONSIDERATIONS

A hydraulic excavator might be selected where digging conditions are considered to be in an intermediate range of severity. Consolidated formations should be drilled and blasted to reduce machine abuse and to maximize productivity. The typical application is to load trucks or a mobile hopper with occasional moves of the machine to different digging faces within the pit area, if required for blending. The diesel drives, the medium speed propel, and the high gradeability, make such machine moves practical.

The hoe vs. shovel configuration selection is based on site conditions and the mine plan. If conditions permit, the shovel appears to be capable of higher production under more difficult digging situations. Graphs 6.7 and 6.8 show the typical cut and dump heights for shovels and hoes. Specific units should be checked (see machine specifications at end of section) against the trucks and/or hoppers to be employed to confirm range limits and proper size matching.

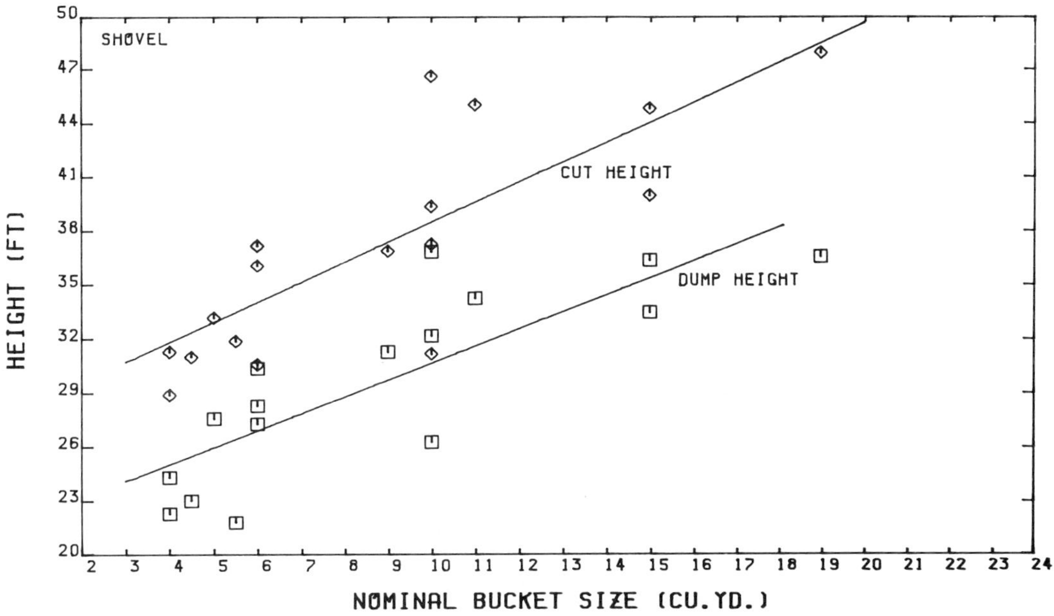

Graph 6.7 Dump height & cut height/nominal bucket size (shovels)

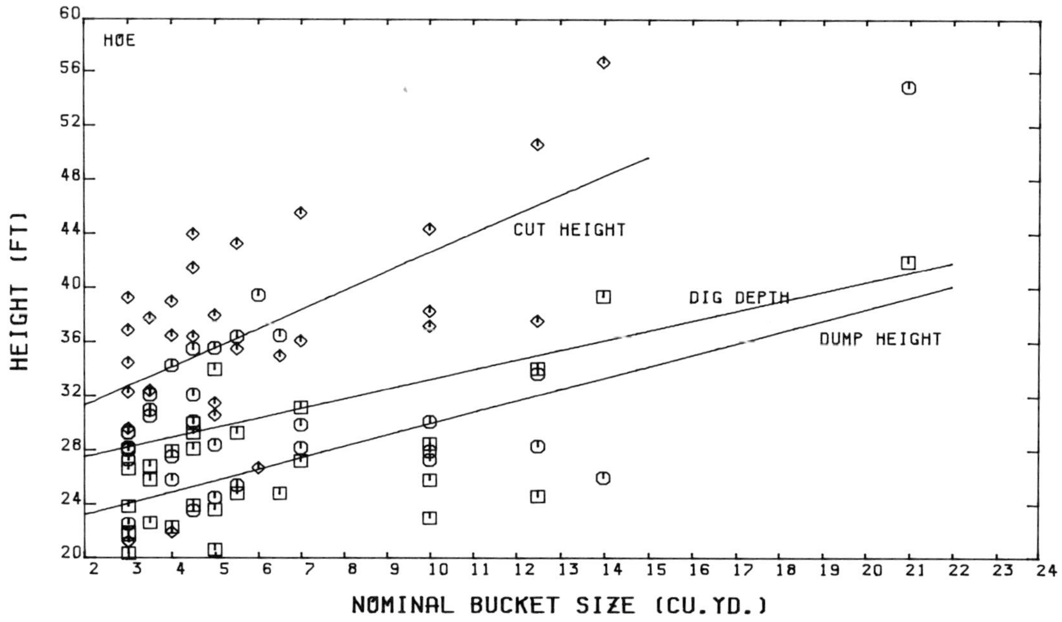

Graph 6.8 Dump height, cut height & dig depth/nominal bucket size (hoes)

As noted earlier, most of the larger machines have been introduced in recent years and have, therefore, accumulated relatively limited operating experience. Good planning suggests that the extent, availability, and the response time of dealer support and training be verified.

Production calculations and cost estimates are considered a separate subject, too extensive to be comprehensively treated in this book. Estimates must be directly correlated with detailed site conditions and matched to specific machine performance characteristics. Equipment manufacturers will provide these estimates for stated conditions, prepared by their sales technical personnel, based on their accumulated experience on similar applications. It must be recognized that such estimates are based on average data extrapolated to meet anticipated average site conditions over the life of the operations. The spread in the averages can be substantial and the correction factors introduced require a considerable amount of judgement. Costs, in particular, vary locally with time and economic conditions which make generalization risky. Some of the considerations entering into these calculations are:
Production

- Material swell
- Material unit weight
- Bucket size
- Bucket fill factors
- Job efficiency
- Swing angle
- Operational delays
- Operating schedule
- Machine stability

Costs

- Machine price
- Salvage value
- Machine life
- Interest
- Insurance
- Taxes
- Fuel consumption

- Lube, oils, greases, filters, etc.
- Operator wages
- Cutting edges, teeth, etc.
- Maintenance

When selecting a hydraulic machine, a limited number of equipment alternatives are offered, dependent on the manufacturer.

- Engines (manufacturer)
- Buckets (size, type, duty)
- Alternate teeth
- Booms and dippersticks (various lengths, extendable)
- Electric motor drive with trailing cable (large sizes)
- Elevated cab
- Alternate crawler shoes
- Wheel mountings (smaller sizes)
- Outriggers (smaller sizes)
- Extendable crawler frames (smaller sizes)

Optional equipment in addition to the basic machine includes:

- Deck leveling device (JI Case–smaller sizes)
- Automatic lubrication
- Cab rock guards
- Wrist cylinder rod guard
- Track guide guards
- Track motor guards
- Swivel guard
- Hydraulic counterweight removal device
- Vandalism protection devices
- Operator comfort/convenience items such as air conditioning, heater, radio, windshield washer/wiper, etc.
- Fire extinquisher/protection system
- Special lighting
- Tinted glass
- Radio telecontrol (P&H 1200)
- Spark arresters
- Starting receptacle
- Fast fuel fill systems
- Fire suppression system

NEW DEVEOPMENTS & TRENDS

Because of their recent development, most of the larger hydraulic machines incorporate the latest technology. The dominant emphasis has been on machines of ever-increasing size, but they may have reached a top size plateau, for a few years, with the new O&K RH300 model. (See Photograph 6.7) The three largest machines currently available are from overseas manufacturers. There seems to have been a special effort by those manufacturers to be the first to have the largest. They were developed in anticipation of a future demand. Rather than a sequence of models of stepped sizes, the largest O&K is more than three times bigger than the next smaller size O&K units, and the larger Demag unit was a two-fold increase over its smaller predecessor. Such size jumps are not uncommon in smaller sizes with proven technology, but these new machines may be a gamble at this point in time. Larger models require new, larger pumps and motors or the use of multiple drive systems. Multiple drives increase the complexity and congestion of the associated hydraulic plumbing.

There probably will be a continued pattern of new model introduction as the traditional U.S. manufacturers progressively expand their product

Photograph 6.7 New O&K RH300, a 30 cubic yard hydraulic shovel

lines to the larger sizes. With increasing size, some of the inherent features attributed to the hydraulic machines may be sacrificed. Multiple diesels introduce added maintenanc costs. Machine weight forces heavier low speed crawler designs and reduction in machine mobility. Fuel costs could well necessitate single motor electric primary drives (now optional) and trailing cables with a further reduction in mobility. Highly sophisticated hydraulic systems are proving to be very expensive. Cycle times are comparable to other types of machines. The added bucket orientation control in the digging phase may be of only limited value if adequate brute power is available.

Efforts continue to modify the front-end geometric relationships to maximize digging forces and optimize the path of the bucket. Caterpillar and, more recently, Komatsu have slave cylinders on the booms which improve machine level crowd capability. Northwest Engineering has a machine which merges some of the mechanical and hydraulic shovel characteristics. In the following chapter on electric shovels, the Marion "superfront" design is discussed. The "superfront" incorporates characteristics of both the cable and hydraulic shovel designs.

Hydraulic system refinements continue, aimed at efficient power utilization and associated reductions in fuel consumption. Since these systems are readily adapted to feedback controls, load sensing, etc., progress is being made in expanded performance monitoring, overload limits, and automation of portions of the cycle. Pressure levels currently appear to have stabilized in the 4000 to 5500 psi range.

With larger volume flows, bigger pipe diameters, and more hydraulic circuits, the required hydraulic plumbing is a major design challenge. Fittings are improving in reliability but accessiblity and failure or leakage identification is still a maintenance problem.

MACHINE SPECIFICATIONS

See Figure 6.14—Hydraulic Shovel Dimensions
See Table 6.2—Hydraulic Hoe Specifications
See Table 6.3—Hydraulic Shovel Specifications

Hydraulic machines are listed alphabetically by manufacturer, and then in ascending order by nominal bucket size. Nominal bucket size is based on the manufacturer's listed general purpose or medium heavy-duty bucket. Bucket range includes all special duty, light and heavy application buckets. Break-out and crowd forces are based on the general purpose nominal bucket as previously described.

Horsepower is based on the manufacturer's rating of the standard engine or engines (where dual engines are standard). In some cases, optional engines are available.

Hydraulic capacity, as listed, is based on the total capacity of the entire hydraulic system, including steering, pilot and auxiliary pumps. Hydraulic pressure is based on the maximum operating pressure of the working circuits.

All dimensions are based on the standard nominal bucket and standard or shortest boom. Clearance radius, cut radius, dump radius and reach at dump are all measured from the center of rotation of the machine. Gradeability is as listed by the manufacturer; here, gradeability refers to continuous, not maximum. Constraints on gradeability listed by manufacturers include traction and engine lubrication requirements.

All specifications, capacities, capabilities and dimensions are based on published manufacturer data. Although the information is believed to be current and the interpretation to be consistent and correct, it is possible that there are some inaccuracies or out-of-date information in the specification charts. These specification charts are not meant to be used as an in-depth analysis, or as a head-to-head comparison of the available equipment, but rather as a general overview of the equipment, available sizes, and approximate operating data. Specific questions relating to performance or purchase should be directed to the manufacturer or authorized distributor/dealer in the user's specific area.

Photograph 6.8 Northwest Engineering 65-DHS hydraulic shovel

Table 6.2
HYDRAULIC HOE SPECIFICATIONS

MAKE	MODEL	NOMINAL BUCKET SIZE (cu yd)	AVAILABLE BUCKET SIZES (cu yd)	ENGINE HP	WEIGHT (lbs)	BREAKOUT FORCE (lbs)	CROWD FORCE (lbs)	FUEL TANK (gal)	COOLANT CAPACITY (gal)
Amhoist	780	4.0	3.5-6	456	171,940	—	—	164	—
Bucyrus-Erie	400-H	4.5	3-6	392	152,470	—	—	235	30
Bucyrus-Erie	500-H	5.5	4-6	630	224,000	—	—	—	—
Caterpillar	245	3.0	2-3.5	325	126,700	—	—	158	21
Demag	H-71	4.0	3.5-7	330	164,022	57,870	56,658	251	—
Demag	H-121	10.0	10-14	675	270,376	101,412	110,230	462	—
Demag	H-241	14.0	10-21	1318	604,000	192,200	192,200	1110	—
Drott	120	3.5	2-3.5	456	134,025	—	—	182	—
FMC	LS-6400	3.5	-	380	138,000-	—	—	—	—
FMC	LS-7400A	4.5	—	456	170,000-	—	—	—	—
Hein-Werner	C-28	3.0	—	318	112,000	—	—	120	—
Hitachi	UH-20	3.0	2-4	300	110,000	50,600	35,200	172	32
Hitachi	UH-30	4.5	2.5-5	400	156,000	58,700	49,900	195	—
Hitachi	UH-801	10.0	10-15.5	800	346,000	110,000	110,000	528	—

Insley	H-3500C	3.5	2-4.5	390	130,000	65,800	48,700	200	—
Koehring	866	3.0	1.5-3	430	129,430	42,900	40,100	165	23
Koehring	1066	4.0	1.5-4	456	161,820	50,311	41,846	218	28
Koehring	1166	5.0	3-6.5	464	179,080	71,593	65,598	296	28
Koehring	1266	6.0	3-7	780	257,700	66,900	60,700	435	28
Koehring	1466	12.5	8-16	898	279,800	84,600	73,600	582	28
Liebherr	R-982	4.0	—	360	—	—	—	—	—
Liebherr	R-991	7.0	4.5-7.5	720	351,832	99,450	99,450	570	—
Northwest	55-DH	3.0	3-6.5	456	126,000	—	—	—	—
Northwest	100-DH	6.5	6.5-13	800	236,960	—	—	—	—
O & K	RH-40	5.0	3-6	480	185,660	67,035	56,890	264	air
O & K	RH-75	7.0	3.5-8.5	714	284,320	110,000	110,000	528	—
O & K	RH-300	21.0	16-21	2352	930,000	225,000	193,000	2900	—
P & H	1200	10.0	6-10	820	350,330	105,820	110,230	581	—
Poclain	300-CK	3.0	2-4	309	127,600	63,200	47,500	187	air
Poclain	400-CK	4.5	3-5	409	165,000	68,300	61,200	249	air
Poclain	600-CK	5.5	3-6	607	271,000	86,000	68,500	462	—
Poclain	1000-CK	12.5	7-17	883	456,200	117,150	109,300	694	—
Warner & Swasey	1900	5.0	3.5-6	614	206,378	—	—	224	—

Table 6.2 (Continued)
HYDRAULIC HOE SPECIFICATIONS

				HYDRAULIC SYSTEM DATA				
MAKE	MODEL	ENGINE MAKE	ENGINE MODEL	PUMP FLOW (gpm)	RELIEF VALVE PRESSURE (psi)	PUMP TYPE	NO. OF CYLINDERS	NO. OF MOTORS
Amhoist	780	Detroit	12V-71N	330	3000	piston	4	2
Bucyrus-Erie	400-H	Detroit	8V-92T	331	5000	piston	4	3
Bucyrus-Erie	500-H	Detroit Cummins	16V-71T VTA-1710-C700	—	—	piston	4	—
Caterpillar	245	Caterpillar	3406	328	4500	piston	4	3
Demag	H-71	Caterpillar	3406-T	170	4266	piston	4	—
Demag	H-121	Detroit Cummins	16V-71T VTA-1710-C700	355	4266	piston	6	—
Demag	H-241	GM	16V-149T	713	4350	piston	6	—
Drott	120	Detroit	12V-71N	300	3000	gear	6	—
FMC	LS-6400	Detroit	-	—	—	piston	—	—
FMC	LS-7400A	Detroit	12V-71N	—	—	piston	—	—
Hein-Werner	C-28	Detroit	8V-71N	344	2550	vane	4	—
Hitachi	UH-20	2-Isuzu	E-120	196	3300	piston	4	4
Hitachi	UH-30	2-Isuzu	E-120T	291	3300	piston	4	—
Hitachi	UH-801	2-Cummins	KT-1150-C450	479	3600	piston	4	—

Insley	H-3500C	Detroit	8V-92T	309	—	gear	4	3
Koehring	866	GM	8V-92N	222	3000	piston	4	2
Koehring	1066	GM	12V-71N	300	3000	piston	4	2
		Cummins	KT-1150-C450					
Koehring	1166	Detroit	12V-71T	475	3000	piston	4	4
Koehring	1266	2-GM	12V-71N	506	3000	gear	4	2
		2-Cummins	KT-1150-C450					
Koehring	1466	2-Detroit	12V-71T	742	3000	piston	4	4
		2-Cummins	KT-1150-C450					
Liebherr	R-982	Cummins	NTA-855-C	203	4000	piston	4	3
Liebherr	R-991	2-Cummins	NTA-855-C	317	4000	piston	4	3
Northwest	55-DH	GM	12V-71N	405	2750	vane	4	—
Northwest	100-DH	GM	16V-92T	868	2500	gear	4	—
O & K	RH-40	Deutz	BK-12L-413F	216	4267	—	5	—
		Cummins	KTA-1150-C525					
O & K	RH-75	2-Cummins	NTA-855-C335	264	4270	piston	5	—
O & K	RH-300	2-Cummins	KTA-2300-C	—	4267	—	7	—
P & H	1200	2-Cummins	KT-1150-C450	607	4090	piston	4	4
Poclain	300-CK	Deutz	F-12L-413	132	5800	piston	4	2
Poclain	400-CK	Deutz	BF-12L-41	168	5800	piston	4	3
Poclain	600-CK	2-Cummins	NT-855-C310	268	5800	piston	4	4
Poclain	1000-CK	2-Cummins	KT-1150-C450	358	5800	piston	4	4
Warner & Swasey	1900	2-GM	8V-71N	296	2750	gear	4	6

Table 6.2 (Continued)
HYDRAULIC HOE SPECIFICATIONS

MAKE	MODEL	MINIMUM LENGTH BUCKET ON GROUND (ft)	LENGTH W/O BOOM (ft)	MAX WIDTH W/O CRAWLERS (ft)	HEIGHT TO TOP OF CAB (ft)	MAX HEIGHT BUCKET RAISED (ft)	GROUND CLEARANCE (in)	CLEARANCE RADIUS REVOLVING FRAME (ft)	CRAWLER DATA: OVERALL LENGTH (ft)	CRAWLER DATA: OVERALL WIDTH (ft)	CRAWLER DATA: STANDARD SHOE WIDTH (in)	GROUND PRESSURE (psi)
Amhoist	780	46.9	24.0	10.8	12.9	37.2	27	13.8	20.3	14.5	35	11.6
Bucyrus-Erie	400-H	—	22.9	13.4	11.9	42.3	—	13.3	19.3	14.7	36	10.5
Bucyrus-Erie	500-H	—	26.9	14.9	13.0	38.9	—	16.0	22.3	15.5	42	11.2
Caterpillar	245	43.3	21.6	12.0	11.8	37.0	29	12.5	18.3	12.3	32	10.5
Demag	H-71	—	21.8	10.5	12.8	40.0	24	12.7	18.9	13.4	28	15.4
Demag	H-121	—	25.9	14.8	18.3	39.8	28	15.6	20.7	15.5	32	18.0
Demag	H-241	—	33.7	20.5	21.6	58.7	35	20.3	27.9	20.3	58	19.0
Drott	120	46.1	21.3	12.0	11.0	33.6	22	12.3	18.0	13.5	36	9.9
FMC	LS-6400	—	—	—	—	—	—	—	18.0	—	—	—
FMC	LS-7400A	—	—	—	—	—	—	—	20.3	—	—	—
Hein-Werner	C-28	—	20.0	11.3	11.2	31.6	16.5	11.9	16.5	11.3	32	7.7
Hitachi	UH-20	43.2	20.5	9.8	10.5	39.3	20	12.2	16.9	11.8	24	12.9
Hitachi	UH-30	—	23.6	10.3	11.8	44.0	26	14.1	19.0	13.3	28	14.5
Hitachi	UH-801	—	30.5	17.4	19.3	44.4	31	19.0	23.0	17.3	32	23.6

Insley	H-3500C	44.0	21.8	12.6	11.3	38.2	22	12.8	18.0	12.5	36	9.8
Koehring	866	—	—	—	11.8	—	18	13.5	18.4	14.2	30	—
Koehring	1066	—	—	—	11.9	—	21	14.0	20.6	14.3	30	—
Koehring	1166	46.6	24.5	14.5	12.6	34.4	21	14.0	20.9	14.7	40	11.0
Koehring	1266	—	—	—	13.4	—	27	16.8	21.4	17.8	48	—
Koehring	1466	52.7	28.3	15.1	13.9	39.6	26	17.0	22.5	16.7	40	15.6
Liebherr	R-982	—	—	—	—	—	34	—	—	—	—	—
Liebherr	R-991	—	30.2	18.1	17.5	46.3	30	19.6	25.7	17.4	32	20.7
Northwest	55-DH	—	21.3	10.3	11.0	34.5	14	13.0	16.7	13.2	36	—
Northwest	100-DH	—	27.0	15.4	13.0	39.1	—	16.1	21.8	20.4	48	—
O & K	RH-40	—	23.0	15.9	10.5	35.1	28	13.8	19.1	14.8	27	17.1
O & K	RH-75	—	28.0	18.9	18.3	42.0	32	17.8	21.2	15.1	31	19.9
O & K	RH-300	—	36.9	24.9	23.0	60.0	40	24.0	27.8	23.5	59	—
P & H	1200	55.4	31.0	18.5	20.0	41.5	34	18.3	25.3	17.9	34	21.6
Poclain	300-CK	36.1	20.9	10.8	11.5	32.3	23	12.3	17.1	13.0	24	14.4
Poclain	400-CK	39.1	24.1	13.3	12.2	36.4	26	14.7	18.7	13.8	24	17.5
Poclain	600-CK	43.7	25.6	17.8	16.2	43.3	32	15.1	21.0	15.6	34	17.1
Poclain	1000-CK	52.5	29.5	20.7	10.8	50.7	34	18.4	21.8	17.2	44	20.8
Warner & Swasey	1900	—	25.0	—	—	40.0	20	15.0	20.0	13.8	40	12.2

Table 6.2 (Continued)

HYDRAULIC HOE SPECIFICATIONS

MAKE	MODEL	MAX SPEED (mph)	GRADEABILITY (%)	SWING SPEED (rpm)	MAX CUT RADIUS (ft)	MAX CUT HEIGHT (ft)	MAX DIGGING DEPTH (ft)	MAX DUMP RADIUS (ft)	MAX DUMP HEIGHT (ft)	REACH AT MAX DUMP HEIGHT (ft)
Amhoist	780	1.8	80	3.8	44.5	36.5	27.5	33.8	22.3	26.2
Bucyrus-Erie	400-H	1.5	—	4.0	49.3	41.5	32.1	38.9	28.1	26.3
Bucyrus-Erie	500-H	—	—	—	56.8	35.5	36.4	50.8	24.8	36.8
Caterpillar	245	1.9	—	4.8	42.4	36.9	28.0	33.1	23.8	22.0
Demag	H-71	1.0	80	6.0	44.1	39.0	25.8	32.9	27.9	24.5
Demag	H-121	1.6	70	4.5	44.2	37.2	27.3	32.5	23.0	19.7
Demag	H-241	1.6	55	3.4	60.2	56.8	26.0	45.8	39.4	29.3
Drott	120	2.8	—	—	48.1	32.4	31.0	39.3	22.6	29.4
FMC	LS-6400	2.1	58	—	47.0	—	32.1	—	25.8	—
FMC	LS-7400A	1.9	58	—	53.3	—	35.5	35.6	30.0	—
Hein-Werner	C-28	3.4	50	6.4	43.1	29.6	29.3	36.8	20.3	26.2
Hitachi	UH-20	2.5	60	6.0	44.2	39.3	27.3	33.5	26.6	28.4
Hitachi	UH-30	1.3	60	4.2	49.3	44.0	30.1	37.7	29.3	26.4
Hitachi	UH-801	1.6	58	4.5	51.3	44.4	30.1	—	28.5	—

Insley	H-3500C	2.2	60	4.2	46.3	37.8	30.5	36.9	26.8	25.6
Koehring	866	1.8	58	4.6	—	21.2	29.3	—	—	—
Koehring	1066	1.9	58	3.6	—	21.9	34.3	—	—	—
Koehring	1166	1.9	58	3.7	44.0	31.5	24.5	37.6	20.6	24.1
Koehring	1266	1.2	58	2.7	—	26.7	39.5	—	—	—
Koehring	1466	1.5	58	4.3	52.0	37.6	28.3	43.2	24.6	29.3
Liebherr	R-982	1.5	90	—	—	—	—	—	—	—
Liebherr	R-991	1.3	60	5.0	52.5	45.6	29.9	43.1	31.2	30.8
Northwest	55-DH	1.5	30	3.8	43.9	34.5	28.2	35.8	21.9	28.6
Northwest	100-DH	0.7	30	3.0	55.8	35.0	36.5	48.7	24.8	38.3
O & K	RH-40	1.5	75	6.5	43.9	30.6	28.4	37.4	23.6	28.3
O & K	RH-75	1.3	65	5.5	50.2	36.1	28.2	43.1	27.2	31.8
O & K	RH-300	1.7	47	—	82.0	—	55.0	—	42.0	—
P & H	1200	1.2	70	4.5	53.5	38.3	27.9	45.9	25.8	32.8
Poclain	300-CK	1.8	72	3.9	35.6	32.3	22.5	35.4	21.7	25.6
Poclain	400-CK	1.8	65	3.5	38.7	36.4	23.5	39.4	23.9	30.5
Poclain	600-CK	1.6	55	3.8	44.0	43.3	25.4	36.1	29.3	26.9
Poclain	1000-CK	1.1	58	2.7	54.3	50.7	33.7	41.3	341	29.5
Warner & Swasey	1900	0.9	60	4.0	49.3	38.0	35.6	42.2	34.0	32.2

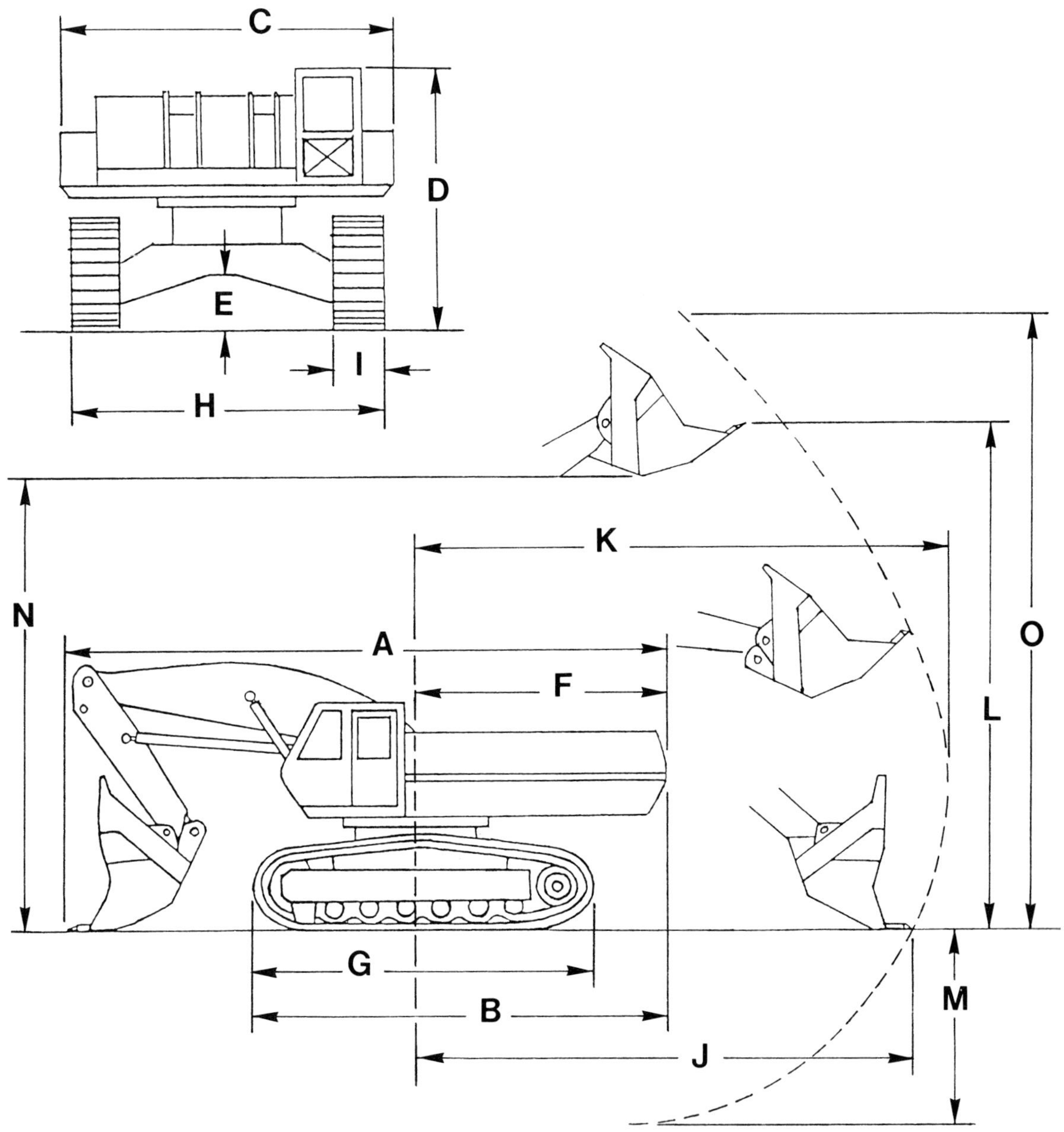

A Minimum Length (ft) (bucket on ground)
B Overall Length w/o Boom (ft)
C Overall Width (ft)
D Height to Top of Cab (ft)
E Ground Clearance (in)
F Clearance Radius of Revolving Frame (ft)
G Length of Crawler (ft)
H Width over Crawlers (ft)
I Width of Shoe (in)
J Max Reach at Level Grade (ft)
K Max Cut Radius (ft)
L Max Cut Height (ft)
M Max Digging Depth (ft)
N Max Dump Height (ft)
O Overall Height (ft) (to top of raised bucket)

Figure 6.14 Hydraulic Shovel Dimensions

Table 6.3
HYDRAULIC SHOVEL SPECIFICATIONS

MAKE	MODEL	NOMINAL BUCKET SIZE (cu yd)	AVAILABLE BUCKET SIZES (cu yd)	ENGINE HP	WEIGHT (lbs)	BREAKOUT FORCE (lbs)	CROWD FORCE (lbs)	FUEL TANK (gal)	COOLANT CAPACITY (gal)
Cat	245	4.0	4-5	325	146,700	95,800	82,700	155	21
Demag	H-71	5.5	5.5-10	330	169,533	88,184	92,593	251	—
Demag	H-121	10.0	7-14	675	270,376	103,620	108,000	462	—
Demag	H-241	19.0	14-28	1318	604,000	204,568	213,560	1110	—
Hitachi	UH-20	4.5	3.5-6	300	114,000	66,000	66,000	172	32
Hitachi	UH-30	6.0	5-8	400	161,000	83,600	78,300	198	34
Hitachi	UH-801	11.0	11-16	800	346,000	137,000	137,000	528	—
Koehring	1166-FS	6.0	6-8.5	464	195,640	84,000	70,000	296	28
Koehring	1466-FS	10.0	10-15	898	303,560	114,500	109,000	582	28
Liebherr	R-982	4.0	3.5-7	360	169,300	70,180	64,900	183	—
Liebherr	R-991	10.0	10-14	688	361,556	121,550	130,390	570	—
Northwest	65-DHS	6.0	5-7	456	210,100	—	—	—	—
O & K	RH-40	6.0	4-8	493	187,870	72,765	72,765	264	air
O & K	RH-75	10.0	6-13	714	284,320	122,000	94,000	506	—
O & K	RH-300	30.0	23-40	2352	930,000	405,000	440,000	2900	—
P & H	1200	15.0	10.5-18	820	326,000	138,890	187,390	581	14
Poclain	300-CK	4.0	4-6	309	127,600	49,300	71,700	187	air
Poclain	400-CK	5.0	4.5-7	410	165,000	98,600	99,800	249	air
Poclain	600-CK	9.0	7-15	607	265,000	136,200	153,400	462	—
Poclain	1000-CK	15.0	9-22	883	425,000	175,300	207,400	694	—

Table 6.3 (Continued)
HYDRAULIC SHOVEL SPECIFICATIONS

				HYDRAULIC SYSTEM DATA				
MAKE	MODEL	ENGINE MAKE	ENGINE MODEL	PUMP FLOW (gpm)	RELIEF VALVE PRESSURE (psi)	PUMP TYPE	NO. OF CYLINDERS	NO. OF MOTORS
Cat	245	Cat	3406	324	4500	piston	6	3
Demag	H-71	Cat	3406-T	170	4266	piston	5	—
Demag	H-121	Detroit	16V-71T	355	4266	piston	5	—
		Cummins	VTA-1710-C700					
Demag	H-241	GM	16V-149T	713	4350	piston	6	—
Hitachi	UH-20	2-Isuzu	E-120	196	3300	piston	7	—
Hitachi	UH-30	2-Isuzu	E-120T	291	3300	piston	7	—
Hitachi	UH-801	2-Cummins	KT-1150-C450	493	3600	piston	—	—
Koehring	1166-FS	Detroit	12V-71T	475	3000	piston	6	4
Koehring	1466-FS	2-Detroit	12V-71T	742	3000	piston	6	4
		2-Cummins	KT-1150-C450					
Liebherr	R-982	Cummins	-	203	4000	piston	5	3
Liebherr	R-991	2-Cummins	NTA-855-C	317	4000	piston	6	—
Northwest	65-DHS	GM	12V-71N	—	—	—	5	—
O & K	RH-40	Deutz	BF-12L-413F	216	4267	—	5	—
		Cummins	KTA1150-C525					
O & K	RH-75	2-Cummins	NTA-855-C	234	4267	—	6	—
O & K	RH-300	2-Cummins	KTA2300-C1200	—	4267	—	7	—
P & H	1200	2-Cummins	KT-1150-C450	607	2190	piston	6	—
Poclain	300-CK	Deutz	F-12L-413	132	5800	piston	4	—
Poclain	400-CK	Deutz	BF-12L-41	168	5800	piston	4	2
Poclain	600-CK	2-Cummins	NTA-855-C310	268	5800	piston	4	3
Poclain	1000-CK	2-Cummins	KT-1150-C450	358	5800	piston	5	—

Table 6.3 (Continued)

HYDRAULIC SHOVEL SPECIFICATIONS

MAKE	MODEL	MINIMUM LENGTH BUCKET ON GROUND (ft)	LENGTH W/O BOOM (ft)	MAX WIDTH W/O CRAWLERS (ft)	HEIGHT TO TOP OF CAB (ft)	MAX HEIGHT BUCKET RAISED (ft)	GROUND CLEARANCE (in)	CLEARANCE RADIUS REVOLVING FRAME (ft)	CRAWLER DATA: OVERALL LENGTH (ft)	CRAWLER DATA: OVERALL WIDTH (ft)	CRAWLER DATA: STANDARD SHOE WIDTH (in)	GROUND PRESSURE (psi)
Caterpillar	245	30.6	21.9	10.2	11.8	34.2	30	12.5	18.4	11.3	24	—
Demag	H-71	31.0	21.8	10.5	12.8	37.0	24	12.3	18.9	13.4	26	15.9
Demag	H-121	37.9	25.9	14.8	18.3	42.3	28	15.6	20.7	15.5	32	18.0
Demag	H-241	49.2	33.6	20.5	21.6	51.2	35	19.7	27.9	20.3	43	22.0
Hitachi	UH-20	29.4	20.6	9.8	10.5	32.8	20	12.2	16.9	10.3	24	13.5
Hitachi	UH-30	34.5	23.6	10.3	11.8	37.2	26	14.1	19.0	13.3	28	14.8
Hitachi	UH-801	44.0	30.6	17.4	19.3	45.1	31	19.0	23.0	17.3	32	23.6
Koehring	1166-FS	34.0	24.5	16.8	17.1	36.9	21	14.0	20.9	14.0	32	14.1
Koehring	1466-FS	38.3	28.3	19.5	20.2	40.8	26	17.0	22.5	16.0	32	20.6
Liebherr	R-982	—	—	—	—	—	34	—	—	—	—	—
Liebherr	R-991	41.4	30.2	18.1	17.5	49.2	30	17.4	25.7	17.4	32	21.3
Northwest	65-DHS	24.3	23.8	10.7	13.8	37.2	11	14.3	19.0	14.5	27	—
O & K	RH-40	32.3	23.0	15.9	12.9	39.5	28	13.8	19.1	15.8	27	17.1
O & K	RH-75	38.4	28.3	18.9	18.3	40.7	32	17.8	21.2	15.1	31	19.9
O & K	RH-300	52.8	36.9	24.9	23.0	—	40	24.0	27.8	23.5	60	27.0
P & H	1200	42.4	31.0	18.5	18.3	45.6	34	18.3	25.3	17.9	34	21.7
Poclain	300-CK	29.0	20.9	10.8	11.5	32.5	23	12.3	17.1	13.0	24	14.4
Poclain	400-CK	33.2	24.0	13.3	12.2	35.8	26	14.8	18.7	13.8	24	17.5
Poclain	600-CK	37.7	25.6	17.8	16.3	41.0	32	15.1	21.0	15.2	24	25.5
Poclain	1000-CK	41.7	29.5	20.7	18.8	45.6	34	18.4	21.8	16.4	34	27.8

Table 6.3 (Continued)
HYDRAULIC SHOVEL SPECIFICATIONS

MAKE	MODEL	MAX SPEED (mph)	GRADEABILITY (%)	SWING SPEED (rpm)	MAX REACH LEVEL GRADE (ft)	MAX CUT RADIUS (ft)	MAX CUT HEIGHT (ft)	MAX DIGGING DEPTH (ft)	MAX DUMP RADIUS (ft)	MAX DUMP HEIGHT (ft)	REACH AT MAX DUMP HEIGHT (ft)
Caterpillar	245	1.9	—	4.8	30.8	32.5	31.3	9.4	—	22.3	18.8
Demag	H-71	1.0	80	6.0	32.2	33.4	31.9	9.4	—	21.8	18.9
Demag	H-121	1.6	70	4.5	36.3	38.9	37.3	13.0	—	26.3	22.5
Demag	H-241	1.6	55	3.4	47.7	50.2	48.0	13.6	—	36.6	32.0
Hitachi	UH-20	2.5	60	6.0	29.0	30.0	31.0	13.1	—	23.0	18.1
Hitachi	UH-30	1.3	60	4.2	33.5	35.1	37.2	16.0	—	27.3	18.2
Hitachi	UH-801	1.6	58	4.5	40.9	42.8	45.1	17.7	—	34.3	23.0
Koehring	1166-FS	1.9	58	3.7	34.6	35.9	30.6	10.1	25.8	—	—
Koehring	1466-FS	1.5	58	4.3	38.9	40.6	31.2	11.6	30.3	—	—
Liebherr	R-982	1.5	90	—	—	—	—	—	—	—	—
Liebherr	R-991	1.3	60	5.0	43.1	45.3	46.7	15.0	—	36.9	26.3
Northwest	65-DHS	1.7	—	—	39.0	39.8	37.2	15.1	—	28.3	24.0
O & K	RH-40	1.5	75	6.5	33.8	35.5	36.1	13.1	29.5	30.4	17.4
O & K	RH-75	1.3	65	5.5	37.2	39.7	39.4	9.7	33.6	32.2	23.3
O & K	RH-300	1.7	47	—	—	—	—	—	—	—	—
P & H	1200	1.2	70	4.5	39.3	41.0	44.9	9.5	—	36.4	20.3
Poclain	300-CK	1.8	58	3.9	26.8	28.6	28.9	7.9	—	24.3	14.7
Poclain	400-CK	1.8	58	3.5	29.2	30.8	33.2	7.6	—	27.6	15.1
Poclain	600-Ck	1.6	55	3.8	32.3	34.8	36.9	7.4	—	31.3	16.4
Poclain	1000-CK	1.1	58	2.7	36.9	39.3	40.0	9.2	—	33.5	20.3

Chapter 7

ELECTRIC SHOVELS

Figure 7.1 Electric Shovel Loading Trucks

The shovel is one of the oldest types of excavating equipment. With time, the machines grew in capacity, steam power was replaced by gas, then diesel fuel and finally, in the larger units used in mining today, by electricity. The basic configuration has changed relatively little but the efficiency and sophistication have increased substantially.

In recent years, smaller shovels below 5 cubic yards in capacity are being replaced by front-end loaders and hydraulic machines. Some 5 to 8 cubic yard diesel/mechanical machines are still utilized in extremely rugged applications.

The two types of electric shovels in service are the loading shovel and the stripping shovel, the primary difference being the size of the machine. Typical units are illustrated in Figure 7.2. Stripping shovels will not be considered as a current product since none has been placed into service in the last ten years. These have been superceded by the large walking draglines. Stripping shovels are worthy of note, however, in that they show the level of development that preceeded the current machines. Many of these older machines are still in operation with the two largest being a Marion 6360, 180 cubic yard (see Photograph 7.1), and a Bucyrus-Erie 3850B, 140 cubic yard (see Photograph 7.2).

The two-crawler electric shovels in the 10 to 70 cubic yard sizes remain a dominant factor in surface mining, both in ore and overburden removal. This text will discuss only these units. They are used primarily for loading large off-highway trucks or, in some cases, mobile hoppers.

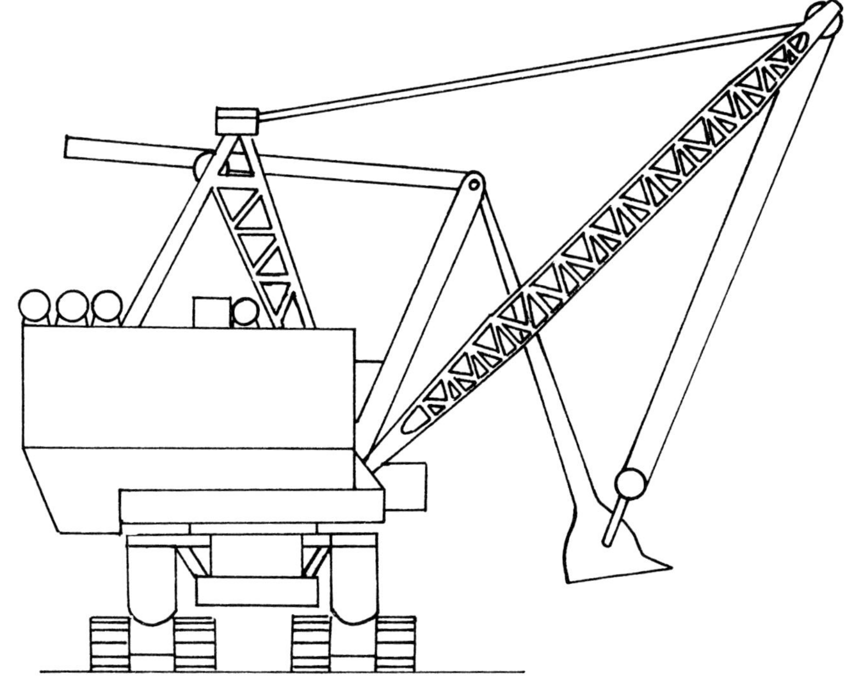

STRIPPING SHOVEL

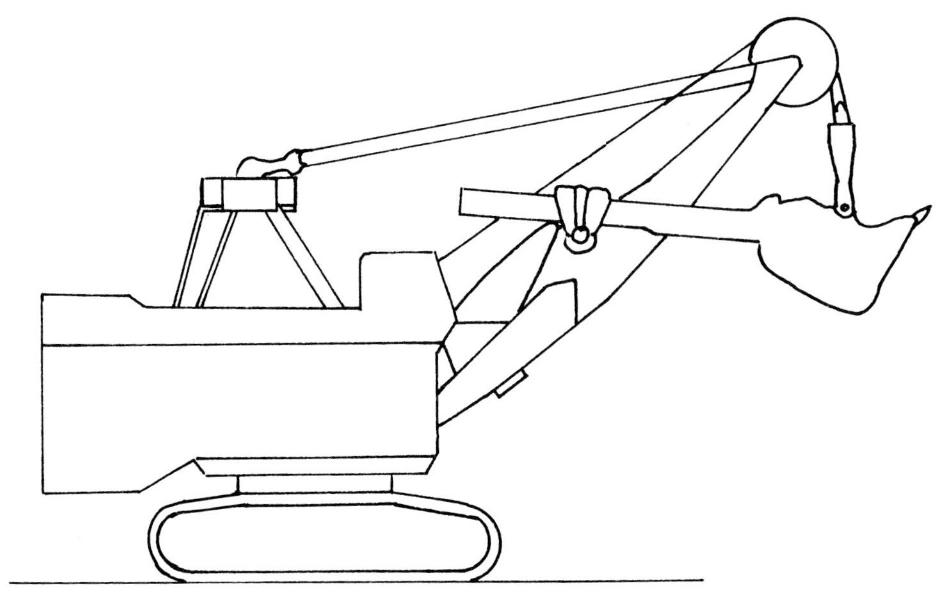

LOADING SHOVEL

Figure 7.2 Electric Shovel Types

Photograph 7.1 Marion 6360, 160 cubic yard, overburden stripping shovel

TYPICAL UNITS

The loading shovels are rugged, high production units with all electric drives and a trailing cable to bring power to the machine. Diesel units are available but not very common. Typical machines have the following features:

- Crawler mounted (two tracks)
- Full revolving upper works
- Cable hoist system
- Cable or rack & pinion crowd system
- Fully enclosed machinery house
- Capacities from 10 to 70 cubic yards
- Total horsepower from 1,000 to 5,500
- Machine weights from 500,000 to 3,500,000 lbs.

Manufacturers

The electric mining shovel market is dominated by U.S. companies who meet international mining demands. There are only three manufacturers: Bucyrus-Erie, Marion and Harnischfeger (P&H). (See Table 7.1) All have been producers of these machines for close to fifty years. Bucyrus-Erie and Marion made the larger stripping shovels in earlier years. Sales and service are generally handled directly through the manufacturers' marketing organization.

Photograph 7.2 Bucyrus-Erie 3850B 140 cubic yard overburden stripping shovel

Table 7.1

ELECTRIC MINING SHOVEL MANUFACTURERS

Bucyrus-Erie Company 1100 Milwaukee Street South Milwaukee, WI 53172	8–60 cu. yd.
Harnischfeger Corporation P.O. Box 554 Milwaukee, WI 53201	6 –75 cu. yd.
Marion Power Shovel Division Dresser Industries, Inc. P.O. Box 505 617 W. Center Street Marion, Ohio 43302	7–40 cu. yd.

BASIC MACHINE OPERATION

A typical operating cycle consists of a digging cut, a loaded swing to the discharge area, the dump, and empty return swing to the digging face. With a swing angle of 70 to 120 degrees, the cycle time usually ranges from 25 to 35 seconds.

- The machine is propelled periodically to reposition it at the digging face. Relocation within the pit requires special procedures for shifting the trailing cable.

- The digging path of the dipper in the face is produced by a combination of hoisting (vertical) motion and a crowd (horizontal) motion. Path geometry is relatively restricted and generally typified by vertical circular arcs.

- Dumping is achieved by the release of a latch securing a bottom door in the dipper which swings open under gravity, allowing the load to fall through.

- The rotating upper works permits swinging the boom and dipper to the right or left for discharge into a hopper, truck or stockpile.

APPLICATIONS

Electric shovels generally have the same applications as hydraulic shovels although the electric units are considered to be particularly suited to more severe digging conditions. They are available in larger sizes and have a proven service record in multi-shift mining operations. Electric shovels also tend to have longer range capabilities.

These shovels are applied in benching operations in either overburden or coal/ore. Discharge is commonly into trucks but can also be into mobile hoppers. The larger models and/or those equipped with long range front ends may be applied in direct spoiling overburden removal operations.

Photograph 7.3 Bucyrus-Erie 195B long range unit loading coal into a bottom dump truck

GENERAL CHARACTERISTICS

The machine's popularity is derived, in part, from its simplicity. The powered functions are illustrated in Figure 7.3.

- Maximum cutting heights from 35 to 65 feet
- Maximum dump heights from 20 to 43 feet
- Crawler propel, electric motor
- Hoist, cable and drum, electric motor
- Crowd, cable and drum or rack and pinion, electric motor
- Upper works rotation, electric motor
- Dipper trip, electric motor (small)

Common features of these machines are as follows:

- High sustained production capability
- Heavy, rugged construction, suited to tough digging conditions
- Minimal preparation (blasting) required
- High reliability and a proven design
- Limited gradeability
- Good stability
- Medium ground pressures of 30 to 50 psi
- Good blending vertically at face
- Limited capability for selective loading
- Low propel speeds (less than 1 mph)
- Limited mobility between faces
- Optional front-end lengths for different digging geometry

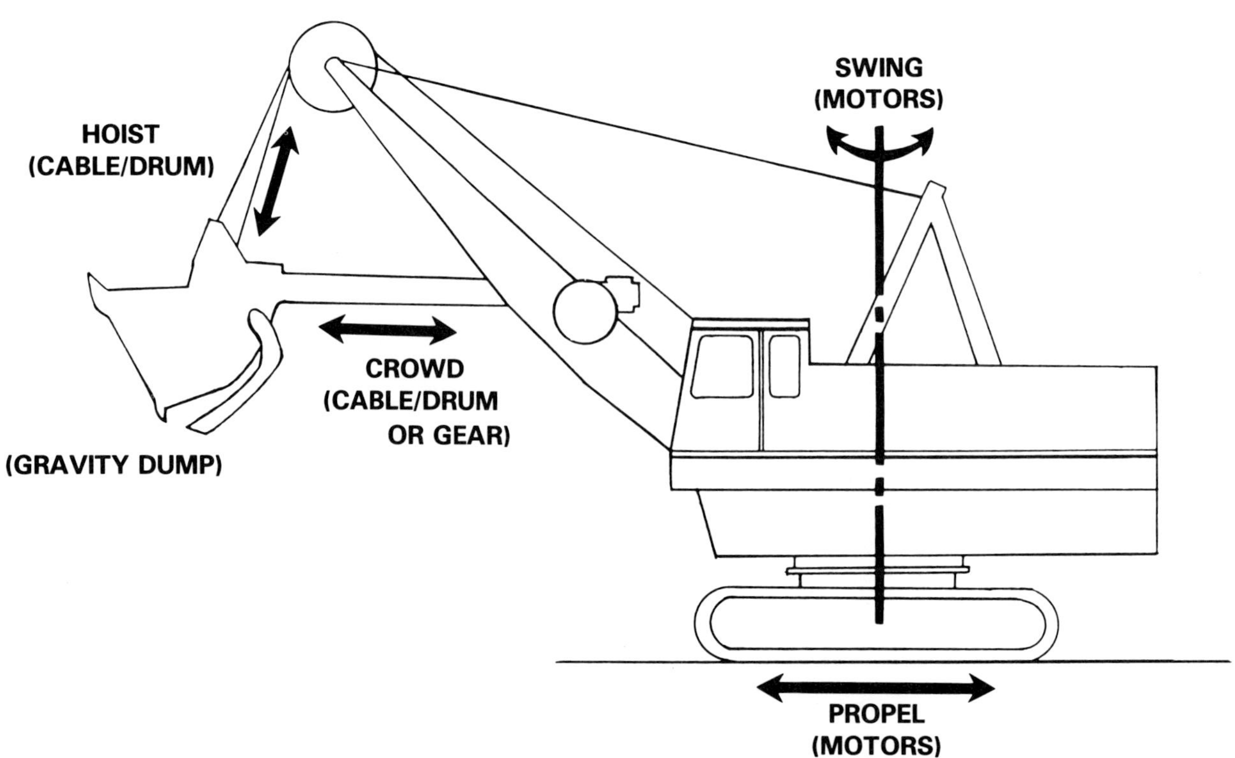

Figure 7.3 Shovel Powered Functions

- High tooth forces on dipper, limited by machine stability
- Relatively fixed digging path
- Can work in close quarters
- Dumping action can be hard on trucks
- Good performance with bad floor conditions
- Limited capability to dig below its floor level
- Support equipment required to maintain working area
- Altitude, temperature or humidity may reduce electrical efficiency. Standard ratings apply to 3000 feet altitude; special options are available for other conditions. Generally, however, the effect of weather is minimal. Electric power has little difficulty with cold weather starts.
- Safe operator location and good operator visibility
- Requires a crew of 1 or 2. Moderate/high operator skill and training is required; low operator fatigue.
- Maintenance is performed in pit.
- Machine normally has a modular construction to allow for knockdown shipment and field assembly.
- Long service life of over 20 years
- High capital cost and low operating cost

Graph 7.1 shows the trend in shovel operating weight versus nominal bucket size. The weight of an electric shovel is about double that of a comparable hydraulic shovel. The counterweight on the electric machine is 3 to 4 times heavier than that on a similar hydraulic unit.

The maximum horsepowers typically available for each of the powered functions are plotted in Graph 7.2. It must be recognized, however, that there are considerable differences in the electrical systems employed on various models. These differences can significantly impact operational torque-speed characteristics and machine performance. Graph 7.3 shows the bail pull (hoisting force) available as a rough indicator of the potential digging forces.

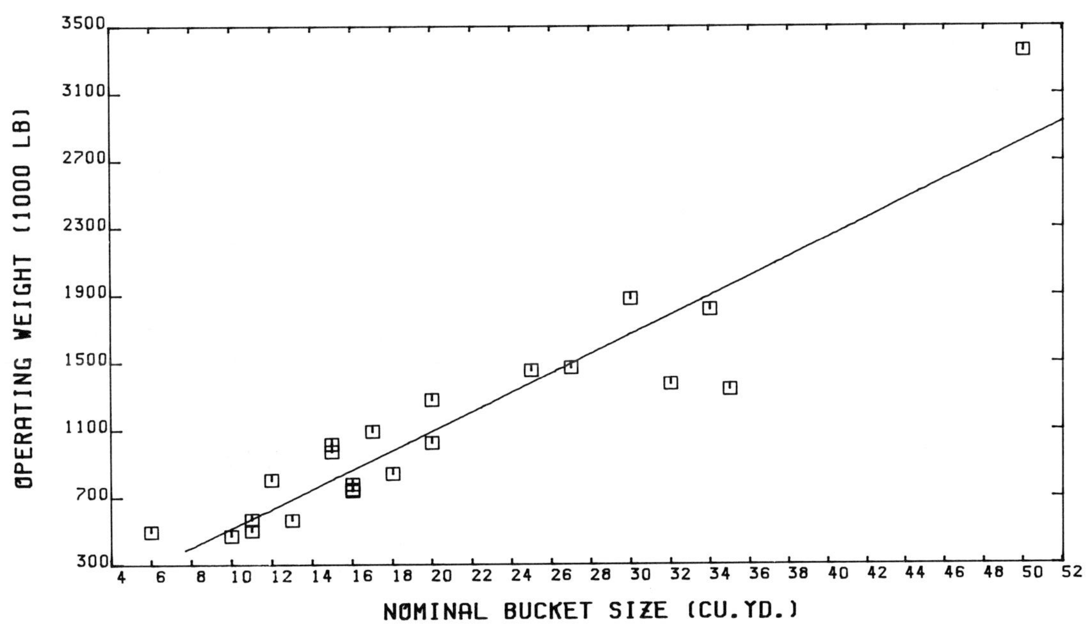

Graph 7.1 Operating weight/nominal bucket size

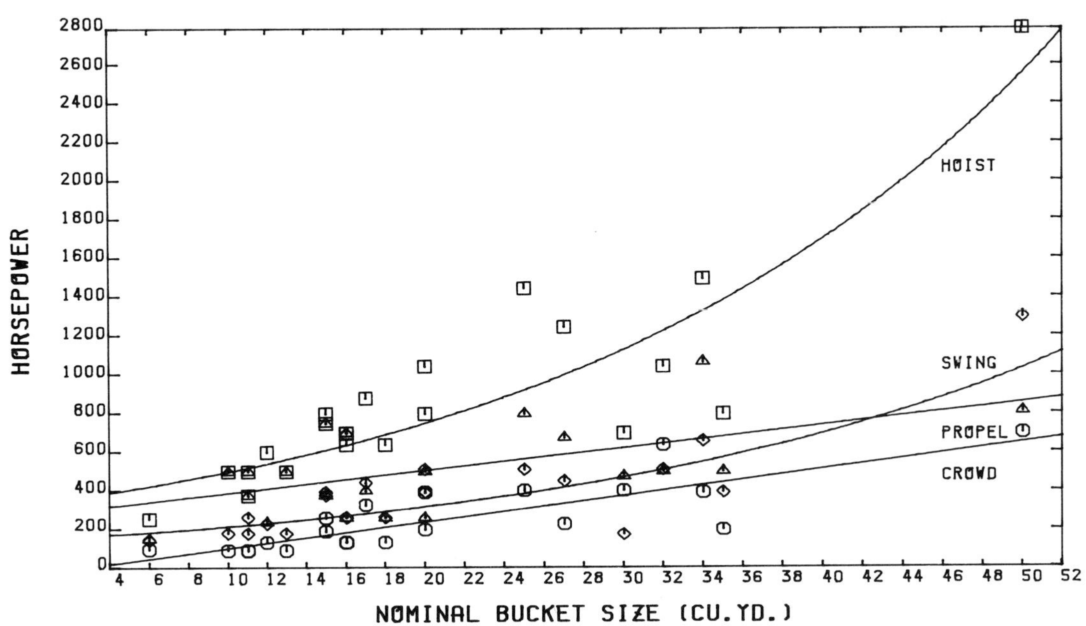

Graph 7.2 Hoist, crowd, swing, propel HP/nominal bucket size

Electrical and diesel horsepower ratings cannot be compared directly, but it is of interest to note that the primary power A.C. motor drive on the electric shovels (Ward-Leonard systems) is of lower horsepower than the diesels on comparable hydraulic shovels. The bail pull on electric shovels is significantly higher than the breakout forces on comparable hydraulic shovels.

The single or two-piece structural boom is pinned to the front of the main deck, and cable suspended from an A-frame mounted slightly to the rear on the main deck. The boom position is fixed and does not articulate during operation. Because of the flexible cable suspension, the boom can be lifted if excessive crowd forces are applied.

The digging path of the dipper is produced by the extension/retraction of the handle (crowd) and by the cable hoisting action. (See Figure 7.4) Hoisting of the dipper is accomplished by means of cables attached to the dipper which pass over sheaves at the boom point and spool on a deck mounted powered drum. The cables must be replaced at regular intervals but are a convenient means of providing a long travel path. They cannot produce any down pressure. The crowd action is produced either by cables or a direct rack and pinion gear drive.

The handle slides in and out of a saddle block which is shaft mounted on the boom. The saddle block acts as a pivot for the handle allowing the dipper to make an approximate arc with relatively limited variation in radius. At the top of the cut, if the handle is extended, the hoist and crowd partially counteract each other. Gravity acting on the dipper, plus the cable angularity, tends to retract and lower the dipper to the start of the dig position with very little power required.

The dipper is often pinned to the handle to facilitate replacement or to allow linkage modifications changing its general orientation. Tooth angularity, with respect to the handle, is fixed during operation. This simplifies the connection and avoids the additional weight associated with a positioning device. The design is rugged and, essentially, maintenance-free. The fixed orientation, however, makes it difficult to attack large rocks or boulders at the toe of the face. (See Figure 7.5) In essence, the electric shovel uses a very simple front-end structure and brute force for its digging rather than the complex, variable digging path approach of the hydraulic shovel.

Dippers for regular duty are roughly square in cross-section, heavily reinforced at the leading edge

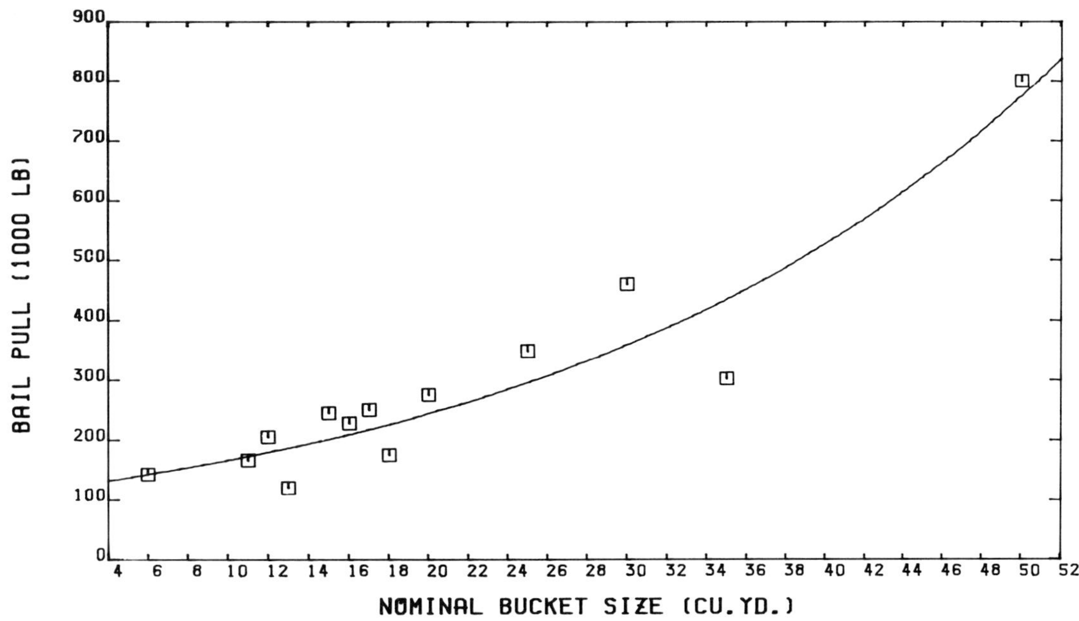

Graph 7.3 Bail-pull/nominal bucket size

and slightly tapered to improve the dirt shedding characteristics. A hinged door at the rear is unlatched to dump the load. The door opens under the force of gravity in an uncontrolled discharge. If there are large boulders in the dipper, it must be lowered close to the truck body or other discharge surface to avoid high impact loads. The door automatically relatches as the dipper is lowered into the start of the dig position. The swing drive, which provides the revolving frame rotation, is composed of two to four deck mounted motor-gear reduction modules which power pinions contacting a large gear fixed to the lower works.

The crawlers permit intermittent repositioning of the machine as the digging face is advanced, periodic relocations during blasting and moves to new operational sites within the pit. They also provide a solid working base to support the machine weight, and to resist digging and rotational forces. The physical size of the machine, plus the trailing cable for power input, restricts speeds and limits mobility. Moves between faces can be facilitated with the diesel drive option and/or use of a tag-along generator set.

A full enclosure is provided to house the deck machinery, all electrical cabinets and accessories. Forced air circulation (filtered) is incorporated to remove heat generated during operation. The operator's station is attached in a high forward position to maximize visibility during digging and dumping.

To aid rapid change-out of defective or worn parts; dippers are pin connected, machinery units are pin connected/bolted, machinery enclosures have removable hatches or panels, drives are located in easily accessible areas, and electric control panels are modular. Static drive components are mounted in pullout drawers for easy access. Indicator trays facilitate trouble shooting, and interchangeable trays permit quick replacement.

Basic maintenance programs involve the following:

- Major electrical components are exchanged with the manufacturer or sent out to specialty shops for rebuilding on an extended schedule.

- Intermittent structural repair welding is performed as required; constant inspection is recommended to identify problems.

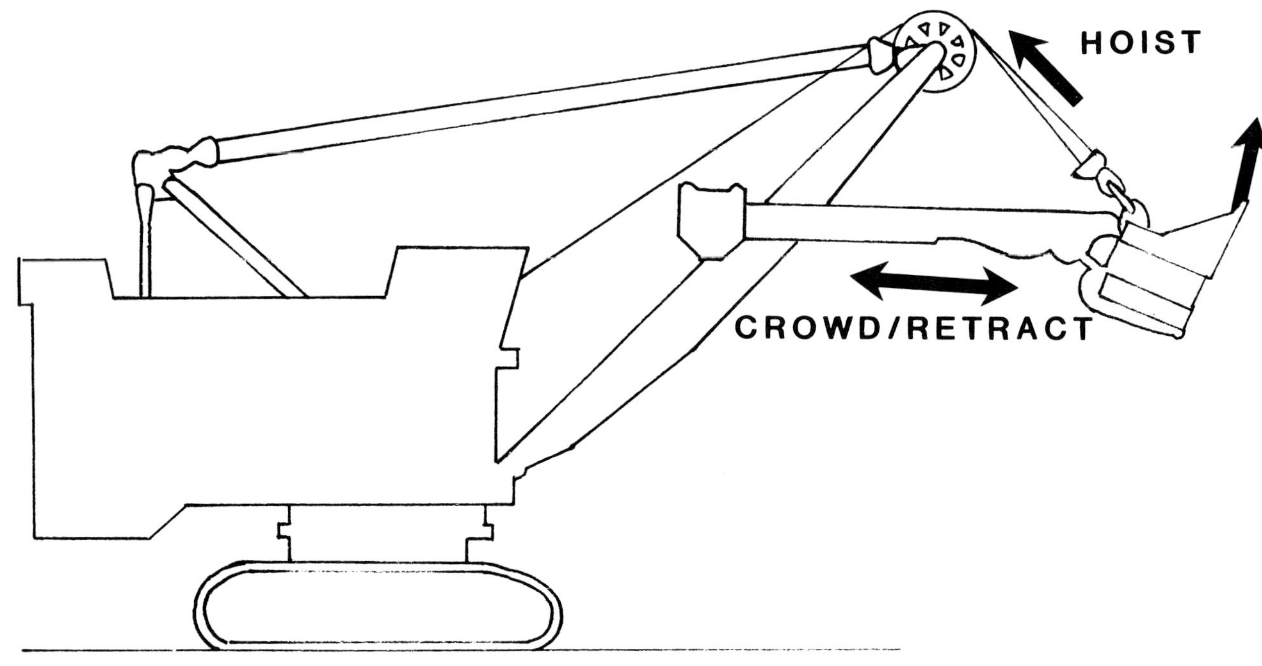

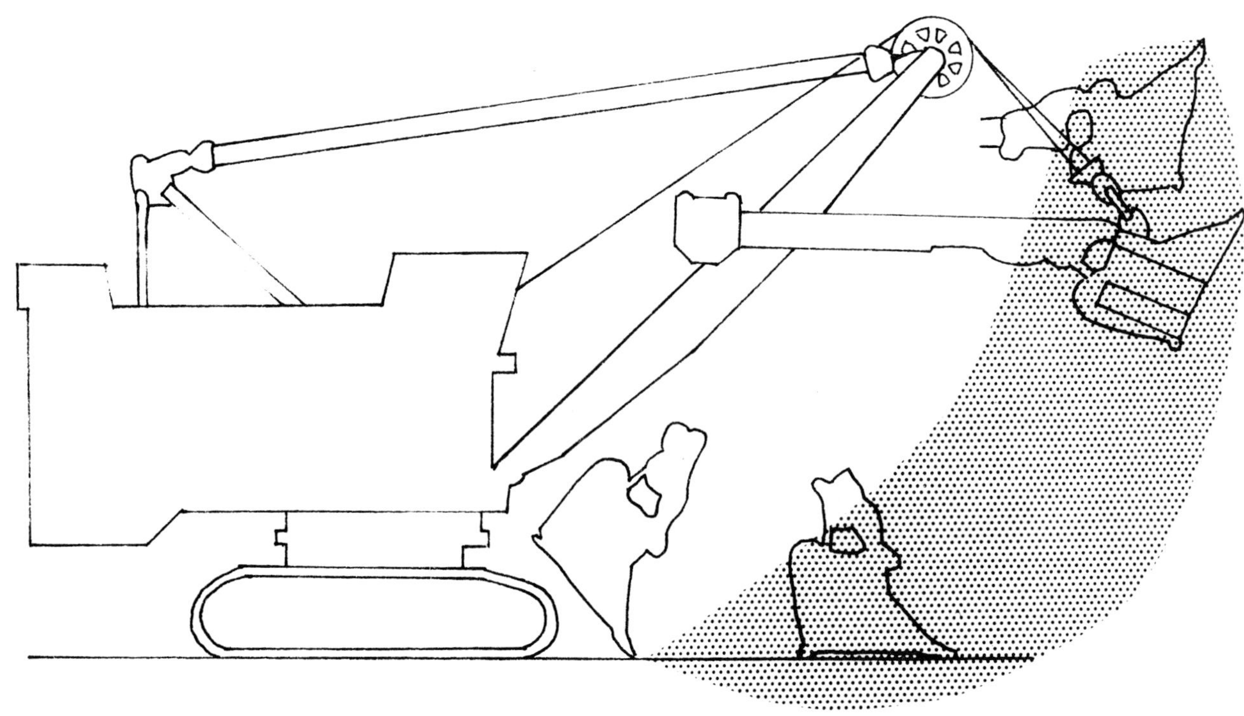

Figure 7.4 Shovel Digging Forces & Digging Profile

- Tooth replacement and dipper repair welding are done as required; full dipper interchange is on a regular basis.

- Cable replacement is on an as-required or programmed schedule.

OPERATIONAL PRACTICES

There is a great deal of accumulated experience in using electric shovels in mining operations. Common practices can be briefly summarized as follows:

- Strongly consolidated materials should be drilled and blasted prior to excavating.

- The shovel should operate on a level, flat digging floor whenever possible.

- Digging downhill permits higher digging thrust forces because gravity adds to the machine's resistance to movement away from the digging face.

- Normally, the digging face should not be higher than the boom point sheave.

- The toe of the bank should be below the rear of the point sheave or shipper shaft.

- Crawlers should be perpendicular to the face to minimize possible damage from materials sliding down the face and to facilitate positioning maneuvers.

- Short frequent moves are desirable to keep the shovel close to the face maximizing the effectiveness of the crowd and hoist forces.

- Hoist and crowd motions should be coordinated for an efficient dipper path up through the bank, starting at the toe and filling in a single pass.

- Penetration of the face should be uniform with a depth sufficient to fill the dipper in 2 or 3 dipper lengths. The dipper should then be retracted clear of the face to minimize travel in the bank.

- To optimize motor life, prolonged stalling of the motors while digging is to be avoided.

- Excessive crowd will jack the boom and lead to reduced suspension cable life.

- Under difficult conditions, the top of the bank should be dug away to ease hoisting through the lower portion.

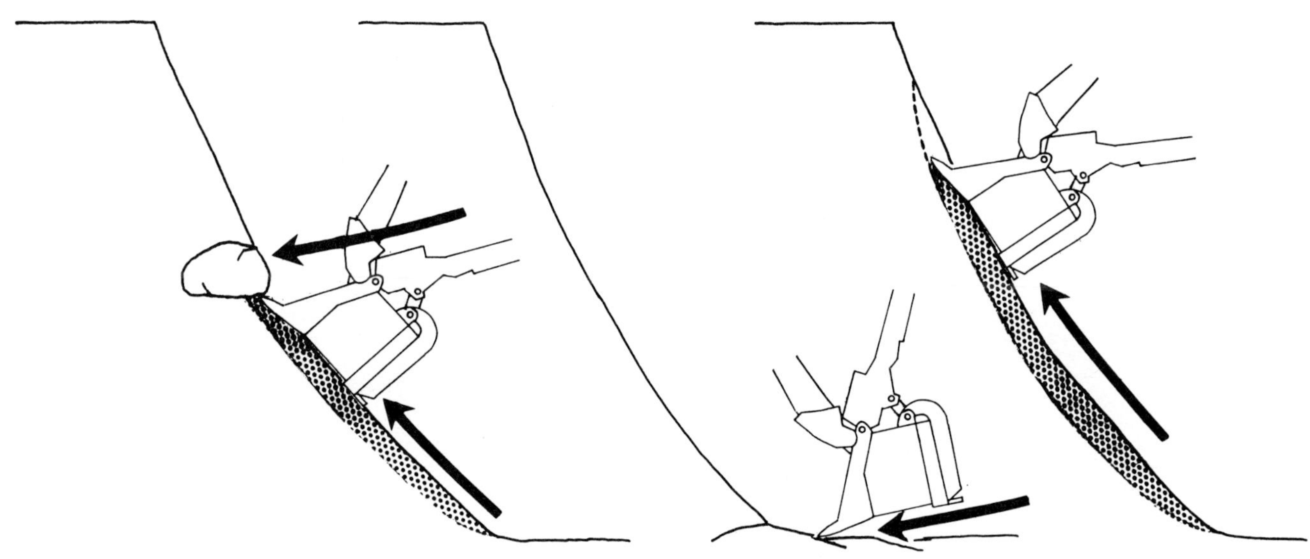

Figure 7.5 Shovel Digging Action

- Boulders or frost caps are removed by digging underneath and then lifting out with the dipper.

- If a hard toe is present, the machine should be moved as close to the bank as practical, to gain maximum benefit of crowd forces and improve tooth orientation for penetration.

- The face can be prepared (while waiting for trucks) by raking the bank with the dipper, with the door open.

- To dislodge a boulder jammed in the dipper, the dipper should be passed through the bank with the door open, or the dipper teeth bumped on the floor.

- Oversized boulders must be cast clear of the operations or carefully set behind the machine. They should not be rolled over the teeth into the truck. (Swinging up to the truck is dangerous.)

- The dipper should not be swung until clear of the bank. Neither should it sweep the floor or face to push rocks aside. These actions produce very high torsional loads in the boom and can also result in bending the handle.

- The swing angle must be kept as low as possible to minimize cycle time.

- The machine is accelerated rapidly at the start of the swing and then plugged (reverse power) to decelerate. The action should be smooth to avoid spillage.

- The truck or hopper to be loaded should be positioned under the boom point sheave. The suspended dipper can be used as a target to spot trucks.

- Good practice requires loading over the rear of the truck to avoid any spillage near the truck driver. The empty/loaded dipper should not be swung over personnel or other equipment.

- The dipper should be lowered close to the truck body (or hopper) prior to dumping to minimize the impact from the discharge.

Photograph 7.4 Marion 151-M diesel machine loading coal into a rear dump truck

- It is desirable to avoid unnecessary crowding or retracting when positioning the dipper over the truck or hopper.

- If there is the potential for face slides, it may be desirable to load the trucks on the blind side (opposite from the operator station) so that the operator has a good view of the face during the entire operating cycle.

- The trailing cable should be clear of the truck maneuvering area.

- When practical, fines should be loaded into the truck or hopper first to provide a cushion for subsequent coarse materials.

- The typical digging cycle should take 25 to 35 seconds: dig 24%, swing loaded 32%, dumping and swinging empty 34%, and bucket positioning 10%.

- The operator should be in front, in the direction of travel, for all machine moves.

- During propel, drive sprockets should be in the rear so that any track belt slack is accumulated along the top.

- Gradual intermittent turns should be made to minimize material build-up on the side of the track (15 to 20 degree increments).

- Care must be exercised in negotiating steep downhill grades. The propel brakes and steering clutches should be checked, the dipper held close to the ground to serve as a possible emergency retarder, and the trailing cable kept on the uphill side of the machine.

- Track lubrication should be checked before and during long machine moves.

- No personnel should climb on board without prior notification of the shovel operator.

To obtain the high production capabilities of the shovel during truck loading, the loading site must be carefully synchronized. Clean-up around the

Photograph 7.5 P&H 2300 digging overburden

machine and positioning of the trailing cable should be timed to minimize the interruption of production. Trucks must also be available and positioned to permit a continuous cycling of the shovel. A signal or communication system should be provided to alert the truck driver when his unit has been fully loaded.

Figures 7.6 and 7.7 show typical shovel/truck loading layouts for different mining sequences. The trailing cable must be supported on standards off the ground, enabling the trucks to be moved into their loading positions. Dual trucks, when they can be employed, will maximize the system production. In all cases, an effort is made to keep the shovel swing time, as well as the amount of time required for truck maneuvering, to a minimum.

The system should minimize truck back-up in areas where there can be spillage from the dipper. The loading area should be sufficiently removed from the face to keep the tires out of the rocks at the toe. Generally, a dozer is available to periodically clean the floor. In the single truck plan, it is assumed that the following truck is located to rapidly move into position as the loaded truck departs.

Figure 7.8 is another alternate operational plan in which the truck is to drive by for its loading cycle. Generally, the angle of swing of the shovel is held to less than 90 degrees, with the center of the machine in line with the highwall rather than the maximum cut depth illustrated. Increasing the depth of the cut reduces the frequency of machine moves because of the wider digging face area. In this arrangement, the trailing cable does not cross a truck driving area but the truck tires are more exposed to dipper spillage. The shovel can readily set oversized material aside without interferring with truck operations. Trucks are not required to back up into the loading position. Production is less than for the dual truck loading pattern.

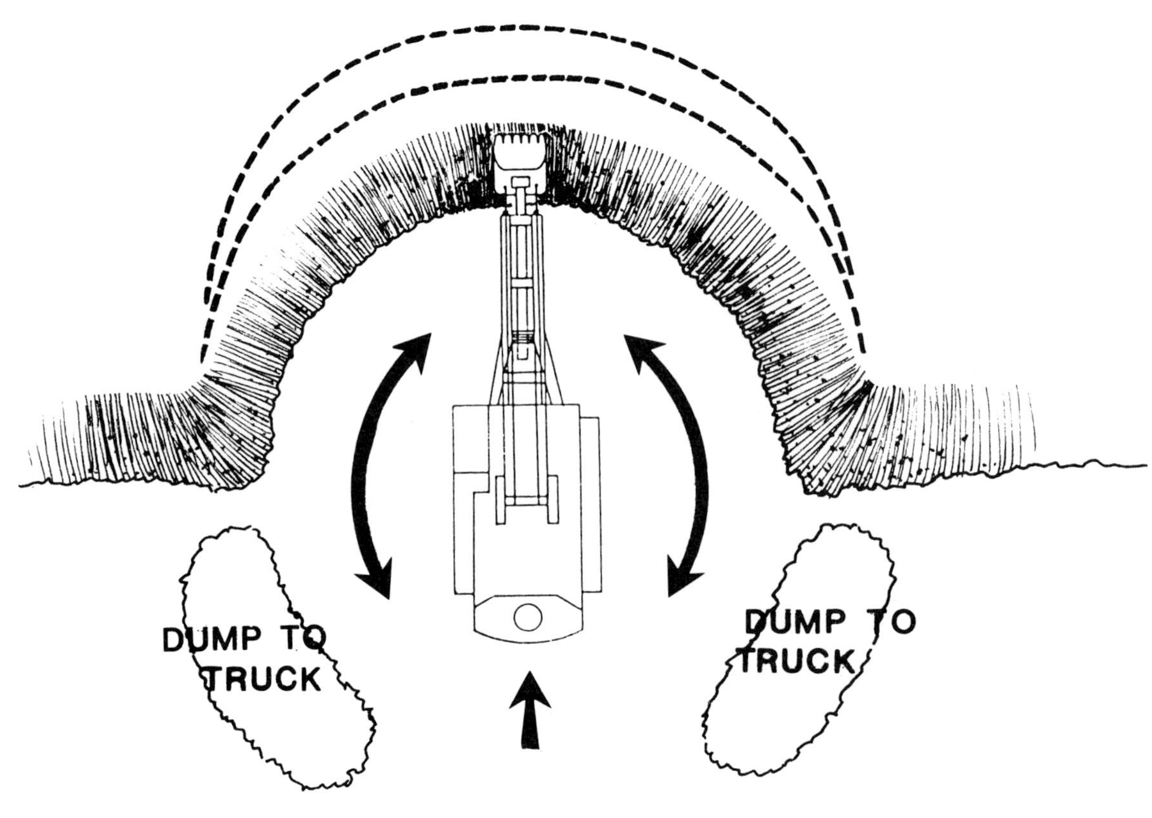

Figure 7.6 Typical Shovel-Truck Loading Plan: Double Truck Back-Up

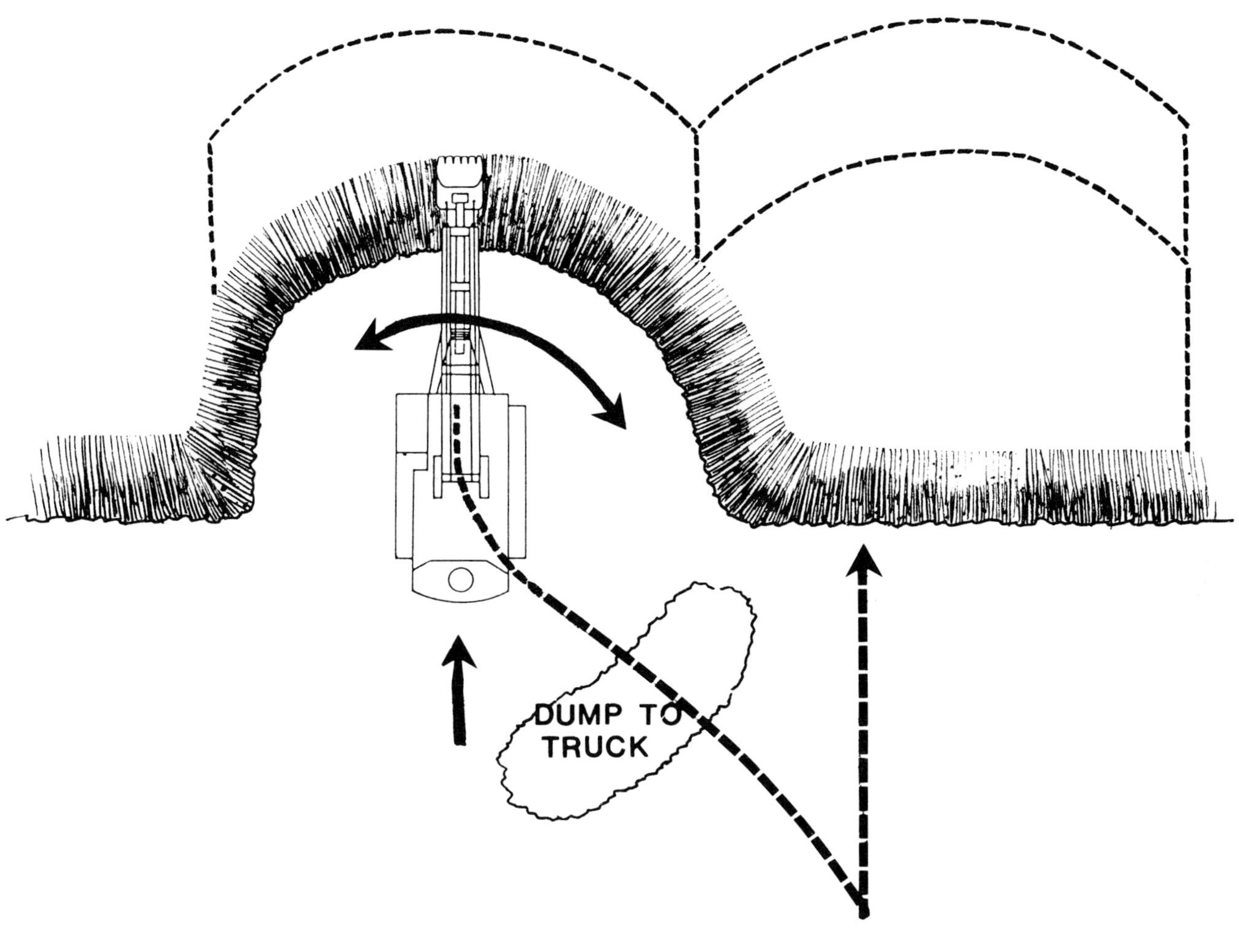

Figure 7.7 Typical Shovel-Truck Loading Plan: Single Truck Back-Up

DESIGN FEATURES

Common nomenclature is shown in Figure 7.9.

There has been considerable design development in the drive system on these machines in recent years. This development, together with mechanical configuration variations, has resulted in rather significant differences in the models from the three manufacturers. The major differences for each of the powered functions will be identified without attempting to assess the very complex impact the various features have on overall machine performance. The size, general arrangment and structural design is similar for all.

The pit power supply is three phase: 60 cycle, 4,160, 6,900 (7,200) or 14,400 volts brought on board the shovel via the trailing cable and high voltage collectors mounted between the upper and lower works.

The electric drives are currently in a design transition with manufacturers having different systems on models within their line. For many years, a Ward-Leonard motor generator system, using D.C. drive motors, was fairly standard (exceptions will be noted later). This system was progressively refined to include static type controls. More recently, static type A.C. to D.C. converters have been introduced as well as A.C. variable frequency systems.

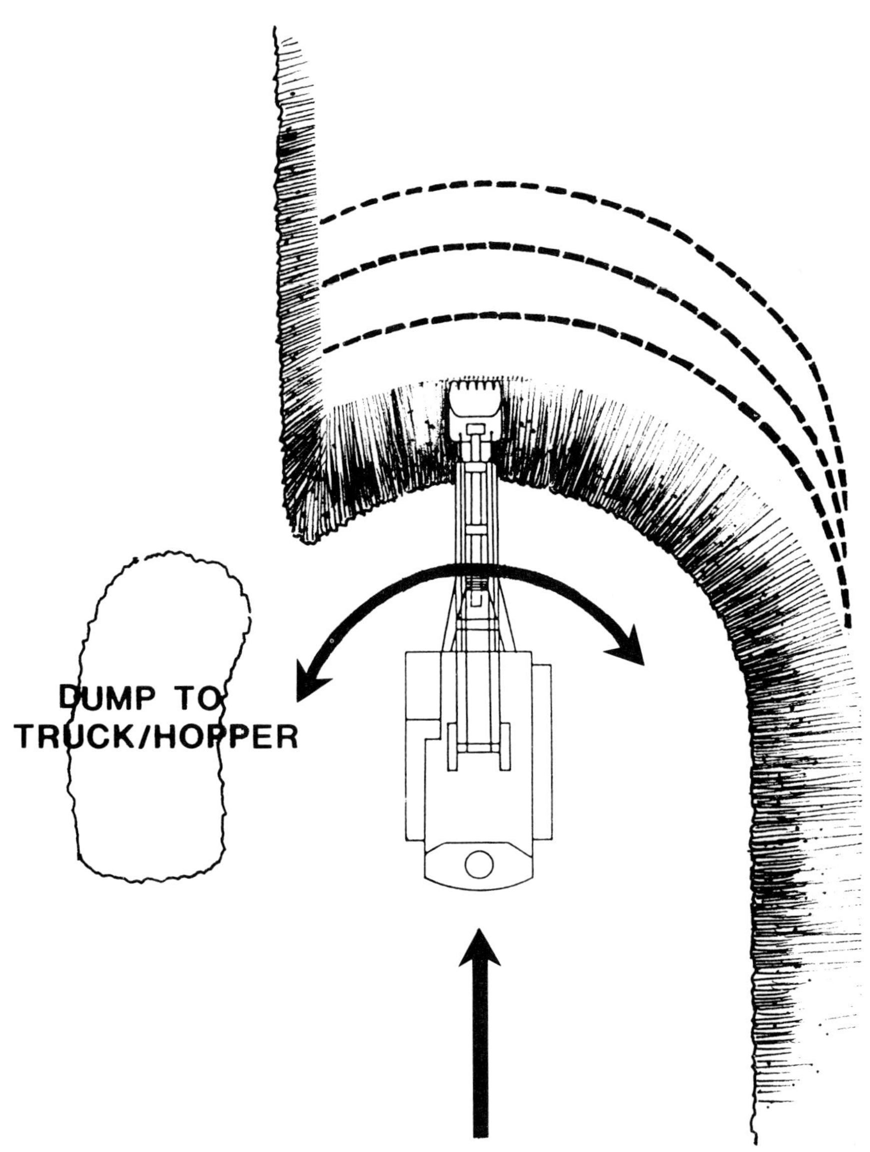

Figure 7.8 Typical Shovel-Truck Loading Plan: Truck Drive-By

At present, all three systems are available on selected models and can be very briefly summarized as follows:

- The Ward-Leonard motor generator system is a direct current, adjustable voltage system. A main A.C. motor, with power supplied through the trailing cable, drives separate D.C. generators for each function. The D.C. generators, in turn, power the armatures of D.C.motors. Individual D.C. mill-type motors drive the hoist, crowd, propel, and swing functions (multiple as required). It uses simplified static SCR (Thyristor) type controls. Motors have armature and field input for optimum control flexibility.

- The Ward-Leonard static generator system employs a tranformer to change high voltage A.C., from the trailing cable, to low voltage A.C. The A.C. power is then converted to D.C. by solid state electronics. D.C. mill-type motor drives are used with simplified static SCR (Thyristor) type controls to power the various functions. Motors have armature and field input for optimum control flexibility. The system has diagnostic aids and modularized construction.

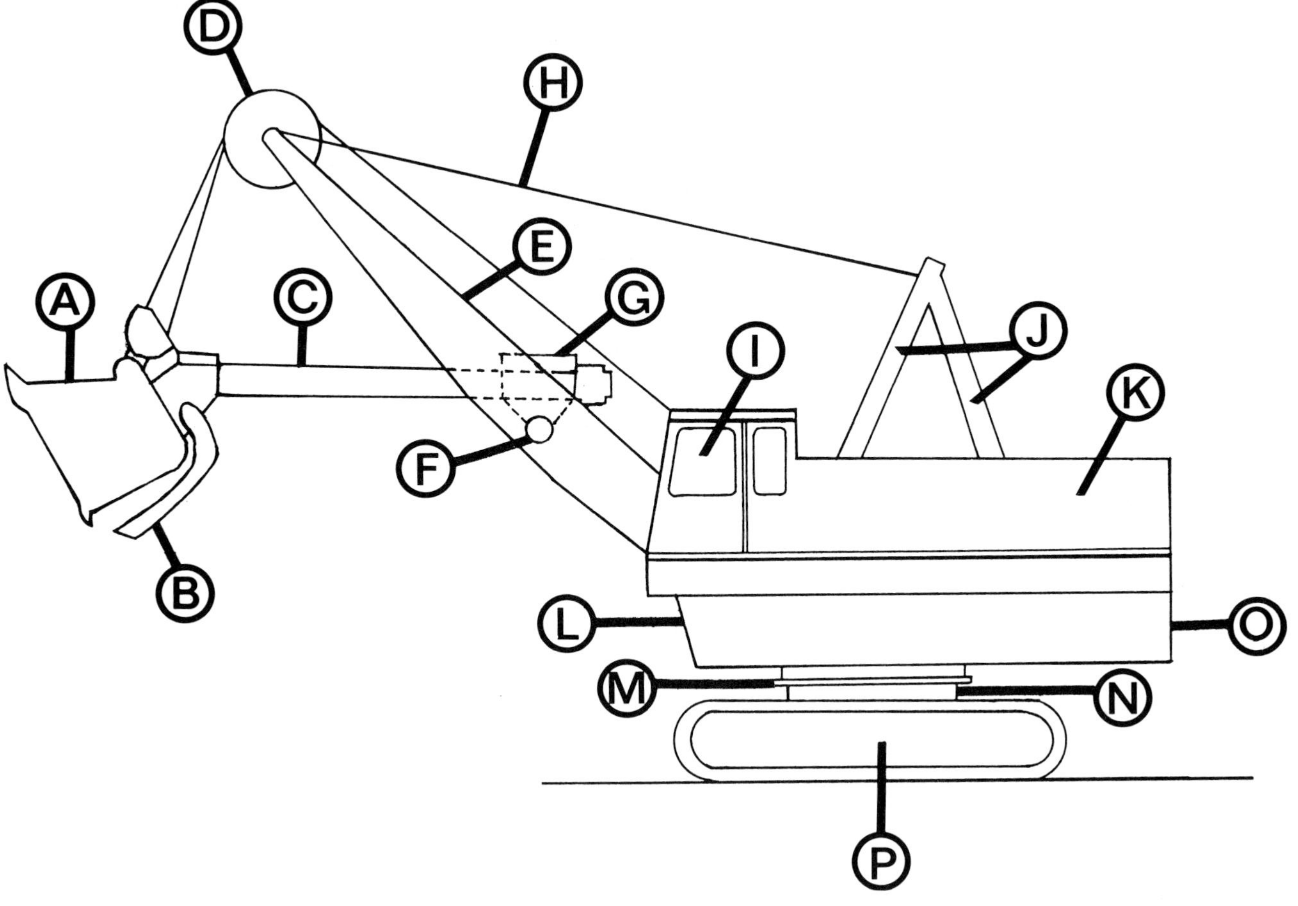

A	Dipper	I	Cab
B	Dipper Door	J	A–Frame
C	Dipper Stick (handle)	K	Machinery House
D	Point Sheave	L	Revolving Frame
E	Boom	M	Swing Circle (roller path)
F	Shipper Shaft	N	Lower Works
G	Saddle Block	O	Ballast Box
H	Suspension Cable	P	Crawler Side Frame

Figure 7.9 Shovel Nomenclature

- Static A.C. variable frequency systems employ a transformer and rectifier to convert to D.C. Inverters then convert to controlled frequency A.C. with separate units for each powered function (solid state). Individual functions are powered by A.C. induction squirrel cage motors with a simplified static type control. The system has diagnostic aids and modularized construction.

The Ward-Leonard motor generator system (still used almost exclusively on draglines) has proven to be a good system with relatively high overall efficiency (75%) and all the necessary control characteristics for excavator service. Power is fully reversible and, during the plugging or braking phase, regenerates usable power for other motions, with excess power returned to the line. A negative feature is the routine maintenance required for the motor generator set mounted across the rear of the revolving frame; this unit requires periodic alignment and balancing, commutators and brush changes. Figure 7.10 shows a typical deck layout for this drive system.

Static D.C. drives, which have no motor generator set, eliminate the associated maintenance and noise, substituting relatively trouble free components with a long life. The system has a faster response without any mechanical inertia. It also significantly reduces the peak current demand with machine start-up. There is a 12 to 15% improvement in overall efficiency. Figure 7.11 shows a typical deck plan for this drive system.

The static A.C. drives also eliminate the motor generator set and secure the advantages noted earlier. The D.C. mill-type drive motors are exchanged for A.C. squirrel cage motors which are cheaper, have higher thermal application ratings and can provide higher speeds and somewhat better accelerations. The A.C. motors have no commutators or brushes, further reducing routine maintenance. This system maintains a constant power factor and has the potential for reduced total power consumption. Power can be regenerated to supply other motions, but excess power cannot be fed back into the line. This system is more tolerant of voltage dips in the incoming power. Figure 7.12 shows a typical deck plan for this drive system.

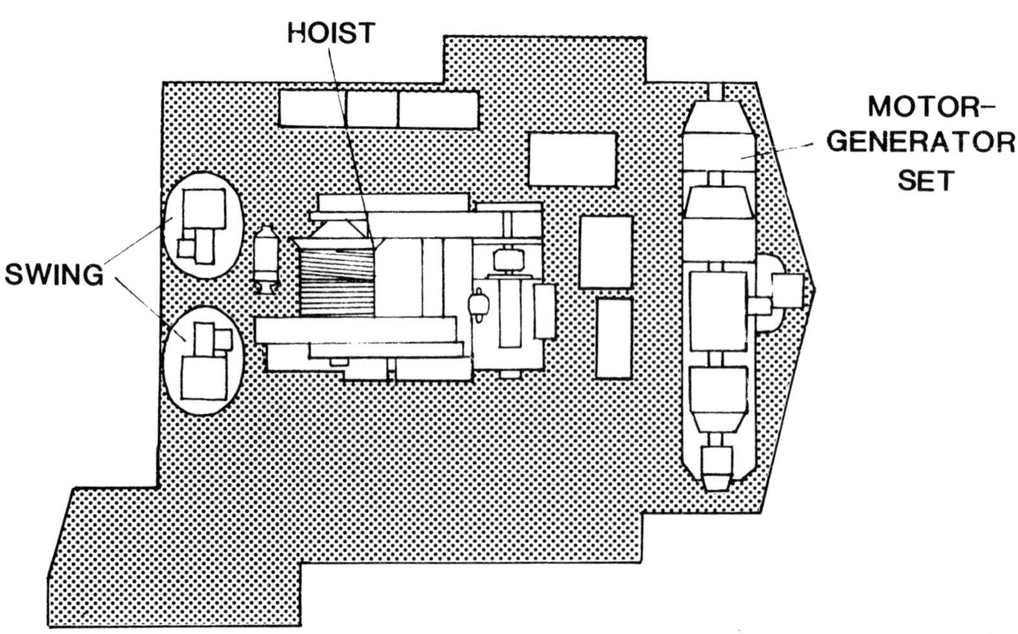

Figure 7.10 Typical Deck Plan — Motor Generator Drive System

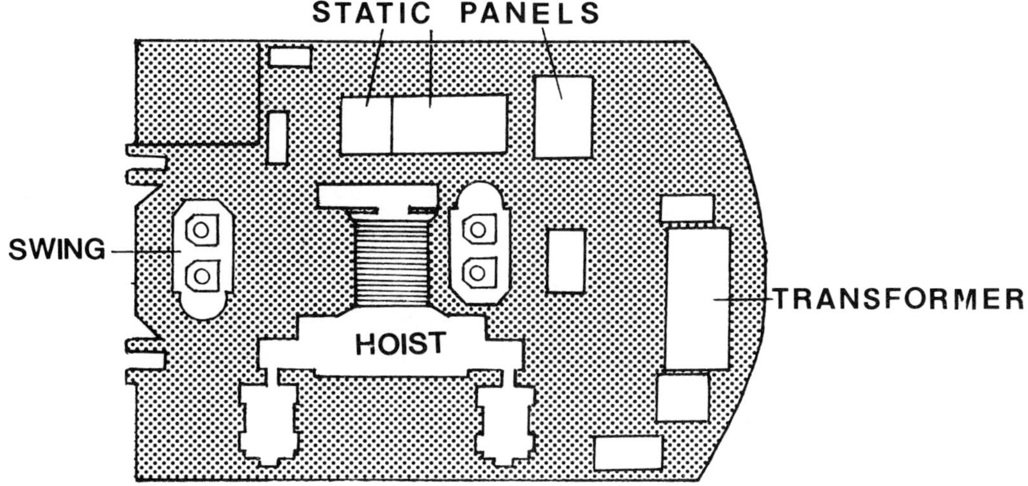

Figure 7.11 Typical Deck Plan — Static D.C. Drive System

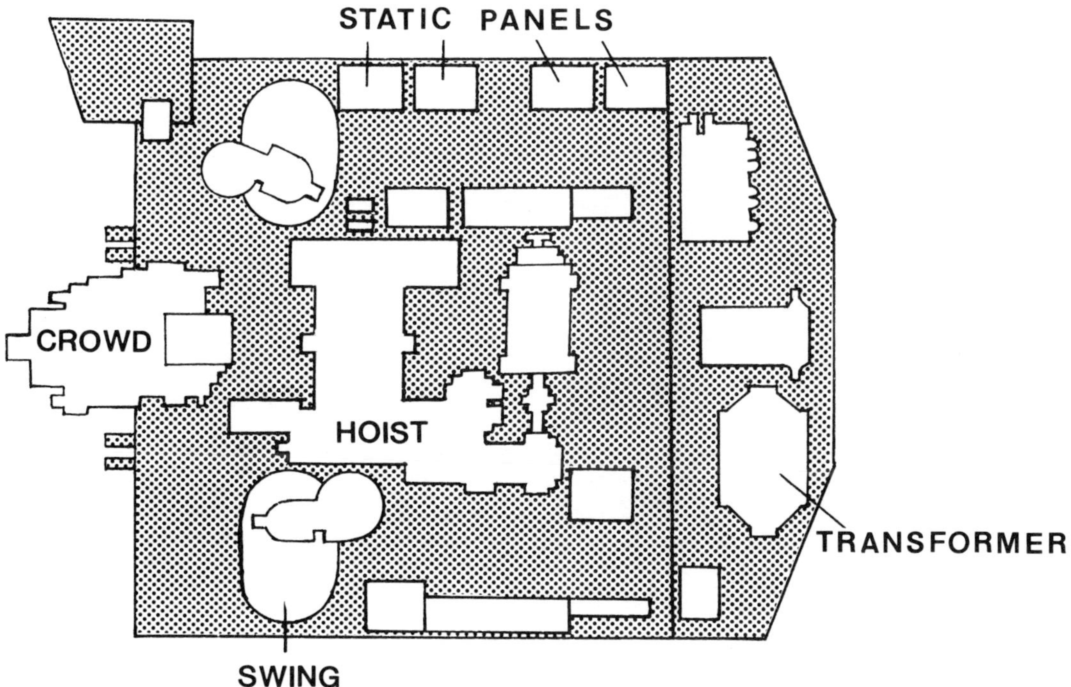

Figure 7.12 Typical Deck Plan — Static A.C. Drive System

There are two different hoist system designs in use today. The magnetorque version appears to be gradually phasing out as the new static drives are introduced on the various models.

- The conventional system (Bucyrus-Erie, Marion, larger P&H models) utilize D.C. or A.C. motors with a direct double gear reduction and drive to the cable drum. The system requires minimal total energy and is good in average digging. It provides excellent control of dipper speed and position. Power lowering is possible and it provides power regenerative energy or braking. In those cases where the hoist is also used for the propel drive, a disconnect clutch is provided.

- Magnetorque (P&H smaller machines) utilizes the main A.C. squirrel cage induction motor which operates constantly at full speed in one direction, driving the cable drum through an eddy-current clutch and double gear reduction. The clutch controls the drive speed by level of excitation; reduced excitation increases slip, reducing speed. It functions as a slipping clutch to absorb shock loads; lowering is dependent on gravity. The operator basically controls torque — high torques at low speeds, low torques at high speeds. The magnetorque system can develop very high torques at constant speed but consumes more energy and generates considerable heat. Power consumption is greater than a Ward-Leonard drive.

Each of the manufacturers has a different approach to the arrangements for the crowd drive system: (See Figure 7.13)

- Bucyrus-Erie uses crowd ropes with power supplied by variable speed D.C. or A.C. motors. Three levels of gear reduction are used to drive a crowd/retrack cable drum. Wire ropes, anchored at the drum, transmit power up the boom, over sheaves mounted on the shipper shaft, then along the handle. The crowd cable passes over a deflection sheave on top (mounted on shock absorber pads). The retract cable is attached to an adjustable take-up screw at the bottom. Wire rope elasticity, combined with cushioned sheaves, reduces shock loads. An electric feedback circuit provides additional crowd cushioning. The machinery is deck mounted, placed in a protected position which adds little to machine swing inertia. The dipper handle is a circular tube and is essentially free to rotate.

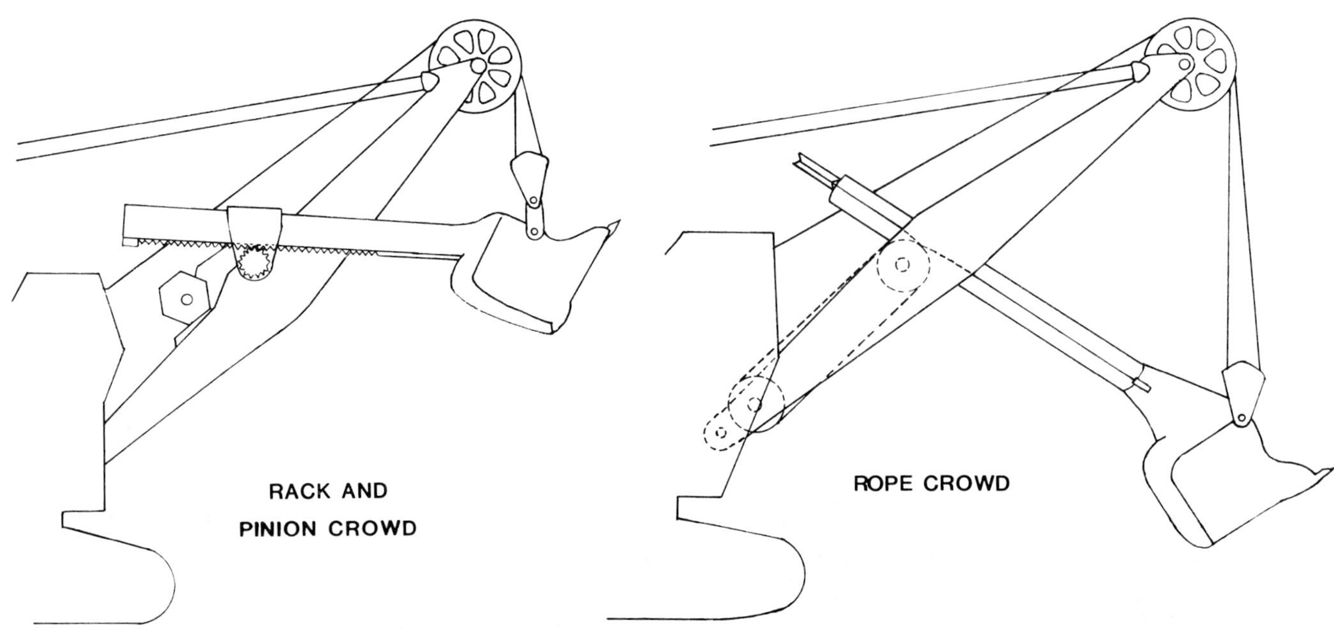

Figure 7.13 Crowd Drive Systems

- Marion has a rack & pinion crowd system with power supplied by a variable speed D.C. or hydraulic motor drive. The motor drives a pinion through a gear reduction which directly engages a rack on the lower side of the handle(s). Shock digging loads transmitted into the rack and pinion and drive are cushioned by an adjustable air clutch. The crowd machinery is located on the boom near the shipper shaft. The machinery is exposed and adds to machine swing inertia. The handle is made of either a single box-sectioned member, or twin box members, designed to straddle the boom.

- P&H employs a power band crowd system which is the rack & pinion design modified to incorporate a multi-belt drive (common back) between the D.C. motor and the final gear reduction, thereby cushioning shock loads. Long belts are used to permit locating the motor lower on the boom to reduce the swing inertia. It uses a twin legged, box sectioned, handle exclusively, with racks on the lower side.

The nature of the crowd drive influences the design of the dipper handle and boom. Any off-center loading of the dipper, while digging, is resisted by the hoist cable and/or the torsional rigidity of the handle. If a bail is used for a central hoist cable connection to the dipper, all of the torque must be carried through the handle. The rectangular box section handles, required by the rack & pinion crowd system, transmit torque to the boom. The handle and lower boom, therefore, must be designed to carry these loads. The rope crowd, with its tubular handle, has only minor torsional load transmitting capability through the handle. Hoist cable connections must be made to either side of the dipper and this is combined with a wider spread between the boom point sheaves to resist the offset loading.

Boom lengths are varied to meet specific mine plan requirements in terms of cutting height and dumping reach. As with all other machines of this type, stability requirements force reduction in dipper size as the range is extended. The structural design must support the torsional loads noted previously, as well as the axial loads introduced by the cables passing over the point sheave and the resulting interaction of the boom suspension cables. Because of the impact on machine swing inertia, the design must also minimize the weight. Bucyrus-Erie has, on some of its models, used a two-piece configuration with a very light upper section, in an

Photograph 7.6 Bucyrus-Erie 295B1 loading a large rear dump truck

effort to reduce overall weight. Harnischfeger includes, in its design, shock absorbing devices in the boom foot connections to the revolving frame.

Shovel dippers (see Figure 7.14) are sized to match the unit material weights to be handled; with ruggedness increasing with the severity of the expected digging conditions. Increasing the width can improve the rate of fill but reduces the penetration forces per unit of width. Reducing the depth may improve the filling. Light duty dippers are wide, approaching that of a loader but with intermediate reinforcing diaphrams to help withstand the high digging forces. The heavy duty versions are narrower and deeper than those for hydraulic shovels. Heavy cast lips of abrasive resistant steels, wrapping up the sides, provide the high strength required. Teeth concentrate the digging forces. The angle between the centerline of the handle and the bottom surface of the dipper and teeth (rake angle) is generally about 65 degrees. As noted earlier, the dipper is tapered to minimize rock jamming. The dipper door is hinged at the back and snubbers are provided to reduce its free swinging. The dipper latch is a simple sliding bar arrangement, activated through a light cable driven by a small electric motor. (See Photograph 7.7) The dipper represents a dead weight load that subtracts from the load that can be raised by the machine (payload); and, hence, its weight must be kept to a minimum.

Enclosed ball or roller bearing swing circles are not available in the sizes required by electric shovels. The swing circle located between the revolving frame and the lower frame is made of lower and upper segmented roller paths, with tapered rollers mounted in a circular carrier. Rollers and rails are of high alloy forged steel, all precision machined. Rollers are grease lubricated or may be provided with self-lubricated cast nylon bushings.

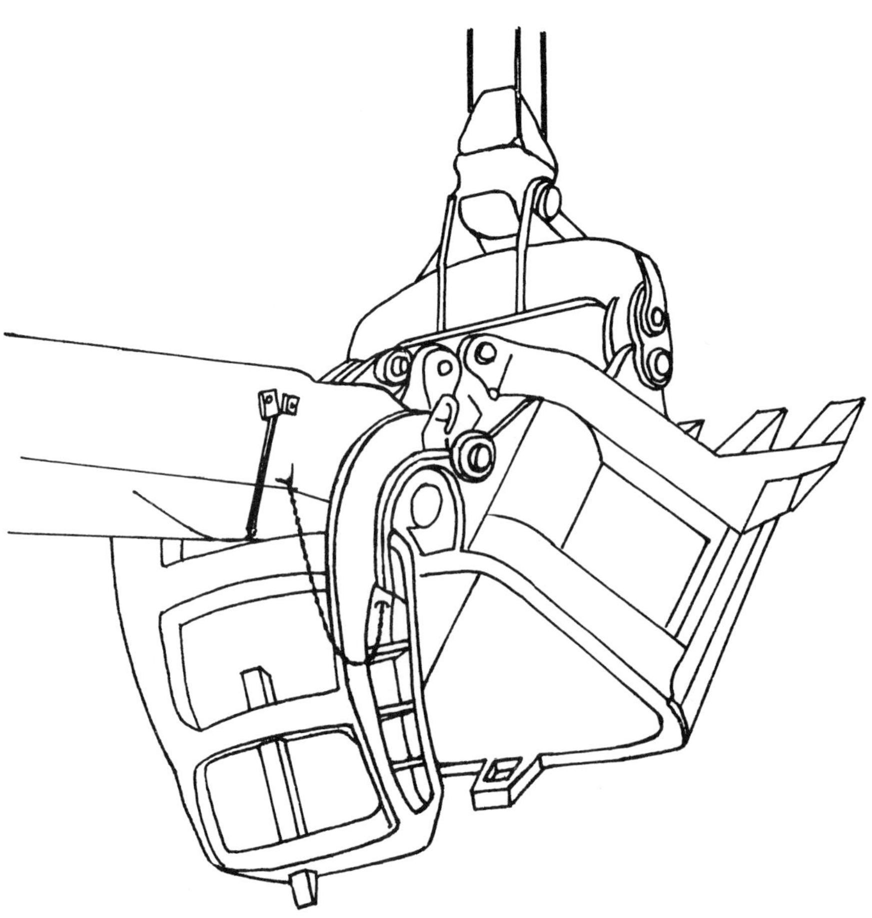

Figure 7.14 Shovel Dipper

Photograph 7.7 Shovel dipper dumping action

The external swing gear is generally a separate, segmented, high alloy casting with machined teeth, mounted outside of the swing circle. Swing pinions, from multiple drive modules, provide a balanced distribution of tooth forces around the ring gear. A center pintle (gudgeon) provides a transverse thrust type bearing which centers the upper works on the lower frame.

All structural units are made of welded high strength steels, with diaphrams to provide the required stiffness. A large ballast box extends across the rear of the revolving frame. This is filled with scrap or other high density materials during field erection, and adds to the machine stability during operation. Most of the gearing is fully enclosed, running in oil, with automatic lubrication for all critical bushings and/or bearings.

At present, there are four different propel systems available:

- Drive from the main machinery deck (selected Bucyrus-Erie and Marion models) is the oldest design. It utilizes the hoist motor, with a drive connection through the center pintle to the lower works. Slow speed bevel gears provide power to each track with intermediate steering jaw clutches and brakes. Tracks can be propelled simultaneously in one direction and/or one disconnected to run free or locked with brakes. The system is low cost because it eliminates a motor. However, it also has low efficiency and the machine must be stopped to engage the clutches.
- The lower works electric drive (P&H, selected models Bucyrus-Erie and Marion) has the drive motor centrally located on the rear of the truck frame. The transmission is a triple reduction, totally enclosed unit, with separately mounted gear cases for each track. Disc or jaw type steering clutches can be energized, while propelling, to power both tracks or either side independently. (See Figure 7.15) Each track has a spring set, air release brake. This system provides improved power utilization, improved accessibility, and less maintenance.

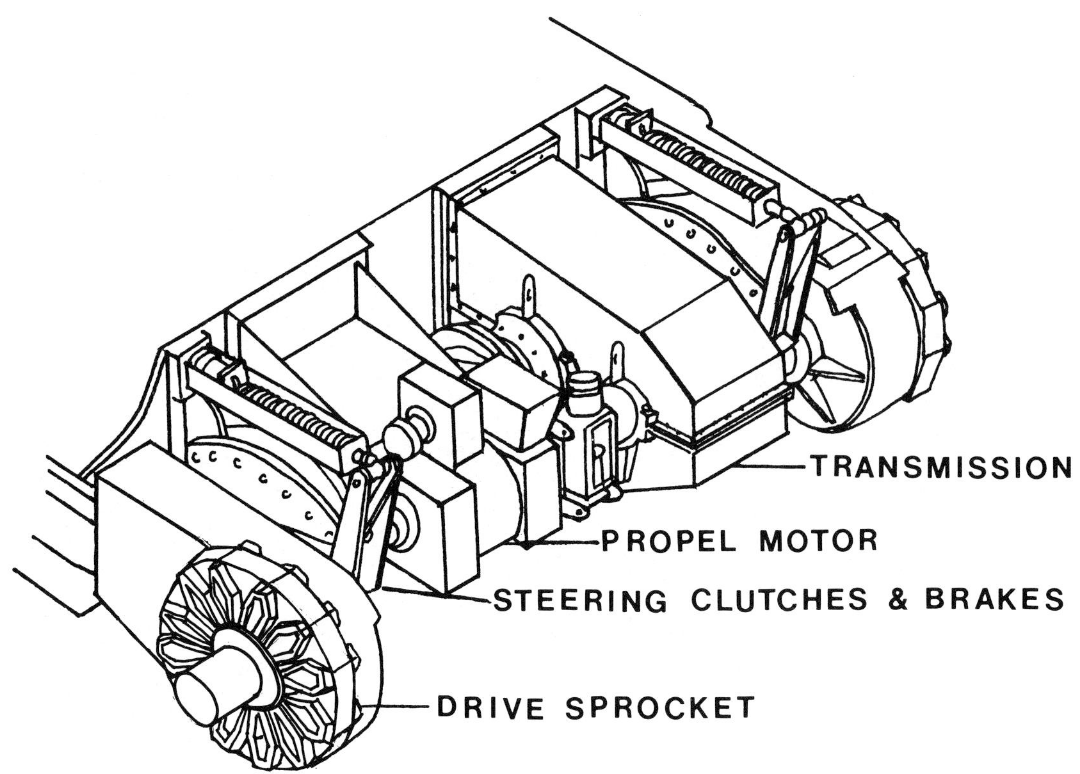

Figure 7.15 Propel Drive

- Lower works hydraulic drive (optional for Marion) has the drive located on the rear of the truck frame. Multiple hydraulic motors provide for individual track drive, so there is spin turn capability. Hydraulic pumps are located on the main machinery deck. There is limited experience with this system.

- Direct D.C. electric motor drive to each track (P&H 5700).

The low speed propel system is designed to provide a rugged working platform for the shovel. Track shoes are heavy, deep sectioned castings without grousers. Shoe width is dictated by site bearing pressure limitations. The track rollers are, in most cases, rigidly mounted to the frame. The new Bucyrus-Erie 395B (see Photograph 7.8) and the P&H 5700, the largest models in their product lines, have bogie mounted lower rollers to reduce the peak loading possible while traversing obstructions. Separate shoe pads are provided for the roller and drive tumbler contact areas. Bucyrus-Erie has introduced a tumbler with driving lugs which extend through the shoe to assure good driving contact regardless of the slack in the track shoe belt.

The electric shovels utilize flexible wire ropes to transmit power for the hosting and/or crowd motions. These ropes are lightweight, reducing the swing inertia and, in conjunction with their driving drums, can provide extended travel lengths at low investment. They are elastic and, hence, introduce a shock absorbing member into the system. In exchange, they must be periodically replaced. Good design and long life requires good lubrication, large bend radii, and smooth, well fitted grooves on the drums and sheaves. Failures normally result from: fatigue from bending over the drums and sheaves, and at the end of fittings; damage from falling rocks; or scuffing from working over rough surfaces.

Marion has produced a unique front end design on their model 204-M which is called the "superfront". It provides some of the features of a hydraulic machine but uses, essentially, a cable system. The front end configuration is quite complex: the lower boom section is articulated, the up-

Photograph 7.8 New Bucyrus-Erie 395B, 34 cubic yard shovel

per boom and handle are replaced by a combination rigid triangular frame and parallelogram members connected to the dipper. A pivoted intermediate crowd arm is rigidly connected to the triangular member, and, via a cable sheave arrangement, to the crowd machinery mounted on the A-frame. (See Photograph 7.9) The cable/drum hoist system pivots the triangular frame to raise the dipper through a block and tackle sheave sequence. Additional cables provide some dipper pitch modification as the hoist is actuated. A great number of pivots, sheaves, struts and cables are required to produce the desired digging action. The claimed advantages are: increased digging forces, variable dipper pitch (limited), a lighter weight front end, and improved level crowd capability.

SELECTION CONSIDERATIONS

Electric shovels are frequently selected where digging conditions are severe. These machines have successfully accumulated the most operating hours in all types of digging conditions around the world. They continue to be the standard to which other units are compared.

Their high mechanical availability in multi-shift operations, and their accompanying high production output meets the needs of larger mines. Digging cycle times of electric shovels are equal to or shorter than other types of loading equipment.

Electric shovels cannot be readily knocked down and moved to other sites, so that the mineral reserves must be sufficient to utilize the long service life of the machine and justify the high cost.

Electric shovels have unusual range (see Graph 7.4) which facilitates mining high benches (40 to 50 feet), loading trucks on an upper bench, and, in some cases, direct spoiling of overburden. They can be readily matched to the largest truck or rail haulage units currently available.

Machines can be periodically relocated around the pit to meet blending requirements or changes in mine plan. Frequent, long distance moves are not

Photograph 7.9 Marion 204M "superfront" shovel

desirable. A diesel drive (optional) will improve the relocating flexibility. Similarly, a skid-mounted diesel generator set, which can be towed behind the shovel, can provide additional move capabilities for a fleet of shovels. (See Photograph 7.10)

These machines are all electric and require a pit power distribution system. If large blast hole drills, draglines, truck trolley assist systems etc., are also to be employed, the power system will be an integral part of the mine planning and investment. A diesel generator set of the type referenced previously can be used for start-up when power has not reached the site.

Most ground bearing pressure requirements can be met with proper selection of the crawler length and track shoe size. The crawlers are adequate for working on bad floor conditions.

All service and maintenance is performed on the machine in the pit. Shop facility requirements are primarily for dipper rebuilding. Major electric motor repairs are made by exchange programs with regional specialty rebuilding shops. Regional parts depots provide service and replacement parts back-up.

The shovels are generally considered to have relatively low ownership and operating costs. Production calculations and cost estimates, however, are considered a separate subject too extensive to be comprehensively treated in this book. Estimates must be directly correlated with detailed site conditions and matched to the specific machine. The manufacturers will provide production and cost estimates for stated conditions prepared by their technical sales staff, based on their accumulated experience on similar applications. It should be recognized that these estimates are generally based on average data extrapolated to meet anticipated average site conditions over the life of the operations. The spread in these averages can be substantial. Some of the considerations entering into these calculations are:

Production

- Material swell

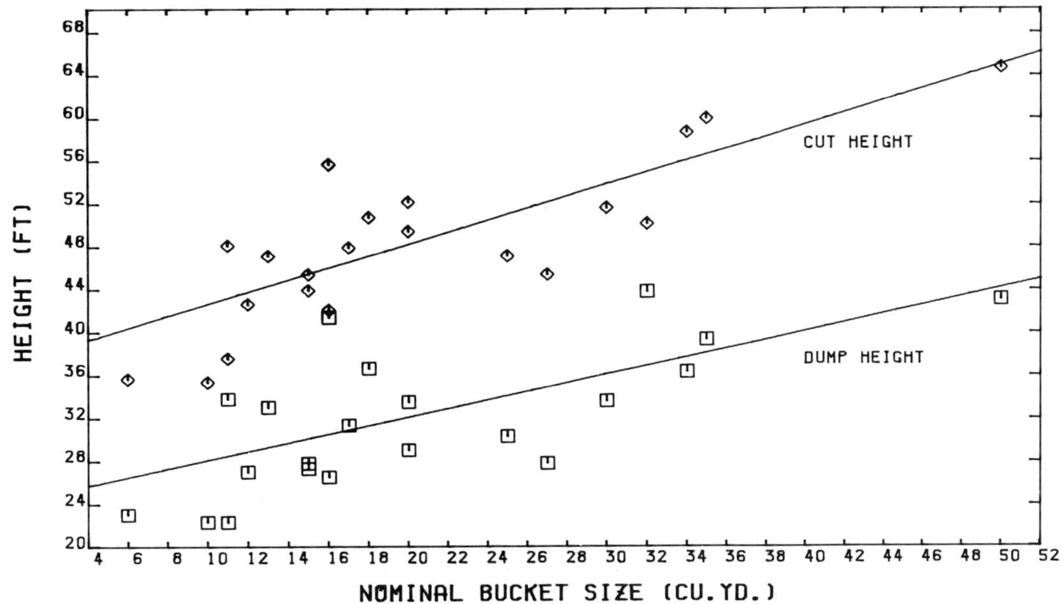

Graph 7.4 Dump height & cut height/nominal bucket size

- Material unit weight
- Dipper size
- Digging conditions
- Swing angle
- Loading conditions (including delays)
- Job efficiency
- Machine availability
- Operating schedule

Costs

- Machine price
- Freight
- Field erection
- Ballast
- Machine life
- Interest
- Insurance
- Taxes
- Power consumption
- Oils, greases
- Teeth and cables
- Crew wages
- Maintenance

Design options available vary with the manufacturer. In general, and depending on the specific model, options include the following:

- Alternate dipper sizes and types
- Boom lengths
- Diesel power
- Electrical
  — Incoming voltage
  — Motor-generator set manufacturer
  — Oversized blowers
  — Low voltage reactor starting
  — Anti-condensation heaters

Photograph 7.10 Skid mounted power supply

- High altitude filtering/ventilation
- Alternate crawlers shoe widths
- Alternate crawler frames
- Elevated cab
- Electric vs. hydraulic propel drive (Marion)
- Colors

Additional optional equipment consists of:

- Special features for low temperature operation:
  - Insulated cab
  - Heaters
  - Heated lube cabinets
  - Special hoses
  - Special steels for selected components
- Expanded automatic lubrication systems
- Auxiliary winch for rope handling
- Power cable reel (see Figure 7.16)
- Skid mounted diesel-electric generator set

NEW DEVELOPMENTS & TRENDS

The change-over to static drives will, undoubtedly, continue until all the product lines have been converted. As noted earlier, the improved efficiency, reduced routine maintenance, and diagnostic displays are desirable features, offsetting the increased cost.

A.C. variable frequency drives may be the next step in the technology evolution, but this will come only after a successful period of field evaluation and cost verification.

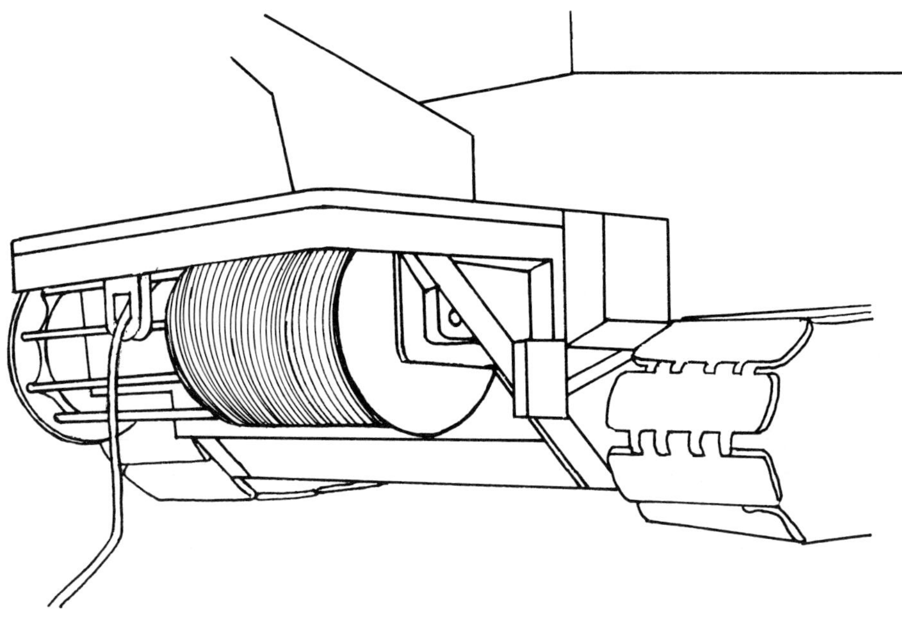

Figure 7.16 Cable Reel

Propel systems with the electric motor in the lower works will most likely be introduced progressively on all machine models. Independent track drive with a separate motor for each track does not appear economically attractive for the immediate future because of the limited need for the added steering capability on these large machines. The P&H 5700, which has such a system, requires two motors to develop adequate tractive power.

Machine sizes have continued to increase, with the P&H 5700 (nominal 50 cubic yard) the largest currently available. (See Photograph 7.11) Since the more popular truck sizes appear to be currently topping out at about 200 to 250 tons, there would appear to be limited demand for larger loading shovel models unless they are to be considered as stripping machines.

The superfront design, with the variable dipper pitch feature, will have to be assessed after added field experience more fully establishes the digging performance and maintenance requirements. An earlier effort by Bucyrus-Erie to produce a hydraulic front end arrangement with a variable pitch dipper was abandoned, in part, because of the design complexity. Special attachments to provide limited hydraulic positioning of the dipper have met with little interest because of the added weight.

Tests have been made of devices to introduce a vibratory motion to the cutting edge and improve shovel digging capabilities. None of these, to-date, have been offered by the manufacturers.

Considerable attention has been directed toward designs which will reduce the field erection time. This has led to increased modular construction and, in some cases, improved quality. The quality of the modular units can be more closely controlled by assembly and tests in the plant prior to shipment.

Gear lubricants are improving which is extending the life of the gear reduction units. Greater attention is being given to all aspects of lubrication in an effort to reduce service and maintenance requirements.

Photograph 7.11 New P&H 5700, 50 cubic yard shovel

MACHINE SPECIFICATIONS

See Figure 7.17 — Electric Shovel Dimensions
See Table 7.2 — Electric Shovel Specifications

Electric shovels are listed alphabetically by manufacturer, and then in ascending order by dipper size. The dipper size was determined by available published data on each machine model, rather than nominal capacity (3000 pound/cubic yard material), or general purpose application. Most machines are available with a range of dipper options; the smaller sizes are tough application rock dippers, and the larger sizes are light material or coal dippers. The specifications of those machines marked with an ( ) are equipped with a coal dipper. In some cases, special application dippers will require different boom lengths and angles than is standard.

Total HP figures are listed when available in the manufacturer's data. Total HP refers to the continuous power rating of the A.C. induction motor. The various component motors are rated at a 460–475 volts (continuous power). Hoist and propel functions sometimes share the same motor(s).

The track shoe width listed is based on the manufacturer's stated standard shoe. Optional shoes with flat track, single or multiple grouser designs and in various widths are usually available as options. Ground pressure will vary depending on the shoe width and design type.

Reach, cut radius, dump radius and boom and frame clearances are measured from the center of rotation of the machine.

All specifications, capacities, capabilities and dimensions are based on published manufacturer data. Although the information is believed to be current and the interpretation to be consistent and correct, it is possible that there may be some inaccuracies or out-of-date information included in the specification charts. These specification charts are provided as a general overview of the equipment, available sizes, and approximate operating data. Specific questions relating to performance or purchase should be directed to the manufacturer or authorized distribution/dealer in the user's specific area.

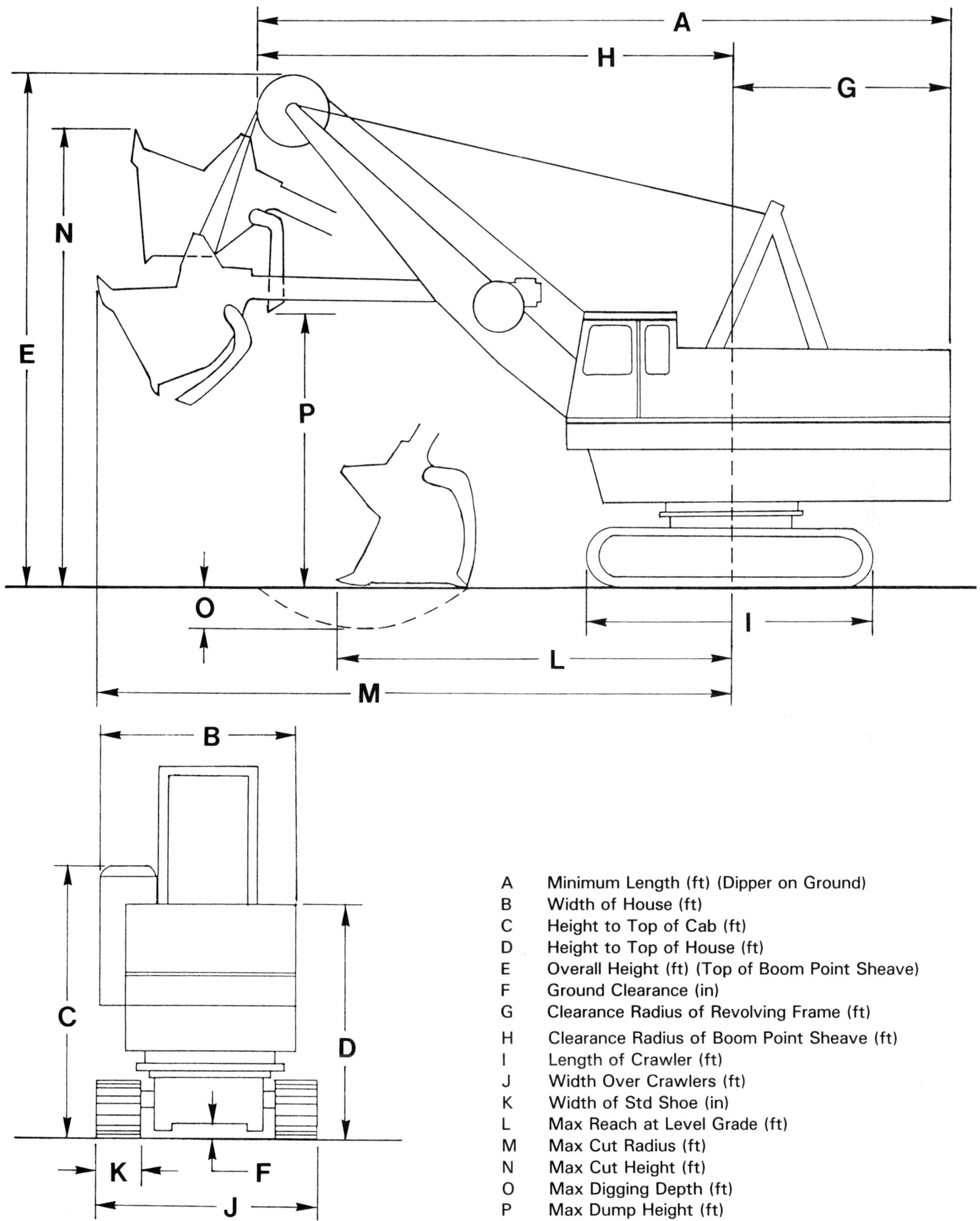

Figure 7.17 Shovel Dimensions

Table 7.2
ELECTRIC SHOVEL SPECIFICATIONS

MAKE	MODEL	NOMINAL BUCKET SIZE (cu yd)	AVAILABLE BUCKET SIZES (cu yd)	MOTOR GENERATOR DRIVE MOTOR TOTAL HP	HOIST MOTOR HP	CROWD MOTOR HP	SWING MOTOR HP	PROPEL MOTOR HP	WEIGHT (lbs)	BAIL PULL (lbs)	POWER SUPPLY (volts)
Bucyrus-Erie	155-B	11.0	9-19	400	375	88	176	375	568,600	166,000	2400/4160
Bucyrus-Erie	155-B()	13.0	13	450	500	88	176	500	563,000	120,000	2400/4160
Bucyrus-Erie	280-B	15.0	8-22	700	750	188	376	750	970,000	245,000	2400/4160
Bucyrus-Erie	195-B1	16.0	13-28	600	640	130	260	255	778,000	228,000	4160/7200
Bucyrus-Erie	195-B()	18.0	18	600	640	130	260	255	841,000	175,000	2400/4160
Bucyrus-Erie	290-B1	20.0	15-34	800	800	195	390	255	1,023,000	275,000	4160/7200
Bucyrus-Erie	295-BII	27.0	20-45	n.a.	1250	225	450	675	1,470,000	—	4160/7200
Bucyrus-Erie	395-B	34.0	25-60	n.a.	1500	390	660	1070	1,812,000	—	4160/7200
Marion	151-M	10.0	7-12	450	500	88	176	500	471,000	—	4160
Marion	151-M()	11.0	7-12	500	500	88	260	500	500,000	—	4160
Marion	191-M-HR	15.0	12-22	n.a.	800	255	390	375	1,015,000	—	—
Marion	181-M()	16.0	8-16	600	700	130	260	700	738,000	—	—
Marion	182-M	16.0	16	600	700	130	260	700	746,000	—	—
Marion	201-M	20.0	16-40	n.a.	1045	390	510	500	1,275,000	—	—
Marion	204-M	32.0	20-40	n.a.	1045	640	510	500	1,372,000	—	—
P & H	1600	6.0	6-14	—	250	95	140	150	494,000	142,000	—
P & H	1900-AL	12.0	10-25	600	600	130	225	235	805,000	205,000	2400/4160
P & H	2100-BL	17.0	17-25	750	880	325	440	400	1,090,000	250,000	4160/7200
P & H	2300	22.0	18-40	n.a.	1120	260	440	360	1,410,000	—	4160/7200
P & H	2800	30.0	24-50	n.a.	1400	400	680	475	1,876,000	460,000	4160/7200
P & H	5700	50.0	50-75	n.a.	2800	700	1300	810	3,350,000	800,000	14,400

Table 7.2 (Continued)
ELECTRIC SHOVEL SPECIFICATIONS

MAKE	MODEL	DRIVE TYPE	CROWD TYPE	PROPEL TYPE	NO. OF SWING UNITS	DRUM DIAMETERS HOIST (in)	DRUM DIAMETERS CROWD (in)	ROPE DATA NO. AND SIZE HOIST (no.-in)	ROPE DATA NO. AND SIZE CROWD (no.-in)
Bucyrus-Erie	155-B	DC/MG set	rope	upper	2	35	30	2-1.38	1-1.50
Bucyrus-Erie	155-B()	DC/MG set	rope	upper	2	44	30	2-1.25	1-1.38
Bucyrus-Erie	280-B	DC/MG set	rope	lower	2	42	34	2-1.75	1-1.75
Bucyrus-Erie	195-B1	DC/MG set	rope	lower	2	40	30	2-1.50	1-1.75
Bucyrus-Erie	195-B()	DC/MG set	rope	lower	2	48	30	2-1.38	1-1.50
Bucyrus-Erie	290-B1	DC/MG set	rope	lower	2	45	42	2-1.88	1-1.88
Bucyrus-Erie	295-BII	static AC	rope	lower	-	—	—	—	—
Bucyrus-Erie	395-B	static AC	rope	lower	2	56	54	2-2.38	1-2.38
Marion	151-M	DC/MG set	gear & pinion	upper	2	42	n.a.	2-1.75	n.a.
Marion	151-M()	DC/MG set	gear & pinion	upper	2	42	n.a.	2-1.75	n.a.
Marion	191-M-HR	static DC	gear & pinion	lower	2	48	n.a.	2-1.88	n.a.
Marion	181-M()	DC/MG set	gear & pinion	lower	2	48	n.a.	2-1.75	n.a.
Marion	182-M	DC/MG set	gear & pinion	lower	2	48	n.a.	2-2.13	n.a.
Marion	201-M	static DC	gear & pinion	lower	2	—	n.a.	—	n.a.
Marion	204-M	static DC	rope	lower	2	50	50	2-2.00	2-2.00
P & H	1600	—	—	—	—	—	—	—	—
P & H	1900-AL	DC/MG set	gear & pinion	lower	2	38	n.a.	2-1.50	n.a.
P & H	2100-BL	DC/MG set	gear & pinion	lower	2	44	n.a.	2-1.75	n.a.
P & H	2300	static DC	gear & pinion	lower	2	50	n.a.	2-2.25	n.a.
P & H	2800	static DC	gear & pinion	lower	2	56	n.a.	2-2.25	n.a.
P & H	5700	static DC	gear & pinion	lower	4	75	n.a.	2-3.00	n.a.

Table 7.2 (Continued)
ELECTRIC SHOVEL SPECIFICATIONS

MAKE	MODEL	MINIMUM LENGTH BUCKET RETRACTED (ft)	WIDTH OF HOUSE (ft)	HEIGHT TO TOP OF CAB (ft)	HEIGHT TO TOP OF HOUSE (ft)	HEIGHT TO BOOM POINT SHEAVE (ft)	STANDARD BOOM LENGTH (ft)	AVAILABLE BOOM LENGTHS (ft)	EFFECTIVE HANDLE LENGTH (ft)	GROUND CLEARANCE (in)	REVOLVING FRAME CLEARANCE RADIUS (ft)	BOOM POINT CLEARANCE RADIUS (ft)	CRAWLER DATA: OVERALL LENGTH (ft)	CRAWLER DATA: OVERALL WIDTH (ft)	CRAWLER DATA: STANDARD SHOE WIDTH (in)	GROUND PRESSURE (psi)
Bucyrus-Erie	155-B	55.6	20.5	23.3	—	38.3	38.0	38-52	21.7	13	18.6	37.0	20.5	17.0	36	36.8
Bucyrus-Erie	155-B()	68.0	20.5	—	17.2	48.3	52.0	38-52	30.5	13	21.0	47.0	24.3	21.5	60	15.8
Bucyrus-Erie	280-B	69.3	25.8	20.8	—	49.3	50.0	—	29.5	18	21.5	47.8	26.3	21.0	42	42.7
Bucyrus-Erie	195-B1	60.8	22.5	19.2	—	42.5	41.5	42-70	24.3	21	20.8	40.0	22.3	20.8	42	41.2
Bucyrus-Erie	195-B()	74.3	22.5	19.2	—	53.3	56.5	42-70	32.8	21	23.5	50.8	22.3	23.8	78	25.0
Bucyrus-Erie	290-B1	68.8	25.8	26.4	—	47.5	47.0	47-80	29.0	23	23.0	45.8	26.3	22.0	42	40.6
Bucyrus-Erie	295-BII	72.3	28.5	27.7	—	50.0	50.0	50-85	28.7	22	23.7	48.6	28.3	23.5	48	47.4
Bucyrus-Erie	395-B	81.8	32.0	31.8	—	58.5	56.0	56	35.9	21	26.5	55.3	33.6	29.1	72	36.2
Marion	151-M	55.5	17.3	—	18.0	38.3	38.0	38	—	6	18.0	37.5	20.8	17.5	36	31.8
Marion	151-M()	60.5	19.2	—	18.0	45.5	45.0	45	—	6	21.0	39.5	20.8	17.5	42	29.0
Marion	191-M-HR	70.6	23.3	—	21.0	48.0	47.0	47	—	26	25.0	45.6	27.0	24.5	42	41.9
Marion	181-M()	71.5	18.8	—	19.8	56.0	57.0	57	—	24	23.0	48.5	30.0	22.5	50	20.0
Marion	182-M	71.5	18.8	—	20.1	56.0	57.0	57	—	11	23.0	48.5	25.8	21.5	55	25.2
Marion	201-M	74.3	22.6	—	—	—	—	—	—	30	24.5	49.8	27.4	24.8	46	49.0
Marion	204-M	68.5	24.3	—	21.6	41.0	n.a.	n.a.	—	30	28.5	40.0	27.4	25.6	55	44.0
P & H	1600	—	—	—	—	—	35.0	—	22.5	—	18.0	—	21.3	17.4	36	—
P & H	1900-AL	63.3	26.4	27.5	25.6	43.0	40.0	—	27.0	21	23.0	40.3	25.0	22.0	42	35.3
P & H	2100-BL	73.1	29.3	28.0	28.8	51.5	50.0	—	30.0	22	24.6	48.5	27.0	24.0	42	46.4
P & H	2300	76.1	34.3	29.6	26.0	52.2	50.0	—	31.8	25	26.0	50.1	28.6	25.2	48	49.7
P & H	2800	78.3	34.3	30.5	29.3	55.3	51.0	—	34.0	28	26.0	52.3	33.3	29.7	56	47.5
P & H	5700	98.3	48.3	42.3	40.8	67.5	62.0	—	41.0	48	37.3	61.0	45.6	40.8	90	39.4

Table 7.2 (Continued)
ELECTRIC SHOVEL SPECIFICATIONS

MAKE	MODEL	MAX SPEED (mph)	CONTINUOUS GRADEABILITY (%)	MAX GRADEABILITY (%)	SWING SPEED (rpm)	LOADED DIPPER SPEED (fpm)	MAX REACH LEVEL GRADE (ft)	MAX CUT RADIUS (ft)	MAX CUT HEIGHT (ft)	MAX DIGGING DEPTH (ft)	MAX DUMP RADIUS (ft)	MAX DUMP HEIGHT (ft)	REACH AT MAX DUMP HEIGHT (ft)
Bucyrus-Erie	155-B	1.0	13	36	3.3	166	34.0	50.0	37.5	7.1	41.8	22.3	40.8
Bucyrus-Erie	155-B()	1.1	17	38	3.2	273	40.3	62.0	47.0	6.0	54.5	33.0	52.0
Bucyrus-Erie	280-B	1.1	14	39	3.0	225	40.5	62.3	43.8	8.5	54.3	27.3	53.5
Bucyrus-Erie	195-B1	0.9	14	39	3.1	201	39.0	55.8	42.0	8.3	48.5	26.5	46.8
Bucyrus-Erie	195-B()	0.9	12	35	3.1	240	43.5	65.8	50.6	8.5	59.5	36.6	57.0
Bucyrus-Erie	290-B1	0.8	12	35	3.0	218	43.0	62.8	49.3	9.3	54.5	29.0	53.5
Bucyrus-Erie	295-BII	0.9	12	34	3.1	207	43.7	64.8	45.3	8.5	55.3	27.8	54.5
Bucyrus-Erie	395-B	0.9	15	37	—	—	50.9	77.4	58.5	9.3	66.8	36.3	64.8
Marion	151-M	1.3	—	—	3.6	—	34.3	50.8	35.3	—	43.5	22.3	41.5
Marion	151-M()	1.1	—	—	3.0	—	36.0	57.0	48.0	—	50.5	33.8	48.5
Marion	191-M-HR	1.4	—	—	3.0	—	42.6	63.5	45.3	—	54.0	27.8	52.8
Marion	181-M()	1.0	—	—	3.0	—	44.5	67.8	55.5	—	60.5	41.3	56.8
Marion	182-M	1.0	—	—	3.0	—	44.5	67.8	55.5	—	60.5	41.3	56.8
Marion	201-M	0.8	—	—	3.0	—	46.8	67.5	52.0	—	57.7	33.5	—
Marion	204-M	0.8	—	—	2.6	—	61.0	67.5	50.0	—	55.0	43.7	35.0
P & H	1600	—	—	—	—	—	30.3	48.5	35.6	—	42.8	23.0	—
P & H	1900-AL	1.0	—	—	3.0	150	39.0	58.5	42.5	—	53.0	27.0	52.0
P & H	2100-BL	0.9	—	—	3.2	148	44.3	65.8	47.8	—	57.5	31.3	55.5
P & H	2300	1.0	—	—	—	—	50.1	70.7	47.0	—	61.3	30.3	60.3
P & H	2800	0.7	—	—	2.6	182	47.4	76.3	51.5	—	68.8	33.6	63.1
P & H	5700	0.8	—	—	—	—	56.0	88.0	64.5	—	77.5	43.0	74.2

Chapter 8

DRAGLINES

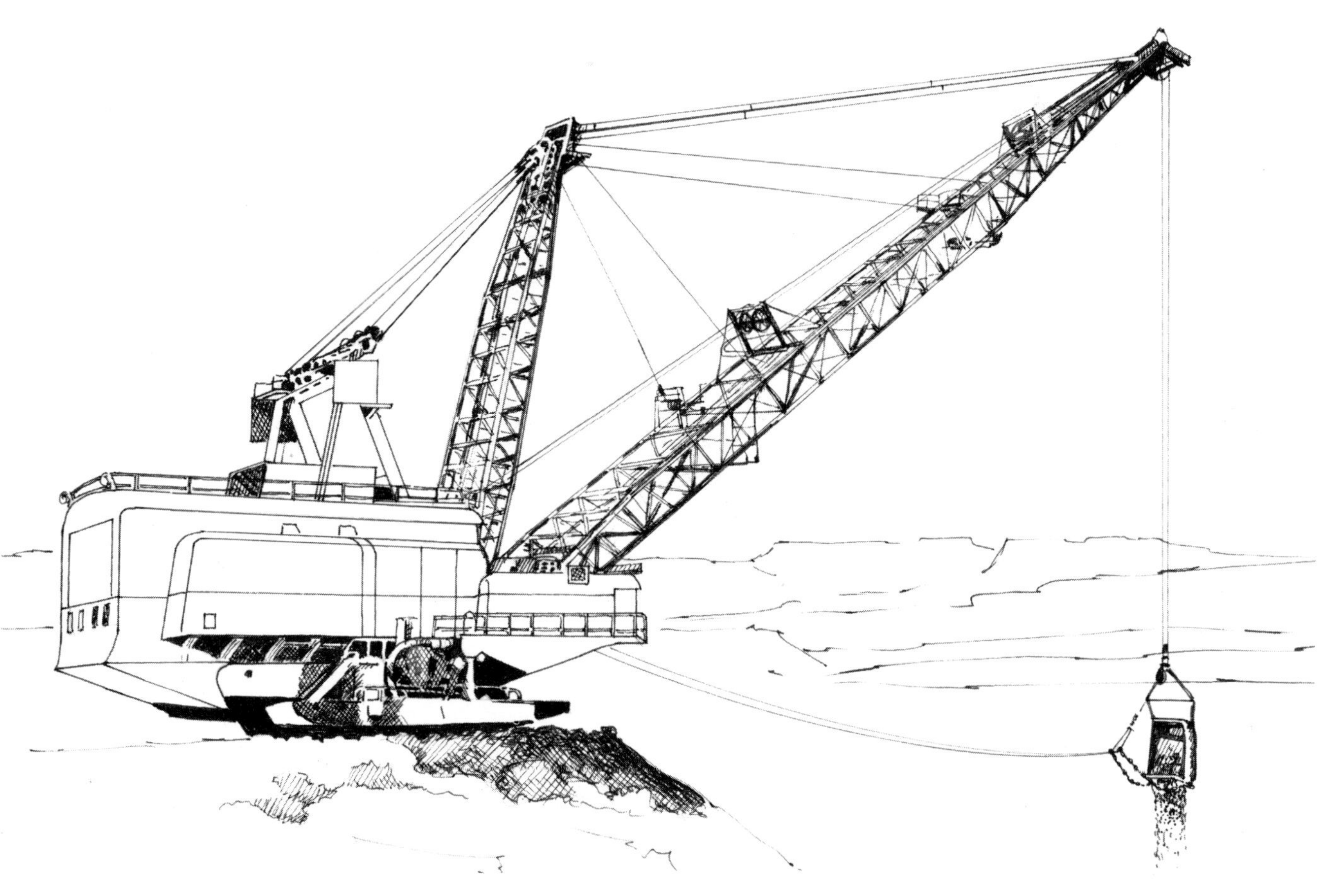

Figure 8.1 Walking Dragline

Through the years, the dragline has remained a unique excavating tool and has experienced a dramatic growth in maximum size. With its long reach and ability to dig to substantial depths below itself, it has had broad applications on many irrigation projects and, in more recent years, in surface mining. The hydraulic hoe has, to some extent, replaced the smaller sized diesel draglines but the larger diesel and/or electric machines retain their popularity. They, along with the bucket wheel excavators, are the largest pieces of mobile equipment currently manufactured. The text analysis will include both crawler mounted and walking machines of approximately 6 cubic yards and larger. Many of the electric mining shovels have been offered with dragline booms; these have not been popular in recent years and are not considered.

Photograph 8.1 American crawler dragline removing overburden

TYPICAL UNITS

The two types of draglines are differentiated by their method of propel. While there is some overlap, the walking draglines generally are larger in size. The high horsepower requirements for the walking machines commonly necessitates an all electric drive system with a trailing cable bringing power to the machine. Both machines operate in the same manner, with their various applications dictated, to a large extent, by their size. These machines are illustrated in Figure 8.2.

Crawler Mounted Draglines

- Diesel engine with mechanical transmission
- Capacities from 6 to 18 cu. yd.
- Horsepowers from 550 to 2000
- Machine weights from 375,000 to 750,000 lb.
- Boom lengths from 90 to 200 ft.
- Propel speeds about 1 mph

Walking Draglines

- Generally all electric drives with a trailing cable
- Capacities from 9 to 220 cu. yd.
- Horsepowers from 1,200 to 24,000
- Machine weights from 1,000,000 to 30,000,000 lb.
- Boom lengths from 175 to 375 ft.
- Propel speed about 0.15 mph

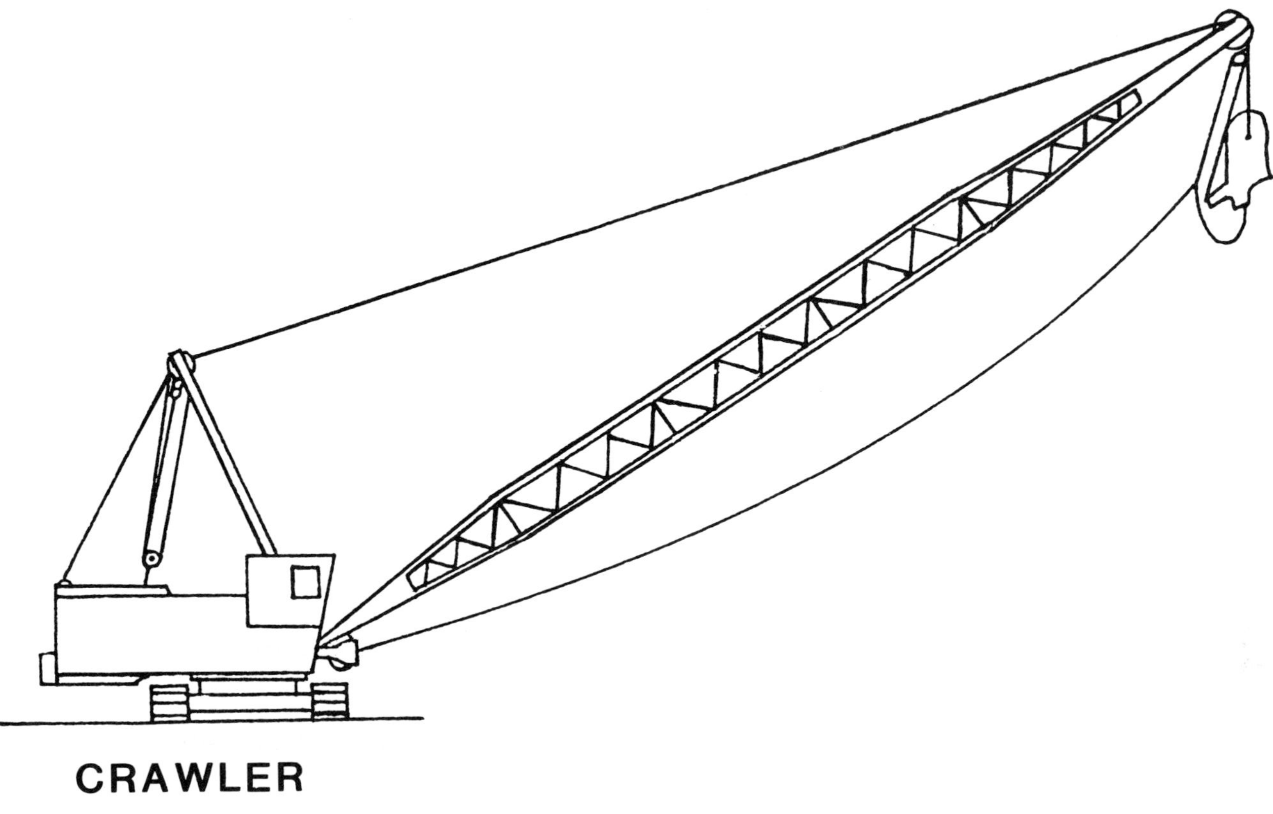

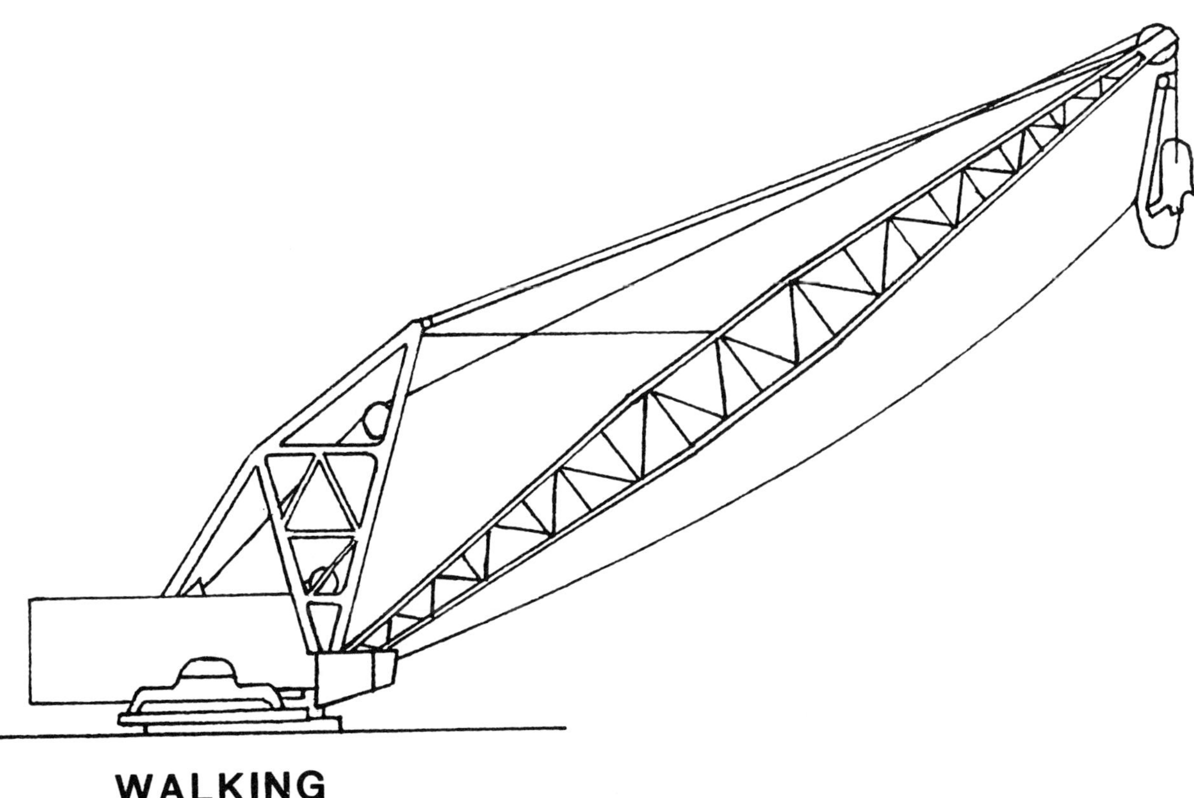

Figure 8.2 Types of Draglines

Manufacturers

Most of the draglines in service in the U.S. were supplied by domestic manufacturers. The crawler mounted units are produced by about eight companies who also manufacture shovels and cranes in similar sizes. (See Table 8.1)

There are four manufacturers of the large electric, walking draglines extensively employed today in mining. Bucyrus-Erie and Marion Power Shovel are the two largest and have, in the past, offered stripping shovels in a comparable size range. The third manufacturer, Page Engineering has supplied machines in the smaller sizes but more recently has expanded into the intermediate range. The fourth company, Ransomes and Rapier Ltd., a British company, discontinued its sales for a number of years and only recently has, again, offered this product line.

The smaller crawler machines are normally sold and serviced through a dealer or manufacturer's representative. Walking dragline sales and service are handled directly by the manufacturer's marketing organization.

BASIC MACHINE OPERATION

The dragline is characterized by a long, lattice boom. This boom gives the dragline substantial reach which allows it to both dig and dump material at a significant radius. Diesel machines can be either crawler mounted or of the walking type. Electric machines are commonly equipped with walking devices. The following brief, operational, explanation considers only the larger units with walking devices.

- Machine mobility is restricted by the large machine size and the trailing cable; long moves require special preparations for shifting the trailing cable and for preparing a roadway. The walking device consists of large rectangular shoes attached by a cam arrangement to either side of the main deck, and located behind the machine center of gravity. To move the machine, the cams are rotated to lower the shoes to the ground and then lift the rear of the machine

Photograph 8.2 Page walking dragline swinging a bucket load of overburden

Table 8.1

DRAGLINE MANUFACTURERS

American Hoist & Derrick Co. 63 S. Robert Street St. Paul, Minnesota 55107	Crawler mounted to 10 cu. yd. 100–160 ft. booms
Bucyrus-Erie Company 1100 Milwaukee Avenue South Milwaukee, WI 53172	Crawler mounted to 12 cu. yd. 50 to 190 ft. booms Walking 9–220 cu. yd. 140–400 ft. booms
Clark Equipment Company Crane Division 1046 S. Main Street Lima, Ohio 45802	Crawler mounted to 8 cu. yd. 100–150 ft. booms
Harnischfeger Corporation P.O. Box 554 Milwaukee, WI 53201	Crawler mounted to 18 cu. yd. 30–200 ft. booms
Manitowoc Engineering Co. Div. of Manitowoc Co., Inc. 500 S. 16th Street Manitowoc, WI 54220	Crawler mounted to 15 cu. yd. 100 to 200 ft. booms
Marion Power Shovel Div. Dresser Industries, Inc. 617 W. Center Street P.O. Box 505 Marion, Ohio 43302	Crawler mounted to 17.5 cu. yd 120–170 ft. booms Walking 9–180 cu. yd. 160–400 ft. booms
Northwest Engineering Co. 201 W. Walnut Street Green Bay, WI 54305	Crawler mounted to 6 cu. yd. 60–110 ft. booms
Page Engineering Company Clearing Post Office Chicago, IL 60638	Walking 11–70 cu. yd. 175–350 ft. booms
Ransomes & Rapier Ltd. Box No. 1 Waterside Works Ipswich 1P2 8HL England	Walking 10–70 cu. yd. 140–350 ft. booms
Weserhutte Aktiengesellschaft P.O. Box 100940 Bad Oeynhausen D. 4970 West Germany	Crawler mounted to 10 cu. yd. 147–187 ft. booms

and slide it rearward 7 to 12 feet. The cam then lowers the machine to the ground, lifts the shoes, and shifts them back to reposition them for another step sequence. A walking step generally takes about 40 seconds. The machine, as it is being moved, rests on the shoes (about 70 to 85% of the weight) and drags the trailing edge of the tub or base (on which the machine normally rests). When the machine is operating, the shoes are carried in a raised position 1.5 to 2 feet off the ground. The unit can be rotated on the tub to "walk" in any direction but the machine actually walks in reverse with the boom pointing in the opposite direction. The walking sequence is illustrated in Figure 8.3.

- Two sets of cables connected to powered drums on the main machinery deck permit vertical (hoist) motion and horizontal (drag) motion of the bucket. The weight of the bucket and its load provide the downward penetrating force while digging. The drag motion pulls the bucket along the surface towards the machine, forcing material into it. The hoist motion is then activated to raise the bucket with sufficient tension kept in the drag cable to keep the bucket orientated so as to retain the load.

- The upper works of the machine is rotated on the base so that the load can be dumped to the right or left. During this swing phase, the bucket is shifted outward to a position under the boom point by slacking up on the drag cable and pulling in the hoist.

- Dumping is accomplished by releasing the tension in the drag cables, permitting the bucket to rotate vertically and discharge the material.

APPLICATIONS

The dragline finds applications wherever the material being excavated needs to be transported only a relatively short distance, or a maximum of approximately 700 feet. If the geologic and operating conditions allow its use, the dragline exhibits advantages when it is considered a digging and transport system, such as:

- Since it digs below itself, it eliminates the often costly vertical lift associated with truck haulage and other systems.

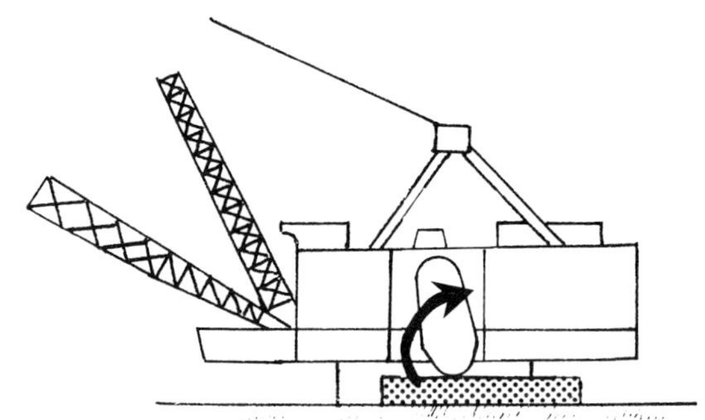

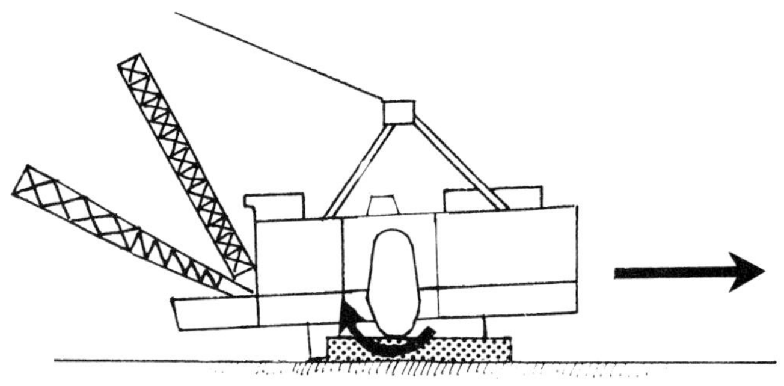

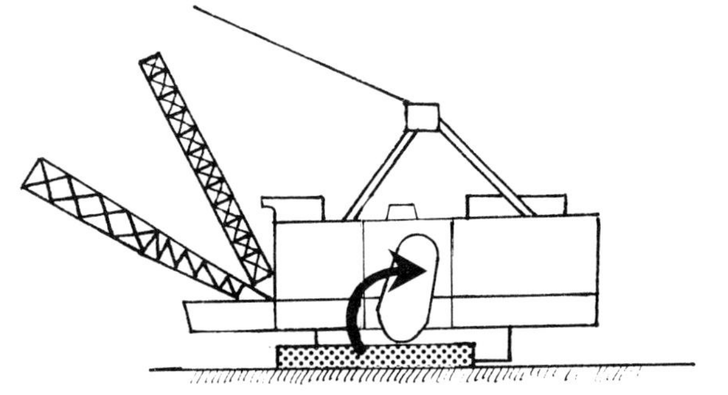

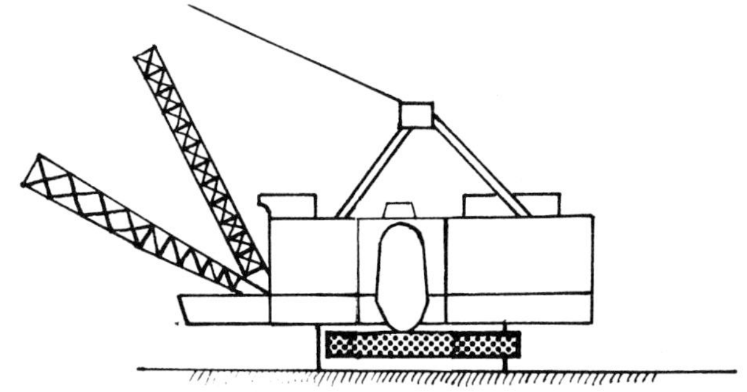

Figure 8.3 Dragline Walking Sequence

Photograph 8.3 Marion 195B crawler dragline

- Because of the dragline's unique capability to rotate and dump at substantial distances, it can eliminate secondary haulage requirements associated with other systems.
- Draglines have the capability to dig and transport material that is blocky or contains boulders.
- Since the dragline is basically stationary during digging and material transport (swinging), it is not significantly impacted by adverse weather conditions.

The two most common large dragline applications involve the strip mining of coal and phosphate. Strip mining techniques usually involve draglines digging long parallel cuts, uncovering the coal or phosphate, etc., and spoiling the waste or overburden into the void created by the previous cut. Draglines are used extensively in phosphate mining where they not only dig and spoil the overburden, but also excavate the phosphate (matrix), thereby eliminating the need for any other equipment in the pit.

The dragline finds applications when deep digging is required and when slope stability or other conditions make operation on the floor of the pit undesirable. Production efficiency can be significantly reduced if the digging face is above the machine level. In sizes above 25 cubic yards, draglines are generally treated as high production units usually operated multi-shift, seven days a week.

If the mining depth exceeds machine digging capabilities, multiple draglines working in sequence can be utilized. Alternatively, a dragline operating bench can be prepared with other equipment and a cross pit conveyor employed to extend the discharge range.

While the larger units are employed primarily for the removal of deep ore deposits or overburden, they can also be used for parting and thin seam removal. The smaller units generally are applied to meet more shallow mining requirements and, in addition, are more frequently assigned to:

- Truck loading of overburden/ore
- Excavation of ponds and basins
- Reclamation leveling
- Removal of partings in multiple coal seam mining

GENERAL CHARACTERISTICS

Although the machine is large and powerful, its operation is simple in terms of the number and complexity of powered functions. These functions are illustrated in Figure 8.4.

- Propel, electric motors
- Hoist, cable and drum (motors)

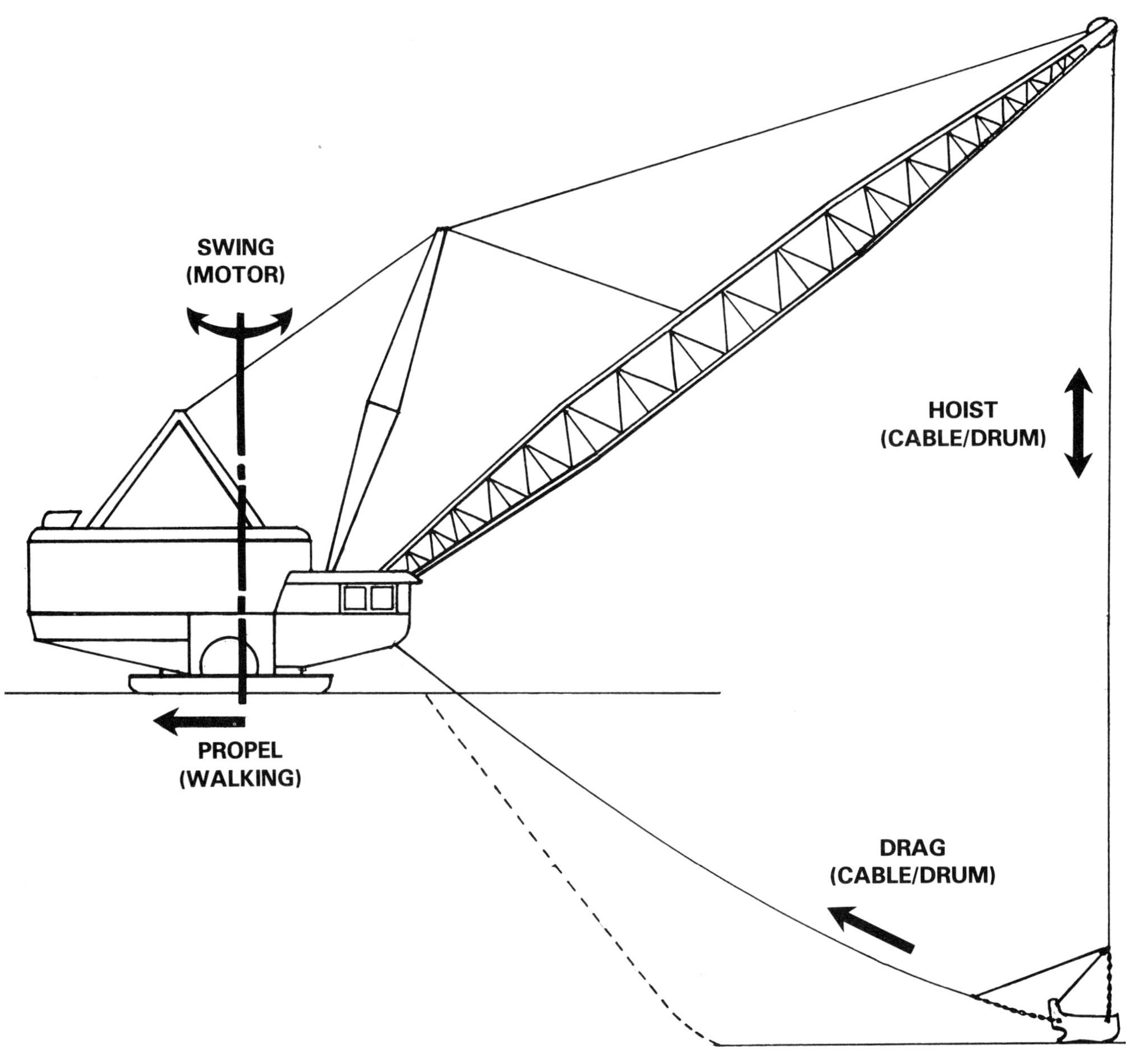

Figure 8.4 Dragline Powered Functions

Photograph 8.4 Bucyrus-Erie 1370W walking dragline dumping

- Drag, cable and drum (motors)
- Upper works rotation, electric motors

Draglines are not as standardized in their designs as some of the other types of equipment. They are produced in lower volumes by smaller companies which have customized their products to meet special segments of the market. In addition, because of the draglines high initial cost, there is a greater effort made to adjust the design and specifications to optimize performance for identified site conditions. The following summaries provide an insight into machine configurations and capabilities. Some models do not fit the general pattern indicated. The unique characteristics of these units will be discussed later in the review of design features.

Crawler Machines

- Diesel powered
- Bucket sizes to 18 cubic yards
- Intermediate length lattice booms — 90 to 200 feet
- Digging depths to 150 feet
- Dumping radius to 180 feet
- Dumping height to 125 feet
- Machine weights 150,000 to 1,400,000 lbs.
- Boom angles of 30 to 60 degrees
- Low ground pressure, from 11 to 20 psi
- Single cable hoist and drag ropes to 2.75 inch cable diameter
- Low speed crawlers
- Good gradeability
- Limited mobility with propel speeds of less than 1 mph

Photograph 8.5 Marion 8200 walking dragline digging overburden

- Demountable upper works for machine relocation
- Full revolving upper works
- Boom can be raised or lowered
- Removable counterweights
- Good operator visibility
- Good performance in medium to hard digging
- One or two man crew
- Relatively high operational skill required
- All maintenance performed in the pit
- Full walk-in machinery house
- Medium service life — to 20,000 hours

Walking Machines

- Commonly electric powered (1,000 to 20,000 HP)
- Bucket sizes from 9 to 220 cubic yards
- Lattice boom from 175 to 375 feet
- Digging depth capability to 200 feet
- Dumping heights to 165 feet
- Machine weights to 30,000,000 lbs.
- Boom angles of 30 to 40 degrees
- Low average ground pressures under tub of 10 to 19 psi
- Multiple cable hoist and drag system with up to 4.5 inch cable diameter

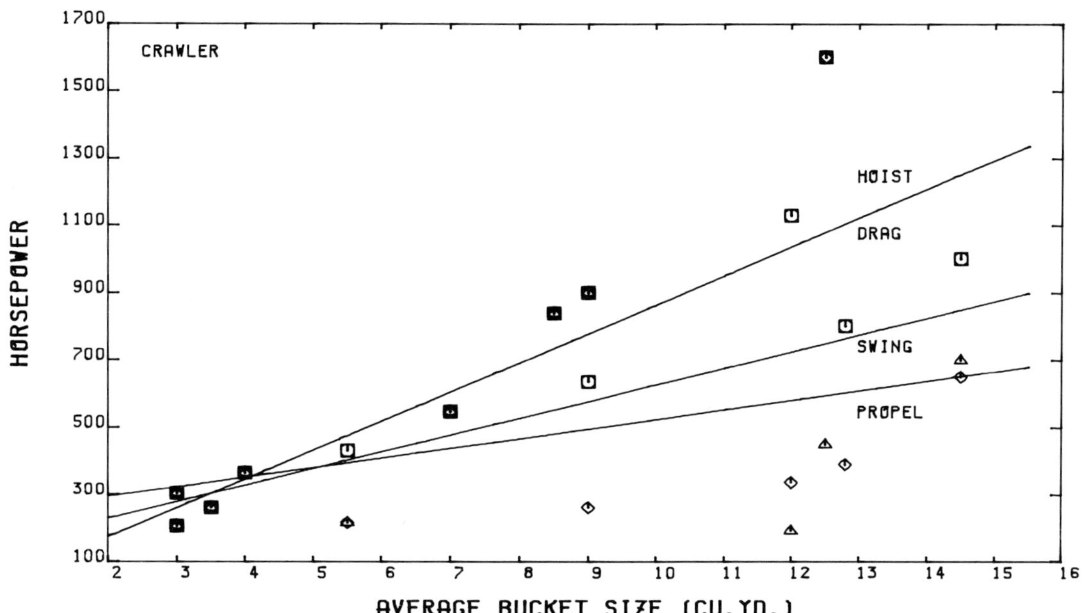

Graph 8.3 Hoist, drag, swing, propel HP/average bucket size (crawler)

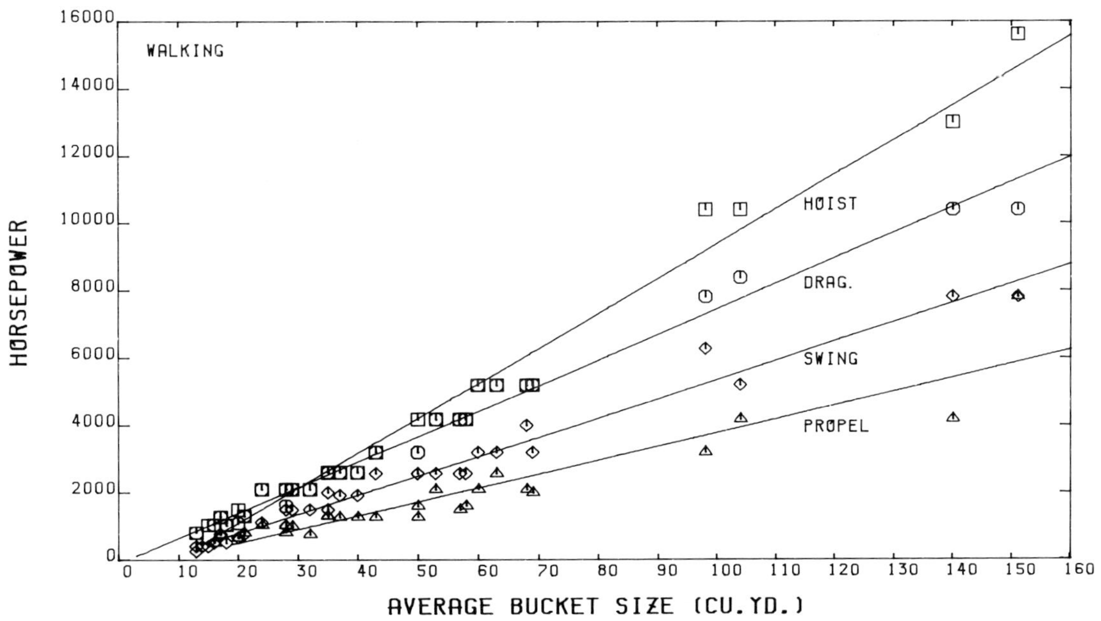

Graph 8.4 Hoist, drag, swing, propel HP/average bucket size (walking)

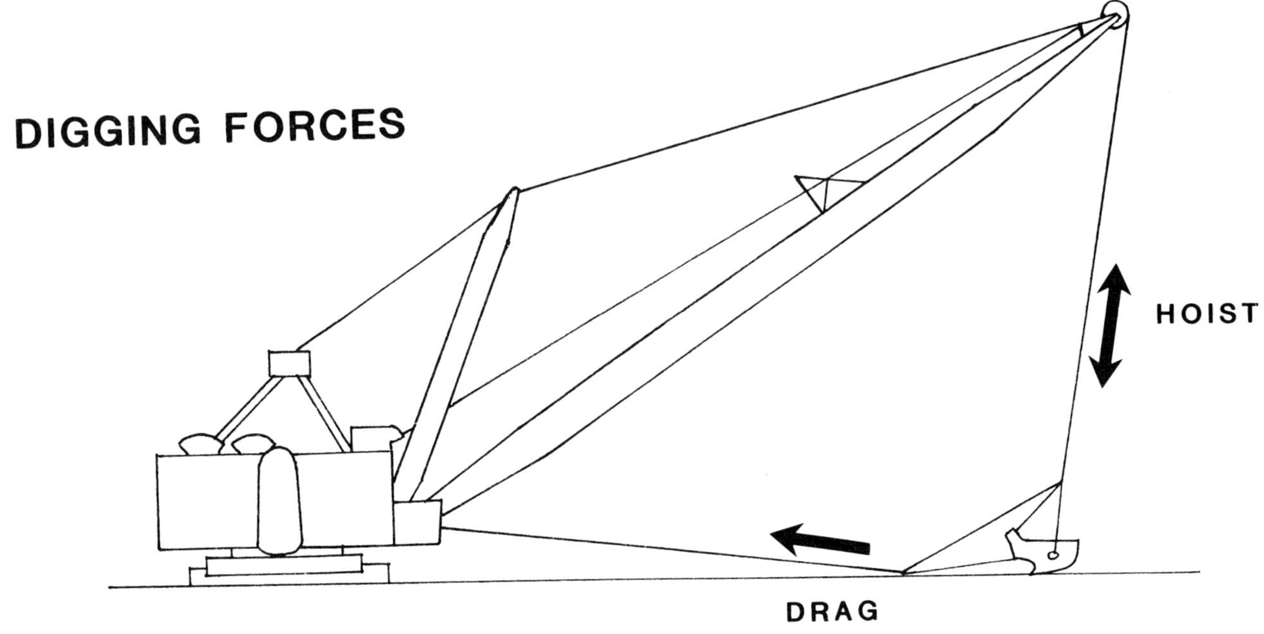

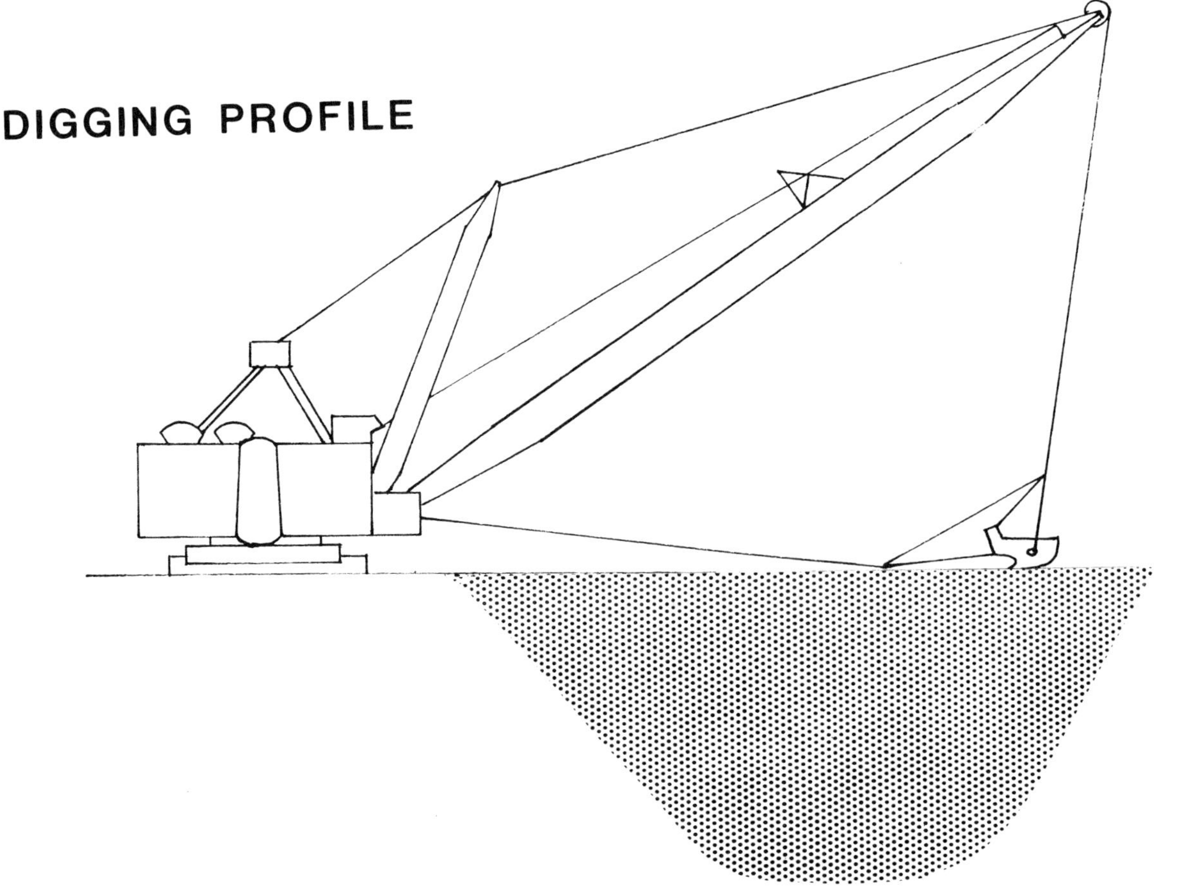

Figure 8.5 Dragline Digging Forces & Digging Profile

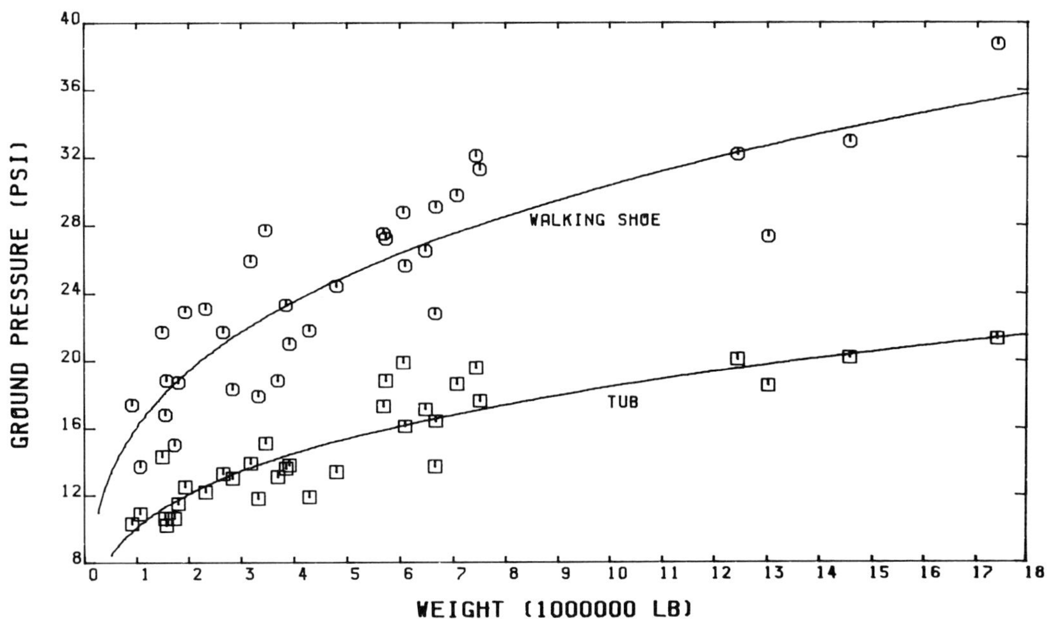

Graph 8.5 Ground pressure/weight (walking)

entirely on the tub. Depending on the ground support conditions and the operating shift of the machine's center of gravity, the ground pressure does vary under the tub. During maneuvering, the dragline is intermittently supported on the walking shoes and a small arc of the tub — ground pressures are then roughly doubled.

Bucket size is a function of production requirements and the unit weight of the material to be handled. The combined weight of the bucket and the excavated material which must be lifted make up the suspended load. Depending on the severity of the digging conditions, bucket weights generally range from 1,600 to 2,400 lbs. per cubic yard of capacity. Suspended load is limited by machine stability. Note that the boom is not subject to digging loads nor is machine stability a factor while digging.

The counterbalancing effect of the machine weight distribution limits the overall load moment around the tipping axis of the dragline. The load moment can be calculated as the operating radius times the suspended load. This means that for any particular machine, as the suspended load is increased, the operating radius must be proportionally decreased. The operating radius can be readily modified by the manufacturer by changing the boom lenth and/or the boom angle. Specifications, as a consequence, commonly show the suspended loads at various operating radii for each model.

The effective operating radius, from a mine planning standpoint, is the distance from the edge of the highwall to the dumping point, assumed to be below the outer edge of the boom point sheave. This distance is called the "reach factor" (operating radius minus 75% of tub diameter, or distance across the shoes plus five feet) and is shown for available walking draglines in Graph 8.6. The dotted lines connecting points represent the range possible for the particular model. On the large draglines, the boom angle and length are essentially integrated into the design and cannot be changed in the field without relatively major modifications.

The walking draglines are rather unique in their design, in that each powered function is an independent module with, in most cases, multiple motor drives. This allows broad componentry standardization in the motor-drive systems and, in the event of a single drive failure, continued operation at reduced power.

Machine size and the inherent limited mobility requires that all service and maintenance activities be performed on the machine while in the pit. To

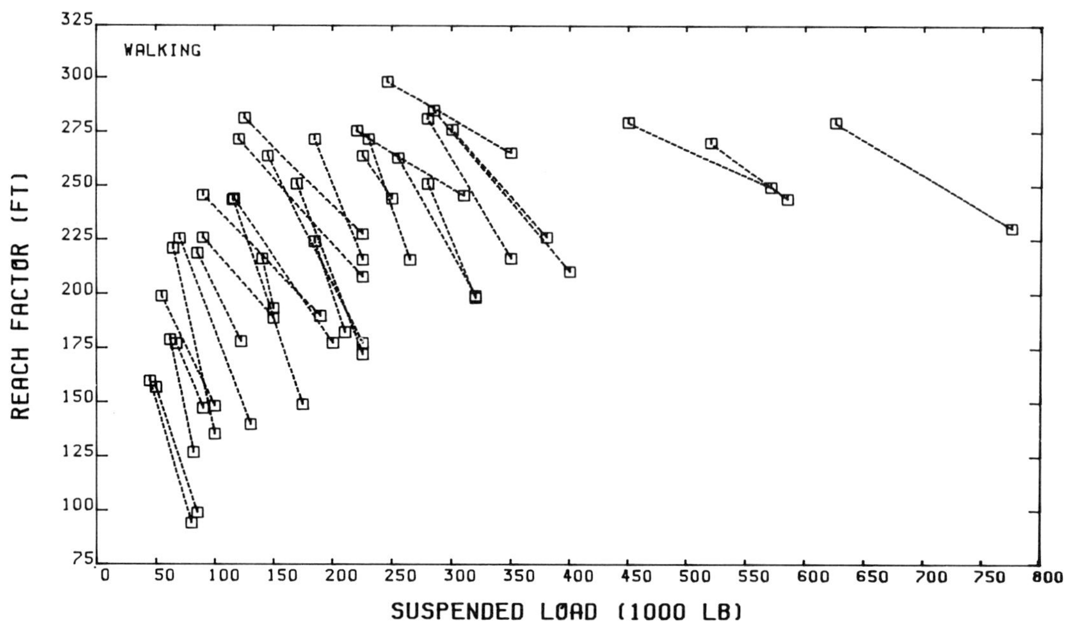

Graph 8.6 Reach factor/suspended load (walking)

facilitate this, all major machinery is located within a house structure equipped with a variety of cranes and hoists to handle the heavy pieces. There are also welding machines for on-board work. Crane rails extending through large side or rear doors, together with roof hatches are provided to remove components and assemblies from the machine. External ladders, stairways and platforms allow full access to the A-frame, mast, and boom structures. A powerful, filtered, forced air ventilation system removes the heat generated by the motors and slightly pressurizes the house.

The degree of fabrication/assembly, and the time required for field erection, increases in proportion to machine size. Rail and/or truck shipment is required to transport the components to the erecting site. This effectively restricts the maximum allowable piece size, and increases the breakdown for shipment and, in turn, lengthens the erection cycle. Rail car shipments range from six to several hundred cars. Erection requires concrete erection pads, heavy lifting equipment, mobile offices, special machining and alignment tools, and, in some cases, temporary protective enclosures for welding. Erection supervision is commonly provided by the manufacturer using either mine personnel or a subcontracted work force. In some cases, the manufacturer provides the erection crew.

The electric draglines require incoming power to the machine. The final flexible link is provided by a 1500 to 2000 foot trailing cable laid on the ground and connected to a movable substation. The substation is a part of a pit power distribution network which serves all the mine electric machines and permits the dragline to traverse the length of the pit.

OPERATIONAL PRACTICES

When used in a mining application, draglines are generally the primary excavating unit, pacing other operations. Optimizing machine performance requires very careful planning and, in some cases, rather complex operational sequencing. This occurs when box cuts, multiple seams, tandem draglines, pitching seams and/or very deep digging are involved. Some situations require rehandling of portions of the excavated material. Detailed practices are well documented in the literature. The manufacturers can be consulted to determine applicable procedures.

The basic operating technique is a side casting sequence involving swing (rotational) angles of 45 to 120 degrees. During filling, the path of the bucket is usually upward at a 30 to 60 degree angle from horizontal. Since the cables provide a flexible connection to the bucket, it must be confined laterally in some manner if a side face is to be shaped. In strip mining, the new highwall is shaped by making an initial key cut each time the machine is advanced. (See Figure 8.6) Because of the machine's position to the far side of the cut, the material is deposited close to the old highwall. Following this cut, the machine is repositioned to remove the remainder of the cut width and the dumping point is shifted further away from the old highwall, permitting a larger volume of spoil to be stacked. (See Figure 8.7) During digging, the machine is periodically repositioned to optimize bucket control for improved filling, for shaping of the digging face, and to maximize reach for disposal.

Production is improved and energy conserved by avoiding multiple passes to fill the bucket. In normal operation, the bucket should be filled in 2 to 2 1/2 bucket lengths of travel. Cycle time is further enhanced by minimizing the swing angle from dig to dump and the vertical hoist distances.

Digging efficiency is reduced and machine abuse is increased when the dragline digs above itself (chop down). It is difficult to control the bucket penetration into the face. High drag cable forces are developed because of the geometric relationships between the bucket and the cables.

Blasting to the full depth of the cut to break up the face is generally desirable prior to excavating. Bucket weight and design (balance), together with the angularity of the drag cable forces, determine its ability to penetrate the material being excavated. Teeth are used to concentrate the forces for improved performance. The drag cables should be kept out of the dirt as much as possible to extend their wear life. Swinging should be minimized while digging to avoid bends in the heavily loaded drag cable.

Dumping occurs during the end of the loaded swing portion of the cycle. At this point, the bucket has considerable inertia so that the material can be cast beyond the boom point if desired, increasing the actual dumping radius.

As noted earlier, only the drag motor is providing power during digging. The hoist cables are essentially slack with all the weight of the bucket and bucket load supported by the ground. Swing and hoist interact for the rest of the cycle, and during the initial acceleration after digging, develop the peak power demand. Drag travel must be reversed during hoisting to hold the load in the bucket. A rapid payout of the drag initiates the dumping action.

A well graded, level, flat bench prepared by a dozer, is required for the walking draglines. This reduces the potential for high local loading of the bottom of the tub and minimizes distortion of the machine structure.

Dirt pushed up in front of the dragline bucket by its digging action produces a roll or bank in front of the machine. This must be periodically pushed back with dozers to reduce drag cable contact and prevent the dirt from penetrating into the roller circle mechanism between the tub and the upper works.

The contact between the tub and the ground must develop enough friction to resist the rotary torques developed by the swinging action. Slippery surfaces, if they exist, must first be removed by a dozer prior to machine repositioning. Highwall stability considerations dictate the set-back of the machine from the edge of the highwall and the digging face.

Care must be taken by the operator to avoid pulling the bucket into the boom ("tightlining") by simultaneously pulling in on both the hoist and drag cables, forcing the bucket up into the boom. Similarly, the bucket must not be pulled up into the fairlead or boom point. Warning devices and mechanical and/or electrical motion cut-offs are available to alert the operator and prevent these actions.

Power costs are determined on the basis of the peak 15 minute demand over a billing period. Operating costs can, therefore, be reduced if efforts are made to minimize prolonged periods of peak demand. Operator recognition of the situation is important. Instrumentation is available to monitor and control peak demand when the 15 minute interval is approached.

The larger machines are commonly scheduled for three shifts a day operation, on a seven day week, with scheduled shut-downs for major repairs.

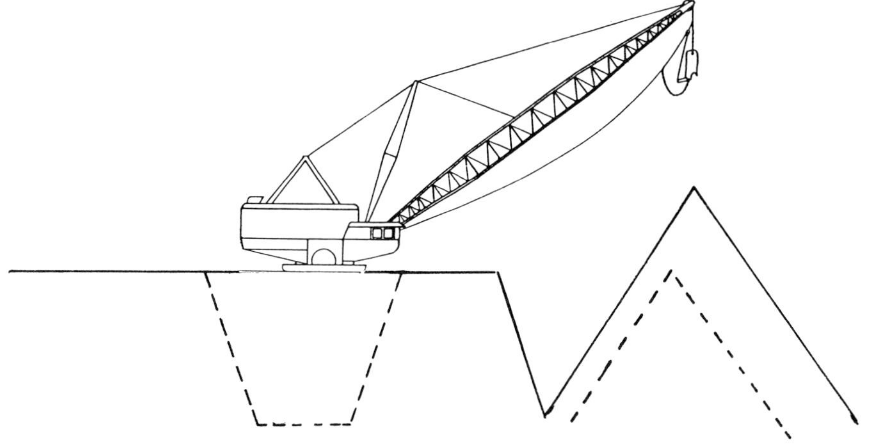

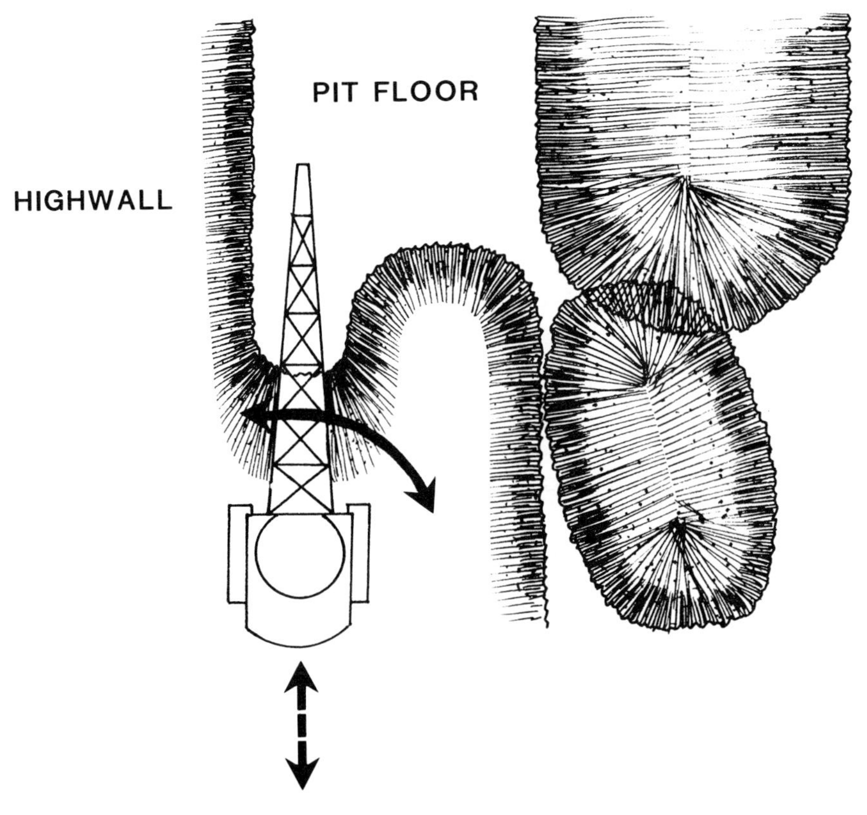

Figure 8.6 Making a Key Cut

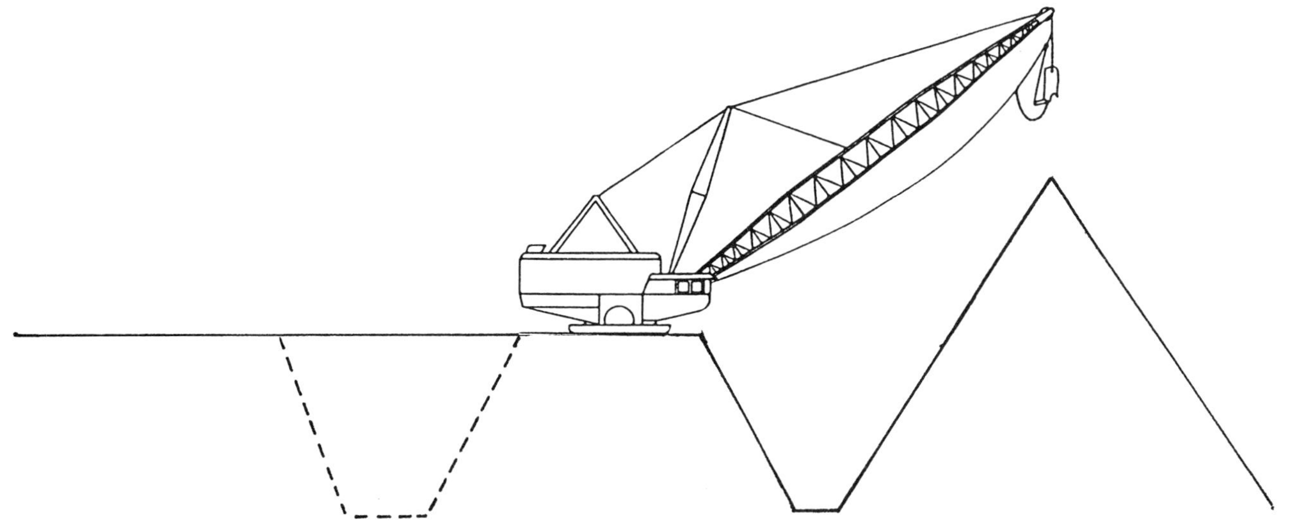

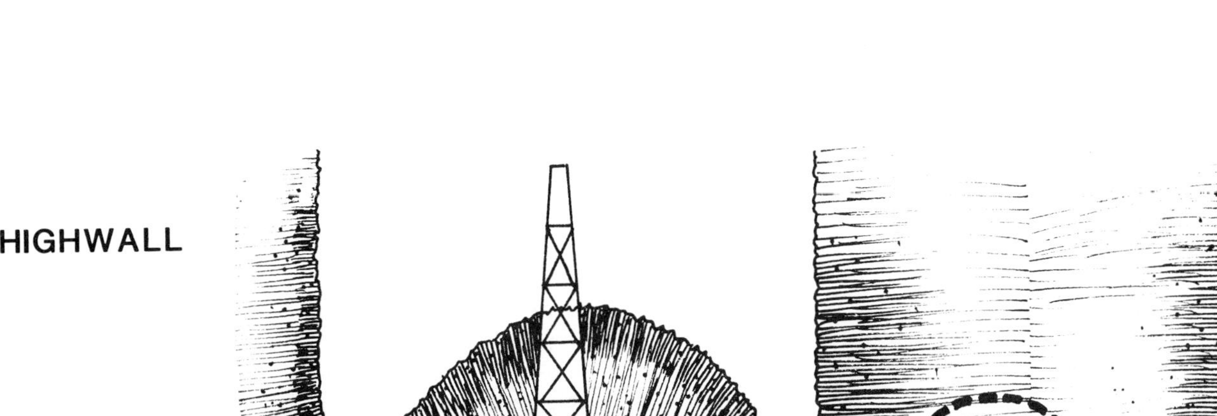

Figure 8.7 Typical Dragline Final Cut

Patrol type inspections of the machinery and structural units are made on a regular basis, with as much service and maintenance as possible performed while the machine is operating. Two buckets are generally provided for each unit to permit rotation, minimizing machine down time and providing ample time fore rebuilding and/or replacing of worn portions.

For the large draglines, cable trucks are often utilized as support equipment. One type mounts a large drum with a spooling device, and operates near the machine to hold extra trailing cable, reducing the length exposed to damage on the ground and simplifying the cable handling. Smaller units with a powered drum are also used to transport lengths of trailing cable to and from the pit for shop repairs. Care is required to avoid damage to the trailing cable from the machine while walking or from other mobile equipment operating in the area.

DESIGN FEATURES

Crawler Machines

Crawler mountings are of a low speed configuration with cast track links. Typically, the track support is rigid. Track side frames are bolted or pinned to the center truck frame (carbody) to permit disassembly for major machine relocations. Drive power arrangements range from a mechanical power train brought down through the center of the machine to direct motor drives mounted on the crawler side frame. The overall structure is usually a wide spread square pattern to provide good machine stability in all directions. Track belt length and shoe width are adjusted to develop the desired ground bearing pressures.

Mechanical drive systems employed with single or twin diesels incorporate torque converters, a chain driven countershaft, and direct gear reductions to the hoist and swing cable drums. High capacity, fan cooled clutches and brakes control the motions. Swing power may be provided by gearing from the countershaft, or by independent electric or hydraulic motor drives. The necessary generators or hydraulic pumps are driven off the main diesel. Some provide for transfer of braking power between the hoist and drag to increase the speed of the motions. It should be remembered that the diesel engine can be replaced by an A.C. induction motor if electric operation is desired.

The boom hoist can be independently powered (hydraulic/electric) or driven off the main countershaft. The hoist, located at the rear of the deck can also position the mast and controls the boom angle through the suspension cables.

The main revolving frame and the truck frame are of a heavy fabricated design with the swing gear and roller circle bolted or welded to the truck frame. The roller circle can be a large ball or roller bearing or a combination hook roller arrangement with pairs of rollers mounted at the front and rear to permit revolving frame rotation and transfer of the load moments. Cast counterweights are bolted to the rear of the revolving frame.

Lattice booms are rectangular in cross section, all welded with angle chords. To facilitate knock down and assembly for machine relocations, the booms are made in 20 to 40 foot sections and bolted together. Aluminum alloy sections are available for some models.

Walking Machines

The basic nomenclature for walking draglines is shown in Figure 8.8.

There are basically four powered functions on draglines — swing, hoist, drag and propel. The actual machinery for these functions is situated on top of the revolving deck or frame. This frame, besides supporting the operating machinery, must be able to withstand the severe stresses encountered while walking, and also must transfer the loads from the boom, mast, A-frame, and fairlead to the roller circle and tub. The revolving frame design differs between manufacturers. Page uses a shallow frame rectangular section design reinforced by a full truss system on all of their models. (See Photograph 8.6) Marion and Bucyrus-Erie employ a deep section box girder frame on most models, with a light truss structure to support the machinery house and service crane system. (See Photograph 8.7) The basic frame of all manufacturers is compartmented for strength with access provided to all areas for inspection.

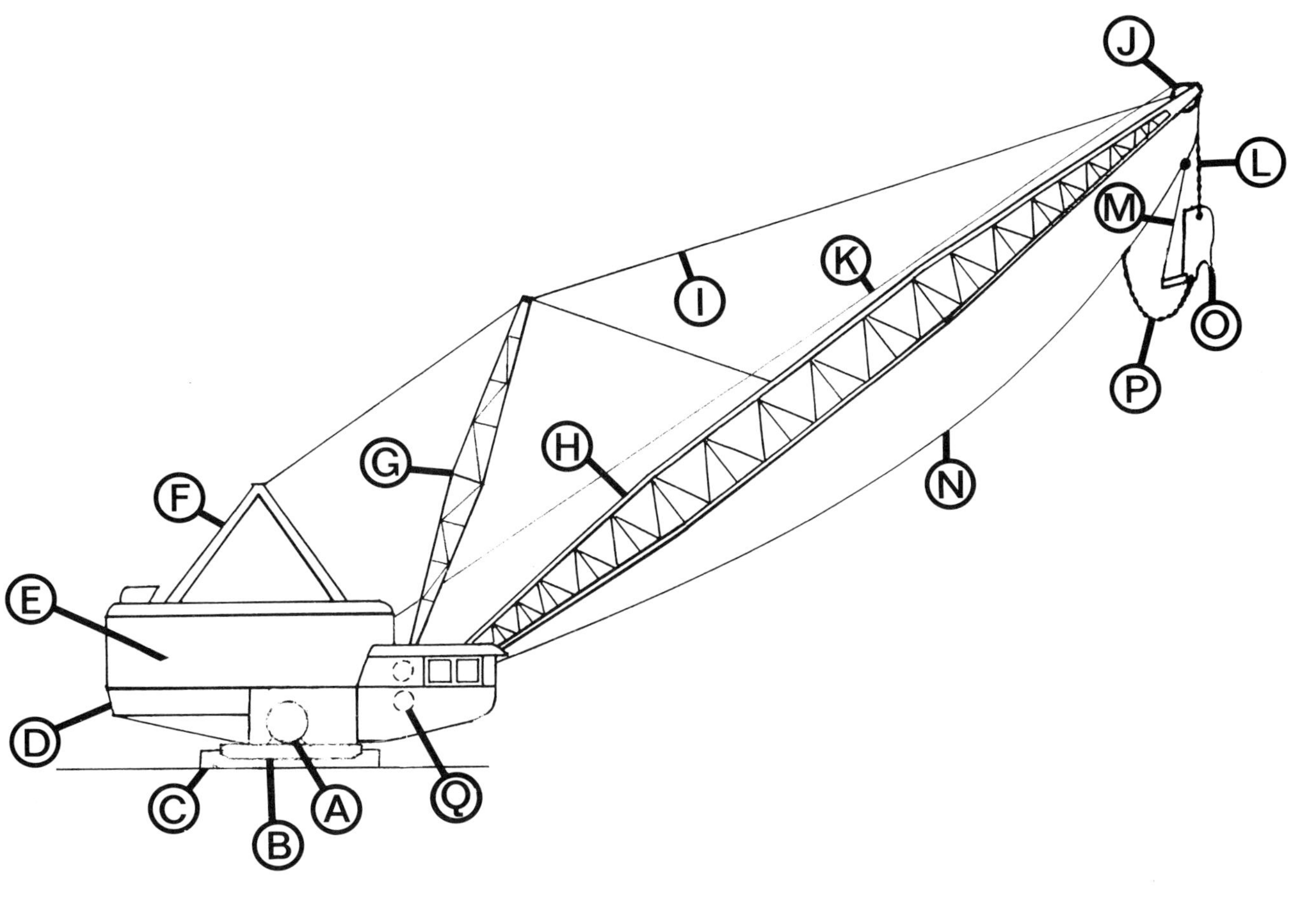

A	Walking Cam	J	Boom Point Sheave
B	Walking Shoe	K	Hoist Cable
C	Tub	L	Hoist Chain
D	Revolving Frame	M	Dump Cable
E	House	N	Drag Cable
F	A-Frame	O	Bucket
G	Mast	P	Drag Chain
H	Boom	Q	Fairlead
I	Suspension Cables		

Figure 8.8 Walking Dragline Nomenclature

The tub is a circular structural unit which accepts the loads transmitted through the roller circle from the revolving deck. The internal structural design, whether rectangular grid or radial pattern must distribute these loads evenly to the total area of the bottom plate. As with the revolving frame, the structure is compartmented with access provided for inspection. The tub's diameter is dictated by allowable ground bearing pressures. The bottom plate outer circumference is protected by thick wear plates to protect it while walking.

The roller circle allows the revolving frame to rotate relative to the tub and ground. The roller circle consists of heavy top and bottom rails which provide the path for cylindrical or tapered rollers.

Photograph 8.6 Truss structure incorporated in Page dragline revolving frame design

Photograph 8.7 Model showing light truss structure for house craneway and A-frame support on Bucyrus-Erie dragline.

Bucyrus-Erie and Marion models utilize tapered rollers, while Page uses a larger diameter rail circle and cylindrical rollers. These assemblies are heavily loaded and must be of high quality steel, properly aligned and well lubricated. Some machines use a partial upper rail, others a full upper rail.

The revolving frame is also connected to the tub by the center pin. It acts as the center of rotation between the revolving frame and tub. Although it does not take any vertical loading, it does serve to resist the horizontal loads, particularly during the walk cycle.

Also mounted between the revolving frame and tub, besides the roller circle, is the swing rack. The swing mechanism commonly consists of multiple swing motors and gear cases; each of these transmit their rotation through a shaft which passes through the revolving frame and whose pinion meshes with the swing rack. Bucyrus-Erie and Marion separate the roller circle from the swing rack, arranging them concentrically on the tub with the external swing rack closer in toward the center pin. Page stacks the roller circle on top of the swing rack, which increases the vertical space between the tub and the revolving frame, improving access for inspection and maintenance. The swing rack, because of its diameter, is segmented for manufacturing and field assembly.

The hoist and drag machinery are module units, self contained and supported in the small to intermediate sizes. Gear reductions are modular and interchangeable; they utilize precision cut spur, helical, or herringbone gear patterns, and are generally splash lubricated. Depending upon the manufacturer, attention has been given to isolating the gear alignments from deflections of the revolving frame. This has been accomplished by mounting the hoist and drag assemblies on separate bases which rest on top of the revolving frame, and also by the use of spherical bearings or bearings mounted in sperical shells on the major gear shafting. Drag machinery may be located in the front or rear, depending upon manufacturer and model.

The fairlead mounted on the front of the revolving frame, guides the drag cable into the machine and assures that there is the proper fleet angle to the drum for proper winding. It also must minimize travel in the dirt and bending of the cable. There are two basic configurations:

- Over and under design: The cable passes under a lower swivel sheave, up and over an upper sheave, and down to the drag drum. (This arrangement may be reversed.)

- Twin roller and deflection sheave design: Twin rollers direct the cable between twin sheaves to the drag drum. The entire assembly is mounted in a swivel frame.

The large dragline booms have a cross section too large to be transported, and thus are assembled as a single piece structure at the erection site. The intermediate and smaller sizes permit partial assembly by the manufacturer prior to shipment. Computer design techniques are applied to produce a structure of minimum weight and yet capable of withstanding the high static and dynamic loading. The boom can consist of a single member with a rectangular or triangular cross section or twin members with either of these cross sections. (See Photographs 8.8, 8.9, and 8.10) Welded steel booms are common with special care given to the welding practices, guaranteeing their quality. The main chords and lacing can be either tubular, H or angle beams. The tubular chords on the Bucyrus-Erie machines are pressurized with an inert gas (60 psi) and monitored to detect any developing cracks. The boom point mechanism usually has a swiveling capability which minimizes wear on the hoist cables.

Marion, on some of the smaller models, offers an optional aluminum boom at an added cost, which reduces the weight about 35%. Special T or WF chords of 6061-T6 alloy are used with square tube lacing. Stainless steel bolts are required throughout for assembly.

The boom support system includes an A-frame and mast. Fixed length bridge cables connect the top and intermediate points along the boom to the top of the mast. Similar cables anchor the mast to the A-frame. Intermediate boom suspension cables may be anchored to a hydraulic cylinder for tension adjustment and/or a vibration dampening device. The mast is a lattice structure similar to the boom, pin mounted at the front of the revolving frame. Its function is to increase the angle the suspension cables make with the boom, reducing the axial compressive loads applied to the boom. The anchoring structure, the A-frame, is a tubular or fabricated plate structure pinned directly to the

Photograph 8.8 Rectangular boom with angle chords

Photograph 8.9 Triangular boom of tubular construction

Photograph 8.10 Twin triangular boom of tubular construction

revolving frame deck or truss structure. In the Marion design, the mast, A-frame and fairlead are combined in a single structure.

The walking mechanism consists of three major components: the shoes, the cam or positioning device and the drive. The shoes are mounted on either side of the revolving frame. They are relatively shallow structural box section members, rectangular in shape with bottom cleats for traction.

The shoe's elliptical operational path, which gives the machine its vertical lift and subsequent horizontal travel, is produced by a cam and/or linkage arrangement. The most common designs are the following: (See Figure 8.9)

- Cam and slide mechanism (Bucyrus-Erie) has a cantilevered drive shaft with a crank arm to the shoe, an anti-friction roller bearing cam arrangement, and a longitudinal hinge and slide base on the shoe.

- Eccentric crank mechanism (Marion) has an outboard bearing for the drive shaft, a crank arm actuator with a ball swivel connection to the shoe.

- Walking spud mechanism (Page) has a cantilevered main drive shaft with double cam eccentrics, one for lift and one for translation, and a ball swivel connection to the shoe.

Good lubrication of these units is important when long walks are required.

The walking drive for the smaller size draglines consists of a cross shaft with a single electric motor drive (alternatively, the drag motor may be used). On the larger models, there are separate electric drives for each shoe, keyed to short shafts, with motion electrically synchronized. It is of interest to

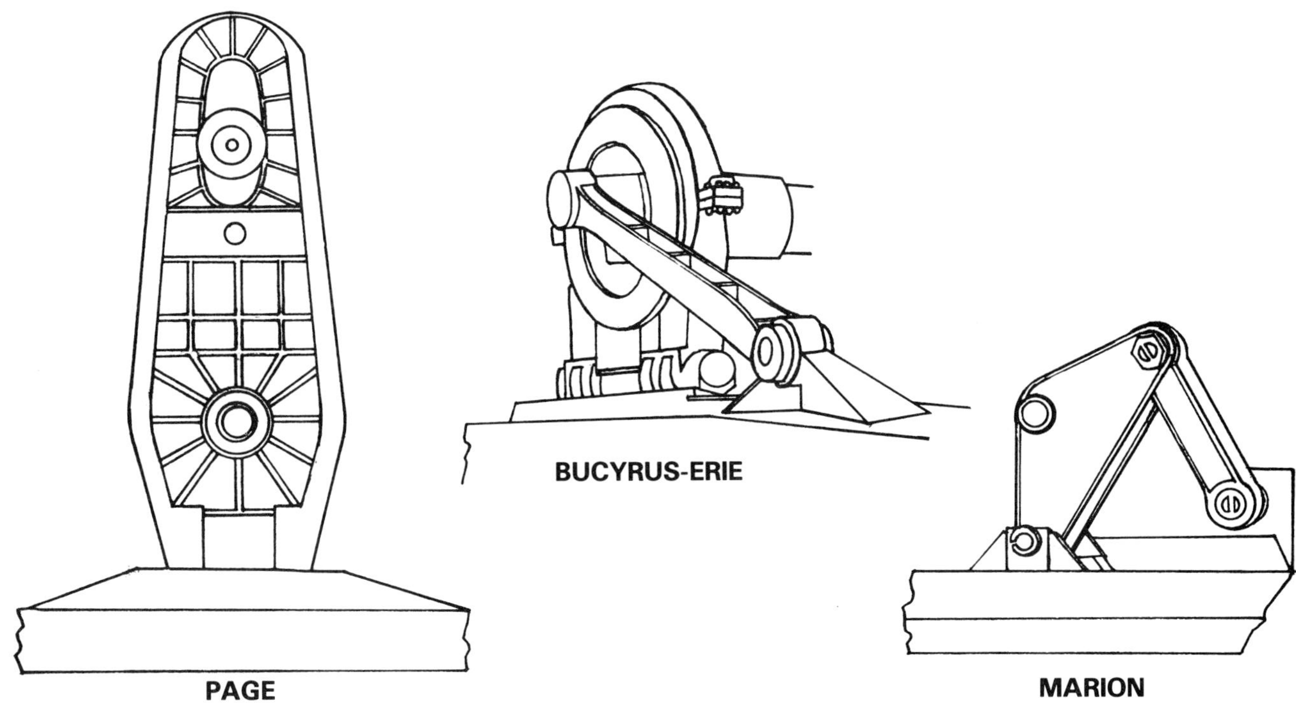

Figure 8.9 Typical Walking Cam Configurations

Photograph 8.11 Hydraulic walking mechanism on Bucyrus-Erie 4250W dragline

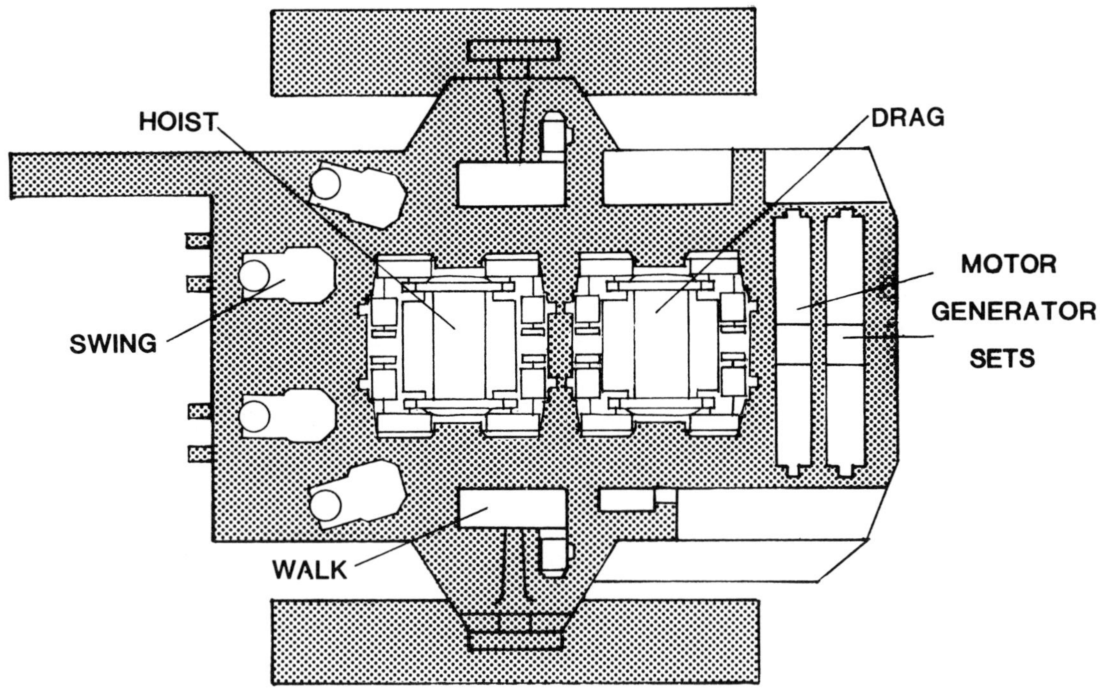

Figure 8.10 Typical Dragline Deck Plan

note that the largest dragline in service has a walking system employing two sets of hydraulic cylinders. One set provides vertical motion and the second set provides the horizontal motion.

Electric drives may be either the static power supply or motor-generator set type discussed earlier in electric shovels. So far, the static system has been limited to the newer, smaller models, with most of the walking machines utilizing motor-generator drives with a Ward-Leonard static control system. On these machines, high voltage (4,160 to 26,000 volts, 60 cycle) A.C. power is brought onboard through the trailing cable to power synchronous or induction motors which, in turn, drive D.C. motor generator sets with a separate generator for each motion. The collector rings and switch gear for the incoming A.C. power are located within the revolving frame around the center journal or in the space between the tub and revolving frame. Power from the D.C. generators is directed to single or multiple D.C. mill motors, driving the various powered functions through gear reductions. There are no clutches or brakes in the main power train. Motors are reversed to decelerate and/or power in the opposite direction. During deceleration, the motors act as generators, feeding power back into the system. Automatic emergency brakes are mounted on the motor shafts, activated in the event of a power failure. Figure 8.10 shows a typical deck layout for the drives and motors.

The Ward-Leonard system is suited to high inertia loads and has good braking characteristics. The D.C. amps control the torque while D.C. volts control the speeds. Maximum (stall) torque occurs at zero speed with the torque dropping off to about 65% of stall as speed is increased. The mill type motors are designed for severe duty, with split frames for accessibility, low inertia armatures, and high insulation ratings. All motors are blower cooled to maximize their thermal capacity.

On the larger models, a centralized main electrical control room is provided with a filtered and pressurized ventilation system. Interconnecting electric cables are laid under walkways for ready access. Switches limit the hoist and drag cable travel and a wide variety of monitoring and detection devices are incorporated for electrical system protection. Thermal detectors on the motors detect

Photograph 8.12 Deck view showing multiple drum drive units and multiple swing drive

excessive temperatures. General Electric and Westinghouse supply the major components with the choice, in most cases, being a customer option.

While the bucket design appears relatively simple, the actual geometry, construction and weight distribution significantly influence performance. Unnecessary weight decreases the potential bucket payload. Bucyrus-Erie and Marion buckets are of cast-weld construction. Due to severe wear and shock loading, most buckets are fitted with wear resistant materials in critical areas. Special abrasion resistant cast steel chain and fittings keep the hoist and drag cable out of the dirt. Optimal loading and material shedding during the dump requires a double tapered body. Hoist trunnions must be located to the rear of the loaded center of gravity to provide proper dumping action, while drag cable forces must act to improve bucket penetration while digging.

The conventional bucket design incorporates a large, heavy arch member over the front of the bucket, above the lip. (See Figure 8.11) Its purpose is to increase strength by effectively tying the sides together. The added weight assists bucket penetration while digging. Page offers an archless design, substantially lower in weight, that performs well in less severe digging conditions. Page also offers a miracle hitch which is an addition to the bucket rigging, modifying the geometry to provide a faster lift out of the front of the bucket when hoisted out of the dirt, minimizing spillage.

SELECTION CONSIDERATIONS

Site conditions, mine plan and production goals dictate the machine requirements. In preparing the detail dragline specifications the following must be considered:

- Bucket size
- Tub diameter in relation to bearing pressures (or track length and width)

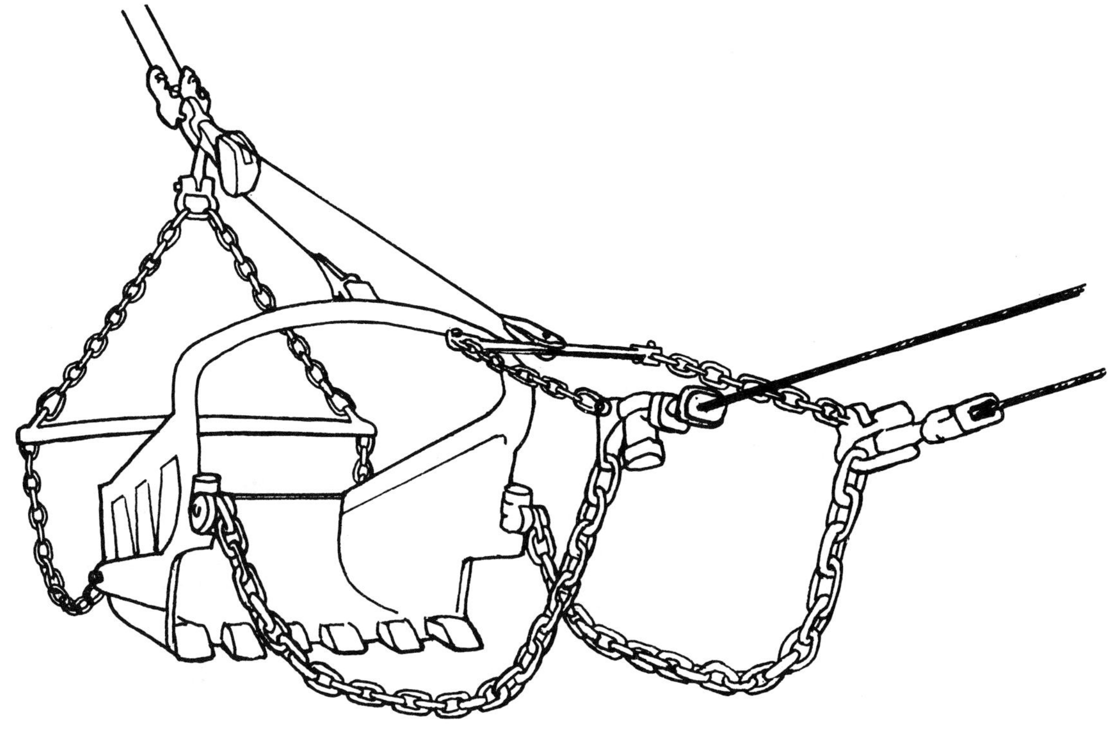

Figure 8.11 Typical Dragline Bucket

- Maximum suspended load at required reach factor
- Maximum digging depth required
- Maximum dumping height required

Hoist and drag cable speeds and swing speed are generally very comparable on units of similar size. Beyond the above direct range and productivity factors , the following should be assessed and, where appropriate, also specified:

- Diesel vs. electric primary power
- Crawler vs. walking propel
- Mechanical, combination or electric motor drives
- Bucket weight and design
- Boom length and angle
- Manufacturer preferred engines, motors, and generators
- Fixed vs. live boom
- General design
- Special servicing and/or maintenance features
- Optional equipment (considered later)
- Power supply

As noted earlier, as the machine size increases the degree of machine customizing possible for specific mining conditions also expands. This, in turn, means engineering studies should be made to properly match the machine to current and future operational needs over the mine property life.

Graph 8.7 attempts to provide a general reference frame of machine power density (horsepower/cubic yard). The variation between available machines is substantial. It must be recognized that nominal bucket size is dependent on a number of variables such as operating radius. Total horsepower and/or the horsepower applied to any particular function can be varied to some degree for most of the models to meet site requirements.

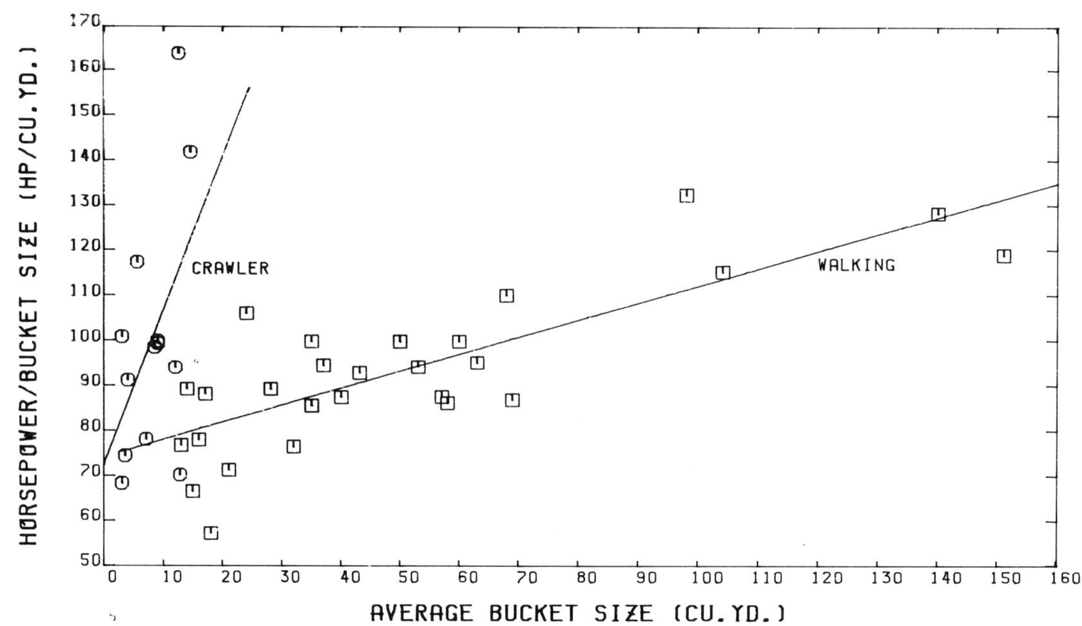

Graph 8.7 HP per nominal bucket size/average bucket size

Some trend lines on machine capabilities for crawler draglines are provided in Graph 8.8. Again, it should be pointed out that there is considerable scatter in the data. The plot reflects the concentration of units in the 7 to 12 cubic yard and the 15 to 18 cubic yard ranges.

The suspended load available on a specific machine is a function of available power and, most importantly, the operating radius, which is tied to the machine stability. Given a specific suspended load, the matching bucket size varies dependent on the unit weight of the material to be handled and the bucket weight. All of these can be varied for each model of walking dragline so that a range of performance specifications is possible. Graphs 8.9, 8.10, and 8.11 illustrate these ranges for digging depth, dump height and dump radius for available machines. Note the resulting considerable overlap between machines, making the selection more complicated. The manufacturer should be consulted to assist in the selection of the appropriate model to best match the range and capacity requirements. (See machine specifications at the end of this section).

While portions of the machine production analysis involve standardized performance factors, the complete analysis requires cut diagrams, machine positioning, mining sequence, etc. This, together with ownership and operating costs, is considered a separate subject and beyond the scope of this book. Some of the production considerations entering into these calculations are:

- Material swell
- Material unit weight
- Fragmentation
- Bucket size
- Reach
- Swing angle (cycle time)
- % rehandle
- Job efficiency

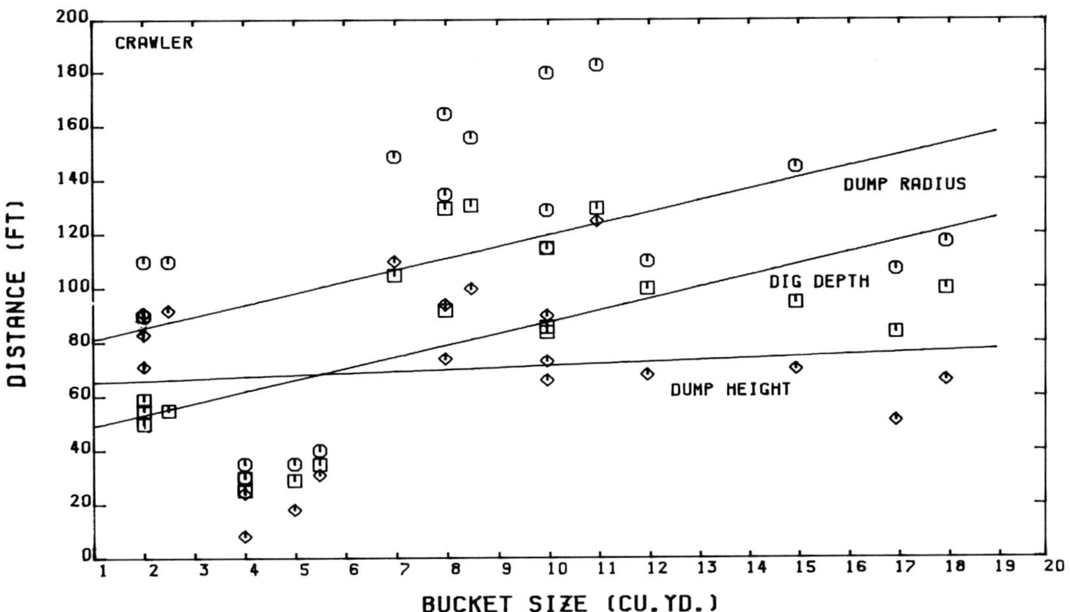

Graph 8.8 Dig depth, dump height, dump radius/average bucket size (crawler)

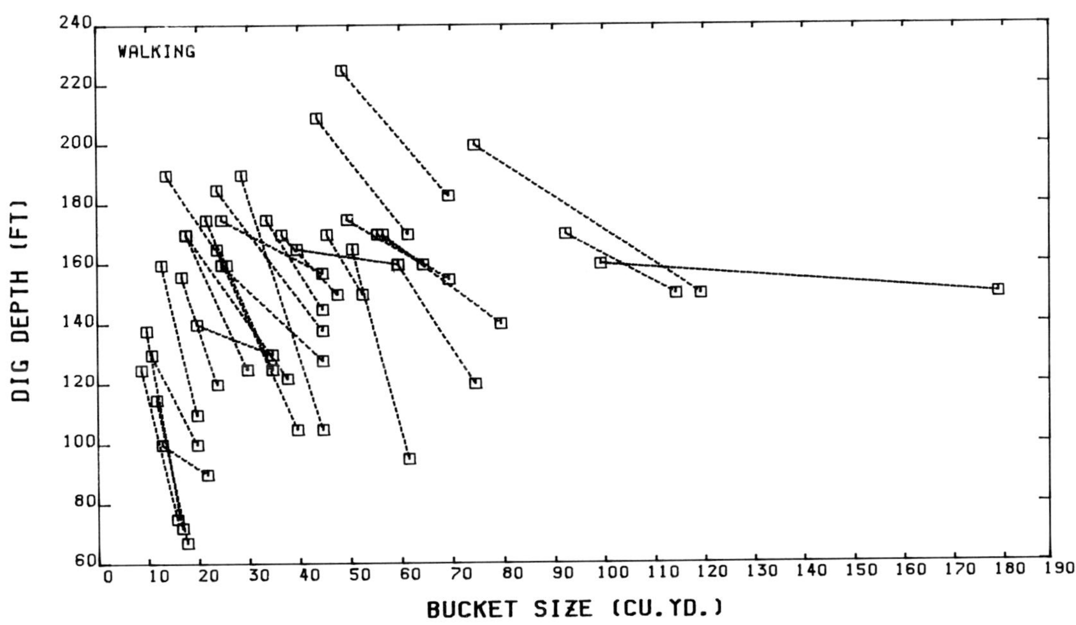

Graph 8.9 Dig depth/bucket size (walking)

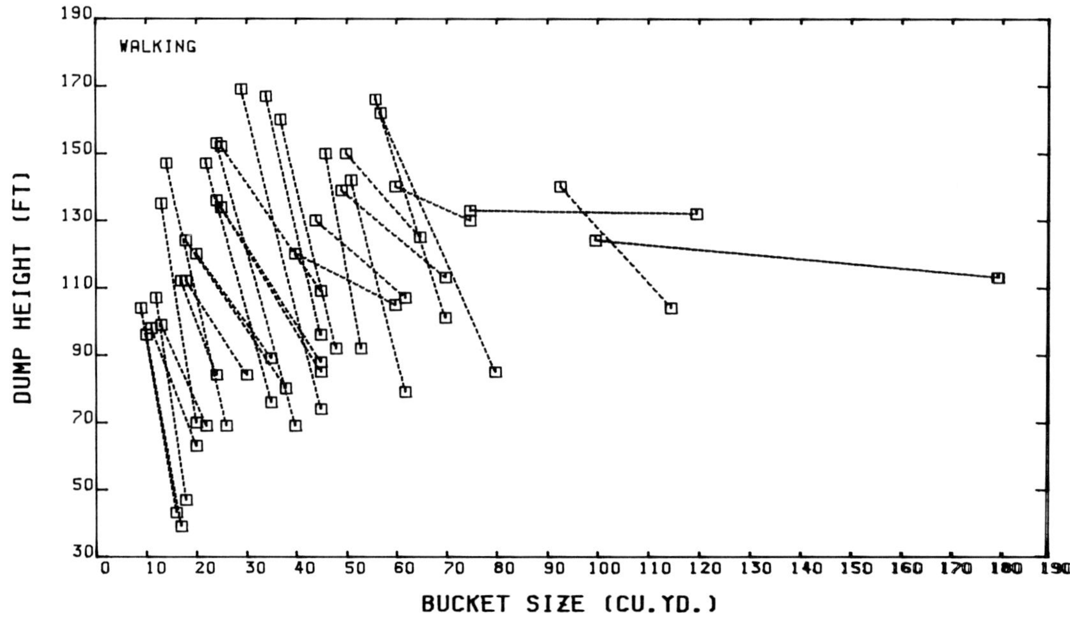

Graph 8.10 Dump height/bucket size (walking)

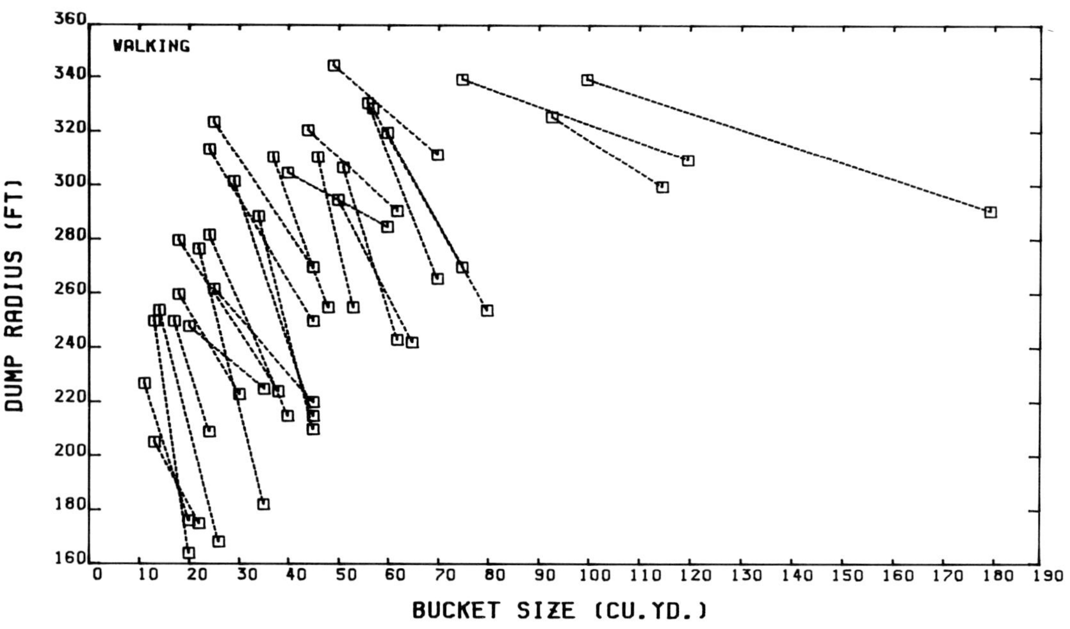

Graph 8.11 Dump radius/bucket size (walking)

- Machine availability
- Operating delays
- Operating schedule

Special material handling situations, mud, toxics, etc., may severely affect production. Some of the costs considerations entering into these calculations are:

- Machine price:
  - Base machine
  - Freight
  - Ballast
  - Erection
- Depreciation (machine life)
- Interest
- Insurance
- Taxes
- Crew wages
  - Operator
  - Oiler
  - Groundman, etc.
- Power (or fuel)
- Lube, oil, greases
- Repairs
- Maintenance
- Supplies

Considerable operating experience has been accumulated over the years so that for preliminary estimates, indices may be used to estimate production.

- Normal stripping mode from highwall — 255,000 cubic yards per cubic yard of bucket capacity per year
- Stripping with extended bench method — 240,000 cubic yards per cubic yard of bucket capacity per year
- Stripping with limited chop down — 225,000 cubic yards per cubic yard of bucket capacity per year
- Stripping with rehandling — 215,000 cubic yards per cubic yard of bucket capacity per year

This cubic yard of material moved per cubic yard of bucket capacity per year is an aggregate number which includes machine availability, efficiencies, etc. These figures shuld be reduced by 10–15% for stripping depths greater than 100 feet.

The optional equipment available differs with the manufacturer.

Crawler Machines

- Buckets
- Diesel engines
- Air compressors
- Alternate crawler widths and lengths
- Hoist and drag drum laggings
- Independent propel
- Counterweight removal device
- Air conditioning
- Heater, defroster, etc.

Walking Machines

- Larger diameter tubs
- Dual operator cabs
- Power factor regulator
- Reduced voltage starting
- Phase reversal protection
- On-board substations
- Unitized power control rooms

- Swing recorder
- Annunciator system
- Tightline limit switches
- Anti-condensation strip heaters
- Automatic lubrication, covering:
  - — Main propel bearings
  - — Rotating frame bearings
  - — Gearing
  - — Cable sprays
  - — Roller circle
- Separate lubrication room
- Bulk lubrication storage
- Communication systems
- Reeving winches
- Cab air conditioner
- Boom hoist
- Welding outlets
- Special steels
- Special lighting equipment
- Stand-by diesel/electric sets
- Special hoists
- Fire control systems
- Data logging system

Note that the buyer usually furnishes ballast and trailing cable.

Photograph 8.13 Bucyrus-Erie 4250W, 220 cubic yard, walking dragline — the largest ever placed in service

Photograph 8.14 Flexible track arrangement on Bucyrus-Erie 300D crawler dragline

NEW DEVELOPMENTS & TRENDS

Possibly as a carry over from the new designs of hydraulic machines being introduced, there has been renewed emphasis placed on designs that can be readily knocked down and reassembled at another site. The goal is to keep part size down to reduce the size of required handling equipment; plan machinery and stuctural units that can mesh together conveniently for a unit shipment; and develop assembly procedures that are simple, reliable, and fast. The emphasis, in this approach, is on the design of modules that can be manufactured, tested and packaged so as to guarantee and control quality.

Some of the future developments in the small to intermediate dragline sizes, may be tied to expanded applications such as their use for reclamation leveling. A dragline in this service would have to propel over very rough and soft ground with relatively frequent moves. The new Bucyrus-Erie 300D appears to have met these requirements with a unique high flotation crawler arrangement. In addition to adding boggies to the track suspension to improve traction and ground adhering characteristics, the track was split into two units on either side. The pivoted twin crawler unit further amplifies the capability of the track to accomodate surface irregularities minimizing highly concentrated loads. To keep the propel drive system and the steering simple and efficient, each track is independently driven by an electric motor.

Each new technology has some inherent advantage which offers opportunities for machine upgrading. In the draglines it has led to mixed drive systems with straight mechanical and/or electric systems giving way to combinations which include hydraulics. Hydrostatic drives are increasingly utilized for track drives and independent functions such as the boom hoist or swing. Similarily, A.C. electrics are appearing for track or swing drives on units with a main torque converter drive to the hoist and drag drums. It must be remembered that hydrostatic and electric drives require no operating clutches and brakes.

The clutches and brakes on the mechanical drives must absorb a great deal of energy in the repetitive reversing duty cycles. Efforts continue to smooth the operation and extend the component life. Considerable progress has been made in modulating clutch engagement and disengagement. Brake service demand is reduced by interconnects between the hoist and drag drums.

At this time, there seems to be no demand for large crawler mounted draglines. Users requiring larger units apparently have applications suited to walking machines.

In the smaller sizes of the electric machines, static D.C. and A.C. variable frequency drive systems are appearing in sizes comparable to the electric shovels. As the components become available at acceptable cost, these will most likely be gradually extended into larger sizes. They offer the opportunity for improved control characteristics, higher reliability and simplified maintenance and reduced power consumption.

Longer booms provide extended operating radii, permitting wider pits, multiple seam applications and/or deeper mining. As a consequence, there has been continuing interest in longer booms, and lengths have increased through the years. At various times booms up to 600 feet have been suggested and considered as technically feasible. Making booms of greater lengths, while maintaining an economical balance with production capacity and overall machine balance, is not an easy task. Partial or full aluminum booms have been in a limited use for many years (machines below 70 cubic yards in size); this is one approach to reducing the structural weight. While the weight saving is significant, it seems that associated problems with available material sizes, strength, complex fabrication techniques, limitations on fasteners, plus high cost are curtailing broad acceptance.

New lubricants are generating considerable interest in three areas; enclosed and open gearing, walking cams, and cables. Experience indicates

Photograph 8.15 Bucyrus-Erie 300D crawler dragline working in overburden

that gear life can be increased and higher loads carried. The heat build-up in walking cams during extended walking periods can be reduced. Operating and suspension cable life can be increased with applied lubricants and those being introduced in the manufacturing process.

Greater attention is being given to hydraulic dampening devices for boom pendants, resilient mounting materials for the roller circle, and polyethylene liners for buckets. Some of the new light weight slippery synthetic materials appear to offer a high potential for replaceable wearing surfaces in buckets if problems associated with fastening techniques can be resolved.

While available for a number of years, computer oriented data logging systems still have found only limited application. With the high ownership and operating costs common to the large draglines, minor improvements in machine availability and/or operational techniques can justify the expenditures. The problem seems to be in getting further acceptance of the recording system's ruggedness and reliability, and developing techniques for data reduction which provides more usable information to operating and maintenance personnel.

Two comprehensive data logging systems for draglines are currently available, one made by the General Electric Company and the other by McDonnell Douglas Electronic Company. Both provide an operator console for inserting information on operational delays, plus feed back displays showing current or accumulated performance such as; yards/hour, % bucket loading, cycle time, swing angle, height/depth of bucket, etc. All functions can be monitored and permanently recorded for statistical analysis, and a wide variety of reports dealing with operator performance, machine productivity, operating conditions, and maintenance activities can be generated. These units incorporate operator alert indicators for static and dynamic tightlining.

In an effort to upgrade dragline operator training, McDonnell Douglas Electronics has, under a U.S. Bureau of Mines contract, developed a sophisticated training unit. This includes a closed circuit television system to provide simulation training similar to that applied in the aircraft industry.

Photograph 8.16 General Electric data logging system operator display and input units

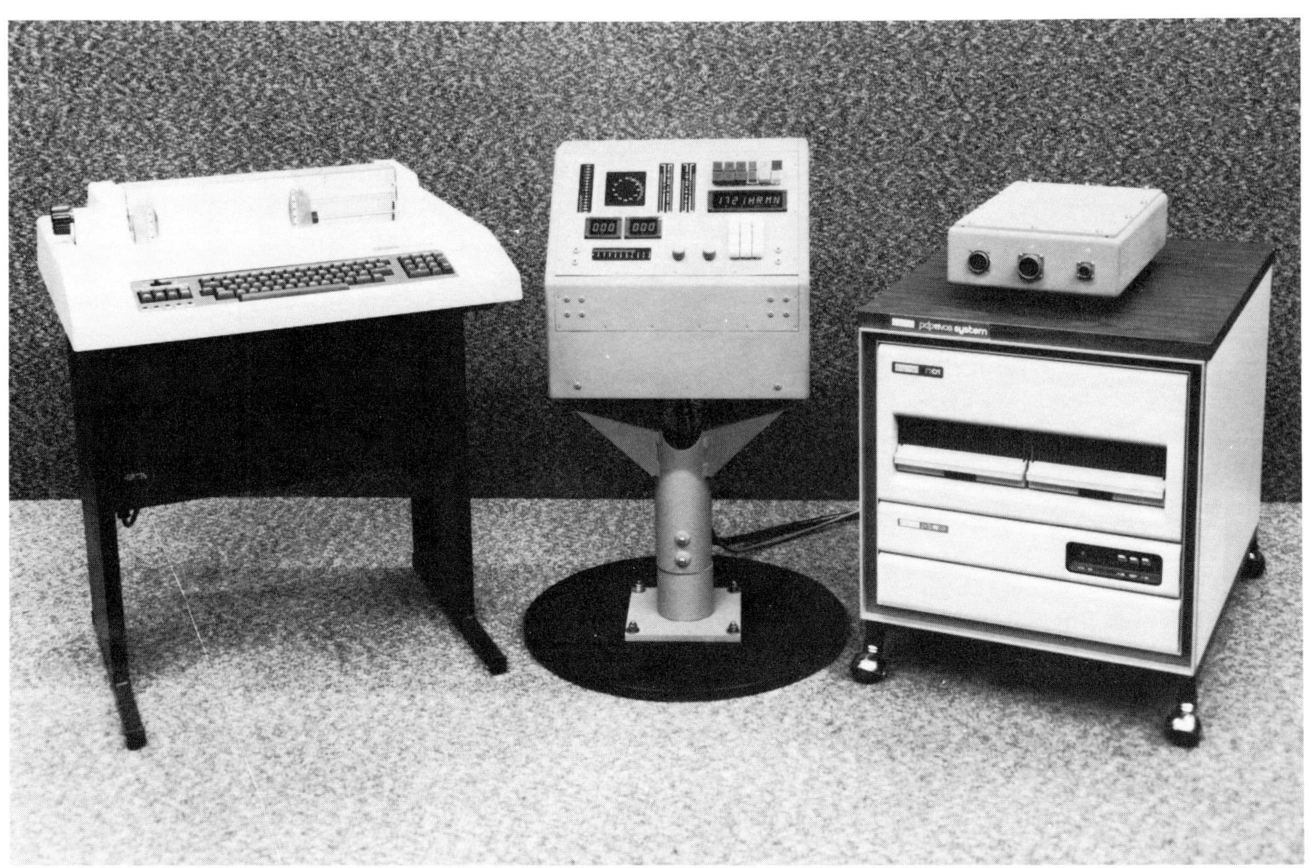

Photograph 8.17 Input station, operator display unit and computer module that are mounted on dragline for McDonell Douglas Electronics data logging system

MACHINE SPECIFICATIONS

Figure 8.12 — Walking Dragline Dimension
Table 8.2 — Walking Dragline Specifications
Table 8.3 — Crawler Dragline Specifications

Dragline specifications are divided into two tables, detailing walking and crawler units separately. In both cases, the machines are listed alphabetically by manufacturer and then in ascending order by bucket size.

Walking Draglines

Total HP, where listed, refers to the aggregate rated HP of the MG set induction motors. The induction motors are rated with incoming power varying between 3/60/3300/6600 and 3/60/4160/7200. Power supply shows the actual rated incoming power for each machine. The other motors are rated at 475 U.D.C., with the listed HP being the total aggregate rating for each function.

Standard operating weight is the total machine working weight including ballast. The amount of ballast is broken out under ballast weight.

The dimensions of the walking shoes are based on the manufacturer's listed standard shoe. In some instances, optional (usually larger) shoes are available. The given shoe ground pressure was either taken directly from the manufacturer's data or calculated from other data at 75% of operating weight/total shoe contact area in square inches. Tub ground pressure, as listed, was either taken directly from manufacturer's data or calculated from the machine specifications at 100% operating weight/total tub bearing area in square inches.

The categories of suspended load, boom-point height, digging depth, dump radius and dump height are all shown as a range of dimensions. This is to allow for changes in operating specifications due to changes in boom length and boom angle. User requirements will determine the actual configuration of any specific machine.

Crawler Draglines

Total horsepower, for all listed crawler machines, refers to the manufacturer's listed rating or net flywheel horsepower of the primary power plant. In all cases, this is a diesel engine(s). Machine functions are then powered either by direct or mechanical linkage from the primary engine, by a hydrostatic pump/motor system, or by electric A.C. or D.C. motors driven by the primary engine through an on-board generator set.

The crawler shoe width listed is the manufacturer's published standard shoe width, and the listed machine ground pressure is calculated on that width. In most cases, optional shoe widths and configurations are available.

The categories of suspended load, boom-point height, digging depth, dump radius and dump height are all shown as a range of dimensions. This is to allow for changes in operating specifications due to changes in boom length and boom angle. User requirements will determine the actual configuration of any specific machine.

All specifications, capacities, capabilities and dimensions are based on published manufacturer data. Although the information is believed to be current and the interpretation to be consistent and correct, it is possible that there may be some omissions, inaccuracies or out-of-date information included in the specification charts. These charts are not meant to be used either as an in-depth analysis, or as a head-to-head comparison of the available equipment, but rather as a general overview of the equipment, available sizes, and approximate operating data. Specific questions relating to performance or purchase should be directed to the manufacturer or authorized distributor/dealer in the user's specific area.

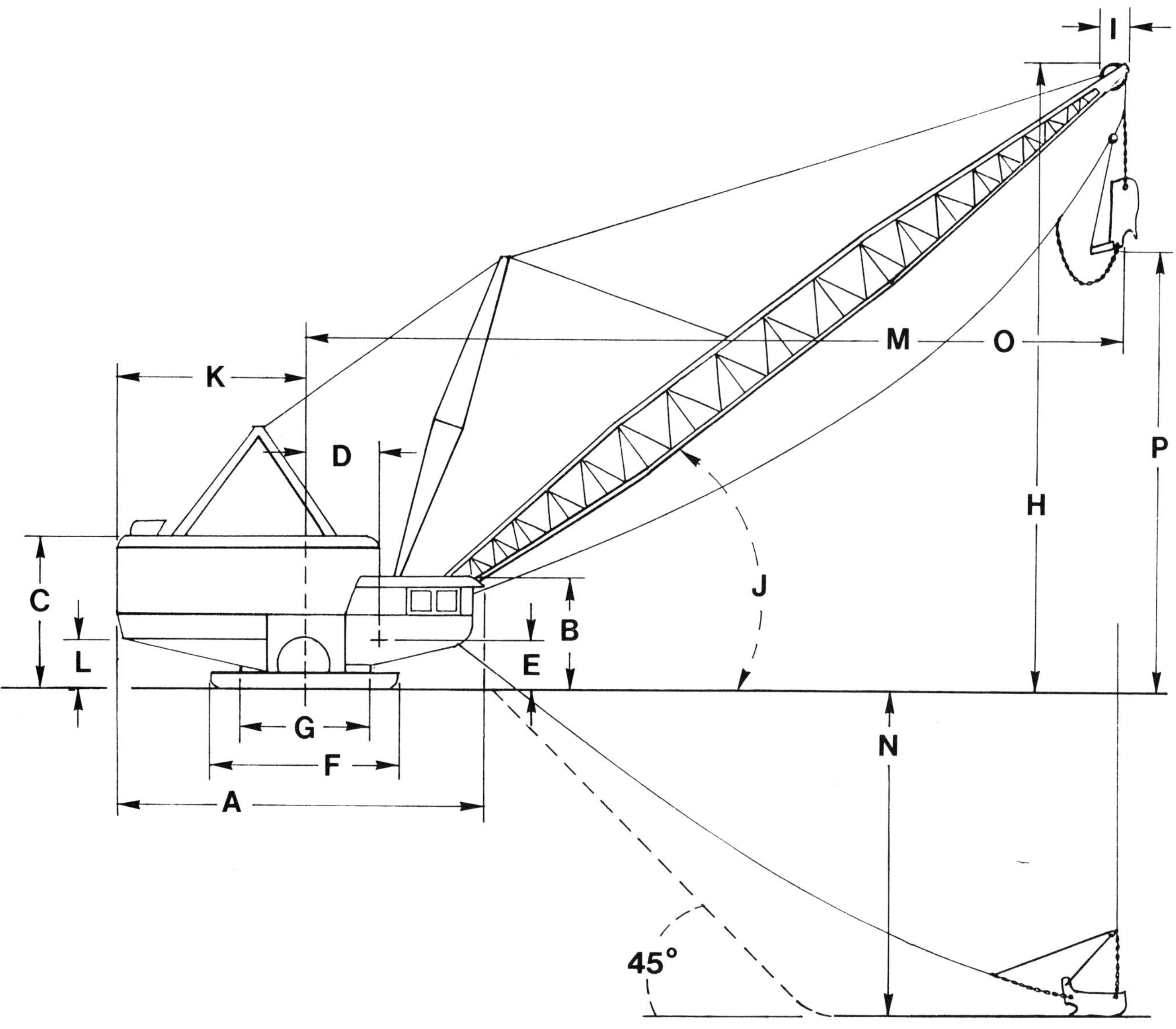

A	Length of House (ft)	J	Boom Angle (degree) (From Horizontal)
B	Height to Top of Cab (ft)	K	Clearance Radius of Revolving Frame (ft)
C	Height to Top of House (ft)	L	Clearance Height of Revolving Frame (ft)
D	Boom Foot Radius (ft)	M	Clearance Radius of Boom Point Sheave (ft)
E	Boom Foot Height (ft)	N	Max Digging Depth (ft)
F	Length of Walking Shoe (ft)	O	Max Dump Radius (ft)
G	Outside Diameter of Std Tub (ft)	P	Max Dump Height (ft)
H	Overall Height (ft) (Top of Boom Point Sheave)		
I	Pitch Diameter of Boom Point Sheave (in)		

Figure 8.12 Walking Dragline Dimensions

Table 8.2
WALKING DRAGLINE SPECIFICATIONS

MAKE	MODEL	AVAILABLE BUCKET SIZES (cu yd)	MOTOR GENERATOR DRIVE MOTOR TOTAL HP	HOIST MOTOR HP	DRAG MOTOR HP	SWING MOTOR HP	PROPEL MOTOR HP	WEIGHT (lbs)	BALLAST (lbs)	APPROXIMATE RANGE OF MAX SUSPENDED LOADS (1000 lbs)	POWER SUPPLY (volts)
Bucyrus-Erie	380-W	9-16	1,000	—	—	250	—	916,000	116,450	45-80	4160
Bucyrus-Erie	480-W	12-18	1,000	1,045	—	390	—	1,545,000	260,000	62-82	4160
Bucyrus-Erie	680-W	14-26	1,600	1,500	1,080	660	660	1,804,000	200,000	70-130	—
Bucyrus-Erie	1260-W	24-40	2,450	2,090	2,090	1,500	750	3,335,000	450,000	116-200	4160
Bucyrus-Erie	1300-W	29-45	3,500	2,600	2,600	1,920	1,280	3,915,000	525,000	145-225	6900
Bucyrus-Erie	1350-W	37-48	4,000	3,200	3,200	2,560	1,280	5,743,000	650,000	185-225	6900
Bucyrus-Erie	1360-W	46-53	5,000	4,180	3,200	2,560	1,280	6,076,900	900,000	230-265	6900
Bucyrus-Erie	1370-W	51-62	5,000	4,180	4,180	2,560	1,500	6,104,700	500,000	255-320	6900
Bucyrus-Erie	1570-W	57-80	6,000	5,200	5,200	3,200	2,000	7,087,700	850,000	285-400	6900
Bucyrus-Erie	2570-W	93-115	12,000	10,400	8,360	5,200	4,180	12,470,000	1,000,000	463-585	6900
Bucyrus-Erie	3270-W	125-176	18,000	15,600	10,400	7,800	7,800	17,435,000	1,800,000	915	—
Marion	7450	9-17	n.a.	800	800	390	800	1,500,000	270,000	40-80	4160/7200
Marion	7500	13-22	1,500	1,300	1,300	510	1,300	1,580,000	350,000	67-108	—
Marion	7620	20-35	2,500	2,090	1,600	1,280	1,045	2,660,000	200,000	110-150	—
Marion	7820	25-45	3,000	2,600	2,600	1,500	1,300	3,860,000	400,000	185-210	—
Marion	8020	40-60	5,000	4,180	3,200	2,560	1,600	5,700,000	800,000	225-250	—
Marion	8050	50-65	5,000	4,180	4,180	2,560	1,600	6,500,000	900,000	280-320	—
Marion	8200	60-75	7,500	5,200	5,200	4,000	2,090	7,450,000	800,000	300-380	—
Marion	8750	75-120	13,000	10,400	7,800	6,270	3,200	13,050,000	1,600,000	400-583	—
Marion	8950	100-180	18,000	15,600	13,000	7,800	4,180	14,600,000	600,000	625-910	—
Page	732	11-20	1,250	1,045	1,045	510	500	1,933,000	340,000	55-100	4160/7200
Page	736	17-24	1,500	1,300	1,300	750	800	2,324,000	430,000	85-122	4160/7200
Page	740	18-30	2,550	2,090	2,090	1,125	1,045	3,184,000	450,000	90-150	4160/7200
Page	840	18-38	2,500	2,090	2,090	1,500	1,045	3,468,000	550,000	90-190	4160/7200
Page	752	24-45	3,000	2,600	2,600	1,500	1,300	4,306,000	580,000	120-225	4160/7200
Page	852	25-45	3,500	2,600	2,600	2,000	1,300	4,823,000	670,000	125-225	4160/7200
Page	852 LR	44-62	5,000	4,180	4,180	2,560	2,090	6,690,000	650,000	220-310	4160/7200
Page	757	49-70	6,000	5,200	5,200	3,200	2,090	7,523,000	860,000	246-350	7200
Ransomes & Rapier	W-700	10-17	1,250	—	—	390	—	1,072,000	180,000	50-85	3300/6600
Ransomes & Rapier	W-800	13-20	1,500	1,280	1,280	750	750	1,730,000	162,000	65-100	—
Ransomes & Rapier	W-1300	22-35	—	2,090	2,090	1,500	1,000	2,840,000	170,000	115-175	—
Ransomes & Rapier	W-2000	34-45	3,500	2,600	2,600	1,920	1,280	3,704,000	374,000	170-225	—
Ransomes & Rapier	W-3000	56-70	6,000	5,200	5,200	3,200	2,560	6,682,000	828,000	280-350	—

Table 8.2 (Continued)

WALKING DRAGLINE SPECIFICATIONS

MAKE	MODEL	DRIVE TYPE	NO. OF SWING UNITS	MAX CABLE SPEEDS		DRUM DIAMETERS		NO. OF ROPES		ROPE SIZE		BOOM POINT SHEAVE PITCH DIAMETER (in)	SWING RACK PITCH DIAMETER (ft)
				HOIST (fpm)	DRAG (fpm)	HOIST (in)	DRAG (in)	HOIST	DRAG	HOIST (in)	DRAG (in)		
Bucyrus-Erie	380-W	DC/MG set	2	—	—	58	45	1	1	2.25	2.25	60	11.7
Bucyrus-Erie	480-W	DC/MG set	2	—	—	46	44	1	1	2.13	2.38	66	23.8
Bucyrus-Erie	680-W	DC/MG set	2	—	—	66	66	1	1	3.00	3.00	95	17.0
Bucyrus-Erie	1260-W	DC/MG set	4	—	—	68	68	2	2	2.50	2.88	85	24.3
Bucyrus-Erie	1300-W	DC/MG set	3	—	—	92	92	2	2	3.00	3.00	85	24.3
Bucyrus-Erie	1350-W	DC/MG set	4	—	—	102	102	2	2	3.00	3.00	120	38.3
Bucyrus-Erie	1360-W	DC/MG set	4	—	—	110	102	2	2	3.25	3.25	120	38.3
Bucyrus-Erie	1370-W	DC/MG set	4	—	—	92	92	2	2	3.50	3.50	120	38.3
Bucyrus-Erie	1570-W	DC/MG set	4	—	—	102	102	2	2	3.88	3.88	120	38.3
Bucyrus-Erie	2570-W	DC/MG set	4	—	—	110	110	2	2	4.50	4.50	144	42.7
Bucyrus-Erie	3270-W	DC/MG set	6	—	—	132	132	4	4	4.38	4.38	145	53.3
Marion	7450	static DC	2	—	—	80	80	1	1	2.00	2.00	—	24.0
Marion	7500	DC/MG set	2	—	—	50	50	1	1	2.00	2.38	60	24.0
Marion	7620	DC/MG set	2	—	—	65	65	2	1	2.38	2.38	90	34.4
Marion	7820	DC/MG set	4	—	—	85	85	2	2	2.75	2.75	90	42.5
Marion	8020	DC/MG set	4	—	—	93	93	2	2	2.75	2.75	90	40.6
Marion	8050	DC/MG set	4	—	—	96	96	2	2	3.25	3.25	120	40.6
Marion	8200	DC/MG set	5	—	—	96	96	2	2	3.75	3.75	120	39.8
Marion	8750	DC/MG set	6	—	—	132	132	4	4	3.25	3.25	90	45.7
Marion	8950	DC/MG set	6	—	—	132	132	4	4	4.00	4.00	120	53.5
Page	732	DC/MG set	2	471	343	60	60	1	1	2.50	2.50	80	30.2
Page	736	DC/MG set	2	514	334	68	68	1	1	2.75	2.75	80	33.9
Page	740	DC/MG set	3	670	444	69	67	2	2	2.25	2.25	80	38.0
Page	840	DC/MG set	4	571	373	69	67	2	2	2.25	2.25	80	38.0
Page	752	DC/MG set	4	589	387	75	75	2	2	2.50	2.50	80	49.8
Page	852	DC/MG set	4	589	387	75	75	2	2	2.50	2.50	80	49.8
Page	852 LR	DC/MG set	4	610	398	98	98	2	2	3.25	3.25	120	48.2
Page	757	DC/MG set	4	671	339	98	98	2	2	3.50	3.50	120	52.2
Ransomes & Rapier	W-700	DC/MG set	2	—	—	55	55	1	1	2.50	2.75	—	24.0
Ransomes & Rapier	W-800	DC/MG set	2	—	—	60	60	2	2	1.88	2.13	—	24.4
Ransomes & Rapier	W-1300	DC/MG set	4	—	—	75	75	2	2	2.38	2.75	—	25.5
Ransomes & Rapier	W-2000	DC/MG set	3	—	—	94	94	2	2	2.75	3.13	—	28.0
Ransomes & Rapier	W-3000	DC/MG set	4	—	—	108	108	2	2	3.50	3.88	—	39.3

Table 8.2 (Continued)
WALKING DRAGLINE SPECIFICATIONS

MAKE	MODEL	WIDTH OF HOUSE (ft)	LENGTH OF HOUSE (ft)	WIDTH ACROSS WALKING SHOES (ft)	BOOM FOOT RADIUS (ft)	BOOM FOOT HEIGHT (ft)	SHOE LENGTH (ft)	SHOE WIDTH (ft)	GROUND PRESSURE (psi)	LENGTH OF STEP (ft)	STANDARD TUB OUTSIDE DIAMETER (ft)	TUB GROUND PRESSURE (psi)
Bucyrus-Erie	380-W	34.0	34.7	41.8	5.3	10.7	23.5	5.8	17.4	7.5	28.0	10.3
Bucyrus-Erie	480-W	30.0	53.5	49.5	16.7	8.5	42.5	6.0	16.8	7.3	36.0	10.6
Bucyrus-Erie	680-W	44.5	48.0	54.5	6.3	15.5	33.5	7.5	18.7	7.3	37.3	11.5
Bucyrus-Erie	1260-W	35.0	62.0	70.0	17.2	17.2	54.0	9.0	17.9	7.3	50.0	11.8
Bucyrus-Erie	1300-W	35.0	71.0	70.0	17.2	17.2	54.0	9.0	21.0	7.3	50.0	13.8
Bucyrus-Erie	1350-W	46.0	86.0	75.3	25.0	14.5	55.0	10.0	27.2	8.5	52.0	18.8
Bucyrus-Erie	1360-W	46.0	86.0	75.3	25.0	14.5	55.0	10.0	28.8	8.5	52.0	19.9
Bucyrus-Erie	1370-W	46.0	86.0	79.8	25.0	14.5	62.0	10.0	25.6	8.5	58.0	16.1
Bucyrus-Erie	1570-W	46.0	86.0	81.3	25.0	14.5	62.0	10.0	29.8	8.5	58.0	18.6
Bucyrus-Erie	2570-W	86.5	99.5	104.5	30.0	16.0	72.0	14.0	32.2	8.5	74.0	20.1
Bucyrus-Erie	3270-W	116.0	123.0	121.5	36.8	37.1	71.0	16.5	38.7	5.5	85.0	21.3
Marion	7450	34.3	59.0	46.0	16.0	10.0	30.0	6.0	21.7	6.0	30.5	14.3
Marion	7500	29.1	55.5	51.2	16.2	7.3	36.5	6.0	18.8	6.2	37.0	10.2
Marion	7620	50.0	72.0	64.0	18.5	11.8	40.0	8.0	21.7	7.0	42.0	13.3
Marion	7820	43.0	75.0	71.0	20.3	13.5	48.0	9.0	23.3	7.0	50.0	13.6
Marion	8020	60.3	88.0	77.0	21.5	12.3	54.0	10.0	27.5	7.0	54.0	17.3
Marion	8050	67.3	89.0	83.0	21.5	12.6	58.0	11.0	26.5	7.0	58.0	17.1
Marion	8200	67.3	94.5	83.0	21.5	15.1	55.0	11.0	32.1	7.0	58.0	19.6
Marion	8750	90.0	115.0	116.0	24.0	22.0	80.0	16.0	27.3	6.0	80.0	18.5
Marion	8950	90.0	116.0	116.5	27.0	23.0	70.0	16.5	32.9	7.0	80.0	20.2
Page	732	—	—	—	20.3	10.0	-	-	22.9	7.0	37.0	12.5
Page	736	—	—	—	23.5	10.5	35.0	7.5	23.1	7.0	41.0	12.2
Page	740	—	—	—	26.3	11.7	40.0	8.0	25.9	6.5	45.0	13.9
Page	840	—	—	—	26.3	11.7	40.8	8.0	27.7	6.5	45.0	15.1
Page	752	—	—	—	32.5	12.0	49.0	10.5	21.8	6.8	56.0	11.9
Page	852	—	—	—	41.3	21.3	49.0	10.5	24.4	6.8	56.0	13.4
Page	852 LR	—	—	—	30.5	11.6	54.5	11.0	29.1	7.2	60.0	16.4
Page	757	—	—	—	35.9	17.3	54.5	11.5	31.3	7.2	61.5	17.6
Ransomes & Rapier	W-700	36.8	48.6	46.8	10.5	13.1	31.2	6.6	13.7	5.2	29.5	10.9
Ransomes & Rapier	W-800	35.0	73.0	56.5	13.0	14.0	40.0	7.5	15.0	6.0	38.0	10.6
Ransomes & Rapier	W-1300	38.0	83.0	65.5	13.3	15.5	45.0	9.0	18.3	6.3	44.0	13.0
Ransomes & Rapier	W-2000	37.8	88.5	74.2	13.5	19.0	55.8	9.2	18.8	7.5	50.0	13.1
Ransomes & Rapier	W-3000	49.4	96.9	95.2	26.3	22.0	65.6	11.8	22.8	7.5	65.6	13.7

Table 8.2 (Continued)

WALKING DRAGLINE SPECIFICATIONS

MAKE	MODEL	MAX SPEED (mph)	SWING SPEED (rpm)	RANGE OF MAX HEIGHTS BOOM PT SHEAVE (ft)	RANGE OF BOOM LENGTHS (ft)	RANGE OF BOOM ANGLES (degrees)	REVOLVING FRAME CLEARANCE RADIUS (ft)	REVOLVING FRAME CLEARANCE HEIGHT (ft)	BOOM FOOT CLEARANCE RADIUS (ft)	RANGE OF MAX DIGGING DEPTH (ft)	RANGE OF MAX OPERATING RADII (ft)	RANGE OF MAX DUMP HEIGHT (ft)
Bucyrus-Erie	380-W	0.40	—	139-81	200-140	30-40	31.0	4.2	5.3	125-75	181-115	104-43
Bucyrus-Erie	480-W	0.15	—	146-95	215-175	30-40	38.0	4.7	16.7	115-67	206-154	107-47
Bucyrus-Erie	680-W	0.17	—	182-111	295-190	30-40	44.0	6.3	6.3	190-160	254-168	147-69
Bucyrus-Erie	1260-W	0.15	—	203-129	302-225	30-38	52.0	9.0	17.3	165-105	282-215	153-69
Bucyrus-Erie	1300-W	0.15	—	219-134	325-235	30-38	57.0	9.0	17.3	190-105	302-215	169-74
Bucyrus-Erie	1350-W	0.14	—	215-157	325-285	30-38	66.0	7.8	25.0	170-150	311-255	160-92
Bucyrus-Erie	1360-W	0.14	—	215-157	325-285	30-38	66.0	7.8	25.0	170-150	311-255	150-92
Bucyrus-Erie	1370-W	0.14	—	212-149	320-270	30-38	66.0	7.8	25.0	165-95	307-243	142-79
Bucyrus-Erie	1570-W	0.14	—	227-157	345-285	30-38	66.0	7.8	25.0	170-140	329-254	162-85
Bucyrus-Erie	2570-W	0.15	—	222-184	360-335	30-38	80.0	14.0	30.0	170-150	326-300	140-104
Bucyrus-Erie	3270-W	—	—	230	330	36	88.0	7.6	36.8	180	311	95
Marion	7450	—	—	—	240-160	34	39.0	4.5	16.0	210-110	226-144	123-55
Marion	7500	—	—	—	240-180	30-32	39.8	4.3	16.2	100-90	205-175	99-69
Marion	7620	—	—	—	300-200	30-35	53.5	5.0	18.5	140-130	248-225	120-89
Marion	7820	0.15	—	—	300-225	30-38	53.5	7.3	20.3	160-128	262-220	134-85
Marion	8020	—	—	—	325-225	30	66.0	8.7	20.5	165-160	305-285	120-105
Marion	8050	—	—	—	340-250	34-39	66.0	8.7	21.5	175-160	295-242	150-125
Marion	8200	—	—	—	350-275	33-36	68.0	11.0	21.5	160-120	320-270	140-130
Marion	8750	—	—	—	375-300	30-33	77.0	15.8	24.0	200-150	340-310	133-132
Marion	8950	—	—	—	400-300	30-33	82.0	15.0	27.0	160-150	340-291	124-113
Page	732	—	—	—	234-175	30	40.3	—	20.3	130-100	227-176	98-63
Page	736	—	—	—	260-212	31	46.8	—	23.5	156-120	250-209	112-84
Page	740	—	—	—	266-223	30	50.0	—	26.3	170-125	260-223	112-84
Page	840	—	—	—	289-225	30	57.0	—	26.3	170-122	280-224	124-80
Page	752	—	—	—	320-247	30	59.0	—	26.3	185-138	314-250	136-88
Page	852	—	—	—	320-258	30	59.0	—	41.3	175-157	324-270	152-109
Page	852 LR	—	—	—	329-294	30	69.7	—	30.5	209-170	321-291	130-107
Page	757	—	—	—	350-312	30	66.0	—	30.5	225-183	345-312	139-113
Ransomes & Rapier	W-700	—	—	—	190-142	30-40	35.4	—	10.5	138-72	179-121	96-39
Ransomes & Rapier	W-800	0.18	—	—	270-180	30-35	55.0	—	13.0	160-110	250-164	135-70
Ransomes & Rapier	W-1300	0.18	—	—	300-201	30-35	66.0	—	13.3	175-125	277-182	147-76
Ransomes & Rapier	W-2000	0.15	—	—	314-245	30-38	68.5	—	13.5	175-145	289-210	167-96
Ransomes & Rapier	W-3000	0.13	—	236-171	346-297	30-38	73.0	—	26.3	170-155	331-266	166-101

Table 8.3
CRAWLER DRAGLINE SPECIFICATIONS

MAKE	MODEL	AVAILABLE BUCKET SIZES (cu yd)	PRIMARY MOTOR TOTAL HP	HOIST MOTOR HP/TYPE	DRAG MOTOR HP/TYPE	SWING MOTOR HP/TYPE	PROPEL MOTOR HP/TYPE	WEIGHT (lbs)	BALLAST (lbs)	APPROXIMATE RANGE OF MAX SUSPENDED LOADS (1000 lbs)	POWER SUPPLY (volts)
American	12210	—	900	900/mech	900/mech	900/mech	900/mech	736,935	96,850	27-47	n.a.
Bucyrus-Erie	61-B	2-4	205	205/mech	205/mech	205/mech	205/mech	152,100	9,500	3-21	n.a.
Bucyrus-Erie	65-D	2-4	303	303/mech	303/mech	303/hyd	303/hyd	146,600	32,000	9-19	n.a.
Bucyrus-Erie	71-B	2-5	261	261/mech	261/mech	261/mech	261/mech	168,190	-	5-24	n.a.
Bucyrus-Erie	88-B	2.5-5.5	365	365/mech	365/mech	365/mech	365/mech	223,600	-	6-29	n.a.
Bucyrus-Erie	300-D	8-12	1130	1130/mech	1130/mech	336/elec	192/elec	801,500	76,000	40-70	n.a.
Clark	2400-B	—	547	547/mech	547/mech	547/mech	547/mech	515,455	59,200	24-42	n.a.
Manitowoc	4600-1	—	646	431/mech	431/mech	215/mech	215/mech	384,100	55,000	—	n.a.
Manitowoc	6400	10-15	2050	1600/hyd	1600/hyd	1600/hyd	450/hyd	1,107,720	140,000	55-80	n.a.
Marion	184-M	8-10	915	635/mech	635/mech	260/elec	635/mech	790,000	149,000	40-60	n.a.
Marion	195-M	8.5-17	900/elec	800/elec	800/elec	390/elec	500/elec	1,085,000	250,000	44-86	—
Northwest	180-DII	—	357	357/mech	357/mech	357/mech	357/mech	224,705	9,880	11-23	n.a.
Northwest	190-DII	—	357	357/mech	357/mech	357/mech	357/mech	247,400	49,500	16-23	n.a
P & H	2355	11-18	3000/elec	1000/elec	1000/elec	650/elec	700/elec	1,420,000	80,000	55-90	4160/7200
Weserhutte	SW-760	7-10	838	838/hyd	838/hyd	838/hyd	838/hyd	518,086	66,138	31-57	n.a.

Table 8.3 (Continued)
CRAWLER DRAGLINE SPECIFICATIONS

MAKE	MODEL	DRIVE TYPE	NO. OF SWING UNITS	MAX CABLE SPEEDS HOIST (fpm)	MAX CABLE SPEEDS DRAG (fpm)	DRUM DIAMETERS HOIST (in)	DRUM DIAMETERS DRAG (in)	NO. OF ROPES HOIST	NO. OF ROPES DRAG	ROPE SIZE HOIST (in)	ROPE SIZE DRAG (in)	BOOM POINT SHEAVE PITCH DIAMETER (in)
American	12210	diesel	1	260	—	—	—	1	1	1.63	—	60
Bucyrus-Erie	61-B	diesel	1	226	184	32	26	1	1	1.00	1.13	42
Bucyrus-Erie	65-D	diesel	1	284	188	22	23	1	1	0.88	1.13	30
Bucyrus-Erie	71-B	diesel	1	236	233	36	35	1	1	1.00	1.25	42
Bucyrus-Erie	88-B	diesel	1	210	210	36	36	1	1	1.25	1.50	48
Bucyrus-Erie	300-D	static AC	2	337	211	48	48	1	1	2.00	2.00	2/60
Clark	2400-B	diesel	1	271	173	46	36	1	1	1.38	1.63	45
Manitowoc	4600-1	diesel	1	—	—	38	38	1	1	1.38	1.63	47
Manitowoc	6400	diesel	2	280	260	50	50	1	1	2.00	2.25	65
Marion	184-M	diesel	2	—	—	38	50	1	1	1.75	2.00	—
Marion	195-M	DC/MG set	2	—	—	60	60	1	1	2.25	2.50	—
Northwest	180-DII	diesel	—	—	—	—	—	1	1	—	—	—
Northwest	190-DII	diesel	—	—	—	—	—	1	1	—	—	—
P & H	2355	DC/MG set	2	—	—	66	66	1	1	2.50	2.75	66
Weserhutte	SW-760	diesel	2	246	130	45	-	1	2	1.65	—	—

Table 8.3 (Continued)

CRAWLER DRAGLINE SPECIFICATIONS

MAKE	MODEL	WIDTH OF HOUSE (ft)	LENGTH OF HOUSE (ft)	HEIGHT TO TOP OF CAB (ft)	HEIGHT TO TOP OF HOUSE (ft)	GROUND CLEARANCE (in)	CRAWLER DATA: NO. OF UNITS	OVERALL LENGTH (ft)	OVERALL WIDTH (ft)	STANDARD SHOE WIDTH (in)	GROUND PRESSURE (psi)
American	12210	20.3	36.2	20.4	18.3	21	2	32.3	25.4	60	17.6
Bucyrus-Erie	61-B	11.8	19.7	12.4	12.4	14	2	15.9	12.8	33	14.5
Bucyrus-Erie	65-D	10.5	20.1	11.8	11.1	16	2	18.5	15.3	36	11.4
Bucyrus-Erie	71-B	12.7	—	—	12.8	16	2	16.8	12.8	36	13.3
Bucyrus-Erie	88-B	12.6	—	15.9	14.3	20	2	17.8	15.3	36	17.1
Bucyrus-Erie	300-D	—	—	—	—	—	4	17.0	24.0	72	14.0
Clark	2400-B	16.8	30.0	17.5	16.3	21	2	25.5	20.2	63	—
Manitowoc	4600-1	17.2	31.4	17.2	13.0	12	2	26.1	21.0	60	—
Manitowoc	6400	—	—	23.8	—	—	2	39.8	—	83	15.4
Marion	184-M	22.0	—	21.8	19.4	23	2	30.0	22.5	66	18.7
Marion	195-M	27.0	—	—	21.0	20	2	32.3	28.5	72	22.3
Northwest	180-DII	12.0	—	—	14.6	—	2	18.1	15.7	36	—
Northwest	190-DII	12.0	—	—	15.6	—	2	21.1	19.5	54	—
P & H	2355	28.0	—	26.8	—	36	2	39.3	33.5	90	19.0
Weserhutte	SW-760	16.1	—	19.7	20.3	35	2	33.8	30.5	59	13.1

Table 8.3 (Continued)
CRAWLER DRAGLINE SPECIFICATIONS

MAKE	MODEL	MAX SPEED (mph)	CONTINUOUS GRADEABILITY (%)	MAX GRADEABILITY (%)	SWING SPEED (rpm)	RANGE OF MAX HEIGHTS OF BOOM PT SHEAVE (ft)	RANGE OF BOOM LENGTHS (ft)	RANGE OF BOOM ANGLES (degrees)	REVOLVING FRAME CLEARANCE RADIUS (ft)	REVOLVING FRAME CLEARANCE HEIGHT (ft)	BOOM FOOT CLEARANCE RADIUS (ft)	RANGE OF MAX DIGGING DEPTH (ft)	RANGE OF MAX OPERATING RADII (ft)	RANGE OF MAX DUMP HEIGHT (ft)
American	12210	1.0	—	—	2.5	114-61	160-100	30-40	26.9	5.7	8.6	167-82	149-87	96-41
Bucyrus-Erie	61-B	1.0	—	—	3.6	102-47	110-60	36-64	14.5	3.5	5.2	55-30	90-35	83-24
Bucyrus-Erie	65-D	0.9	—	—	3.5	89-31	100-50	30-59	14.6	3.4	4.0	50-25	90-30	71-8
Bucyrus-Erie	71-B	1.0	—	—	3.3	108-41	120-60	30-60	16.0	4.3	5.5	59-29	110-35	91-18
Bucyrus-Erie	88-B	0.8	—	—	3.1	109-52	120-70	30-60	17.8	4.6	5.5	55-35	110-40	92-31
Bucyrus-Erie	300-D	0.5	—	—	2.8	119-93	190-130	34-39	29.5	8.1	5.3	130-100	165-110	94-68
Clark	2400-B	0.7	—	—	2.0	—	150-100	30-45	23.8	5.2	6.3	156-101	136-77	98-38
Manitowoc	4600-1	1.3	—	30	2.4	98-58	140-100	30-40	24.2	3.5	6.2	113-75	129-85	78-38
Manitowoc	6400	0.8	—	30	2.1	115-95	200-160	30	33.7	—	6.5	115-95	180-145	90-70
Marion	184-M	—	—	—	—	—	150-140	34-41	26.1	5.0	8.3	92-84	135-115	74-66
Marion	195-M	1.5	—	—	2.6	—	170-130	32-43	28.0	6.4	9.4	131-84	156-107	100-51
Northwest	180-DII	0.7	—	—	2.6	—	110-60	20-45	16.8	4.8	5.5	55-30	110-50	—
Northwest	190-DII	0.7	—	—	2.6	—	110-60	20-45	17.6	4.9	6.2	55-30	110-50	—
P & H	2355	1.2	—	—	2.2	145-86	200-140	30-40	37.0	9.5	7.1	130-100	183-117	125-66
Weserhutte	SW-760	0.9	—	25	2.5	134-100	187-147	37-40	27.0	7.9	7.1	105-86	149-129	110-73

Chapter 9

WHEEL EXCAVATORS

Figure 9.1 Medium Size Bucket Wheel Excavator

Wheel excavators dig with a rotating bucket wheel; this wheel discharges the material onto a belt conveyor. The material is transported on this conveyor or a series of belt conveyors until it is discharged from the machine.

Wheel excavators have been used, in limited numbers, for continuous excavation of unconsolidated materials starting back in the mid 1920's. Interest in the machines has been much greater overseas with the Germans, in particular, performing extensive application studies and machine development. Overall use within the United States has been very limited.

There are three types of wheel excavators: large stripping wheels, medium-sized machines of conventional design, and small diameter fixed wheels. (See Figures 9.2 and 9.3) The last large stripping wheel went into service in the United States in the early 1970's. Presently, this type is of only limited interest and will be considered only briefly. The current U.S. application for these large wheels is in prestripping for the large stripping shovels; this is different than the full face mining operations overseas. Six to eight medium-sized wheels have been placed in service within the last five to eight years and appear to have future mine application potential. The small, fixed wheels are a new U.S. development and most are still in the prototype stage. These last two wheel excavator designs will be considered in some detail. Because of the large variation in basic configuration, emphasis will be placed on general machine characteristics and less on detailed design features.

There will be no discussion of reclaimer type wheels or those special machines designed for digging substantially below the level of their crawlers. The discussion will be further limited to mining size units of approximately 500 cubic yards/hour capacity and larger.

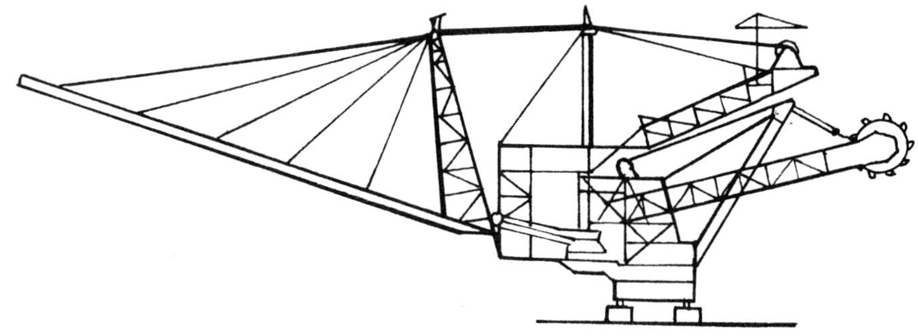

LARGE STRIPPING WHEEL (DIRECT DISPOSAL)

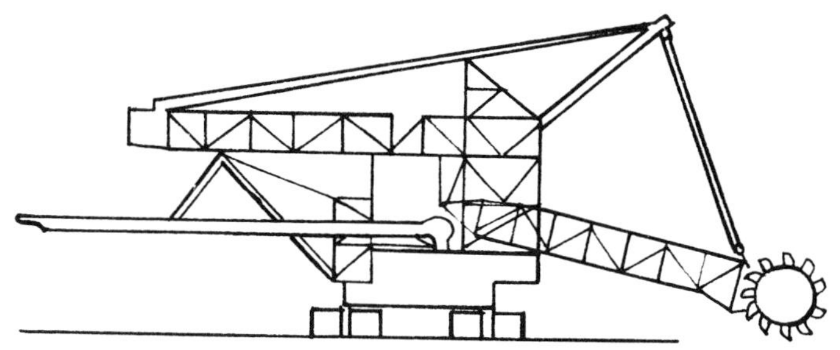

LARGE STRIPPING WHEEL (DIRECT DISPOSAL)

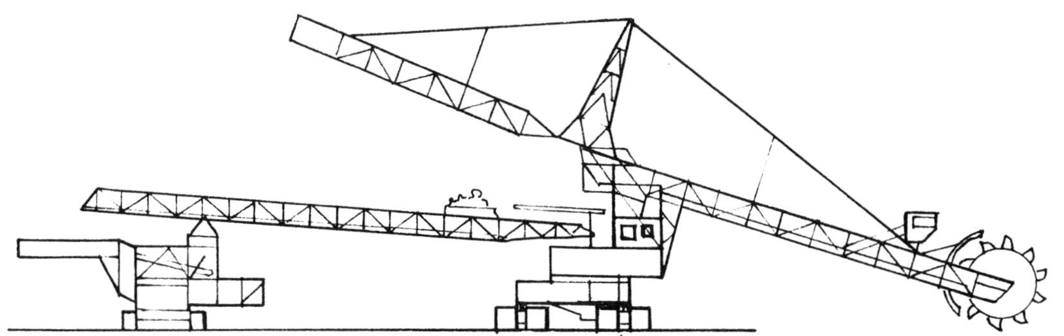

LARGE STRIPPING WHEEL (BELT LOADING)

Figure 9.2 Types of Bucket Wheel Excavators — Large

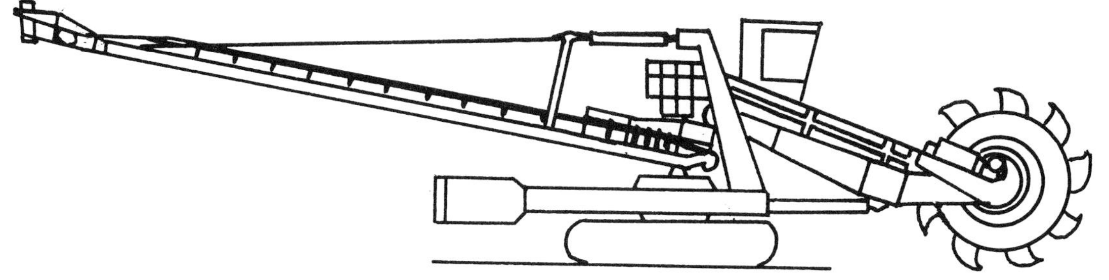

MEDIUM SIZE WHEEL

SMALL FIXED WHEEL

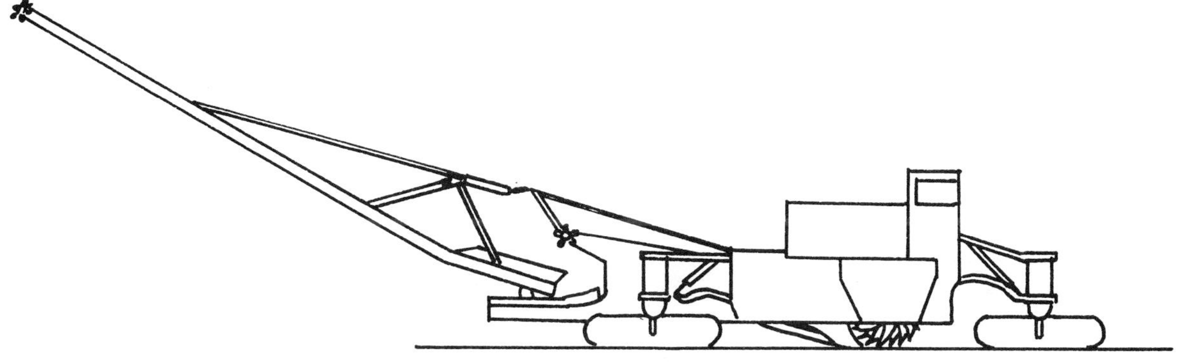

SMALL DRUM TYPE WHEEL

Figure 9.3 Types of Bucket Wheel Excavators —Medium & Small

TYPICAL UNITS

The three broad classes of machines available have significantly different application requirements and specifications. All of the stripping wheels are highly customized to meet specific mine requirements, so that the sizes reflect existing wheel excavators rather than available models. Only units in the United States will be considered, but it should be recognized that much larger machines are in service overseas.

Large Stripping Wheels

- 21.6 to 38 ft. wheel diameters
- All electric drives with a trailing cable
- 2700 to 7600 cu. yd./hr. capacity
- 75 to 104 ft. digging height
- Twin crawler units in each corner

Medium-Size Wheels

- 10 to 25 ft. wheel diameter
- 250 to 1600 HP diesel/electric
- 400 to 3500 cu. yd./hr. capacity
- 20 to 50 ft. digging height
- Two crawler mounting

Small Fixed Wheels

- 3 to 16 ft. wheel (drum) diameter
- 200 to 1600 HP diesel/hydraulic
- 1000 to 2800 cu. yd./hr. capacity
- .5 to 12 ft. digging height
- Two to four crawlers

Photograph 9.1 Bucyrus-Erie bucket wheel excavator removing top layer of overburden

Photograph 9.2 Bucyrus-Erie 654 wheel excavator loading side by side bottom dump trucks

Manufacturers

The manufacturers of bucket wheel excavators are listed in Table 9.1. Please note that this is not a comprehensive list of worldwide manufacturers, but does include those major companies that are marketing in the United States. All three types of wheel excavators are included in the listing.

BASIC MACHINE OPERATION

Loaders, shovels and draglines are cyclic excavators because the actual excavating phase is interrupted by a swing to dump operation. The bucket wheel can excavate continuously.

Multiple small buckets, mounted on the circumference of a rotating wheel structure, take progressive slicing side cuts as the wheel is swung back and forth across the digging face (slewing). Wheel rotation produces an upward cutting action. Material collected in each bucket is initially retained by a fixed internal ring segment that acts as the back (or bottom) of the bucket. This ring segment only closes off the buckets behind the digging sector of the wheel where actual cutting occurs. As the wheel rotates, the material in each bucket falls out when it reaches the top position where there is no closing ring. There is a relatively large discharge zone for each bucket to dump its load while the wheel continues to rotate, and there is sufficient overlap between succeeding buckets to effectively produce a continuous flow.

The wheel is raised or lowered to take successive horizontal arc type cuts across the face, shaping the profile of the face as desired. The bucket penetration into the face for each pass is set by positioning the machine. Cut depth is controlled by swing speed. Actual effective output is about one half the theoretical capacity.

The material discharging from the buckets slides down a slope sheet (fixed deflecting plate) which directs it onto a belt conveyor roughly parallel with the wheel. (See Figure 9.4) The wheel belt conveyor carries the material to the center of rotation of the machine and transfers it to an intermediate conveyor or directly to a discharge conveyor. The

Table 9.I

BUCKET WHEEL EXCAVATOR MANUFACTURERS

Barber-Greene Company 400 N. Highland Avenue Aurora, Illinois 60507	16′ wheel diameter 1500–2000 ton/hr.
Buckau R. Wolf D4048 Grevenbroica Post F.69 West Germany	18′–70′ wheel diameter 1500–20800 cu. yd./hr. 30′–164′ cut height
Bucyrus-Erie Company 1100 Milwaukee Avenue South Milwaukee, Wisconsin 53172	24′–30′ wheel diameter 7000 cu. yd./hr.
CMI Corporation Box 1985 Oklahoma City, Oklahoma 73101	2.5′ wheel diameter 250–400 ton/hr. .5′ cut height
Demag Lauchhammer 4 Dusseldorf-Benrath Forststrasse 16 Germany	7.2′-45.9′ wheel diameter 110–7100 cu. yd./hr. II.3′– 92′ cut height
Easi-Miner Huron Manufacturing Corp. P.O. Box 1398 Huron, South Dakota 57350	to 7′ wheel diameter 1000–2500 ton/hr. 1′–2.5′ cut height
Eisenwerk Weserhutte AG 4970 Bad Oeynhausen West Germany	9′–14.5′ wheel diameter 135–1780 cu. yd./hr. 18′–27.8′ cut height
Fried Krupp GMBH 414 Rhinhausen Franz-Schubert-Strasse 6 West Germany	16′–57′ wheel diameter 940–12230 cu. yd./hr. 32.8′–128′ cut height
Mechanical Excavator Inc. 2960 Marsh Street Los Angeles, California	to 25′ wheel diameter to 3500 cu. yd./hr.
Orenstein & Koppel AG D-2400 Luebeck 1 Einsiedelstrasse G West Germany	17′–71′ wheel diameter 1470–20300 cu. yd./hr. 26′–167′ cut height

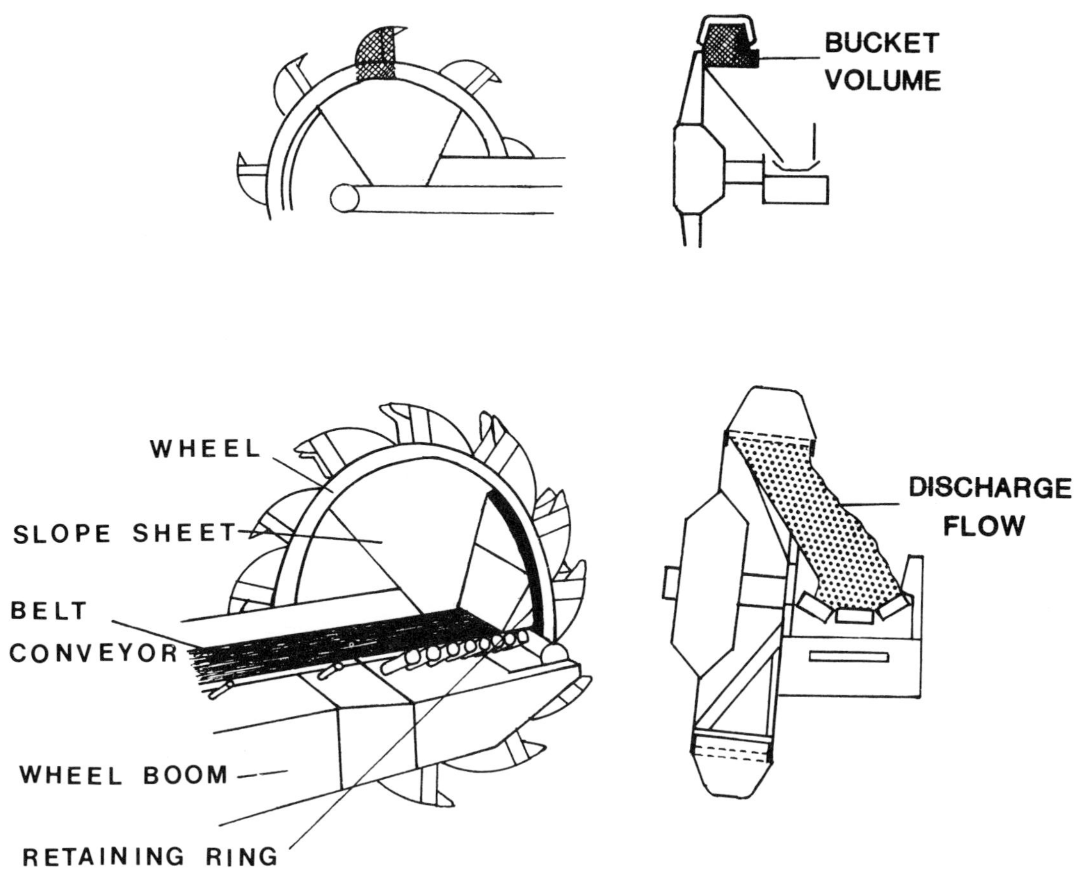

Figure 9.4 Wheel Volume and Discharge Path

discharge conveyor can be mounted on either a swinging boom, or on a bridge conveyor riding at the far end on a pit conveyor frame.

The discharge boom or bridge can be positioned (rotated) independently of the wheel boom and the lower works crawler units. Discharge can be to stockpile or spoil but generally is to a shiftable belt conveyor system running parallel to the direction of machine advance.

Some of the small fixed wheel excavators utilize a rotating drum type cutting head similar to that found on underground continuous miners. The drum has a forward rotation so as to cut downward, pushing material to the rear, thereby assisting the propel motion. The cutting drum extends across the full width of the machine and makes a thin skimming type cut as the machine is propelled. Cut depth is controlled by vertical positioning of the forward crawlers. Shovel and/or pickpoint teeth, mounted on the periphery of the drum, provide the cutting action and help to push the broken material up a following gathering blade to feed a belt conveyor. A double spiral scroll may be added to the drum to help move the material from the ends towards the center to improve the feed to the conveyor. The pick-up conveyor transports the material to the rear of the machine and deposits it on a discharge conveyor. The discharge conveyor is integral with a rotatable cantilevered boom, so that the material can be discharged to trucks running parallel on either side.

APPLICATIONS

There are currently very few bucket wheel excavators in service in the United States. They have been used for:

- Overburden excavation with direct spoiling — full depth (large stripping wheels)

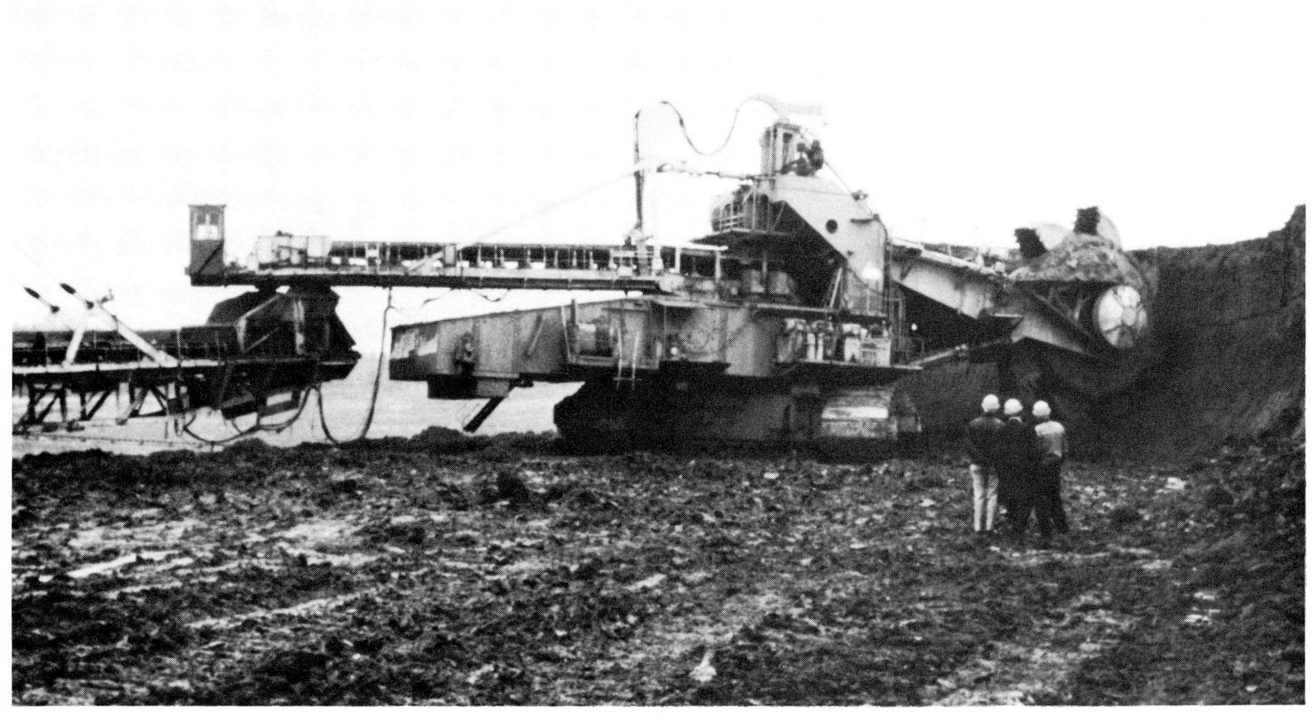

Photograph 9.3 Demag Lauchhammer medium size wheel excavator removing overburden and discharging to a mobile transfer conveyor

- Overburden excavation with direct spoiling: prestripping for stripping shovels (large stripping wheels)
- Overburden excavation with conveyor or truck loading: prestripping for a large dragline or stripping shovel (medium size wheels)
- Large earthmoving projects (medium size or small fixed wheels)
- Coal excavation with conveyor or truck loading (medium size or small fixed wheels)
- Reclamation leveling (small fixed wheels)
- Topsoil removal (small fixed wheels)

At present, most of the large stripping wheels placed in prestripping operations 10 to 30 years ago continue in service. A few medium size wheels have been introduced within the last 4 to 6 years, prestripping for larger excavators. Small fixed wheels are loading coal into trucks. All are operating in relatively uniform, unconsolidated formations.

Dependent to some extent on the sizes of the buckets, the wheels cannot effectively handle large boulders, frozen blocky material, or stumps and debris, etc. This means that special attention must be given to the nature of the formation. The slicing action combined with the small buckets, on the other hand, results in a relatively uniform sized discharge suited to transport on belt conveyors. Wheels, therefore, are typically applied in operations employing belt conveyors for the haulage system of either overburden or coal/ore.

The small drum type wheels are suited to the removal of thin seam deposits and/or parting. No drilling or blasting is required. The discharge is relatively small and uniform in size, which may reduce the need for subsequent processing. (See Photograph 9.5)

Photograph 9.4 Barber Greene WL-50 small fixed wheel excavator, in coal, loading bottom dump trucks

Photograph 9.5 Easi-Miner (manufactured by Huron Manufacturing Company) small fixed drum type excavator, in coal, loading bottom dump trucks

GENERAL CHARACTERISTICS

The bucket wheel excavator is a relatively complex design with multiple powered operating functions. They are illustrated in Figure 9.5.

- Crawler propel, motors
- Bucket wheel rotation, motors
- Wheel boom elevating, hydraulic cylinders
- Wheel belt drive, motors
- Discharge belt drive, motors
- Discharge boom elevating, hydraulic cylinders
- Discharge boom rotation, motors
- Upper works rotation, motors

Where motors are indicated as the power source, they generally are electric motors if the BWE is electric powered and hydraulic motors if the BWE is diesel powered.

The three types of wheels have quite different characteristics; relatively little is common other than the basic excavating mode. They will, therefore, be treated separately with brief summaries of the key features. Specific units may deviate from the indicated general pattern.

Large Stripping Wheels

- Continuous output is provided by multiple buckets depositing on a belt conveyor.
- Can dig medium hard consolidated materials without prior blasting; generally applied in materials with a low cutting resistance (determined by site tests).
- Cannot economically handle large boulders, frozen ground, consolidated hard formations.

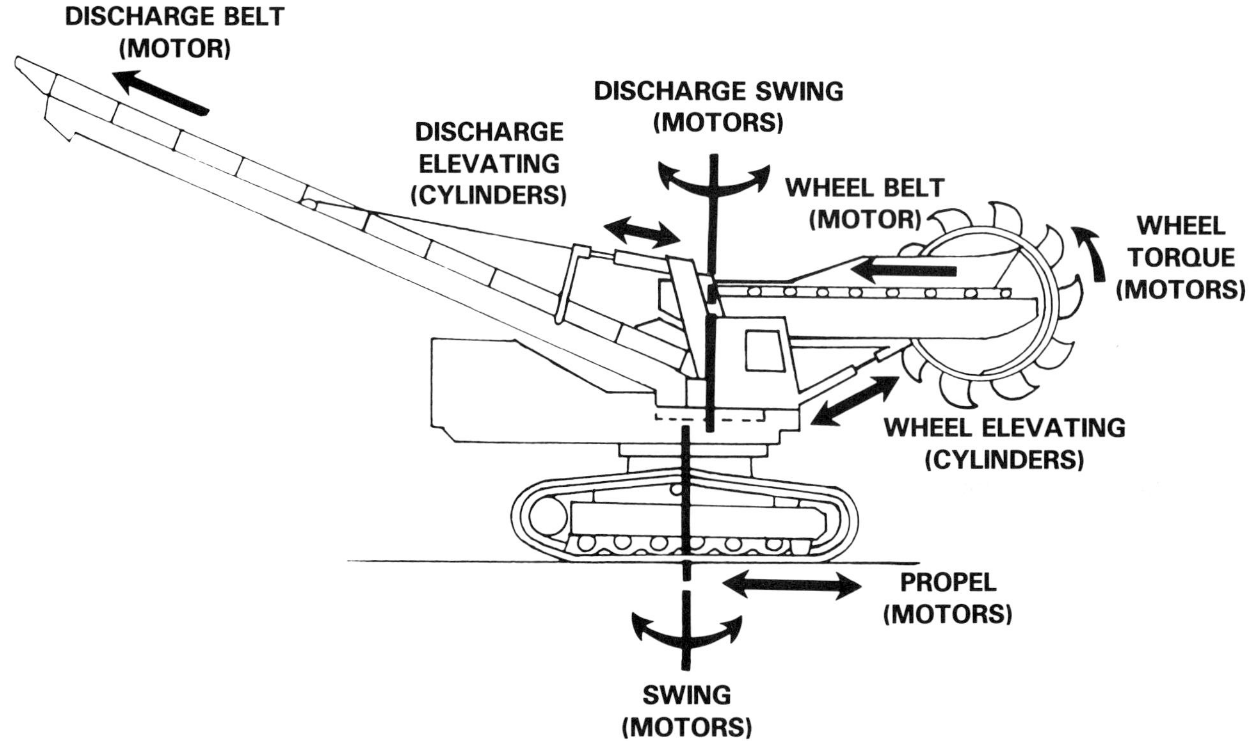

Figure 9.5 Wheel Powered Functions

- Machine configuration changes for direct spoiling, discharge to conveyors or a combination for selective mining.
- Highly customized design
- Vertical ranging wheel for cutting faces 30 to 100 feet high
- Large discharge radius
- Crowd-less design (the boom does not move in and out)
- Cell-less digging wheel design (not compartmented)
- Variable digging wheel speeds
- High speed belt conveyors on the machine (wheel belt, transfer belts, discharge belt)
- Multiple crawler units (3 to 4) with multiple track belts on each (6 to 12)
- All electric D.C. drives with trailing cable for power supply
- Automatic machine leveling
- Removable buckets and teeth
- Adaptable to automatic dig sequencing
- Relatively uniform power consumption
- Very low propel speeds and gradeability
- Limited pit mobility
- Medium ground pressures (10 to 20 psi)
- Produces a sized product for discharge
- Can discharge material above or below its working level.
- Operator has good visibility of wheel; requires auxiliary means (TV systems) to view discharge
- Requires a level operating bench
- Auxiliary equipment required for floor clean-up
- Two to four man crew required
- Maintenance performed in pit
- Fair machine availability
- Good blending capabilities across the face
- Can have some operating problems with high winds
- Long service life (25 years plus)
- Very high initial cost
- Long delivery cycle
- Extended field erection time (1 to 3 years)

Medium Size Wheels

- Continuous output is provided by multiple buckets depositing on a belt conveyor.
- Can dig medium hard consolidated materials with or without prior blasting; generally applied in materials with low cutting resistance.
- Cannot economically handle large boulders, frozen ground, consolidated hard formations.
- Vertical ranging wheel for cutting face to 50 feet high
- Crowd-less design
- Cell-less design
- Swingable discharge boom conveyor
- Medium speed belt conveyors on machine (wheel belt, discharge belt), 600 - 800 fpm
- Two crawler mounting, independent hydraulic drives
- Medium propel speeds (0.2 to 0.8 mph), good gradeability
- Low to medium ground pressures (10 to 20 psi)

- Diesel hydraulic drives or single motor primary electric drive
- Produces a sized product for discharge
- Swinging or fixed discharge boom
- Smooth, shock-free discharge
- Can discharge material above or below its working floor.
- Operator has good visibility of wheel and a limited view of the discharge.
- Auxiliary equipment is required for floor clean-up.
- Crew size is dependent on machine size.
- Maintenance is performed in the pit.
- Good blending capabilities across the face
- Limited operator skill required
- Intermediate length service life
- High initial cost
- Extended delivery and erection cycle

Small Fixed Wheels

- Continuous output is provided by either multiple buckets or a cutting drum, with material pushed up onto a belt conveyor.
- Can unconsolidated material in the case of the wheel, and consolidated materials without blasting in the case of the drum cutter.
- Cannot economically handle boulders
- Vertical cutting height from 6 inches to 12 feet
- Fixed non-elevating digging wheel/drum
- Cut depth controlled by elevation of main frame with respect to crawlers

Photograph 9.6 Weserhutte medium size wheel excavator showing maximum digging height

Photograph 9.7 Easi-Miner excavator, small fixed wheel

- Cutting rate dictated by machine propel speed (drum speed adjusted accordingly)
- Diesel/hydraulic drives
- Non-rotating upper works
- Crawler mounted (3 or 4 tracks)
- Good pit mobility
- Low to medium propel speeds (to 2.5 mph)
- Little operator skill required
- No auxiliary equipment required
- Relatively short life
- Lost cost
- Limited service support available
- Potential dust problem (drum type)

To a large degree, the large stripping wheels are made to order, so that product literature is essentially limited to manufacturers' general capabilities and brief specifications for machines built over the years. This would not be comparable to the data available on other products, so no attempt has been made to prepare specification tabulations for these machines.

The medium size wheels are a standardized product but the range of models in mining sizes and the number of the manufacturers is limited. The data readily available on the various units cannot be correlated to provide a comprehensive overview of general characteristics. Graphs 9.1 through 9.5, however, provide a frame of reference.

Small fixed wheels are in a development period with each machine placed in service being an upgraded version of the previous unit. Analysis of the machine specifications would be misleading. It is of interest to note, however, that the drum type units are a variation of the machines developed as highway concrete trimmers.

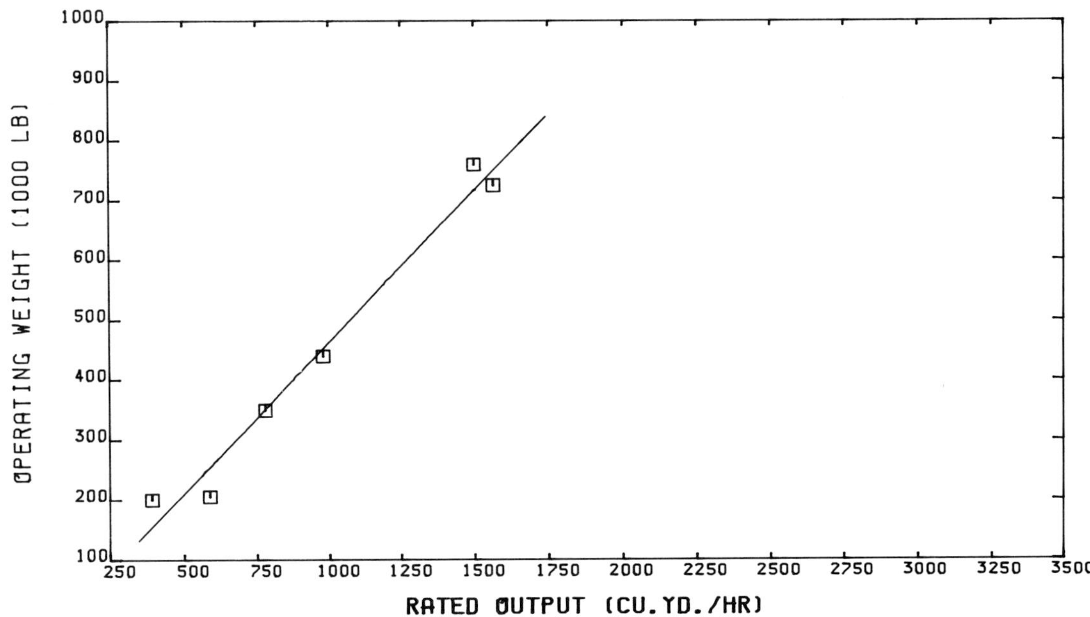

Graph 9.1 Operating weight/rated output

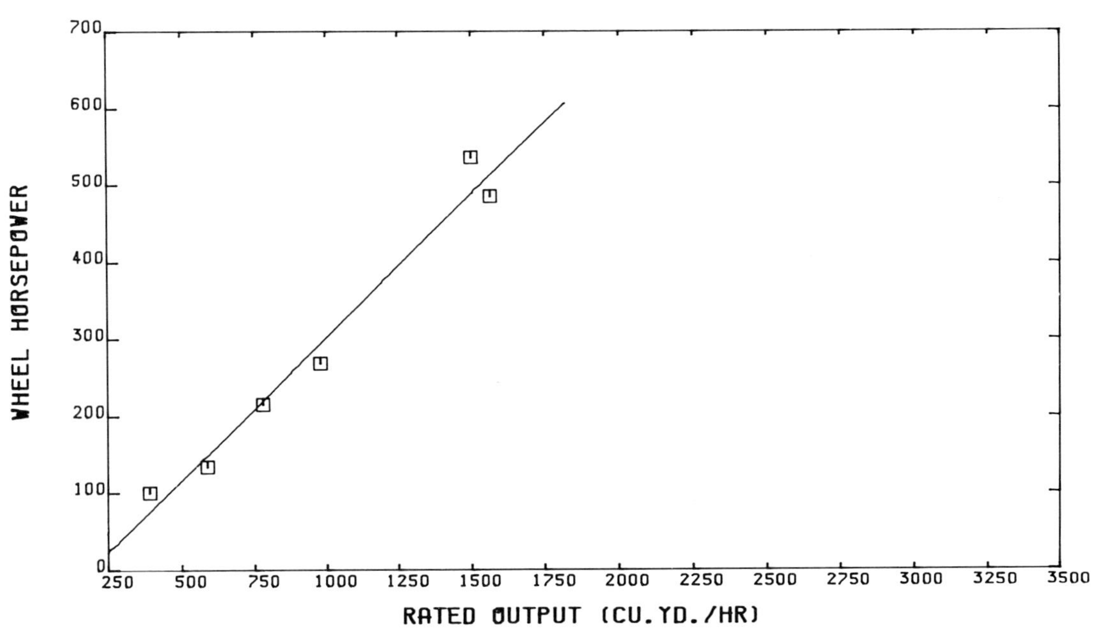

Graph 9.2 Wheel HP/rated output

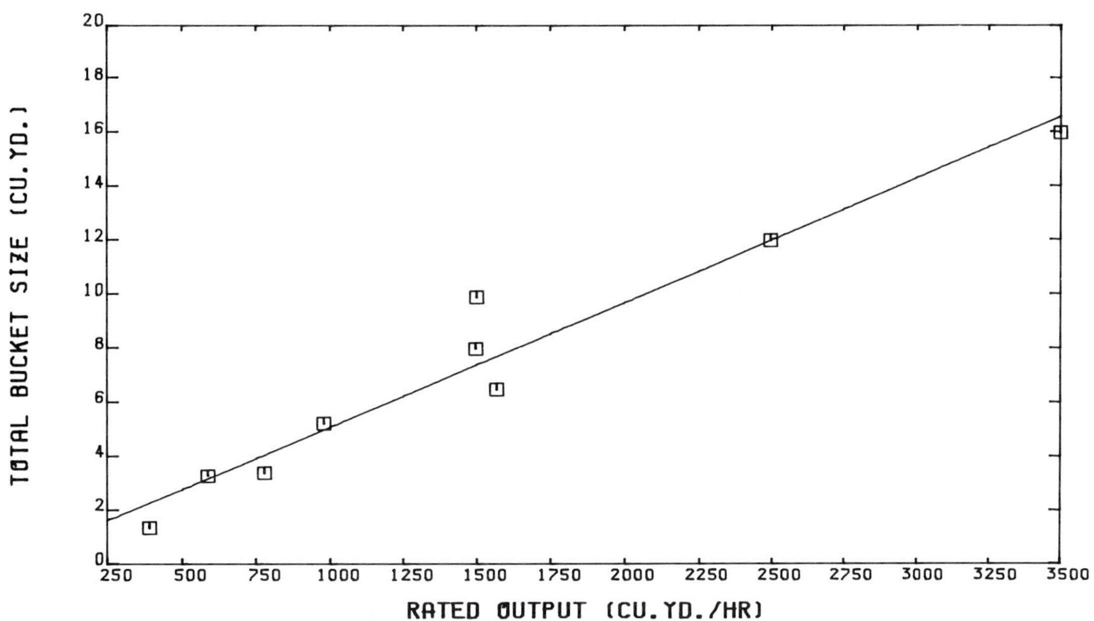

Graph 9.3 Total bucket size/rated output

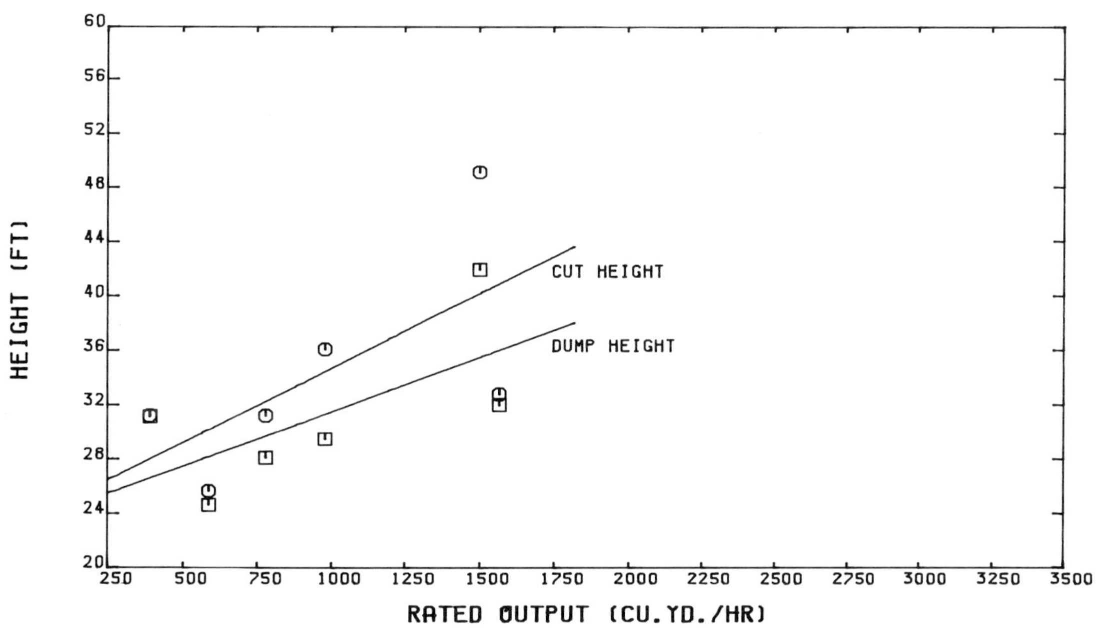

Graph 9.4 Wheel diameter/rated output

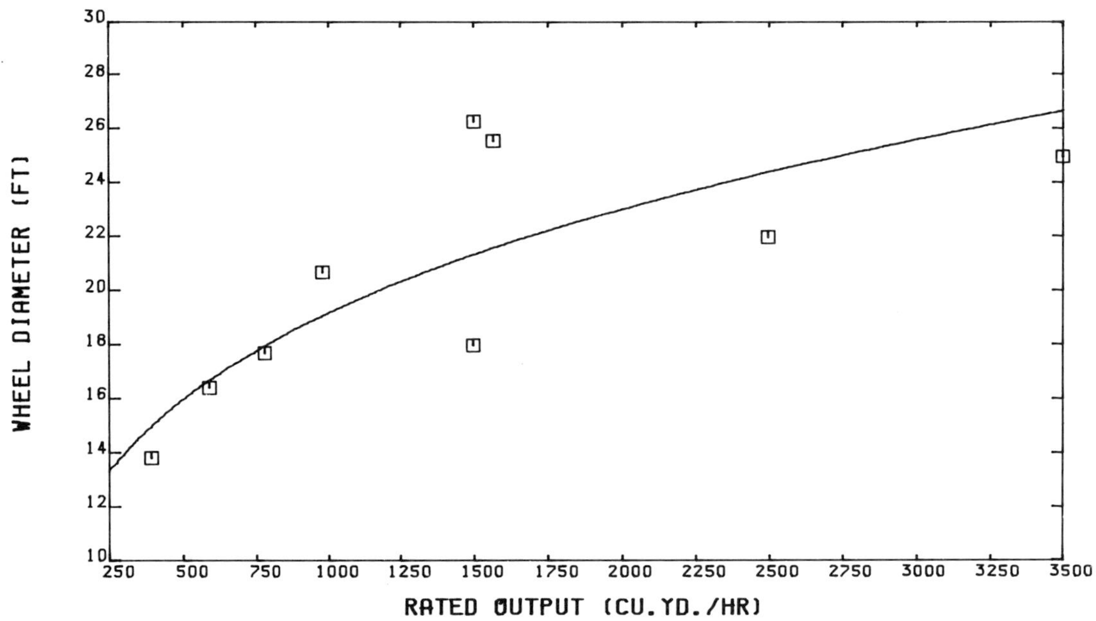

Graph 9.5 Dump height & cut height/rated output

OPERATIONAL PRACTICES

The material face is excavated by swinging the wheel back and forth through an arc with each pass at a selected cut depth, and cutting alternately in both directions. The slice height (or depth) is normally about 0.6 to 0.7 times the wheel diameter. Wheel speed is dictated by the material characteristics and the swing speed.

The swinging arc of the machine in each successive cut is sickle shaped, tapering at the ends as the angle increases from the direction of advance. As a result, the cut depth decreases and output is reduced unless swing speed is proportionally increased. This can mean a 20 to 30% loss in production. The large machines are, therefore, automated to adjust swing speed to maintain a constant output.

Nominal swing speeds of 60 to 70 feet/minute with a maximum of 140 feet/minute are common. The swing inertia dictates the the braking forces required when reversing cutting direction and the energy to be absorbed when striking an obstacle.

There are two basic types of cuts made: terrace and dropping. (See Figure 9.6) In the terrace cut, the face slope is established by a series of small bench type cuts made by the forward arc of the wheel. The machine is propelled forward after each swing pass to provide the cut depth, and this is repeated until the desired terrace depth is achieved. The machine is then repositioned (by propelling backward) for the start of the next terracing cut sequence.

With the dropping cut, the final slope of the face is defined by the vertical arc of the wheel about its boom pivot on the machine main frame. Cuts are made by the bottom arc of the wheel progressing from the top to the bottom of the face as the wheel is swung from side to side. After a cut has been made over the entire face, the machine is repositioned by advancing the machine to the selected cut depth.

Production is approximately the same with each cutting procedure. The dropping cut, which produces a relatively steep face, is only applicable where the bank is adequately stable. It requires less machine travel. The forward and reverse propelling sequence for the terrace cut can tear up the

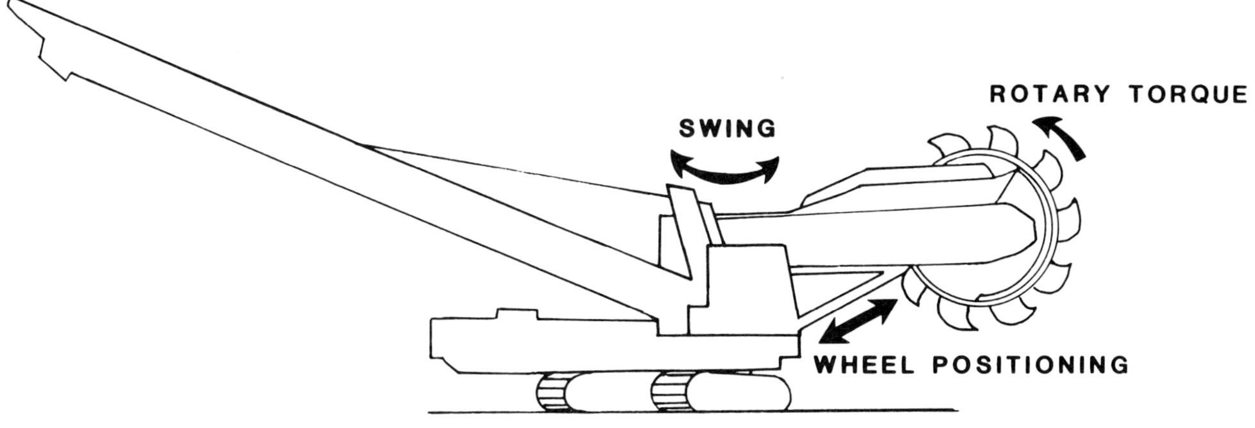

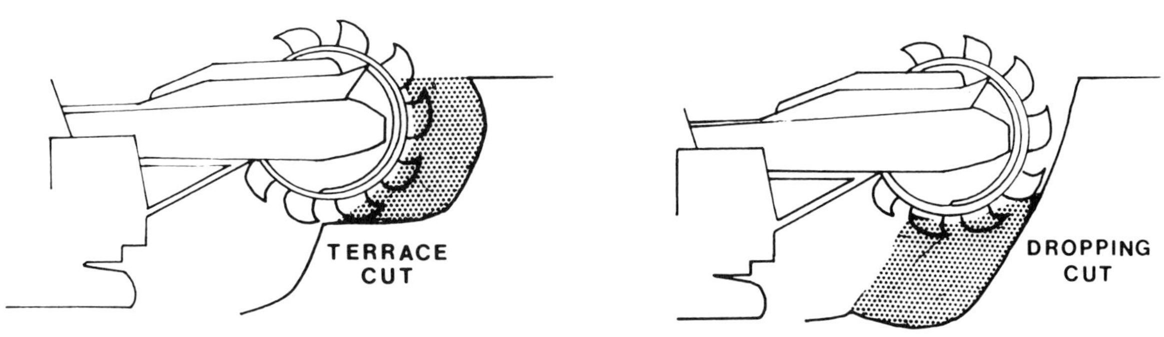

Figure 9.6 Wheel Digging Forces & Digging Profile

floor under marginal conditions. The dropping cut normally requires more power except when the excavating is in a horizontally stratified formation.

Although the wheels have some capability to excavate below their own level, in the newer mines they are rarely applied in this manner. In some cases, the medium size wheels are in benching operations where the wheel digging height equals or exceeds the face height, so that the wheel is not being elevated.

Mining is either by the full block or lateral block method of excavating. Using the full block method, the wheel travels parallel to the highwall, removing a block which is a function of the wheel cutting radius and the swing arc. Optimum block width is normally achieved with a swing angle of 80 degrees to the highwall side and 45 to 50 degrees on the discharge side. This is the most common operating mode.

In the lateral block method, the machine also travels parallel to the highwall but with the crawlers offset. Long deep benches or terraces are cut along the highwall, starting at the top. The angle of swing into the highwall is from 10 or 15 degrees through 90 degrees to the direction of travel. This generally requires a machine equipped with a long boom. There is significantly more travel time required than with the full block method, reducing digging time and increasing crawler wear. However, the lateral block method can be used for selective mining.

Selective mining, working around face boulders, frozen lumps, and/or stoppages from overload cut-offs can substantially reduce production.

As noted earlier, the machines lend themselves to automated sequencing, such as the control of swing speed to provide a uniform digging output. In addition, swing arcs can be programmed to produce automatic reversals and automatic propel advances to the desired cut depth after each pass. In hard digging, power demand monitoring can reduce swing speeds and limits can be set on swing acceleration/decleration.

Although the medium size wheels can be used to load trucks, the advantage of its continuous output capability is lost unless the trucks are always available. Truck delays are minimized if the discharge point is equipped with a double chute and switching gate allowing the flow to be shifted without interruption when sequentially loading trucks in parallel positions. (See Figure 9.7) Obviously, to achieve the ideal, the truck fleet must be matched to wheel output and the truck scheduling must be well supervised.

Dependent on the mining depth, machine size, and the overall mining plan, the wheel can directly spoil or stockpile material. Generally this requires a special, long discharge boom and is most commonly employed with the lateral block method of excavating. (See Figure 9.8)

Most wheels are employed to load belt conveyor systems. The conveyor system can be either an across the pit or an around the pit configuration. The latter is the most common and incorporates a shiftable pit belt conveyor section parallel to the highwall (face conveyor) that accepts the wheel discharge as it advances down the pit.

If the wheel has a relatively long discharge radius, the wheel can discharge directly onto the pit belt conveyor. (See Figure 9.9) The belt must be periodically shifted (advanced with the highwall) to keep the belt within the discharge range of the machine. Since the belt location can only range between the crawlers and the machine discharge radius, the belt must be shifted rather frequently. Each move involves auxiliary equipment, a work crew and lost production time.

To increase the discharge range of the wheel to the shiftable pit conveyor, a mobile transfer conveyor (band wagon) is commonly added to the system. (See Figure 9.10) The angular orientation of the transfer conveyor can be varied to meet changing distances between the digging face and a spoil area or stockpile. In some cases, multiple transfer units are used to meet special site requirements.

The primary purpose of the mobile transfer conveyor is to provide, in effect, an extensible link between the wheel and a shiftable face conveyor. This is achieved by the positioning of the transfer conveyor to a varying angular orientation with the wheel discharge conveyor. (See Figure 9.11) By this means, the frequency of relocations for the shiftable conveyor can be reduced substantially. Note that the transfer conveyor can be located on a bench above or below the wheel and can, in some instances, discharge to a conveyor on yet another bench.

The wheel discharges into a small hopper on one end of the transfer unit and the transfer conveyor, in turn, discharges to a hopper car. This hopper car rides on the pit conveyor frame and is periodically repositioned as the digging face advances. Special transfer units with high vertical angling capabilities may be used at the end of the bench to tranfer the material up to a main conveyor extending across the end of the pit.

The mobile transfer conveyor can be a rather complex machine, generally powered in a similar manner to the wheel. The transfer units are mounted on rotating upperworks, and means are provided to elevate or depress the ends as required.

A wide variety of pit belt systems are possible. Conveyors can be placed parallel to each other on multiple benches with the entire series advancing with the digging face. A conveyor can be located on an intermediate bench and fed by multiple wheels operating on benches both above and below. These systems generally require a wide, well prepared bench for machine maneuvering. Dozers are required for bench cleaning and belt shifting.

High rainfall can cause operational problems if materials with a high clay content are handled. The water can increase the stickiness and the adhesive properties of the material, resulting in bucket plugging, coating of the belts, and loss of friction between the belt and drive pulleys.

Low temperatures often cause materials to stick to the buckets and belts. In some applications, propane heaters have been mounted on the wheel boom to heat the back of the buckets in an attempt to minimize the build-up of material. Freezing conditions may also signficantly increase both the cutting forces required and the formation of slabs or chunks of oversized material.

High winds striking the belts, particularly on the large machines, can cause belt misalignment, spillage, and belt damage due to abrasion against the side skirts. Operator visibility may also be impaired by the accompanying dust conditions.

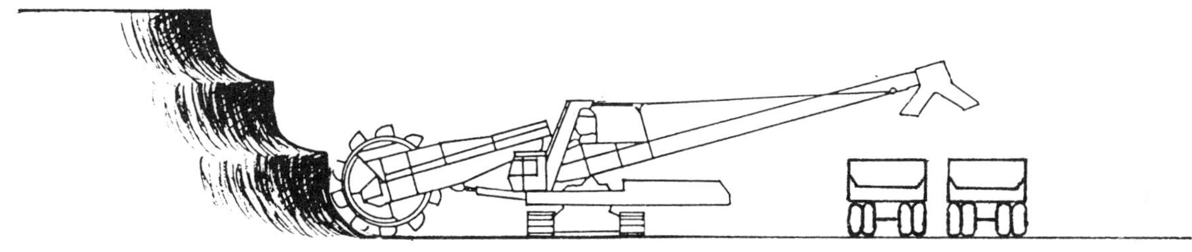

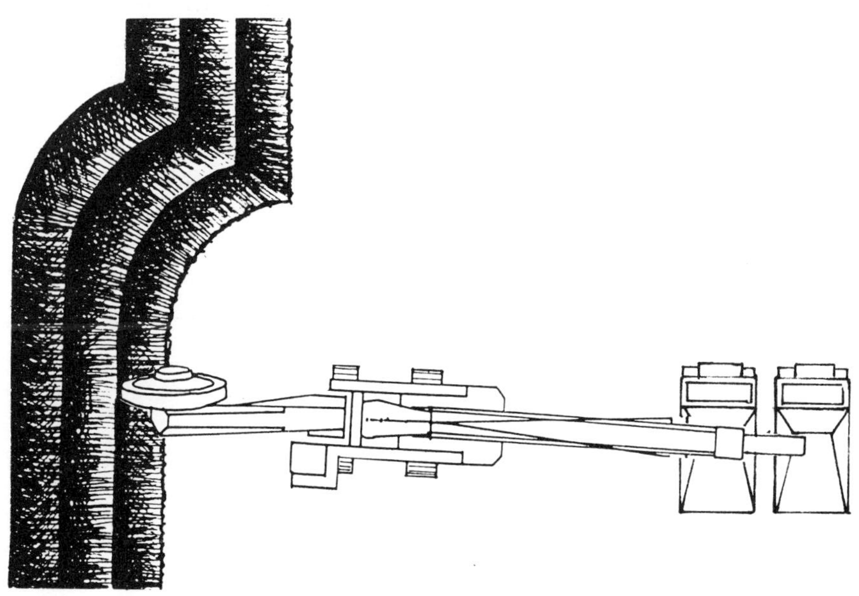

Figure 9.7 Wheel Direct Loading Trucks

Wheel cutting speeds may have to be reduced if hard formations are encountered. Striking such materials transmits impact loads to the wheel drive and wheel boom structure. The repeated impact by the teeth on a hard obstruction can set up a vibration of the entire wheel boom which, if continued for prolonged periods, results in weld failures. Similarly, boulders, rocks or chunks of material discharged from the buckets which strike the conveyor belt, severely load the conveyor idlers and support members. If the rocks bounce, they cause additional dynamic loads and damage to the chute and belt skirts.

If the wheel and/or mobile transfer conveyor are electrically powered, the power cables are brought down the pit supported on the side of the conveyor structure. The hopper car also serves as an electrical station, which incorporates a cable spooling drum, transformers and switch gear, etc. Short power cables from the wheel and mobile transfer unit are connected into the hopper car power supply.

Mine operational procedures have not as yet been fully developed for the small fixed wheels. While they could be used to load conveyor belts, the substantial travel necessary during excavating in-

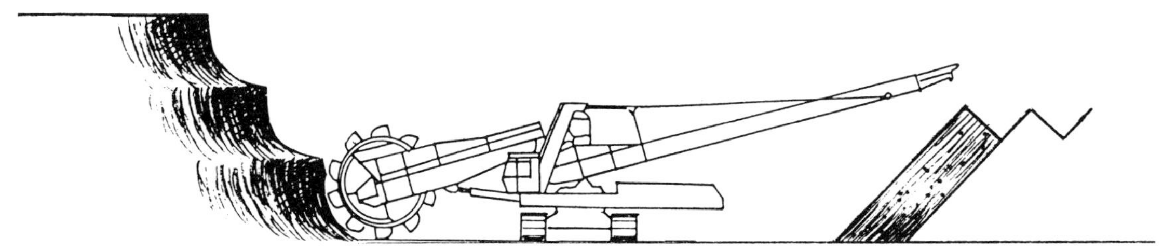

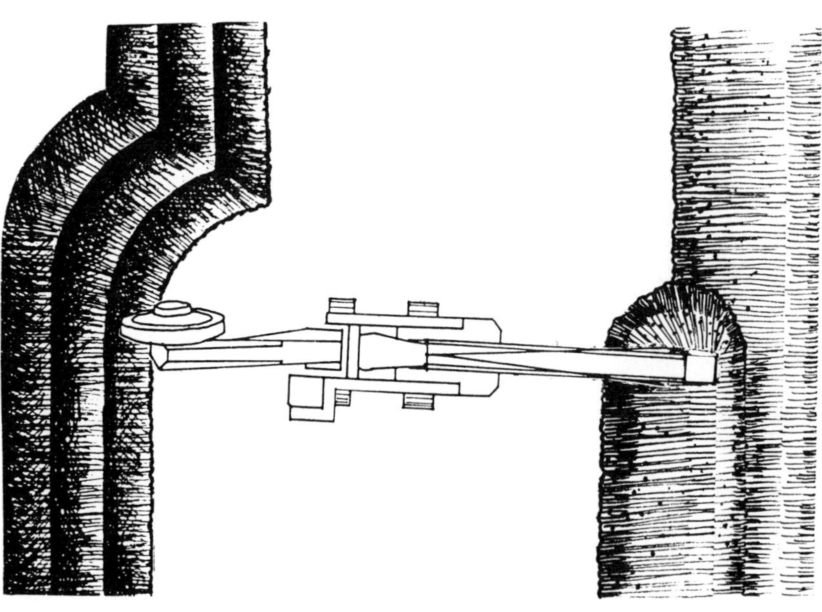

Figure 9.8 Wheel Direct Spoiling, Utilizing Long Discharge Conveyor

troduces some unique problems. Special high mobility transfer conveyor units may be required for this type of haulage system. To-date, such wheels have been applied only to truck loading.

The Barber-Greene continuous excavator has a side mounted wheel. (See Photograph 9.8) With this configuration, cut depth is established by the transverse face overlap selected and maintained by steering the unit as it propels forward. It excavates in only one direction and, after making a cutting pass down the length of the face, loops back for the start of the next cycle. Different from the drum units, however, it can dig to depths of approximately 12 feet (two-thirds of the wheel diameter). The block removed is dependent on the shape of the bank face. Propel speed is adjusted to match digging conditions, with the operator keeping the conveyor full. The buckets are shallow, restricting the size of material which can be handled. Since the machine is operating close to the excavated face, the bank must be relatively stable. No floor clean-up is usually required.

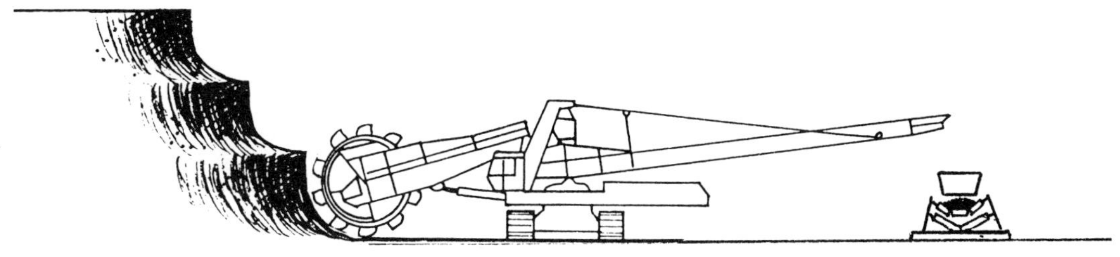

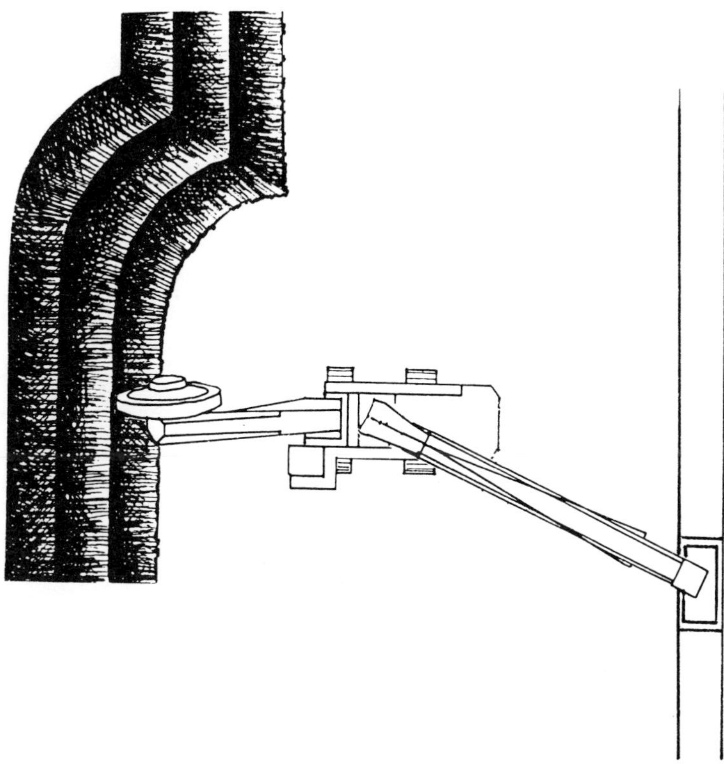

Figure 9.9 Wheel Direct Loading Shiftable Conveyor

Small drum type wheels take parallel, shallow cuts progressively back and forth along the bench or pit. Long cuts minimize lost time related to machine turn-around. Long straight cuts facilitate drive-by type truck load-out, with the trucks operating on the floor established by the prior cut. While being loaded, the truck propel speed may have to be coordinated with the wheel advance.

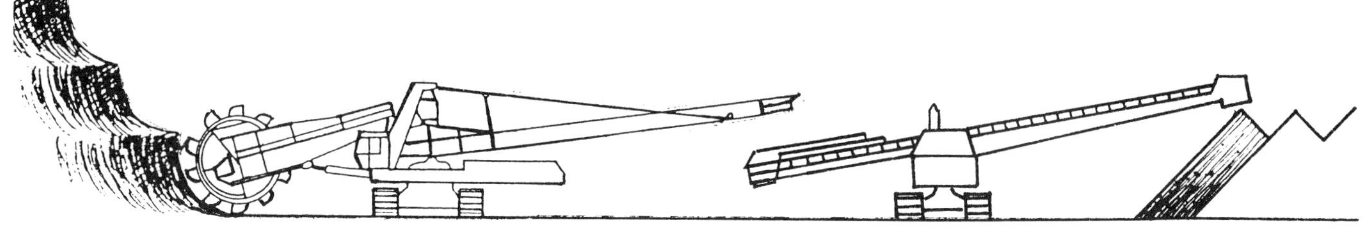

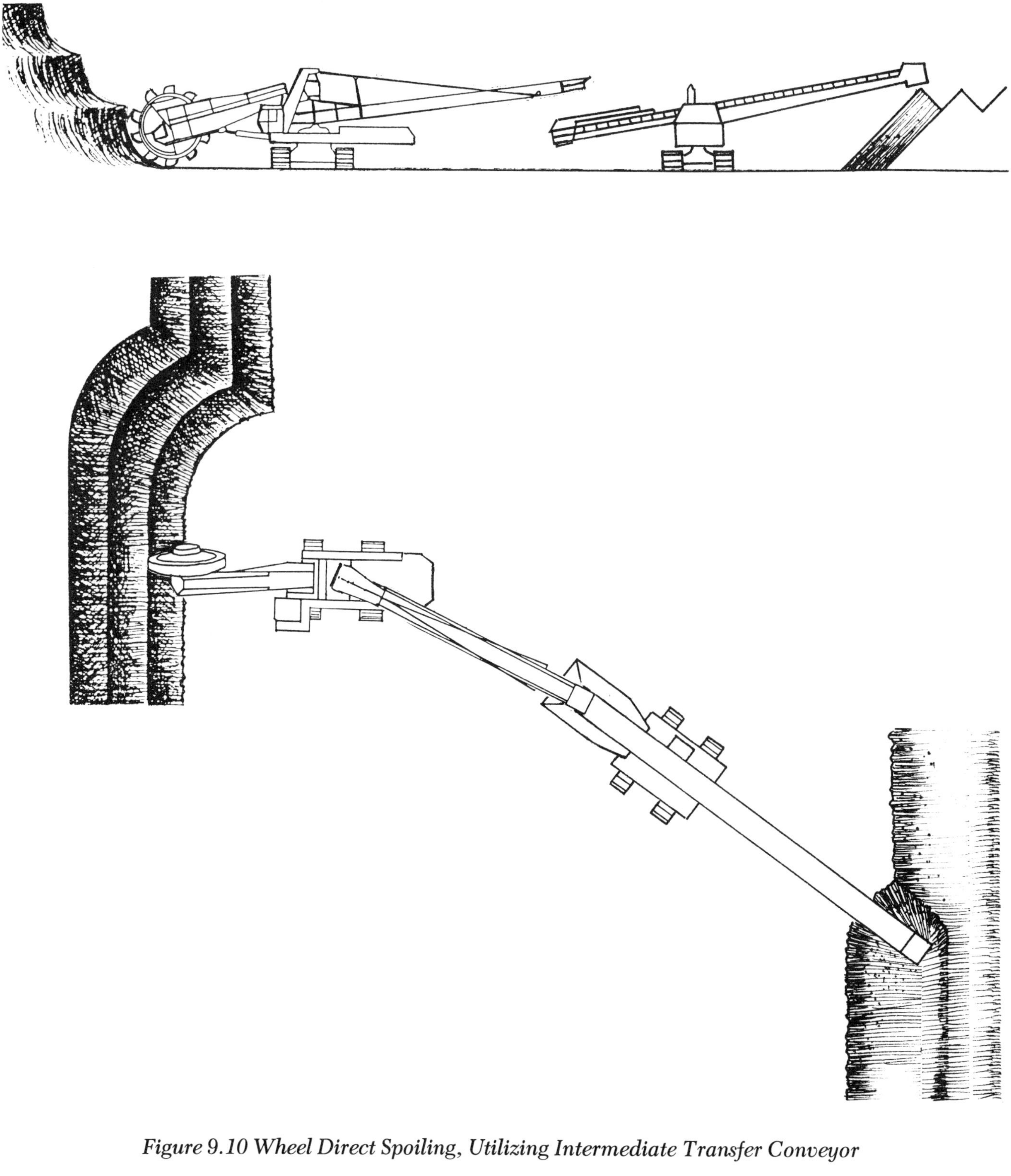

Figure 9.10 Wheel Direct Spoiling, Utilizing Intermediate Transfer Conveyor

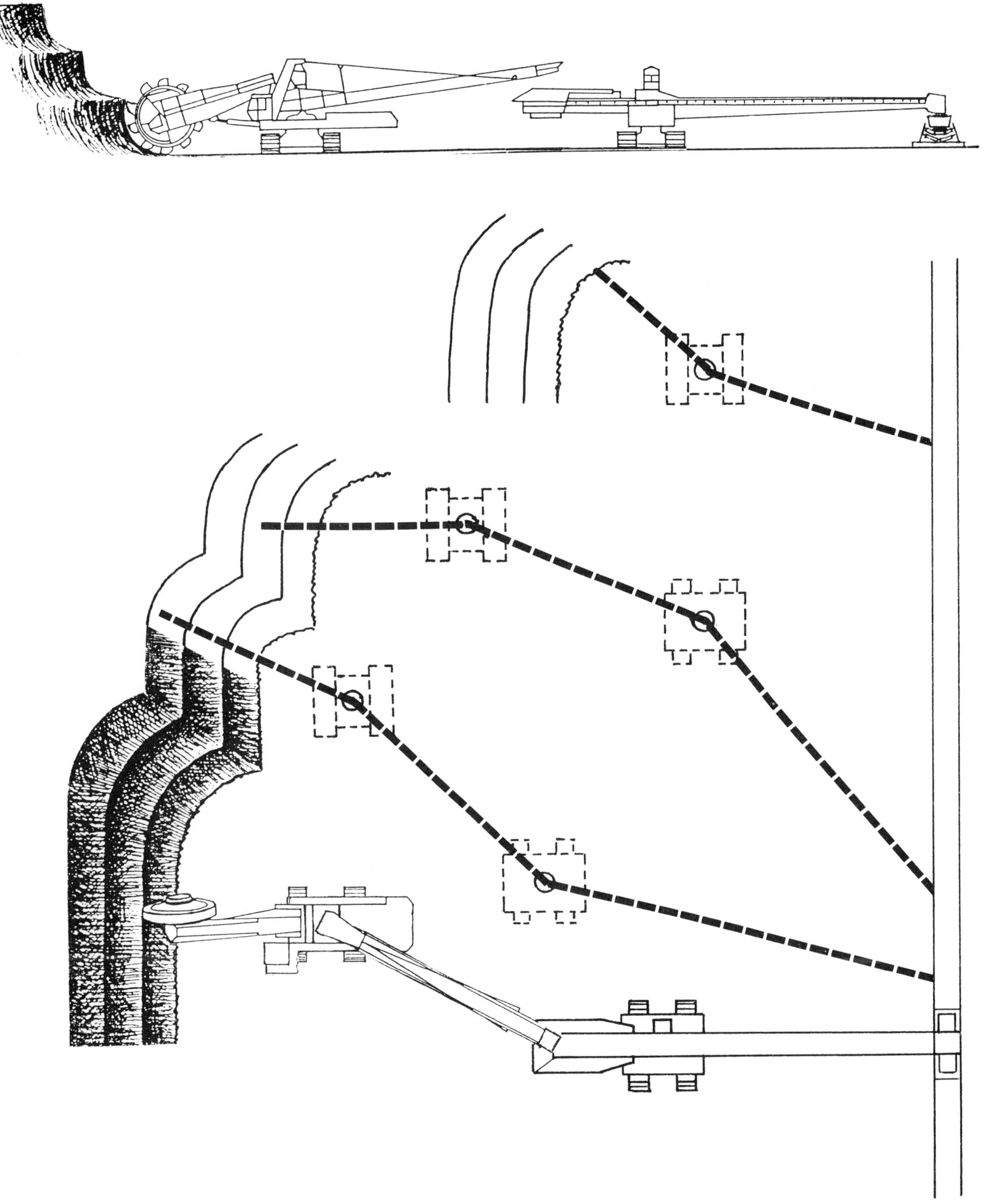

Figure 9.11 Wheel Loading Conveyor, Utilizing Intermediate Transfer Conveyor

Photograph 9.8 Barber-Greene excavator illustrating side-mounted wheel configuration

DESIGN FEATURES

See Figure 9.12 for the general wheel nomenclature.

The structural design and the hydraulic or electric drive systems are similar to those discussed earlier for hydraulic machines, electric shovels and walking draglines. Only those features unique to the wheels will be considered, with the emphasis placed on the medium size wheels.

The popular design for the cutting wheel is a cell-less configuration, meaning that the wheel is not compartmented. The buckets have no backs (bottoms). The dirt is retained by a backing ring in the section of the wheel where the digging occurs. At the top of the wheel, the material is free to discharge down a slope sheet (chute) to a conveyor belt. This design is the most simple and permits a maximum number of buckets to be mounted with optimum material discharge characteristics.

Machine output is a function of bucket size and number. Increasing the number of buckets on the wheel reduces the fluctuations in the drive torque, but the number is limited by their ability to fill and discharge properly. Maximum lump size can be controlled and digging characteristics enhanced by the addition of precutters between buckets. The precutter is, essentially, a narrow blade, shaped similar to the front of a bucket, which prepares the material for the succeeding bucket. The shape of the bucket is dependent on the material to be handled. Flat and wide designs are used for sticky materials; materials with a high swell require larger buckets.

Buckets are pinned to the circumference of the rotating wheel structure to permit interchange for maintenance and to allow for bucket changes dictated by different digging conditions. Some buckets have been lined with slippery, abrasive-resistant synthetics to assist in the discharge of sticky materials. Alternatively, buckets can be equipped

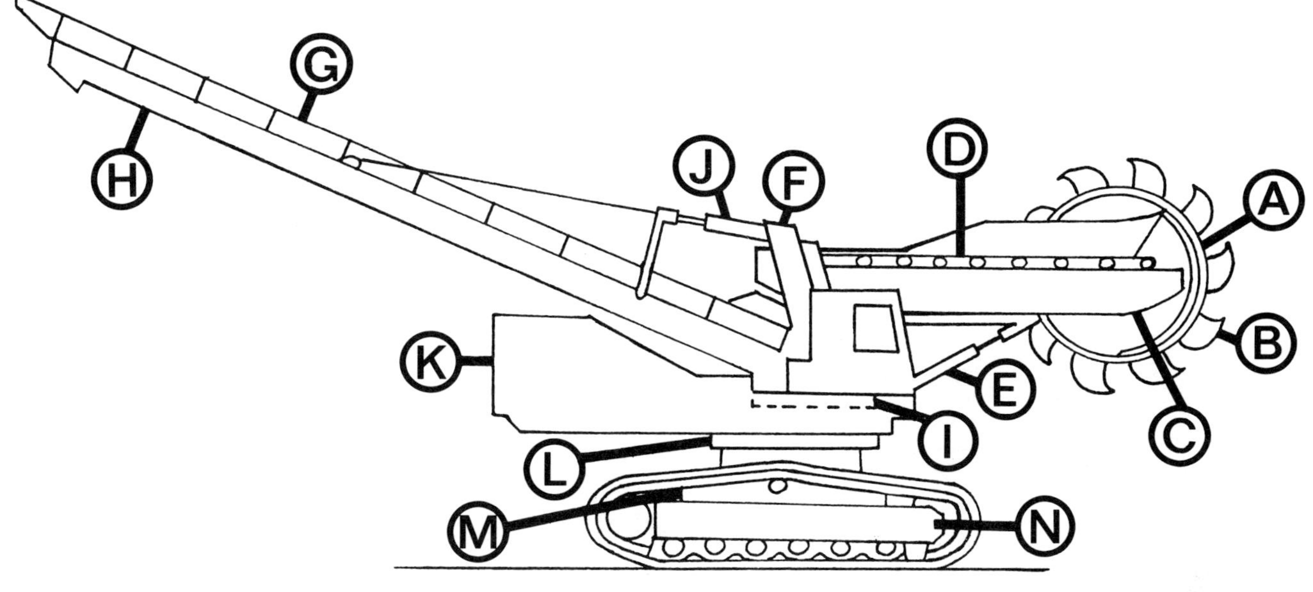

A Wheel
B Bucket
C Wheel Boom
D Wheel Conveyor
E Wheel Elevating Cylinder
F Gantry
G Discharge Conveyor
H Discharge Boom
I Discharge Swing Bearing
J Discharge Elevating Cylinder
K Revolving Frame
L Swing Circle
M Undercarriage
N Crawler Side Frames

Figure 9.12 Wheel Nomenclature

with chain backs (multiple strands of chain) which flex, helping to dislodge sticky materials during the dump phase.

Buckets are generally equipped with teeth to concentrate the cutting forces. Since the wheel cuts primarily on the sides of the buckets, corner teeth and side cutter designs are important.

Overall digging wheel weight is substantial and is a critical design factor in determining the amount of counterweight required to balance the machine. In large machines, this must be given careful consideration because of the large shift in the machine's center of gravity when the wheel is allowed to rest on the ground. The digging wheel drive, which is located in the hub or immediately adjacent to the wheel, adds to the total weight. In the large models, D.C. motors provide the power, driving through conventional gear reductions (see Photograph 9.9); and, in the medium size units, hydraulic motors drive through compact planetary gear reductions. Wheel speeds commonly range from 4 to 11 rpm, depending on diameter and face conditions. Typical drive ratings are: 0.2 to 0.3 KW/cubic meter in light soil; 0.3 to 0.4 KW/cubic meter in average soil; and, 0.5 to 0.7 KW/cubic meter in soft stone. The drive system requires overload or slipping devices to reduce the severe shock loads introduced when obstructions are encountered in the face.

The transfer of material from the buckets to the wheel boom conveyor is a troublesome design area. Sticky materials require steep slope sheets (chutes), but these permit any rocks or boulders to virtually free-fall, impacting and bouncing on the conveyor. This shortens belt and conveyor idler life and increases the dynamic loads on the wheel boom. Ideally, the discharging material would be

Photograph 9.9 General front-end configuration of medium size wheel excavator

deflected by the slope sheet and slide down onto the conveyor. Obviously, to meet varying face conditions, slope sheet design must be a compromise. On the large machines, a number of different approaches have been used, including powered rollers, short belts, rotating conical disc, etc. All of these attempts increase digging wheel assembly weight and complexity. The medium size wheels produced by MX (Mechanical Excavator) employ an angled wheel to improve the discharge flow pattern from the wheel onto the conveyor. (See Photograph 9.10)

The wheel boom, together with its cable suspension system (or elevating cylinder) must support the weight of the digging wheel assembly, plus the digging forces, the weight of the material in the buckets and the loaded belt. The design is complicated by the potential for induced vibrations as individual buckets repeatedly strike a hard fault or obstruction in the face.

Early boom designs included a crowd mechanism which could move the wheel boom in and out to change the cutting depth. This reduced the number of machine moves required during operation and permitted adjusting the cut during swing to produce a constant cut depth regardless of the swing angle. Movement of the cutting wheel and boom assembly, however, required a costly, heavy design configuration, increased overall machine weight 10 to 35%, and complicated the transfer from the wheel belt to the discharge belts. In most of the newer designs, this feature has been eliminated.

In contrast to the cyclic excavators, the bucket wheels require several large roller circles or bearings to permit rotation of major structures. The digging wheel itself, depending on its size, requires a large bearing which is subjected to all of the digging loads. The cantilevered discharge boom, with its conveyor, rotates independently on another large bearing. And the main revolving frame rotates on the lower crawler support structure.

On the medium size machines, the crawler design is very similar to that on hydraulic excavators. The crawler systems employed for the large units vary significantly, in some cases pro-

Photograph 9.10 Mechanical Excavator's angled wheel mount on boom

viding hydraulic leveling of the main structure. They commonly have a three point support configuration with anywhere from 6 to 12 sets of tracks. The steering arrangements are relatively complex.

From the point of bucket discharge, the material is transported within the machine on a series of belt conveyors. There are two separate conveyors on a typical medium or small wheel; more on the large machines, depending on basic design and size. The problems associated with belt abuse in the digging wheel loading area were noted earlier. Each transfer point from one belt to another poses a similar problem. The transfer from the wheel belt to the discharge belt can be particularly severe because of the height of the material drop from one belt to another, necessitated by machine geometry and construction. The angle (horizontal) between transferring belts can vary substantially, making it difficult to optimize the material flow. Deflection plates, special idlers, etc., are required to minimize impact loads on the belts and supporting structures.

Maximum belt angles (vertical) must be limited to prevent damage and jamming from rocks rolling back down. In some cases, dirt belts underneath the main belts are required on the large machines to collect spillage and protect other machinery and/or personnel. Belt drives must permit start-up and shut-down with a loaded belt. On large belt installations, acceleration belts are required, together with sequential starting and shut-down capabilities. Belt tensions must be maintained to minimize sag which would increase the impact loading on the idlers.

Cutter drums on the small fixed wheels are presently of two types. The earlier machines employed pickpoint cutters generally mounted on a scroll which directed the flow of material to the center of the machine. This approach parallels the practices on underground miners and highway concrete trimmers. (See Photograph 9.11) Subsequent experiences suggest that heavy arms capped with shovel teeth are effective in softer formations. (See Photograph 9.12)

Photograph 9.11 Drum cutter head with pick-point type teeth

Photograph 9.12 Drum cutter head with shovel type teeth

SELECTION CONSIDERATIONS

Bucket wheel excavators are capable of high production under suitable formation and operating conditions. Wheels are generally considered for large flat deposits of uniform unconsolidated materials, such as unconsolidated alluvium, soil, or glacial till.

Wheel applications should recognize the following constraints:

- Large boulders or blocky material cannot be handled.
- Hard consolidated materials require excessive power and abuse the machine (should have a low cutting resistance).
- Sticky materials introduce problems with build-up in buckets, on belts, and at belt transfer points.
- Abrasive materials can produce excess wear on teeth and buckets and are hard on the conveyor belts.
- The digging face should be relatively stable.

Wheel type, size, and range are dictated by production requirements and mine plan. It must be remembered that actual output is commonly 45 to 50% of the theoretical capacity. Power requirements, electric vs. diesel, are related directly to machine size with all units larger than the medium sizes commonly being fully electric and requiring a pit power distribution system. Optional single electric motor main drives are available for the medium size wheels.

The anticipated handling of the wheel excavator discharge, whether direct spoiling, truck loading or belt conveyors, has a significant impact on the basic machine configuration. Direct spoiling usually requires a machine with an extended discharge radius and dumping height. Truck loading will not take advantage of the continuous output characteristics of the wheel unless special discharge techniques are applied and a well-matched and scheduled truck fleet is utilized.

The wheels can be integrated into a continuous haulage system using a belt conveyor either through, around, or across the pit. A typical around the pit system includes a considerable number of major elements:

- Wheel excavator(s)
- Transfer conveyor(s)
- Face conveyor
- Side conveyor
- Spoil conveyor
- Tripper car
- Spreader
- Additional special auxiliary equipment may be required for belt handling and for shifting of large belt drive stations.

Each of these additional units in a total bucket wheel system can be considered a major piece of equipment, with an associated percent availability (reliability) which varies with its complexity. Total system availability is the product of all of these multiplied together. Care should be exercised, therefore, to provide excess capacity in the system so as to allow for down periods and/or to provide alternative back-up units. These systems require a very high initial investment with limited flexibility for change after installation.

With the increased emphasis placed on integrating land reclamation with the basic mining operation, the wheel-conveyor systems offer some advantages. In the final stage of material discharge, the system employs a spreader which commonly has the capability to deposit the material relatively uniformly, minimizing subsequent rough grading requirements.

While it is hard to assess overall system manpower requirements, there is evidence that reductions are possible with a full continuous haulage system. A fairly high degree of automation is possible. Personnel are shifted, significantly, from operating to servicing and maintenance activities, making a comparative analysis difficult. There is very limited mining experience for such systems designed and/or installed within the last ten years in the United States.

Production and ownership and operating estimates for the bucket wheel excavator are considered beyond the scope of this book. The variety of wheel configurations and haulage systems, together with the limited comparative data available on U.S. operations, make these analyses very difficult. Particular attention should be given to the unique delays of these systems, which are associated with the periodic relocation/extension of any belt system as mining advances.

As noted earlier, when the machine size increases, the degree of machine customizing to match specific mining conditions also expands. This, in turn, means engineering studies should be made to match the machine to current and future operational needs over the life of the mine property. Machine specification data, similar to that available for the other types of mining equipment, are not available for any but the medium and small sized wheels. The manufacturers will prepare such information on request after consultation on site requirements.

Optional equipment lists for the wheels are generally not available. Any special requirements and/or features are included in the discussion of the basic machine specifications with the manufacturer.

NEW DEVELOPMENTS & TRENDS

Increased use and future development of these units would appear to be primarily application oriented; that is, increased emphasis on recognition of formation or deposit restraints, and/or increased flexibility and reliability in the total haulage system. The latter refers to those cases where the wheel is discharging to a conveyor system. The special capability of the wheel's continuous output can only be effectively utilized if the overall transport system, which starts at the wheel, has a high reliability. Unfortunately, such total systems are costly and include a number of fairly complex components, following in series, making it difficult to achieve the necessary performance levels. Recent U.S. Bureau of Mines studies of the direct spoiling stripping wheels in the United States showed an overall average machine availability of about 65%.

The large wheels have been refined through the years as belt conveyor technology advanced and better power systems became available. Cutting forces have increased significantly in recent years as more power has been added to the wheel drive. This has been accompanied with only minor increases in overall machine weight.

There would seem to be little need for wheels larger than those currently in existence. Further development of large units is dependent on the future opening of new, very large mines with flat deposits in materials suited to wheel excavating.

Undoubtedly, efforts will continue to increase the wheel's digging versatility by increasing its ruggedness and improving frame and drive shock isolation. But the basic configuration, with its closely spaced small buckets, traveling in a fixed path, in a high inertia mounting, will always restrict its use in severe or varying digging conditions.

The medium size wheels are a relatively new option in mine planning. Applications are, again, limited by formation restraints and the overall haulage system commitment. Current machine sizes meet mine requirements. The basic drive technology, and much of the componentry, is the same as that for the hydraulic excavators and is being continually refined by that product.

The future of the small fixed wheels seems dependent on two parallel requirements. The first is design refinement to provide machine reliability equal to other available equipment. In this process, the optimum configuration — wheel or drum — should be resolved. The second is the development of mining operational practices which utilize the unique characteristics and capabilities of these units. They can load trucks or conveyor belts; but because of the distances they travel while excavating, any mating belt system will require some new innovative approach. In many ways, the final de-bugging and/or acceptance of this new tool is being delayed by the lack of a clear picture of how it should be used. With a better understanding of the use, the machine could be tailored to maximize production.

There have been a number of government sponsored studies aimed at evaluating the feasibility of small wheels. An experimental forward rotating wheel mounted across the front (full width) of a crawler driven unit was tested briefly in coal. Preliminary designs have been proposed for vertical side mounted cutting drums to cut along a face similar to the procedures followed by the Barber-Greene wheel. Another variation of this approach envisioned a vertical blade with teeth mounted on pneumatic (or hydraulic) impact hammers to improve penetration characteristics. All assumed a wheel or drum with limited elevation capabilities, track mounting, diesel drive, with cutting advance provided by machine propel.

The Unit Rig and Equipment Company did considerable work and, for a while, offered a full width front mounted small wheel. This machine was unique in that it was rubber tire mounted. It also had a positive discharge arrangement for the buckets, so that the discharge was off the back of the wheel onto a parallel conveyor running transverse to the machine.

It is of interest to note that this same concept of taking continuous thin slices of material and putting it on an internal conveyor for discharge into trucks is also being successfully done with Holland Loaders (manufactured by Holland Loader Company). These units use two coupled dozers to provide a high tractive effort to pull a scraper type cutting blade through the ground. The material passing over the blade is directed onto a belt conveyor which elevates it and then transfers it to a second fixed right angle discharge conveyor. Blades can be mounted horizontally or vertically. In uniform, unconsolidated formations, these units are capable of very high production.

MACHINE SPECIFICATIONS

To date, sales of bucket wheel excavators in the United States have been very limited. There is little readily available published information on machine specifications for current models. Information is commonly obtained through direct discussions with the manufacturer. Machine specifications are modified and adjusted to meet specific site requirements. The brief specifications which follow are restricted to a representative group of medium size wheels.

See Figure 9.13 for medium size wheel dimensions.

Machines are listed alphabetically by manufacturer, and then in ascending order of approximate effective output (BCY/hr) in medium digging conditions. Effective output is based on published manufacturer data. Bucket size (multiple buckets per wheel) is rated in cubic feet rather than cubic yards as with other excavators.

Wheel HP refers to the rated HP of the bucket wheel drive. Primary power can be either incoming A.C. or diesel. Engine make includes either engine manufacturer of the diesel power, or the drive type for the electric configurations. Engine model shows the diesel engine model or the incoming power supply for electric units. Engine HP includes both the total number of motors involved, and the aggregate HP of those motors.

Machine height refers to the highest fixed point on the machine. Dump reach is measured from the center of rotation of the revolving frame to the end of the discharge belt. Dump swing angle is the total arc, side to side. Cut height is based on manufacturer's published data. Cut reach is measured from the center of rotation of the revolving frame to the outside edge of the bucket wheel.

Specifications, capacity, capabilities and dimensions are based on published manufacturer data. Although the information is believed to be current and the interpretation to be consistent and correct, it is possible that there are some inaccuracies or out-of-date information in the specification charts. In addition, these specification charts are not meant to be used either as an in-depth analysis, or as a head-to-head comparison of the avilable equipment, but rather as a general overview of the equipment, available sizes, and approximate operating data. Specific questions relating to performance or purchase should be directed to the manufacturer or authorized distributor/dealer in the user's area.

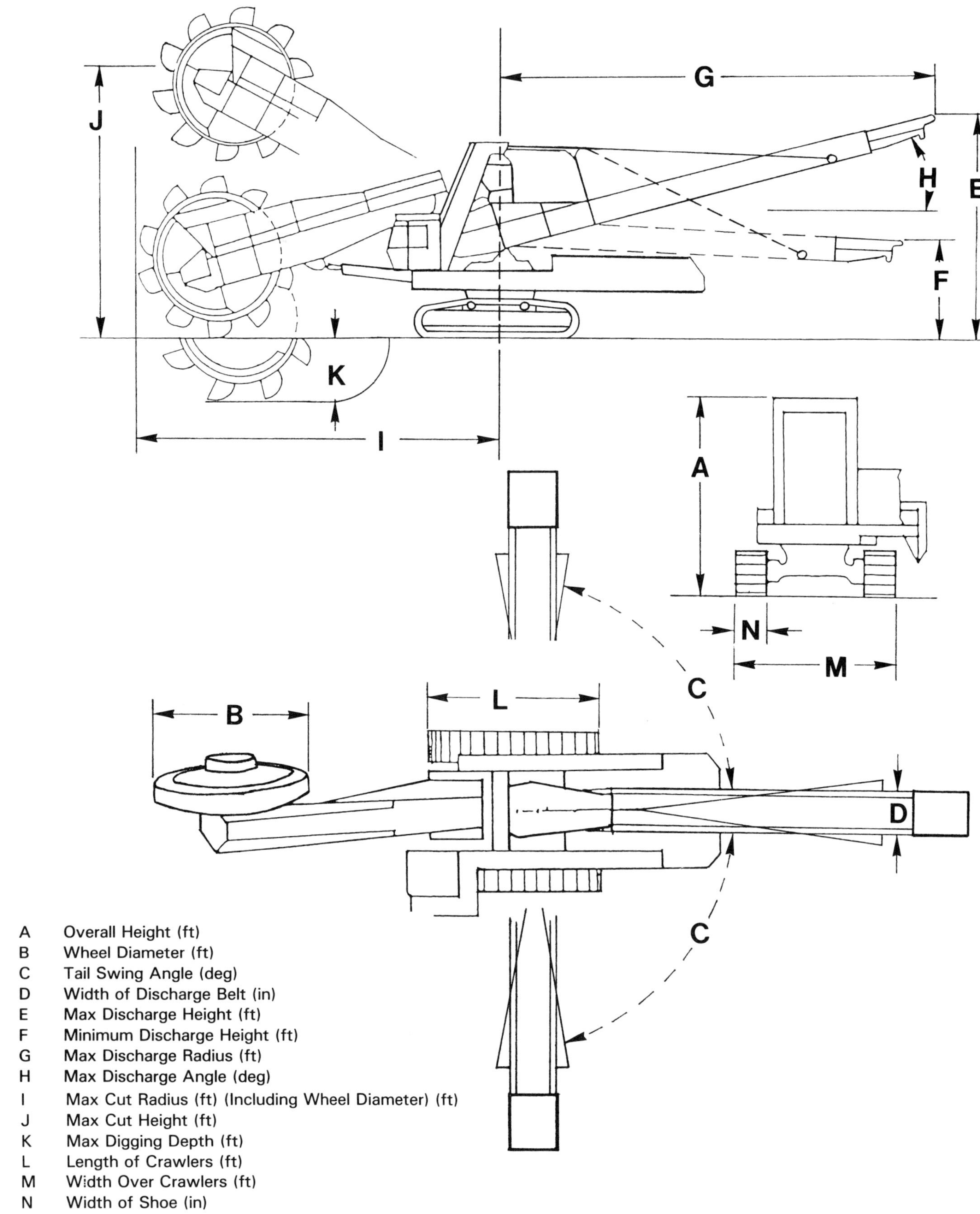

Figure 9.13 Wheel Dimensions

Table 9.2
BUCKET WHEEL EXCAVATOR SPECIFICATIONS

MAKE	MODEL	APPROXIMATE AVERAGE RATED OUTPUT (BCY/hr)	NOMINAL BUCKET CAPACITY (cu ft)	NO. OF BUCKETS ON WHEEL	WHEEL DIAMETER (ft)	RATED HP BUCKET WHEEL DRIVE	WEIGHT (lb)
Krupp	100	390	6.0	6	13.8	100	200,000
Krupp	C-300	780	11.4	8	17.7	215	350,000
Krupp	C-500	1570	17.5	10	25.6	485	725,000
Mechanical Excavator	1000	1500	27.0	8	18.0	—	—
Mechanical Excavator	2000	2500	40.5	8	22.0	—	—
Mechanical Excavator	3000	3500	54.0	8	25.0	—	—
O & K	SH-250	589	8.8	10	16.4	134	205,000
O & K	SH-400	981	14.1	10	20.7	268	440,000
O & K	SH-630	1504	22.3	12	26.3	536	760,000

MAKE	MODEL	ENGINE DATA: TYPE: DIESEL OR ELECTRIC	ENGINE DATA: DIESEL MAKE OR ELECTRIC DRIVE TYPE	ENGINE DATA: DIESEL MODEL OR ELECTRIC POWER (volts)	ENGINE DATA: NO. OF ENGINES / TOTAL HP	MAX SPEED (mph)	TRAVEL GRADEABILITY (%)	EXCAVATING GRADEABILITY (%)	FUEL TANK (gal)	COOLANT (gal)
Krupp	100	electric	—	380/500	10/260	0.2	10	5	n.a.	n.a.
		diesel	—	—	1/250				—	—
Krupp	C-300	electric	—	3000/6000	14/500	0.2	10	10	n.a.	n.a.
		diesel	—	—	1/440				—	—
Krupp	C-500	electric	—	3000/6000	16/1130	0.2	10	5	n.a.	n.a.
		diesel	—	—	1/1000				—	—
Mechanical Excavator	1000	diesel	G.M.	8V-71N	2/-	—	—	—	—	—
		electric	—	—	-/-				n.a.	n.a.
Mechanical Excavator	2000	diesel	G.M.	12V-71N	2/-	—	—	—	—	—
		electric	—	—	-/-				n.a.	n.a.
Mechanical Excavator	3000	diesel	G.M.	16V-71N	2/-	—	—	—	—	—
		electric	—	—	-/-				n.a.	n.a.
O & K	SH-250	diesel	Deutz	—	2/354	0.7	10	5	—	air
		electric	hydraulic	380/500	-/335				n.a.	n.a.
O & K	SH-400	diesel	Deutz	—	2/680	0.8	10	5	—	air
		electric	hydraulic	380/500	2/670				n.a.	n.a.
O & K	SH-630	diesel	—	—	4/1360	0.8	10	5	—	air
		electric	hydraulic	380/500	7/1275				n.a.	n.a.

Table 9.2 (Continued)

BUCKET WHEEL EXCAVATOR SPECIFICATIONS

MAKE	MODEL	LENGTH WITHOUT DISCHARGE CONVEYOR (ft)	LENGTH WITH DISCHARGE CONVEYOR (ft)	MAX FIXED HEIGHT (ft)	OVERALL WIDTH REVOLVING FRAME (ft)	GROUND CLEARANCE (in)	CRAWLER DATA: NO. OF UNITS	CRAWLER DATA: OVERALL LENGTH (ft)	CRAWLER DATA: OVERALL WIDTH (ft)	CRAWLER DATA: STANDARD SHOE WIDTH (in)	GROUND PRESSURE (psi)
Krupp	100	59.4	103.0	32.0	20.8	22	2	17.5	17.4	51	13.0
Krupp	C-300	67.2	105.5	26.9	23.5	28	2	21.8	21.3	51	16.3
Krupp	C-500	90.1	114.4	39.0	30.0	30	2	30.4	29.7	89	13.2
Mechanical Excavator	1000	—	90.0	27.0	—	—	2	23.5	23.5	56	—
Mechanical Excavator	2000	—	105.0	38.0	—	—	2	26.4	24.3	56	—
Mechanical Excavator	3000	—	105.0	31.0	—	—	2	26.3	26.8	56	—
O & K	SH-250	53.2	82.0	21.3	21.2	20	2	16.7	16.1	35	18.5
O & K	SH-400	73.3	112.0	31.8	25.1	30	2	23.4	21.0	55	18.6
O & K	SH-630	95.1	144.4	42.0	35.8	28	2	29.9	32.3	89	14.2

MAKE	MODEL	MAX WHEEL SLEW SPEED (fpm)	BELT WIDTH (in)	BELT SPEED (fpm)	MAX REACH DISCHARGE BELT (ft)	SWING RANGE DISCHARGE BELT (degrees)	MAX DISCHARGE HEIGHT (ft)	MAX DIGGING DEPTH (ft)	MAX CUT HEIGHT (ft)	MAX CUT REACH (ft)
Krupp	100	14-72	36	650	65.6	180	31.2	1.3	31.2	37.4
Krupp	C-300	16-82	48	650	65.6	200	28.1	1.2	31.2	40.0
Krupp	C-500	20-98	54	650	65.6	180	32.0	2.0	32.8	48.9
Mechanical Excavator	1000	—	54	—	—	—	—	—	—	40.0
Mechanical Excavator	2000	—	72	—	—	—	—	—	—	45.0
Mechanical Excavator	3000	—	84	—	—	—	—	—	—	45.0
O & K	SH-250	20-77	39	709	49.2	180	24.6	1.6	25.6	32.8
O & K	SH-400	23-91	47	787	65.6	180	29.5	2.0	36.1	46.4
O & K	SH-630	23-82	55	866	82.0	210	42.0	3.9	49.2	62.3

Chapter 10

ROTARY BLAST HOLE DRILLS

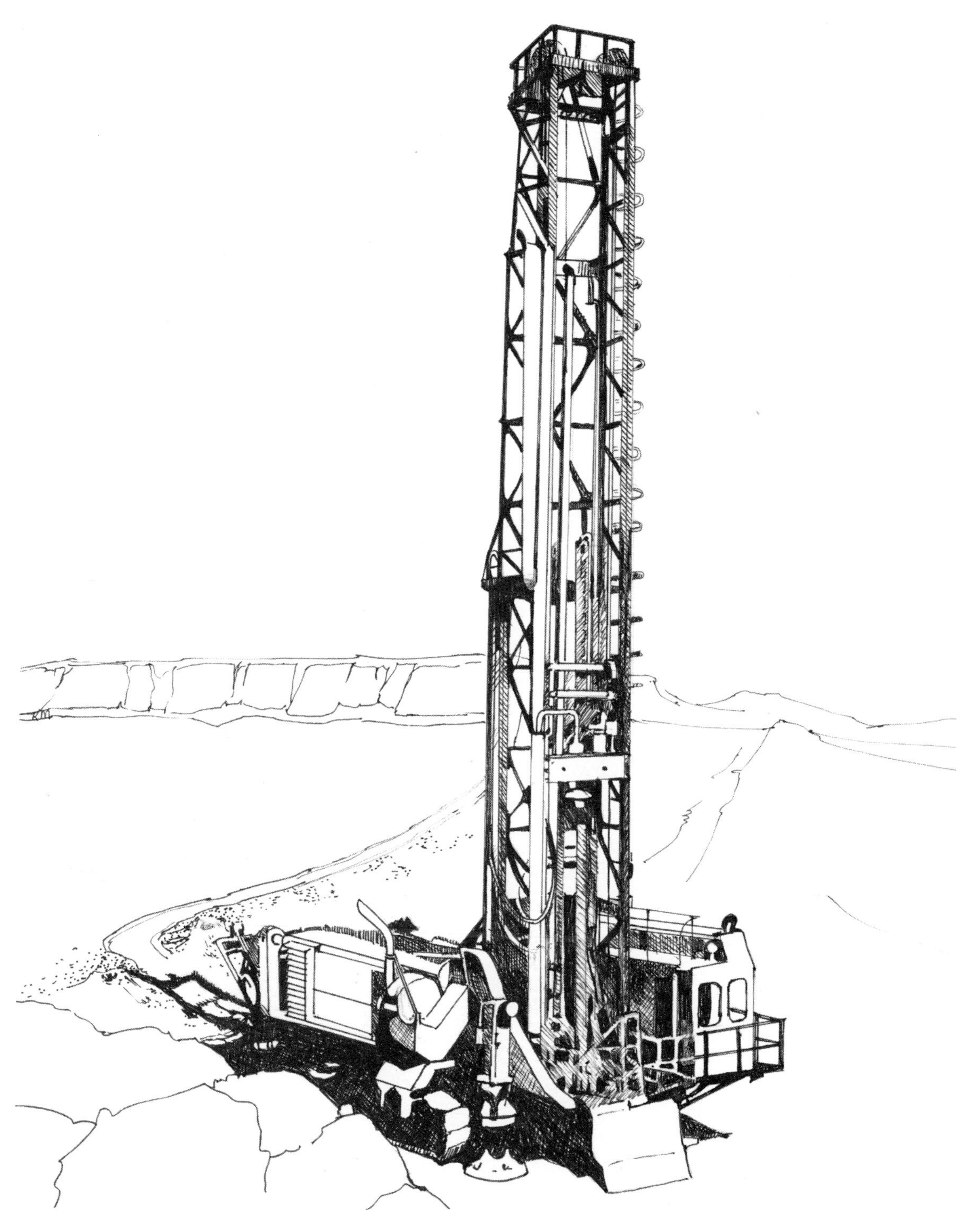

Figure 10.1 Drilling Overburden

Most of the excavating machines discussed in other chapters of this manual incur excessive maintenance costs and/or loss of productivity when operating in hard, consolidated geological formations. To overcome this problem, explosives are utilized to break up the material to the extent that efficient and economic operation of the digging equipment is possible. The explosive is placed in holes drilled into the formation to depths roughly equal to the depth of the planned excavation. Holes are drilled in a specific and predetermined pattern, such as 30 foot centers. They have a diameter that will accomodate the characteristics of the explosive being used, which, in turn, is matched to the power required to produce the desired fragmentation in the specific formation.

Blast hole drills are used in surface mining operations to produce the hole for explosive placement. While small percussion (pneumatic and hydraulic) and jet piercing machines are employed on a limited basis, this discussion will be restricted to the more popular rotary units which have replaced the churn drills of the past. Both crawler and wheel mounted units for vertical and angle drilling will be considered.

Early designs were termed churn drills and functioned in a manner similar to a pile driver; where a heavy wedge type tool was repeatedly raised and dropped to produce penetration. Water was added to flush out the generated chips. The introduction of tricone roller bits, which could produce an effective, continuous cutting action under heavy down pressures and high torques, resulted in the development of the present rotary drills. Drill size and production capability have increased, paced by the continued design improvements of these bits and their availability in larger sizes.

The equipment employed to transport explosives and to load the holes is not discussed in this book. It should also be noted that drills of the type used for mineral exploration, shallow oil/gas wells and water wells drills are beyond the scope of this book.

Photograph 10.1 Gardner Denver GD-25 crawler mounted unit operating in a diamond mine

Photograph 10.2 Marion M-4 crawler blast hole drill setting up to drill overburden

TYPICAL UNITS

Blast hole drills can be conveniently separated into two categories, crawler mounted and wheel mounted. Typical units with their masts lowered in the carry position are shown in Figure 10.2. Power is normally supplied by a diesel engine, but on the large crawler mounted units power is electric with a trailing cable. The choice between the two units is dictated by mine application requirements.

Crawler Mounted

- Hole sizes to 22 inches in diameter
- Machine weight from 33,000 to 300,000 lb.
- Diesel or electric primary drives
- Electric hydraulic, and/or mechanical auxiliary drive units

Wheel Mounted

- Hole sizes to 12 1/2 inches in diameter
- Machine weight from 30,000 to 120,000 lb.
- Commercial truck chassis: 2, 3, or 4 axle
- Gas or diesel drives: 1 or 2 engines
- Electric, hydraulic, and/or mechanical auxiliary drives

Manufacturers

While one foreign manufacturer has been included in the tabulations, the U.S. manufacturers are the primary suppliers. The number of manufacturers decreases as the machine size increases. There is an unusual spread in the size of the manufacturing companies which range from very small to

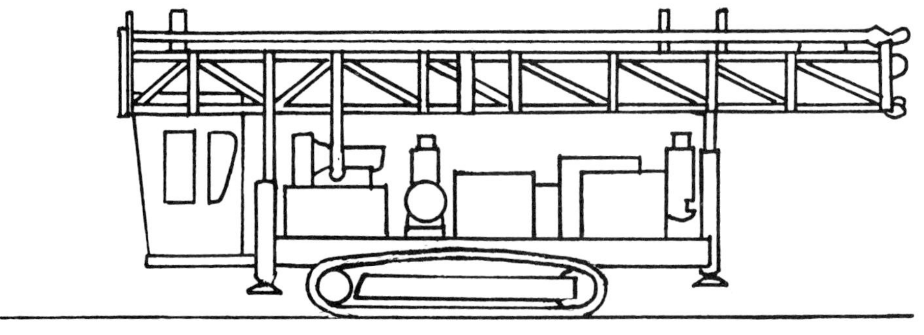

SMALL CRAWLER – DIESEL

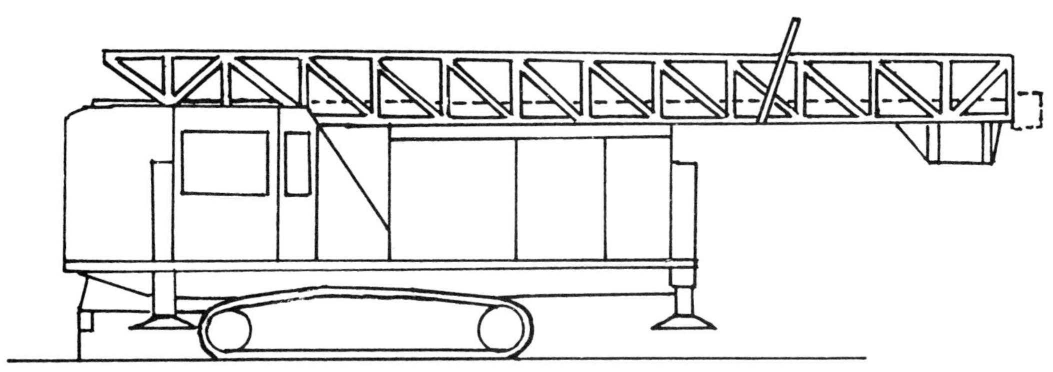

LARGE CRAWLER – ELECTRIC

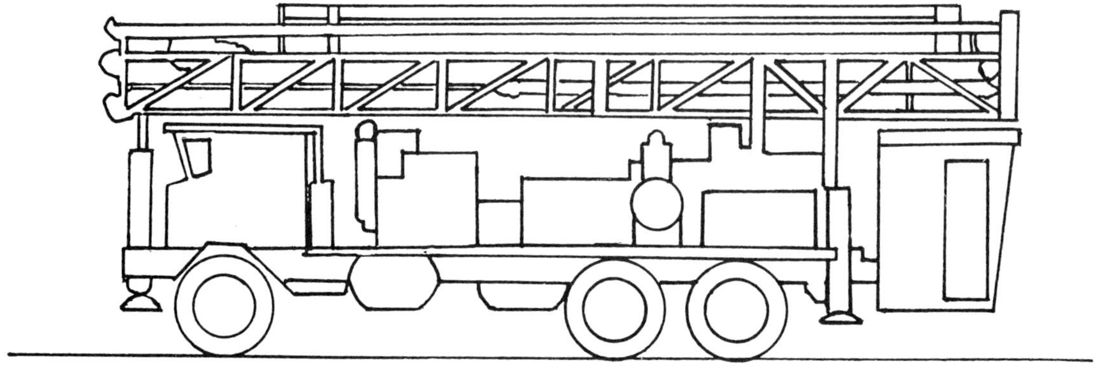

RUBBER TIRED – DIESEL

Figure 10.2 Types of Rotary Blast Hole Drills

Table 10.1

ROTARY BLAST HOLE DRILLS MANUFACTURERS

Acker Drill Co., Inc. P.O. Box 830 Scranton, Pennsylvania 18501	to 4.75 in. holes to 16,000 lbs. P.D. crawler, truck, trailer mtd.
Becker Manufacturing Div. Drill Systems Inc. 616 58th Ave. S.E. Calgary, Alberta Canada T2P 1X3	to 45,000 lbs. P.D. crawler, truck mtd. diesel
Bucyrus-Erie Company 1100 Milwaukee Avenue South Milwaukee, Wisconsin 53172	6–17.5 in. holes 35,000-110,000 lbs. P.D. crawler, truck mtd. diesel, electric
Chicago Pneumatic Drill Division P.O.Box 1225 Enid, Oklahoma 73701	6.25–9.75 in. holes 30,000–50,000 lbs. P.D. truck, crawler mtd. diesel
Davey Rousselle Drill Rig Division 2310 West 78th Street Chicago, Illinois 60620	2.375–7.375 in. holes 17,000–30,000 lbs. P.D. crawler, truck mtd. diesel
Driltech Inc. P.O. Box 338 Alachua, Florida 32615	to 50,000 lbs. P.D. crawler, truck mtd. diesel
Gardner-Denver Company P.O. Box 47114 Dallas, Texas 75247	3.5–17.5 in. holes to 130,000 lbs. P.D. truck, crawler mtd. diesel, electric, air
Haus Herr D 4322 Sprockhovel West Germany	2.5 –17.5 in. holes to, 121,000 lbs. P.D. truck, crawler mtd. diesel, electric
Ingersoll-Rand Rock Drill Division Orrville, Ohio 44667	1.75–12.25 in. holes to 100,000 lbs. P.D. crawler, truck mtd. diesel, electric, air
Joy Manufacturing Co. Robbins Drill Division 300 Fleming Road P.O. Box 6505 Birmingham, Alabama 35217	4–15 in. holes to 120,000 lbs. P.D. crawler, truck mtd. diesel, electric
Marion Power Shovel Co. 617 W. Center Street Marion, Ohio 43302	9.875–15 in. holes to 120,000 lbs. P.D. crawler mtd. diesel, electric

Table 10.1 (Continued)

ROTARY BLAST HOLE DRILLS MANUFACTURERS

Portadrill Inc. 3811 Joliet Street Denver, Colorado 80239	4.75–12 in. holes to 45,000+ lbs. P.D. truck, crawler mtd. diesel
Reed Tool Company Drilling Machinery Division P.O. Box 998 Sherman, Texas 75090	3.5–9.875 in. holes 20,000–50,000 lbs. P.D. truck, crawler mtd. diesel
Schramm Incorporated 800 E. Virginia Avenue West Chester, Pennsylvania 19380	to 9 in. holes 14,700–49,000 lbs. P.D. diesel

large. Many concentrate on one segment of the market and do not offer a complete line. The truck mounted units are, in some case, adaptations of the company's water well drill product line. Several of the companies are also major suppliers of air compressors. Two of the companies (Bucyrus-Erie and Marion) are manufacturers of excavating equipment. For a listing of manufacturers see Table 10.1.

BASIC MACHINE OPERATION

The purpose of the blast hole drill is to drill the hole into which explosive charges will be placed for subsequent blasting.

- The machine is propelled and maneuvered into position where the hole is to be drilled (previously determined by selection of a drilling pattern).
- The mast is raised into position (for short distance moves, the mast is kept in the raised position); the mast supports the drilling mechanism.
- The machine is leveled into a horizontal orientation so that holes drilled are consistently vertical (or at a specific angle); while drilling, the wheels (and/or crawlers) are raised off the ground.
- Drilling is performed by forcing the rotating drill bit into the ground with a high down pressure and rotary torque.
- With increasing depth, the drill pipe on which the cutting bit is fastened is extended by adding additional lengths of drill pipe.
- Chips formed by the cutting action of the bit are removed from the hole by high velocity air fed down through the center of the drill pipe; the material chips are carried up the circular space between the hole and the pipe.
- Blown out chips and fines are collected on the surface, around the top of the hole with the air borne particles passed through a dust collection system.
- When the hole is completed to the desired depth, the bit is retracted with the drill pipe progressively removed; the machine then starts the relocation sequence.

Photograph 10.3 Schramm three axle, wheel mounted blast hole drill working in a quarry

APPLICATIONS

Blast hole drilling and subsequent blasting is an effective means of loosening and fragment the bank prior to excavation. In some cases, the blasting can also physically displace the material, directly assisting in the removal process as well as shaping the face to match excavator characteristics. The amount of drilling required is dependent on the hole spacing and depth pattern selected. This, together with the hole size, must be matched to the capabilties of the explosives and the geology of the formation. The degree of fragmentation required varies with the excavating equipment. While excessive fragmentation is not cost effective, an insufficient amount can also significantly reduce excavating productivity and result in excavator abuse with correspondingly high maintenance costs and downtime. The optimum level is based on past experience and continued evaluation of performance.

Drills are, in most cases, special purpose machines applied only to blast hole drilling. The large crawler units are utilized in overburden only, while the smaller wheel and crawler units may be used in overburden and coal/ore. Wheel units are employed when the terrain permits and high mobility is required for moves between pit areas.

Drills are capable of dry drilling relatively large diameter holes to medium depths in soft to hard formations. Hole alignment is normally within 30 degrees of vertical. The machines used in surface mining are physically large units, too big for use in undeground mining or tunneling.

GENERAL CHARACTERISTICS

Insight into the functioning and complexity of the drill is possible by considering the functions that are performed by the machine. (See Figure 10.3)

- Propel system, motors
- Mast raising and lowering, hydraulic cylinders
- Machine leveling, hydraulic cylinders
- Bit rotation, motors
- Bit down pressure, hydraulic cylinders or motors
- Tool and pipe changing, hydraulic cylinders
- Chip removal, compressed air
- Dust control, optional systems

There are rather significant differences in the features of the blast hole drills from different manufacturers. There is some commonality which can be summarized for two basic groups of machines, the larger crawler units and the carrier (truck) mounted units.

Large crawler units:

- Hole sizes from 9 to 22 inches
- Maximum drilling depths from 150 to 200 feet
- Maximum pulldown from 40,000 to 150,000 pounds
- Maximum bit rotational speeds from 80 to 200 rpm
- Maximum torques from 3500 to 25,000 ft.-lb.
- Maximum feed speeds (down) from 3 to 65 fpm
- Maximum hoist speeds (up) from 74 to 155 fpm

Photograph 10.4 Joy-Robbins PRT-60 four axle, wheel mounted blast hole drill, angle drilling

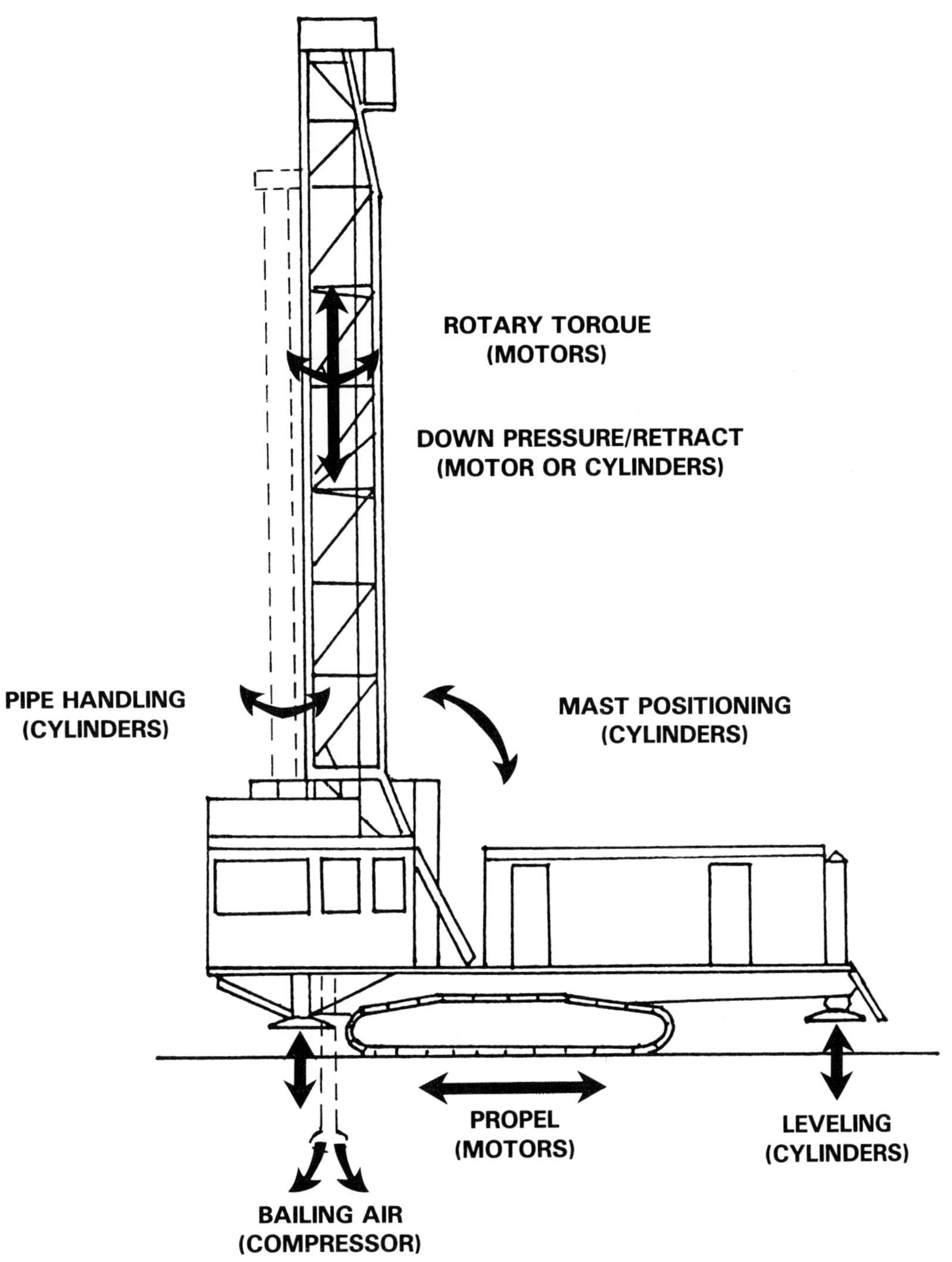

A	Mast	F	Bit
B	Pipe Rack	G	Cab
C	Dust Collection System	H	Machinery House
D	Top Drive	I	Leveling Jacks
E	Drill Pipe	J	Crawlers

Figure 10.3 Drill Powered Functions

Photograph 10.5 Bucyrus-Erie 60R drilling overburden

- Bailing air pressures from 40 to 125 psi
- Drill pipe lengths from 25 to 65 feet
- Maximum propel speeds from .7 to 1.1 mph
- Low ground pressures
- Four leveling jacks
- Optional dust control system
- Main drive, electric with trailing cable
- Auxiliary drives, electric and/or hydraulic
- Rugged construction
- Enclosed operator station
- Enclosed pressurized and filtered machinery house
- Good mobility under poor ground conditions
- Proven designs
- Maintenance performed in pit
- One man operation
- No support equipment required
- Long service life
- Short field erection cycle
- High initial investment

Truck mounted units:

- Maximum hole sizes 6 to 12 inches
- Maximum drilling depths 100 to 160 feet
- Maximum pulldowns 20,000 to 70,000 pounds

Photograph 10.6 Joy-Robbins RR-15 with mast in carry position

- Maximum bit rotational speeds from 24 to 200 rpm
- Maximum rotary torques from 3,000 to 11,500 ft.-lb.
- Maximum feed speeds (down) from 5 to 60 fpm
- Maximum hoist speeds (up) from 25 to 130 fpm
- Bailing air pressures 50 to 250 psi
- Drill pipe lengths 20 to 32 feet
- Propel speeds to 50 mph
- Three leveling jacks
- Optional dust collection systems
- Main drill drive, gas or diesel engine
- Independent truck engine (in most cases)
- Auxiliary drives, direct mechanical or hydraulic
- Optional operator cab
- Open machinery deck
- Three or four axle carriers, tandem rear axle drive
- Carrier width from 8 to 10 feet
- Roadable with special permits in some states
- Limited off-highway mobility
- Relatively large turning radius
- Proven designs
- Maintenance performed in the shop
- One man operation
- No support equipment needed
- Medium service life
- No field erection required
- Low to medium initial investment

The small diesel powered crawler mounted machines tend to fall somewhere between the above two types in terms of mobility and general performance capabilities.

Graphs 10.1, 10.2, 10.3, and 10.4 show the key machine parameters in relation to the maximum bit diameter for crawler and wheel mounted drills. The indicated horsepowers are comparable for both types (large electric units omitted). The crawler machines are heavier and this, in turn, permits higher maximum pulldown forces which are dictated by the machine weight distribution and are about 50% of operating weight. (See Graph 10.5) Maximum torques are comparable for equal sized machines of both types, although there is a substantial spread among the various models.

Too much emphasis should not be placed on maximum drilling depth capabilities. The blasting (hole) depth normally does not exceed the highest digging face, which would be roughly 200 feet for the largest draglines. The drilling depth is established by the addition or removal of lengths of drill pipe. Maximum hole depth is a function of mast height, which limits both the length of the drill pipe and the amount of drill pipe that can be efficiently stored and handled in and out of the pipe racks on the machines. Mast height and pipe storage capacity can be matched to mining requirements.

Drill pipe storage is provided, most commonly, by a vertical, rotatable carousel holding fixture mounted on the front of the mast. The carousel can be pivoted and indexed into position so that individual pipes are aligned for assembly or removal. Three to six pipes can generally be stored and up to two carousels can be installed.

Actual drilling is interrupted by drill pipe changes. It requires on the average of 3 to 10 minutes to make pipe changes. To minimize the lost time, the pipe handling is semi-automated with all functions being powered to facilitate handling of the heavy pieces. Auxiliary hydraulic cylinders are provided to make up the drill pipe and bit joints. The drill pipes are hollow to provide an air passage and are usually 1 1/2 to 2 inches smaller that the bit

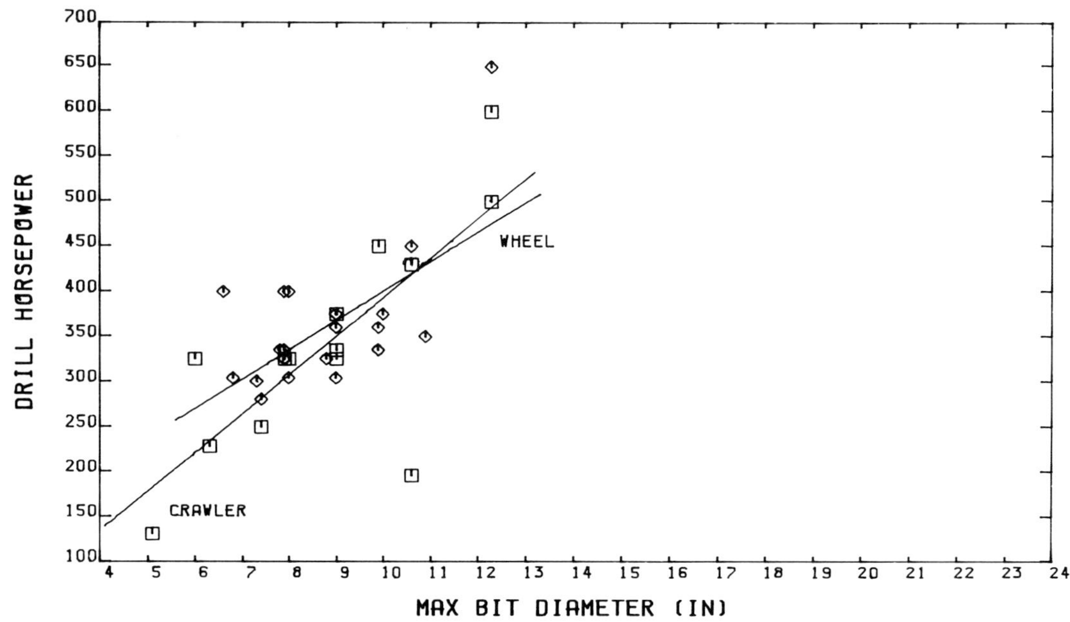

Graph 10.1 Drill rotary hp/max bit diameter

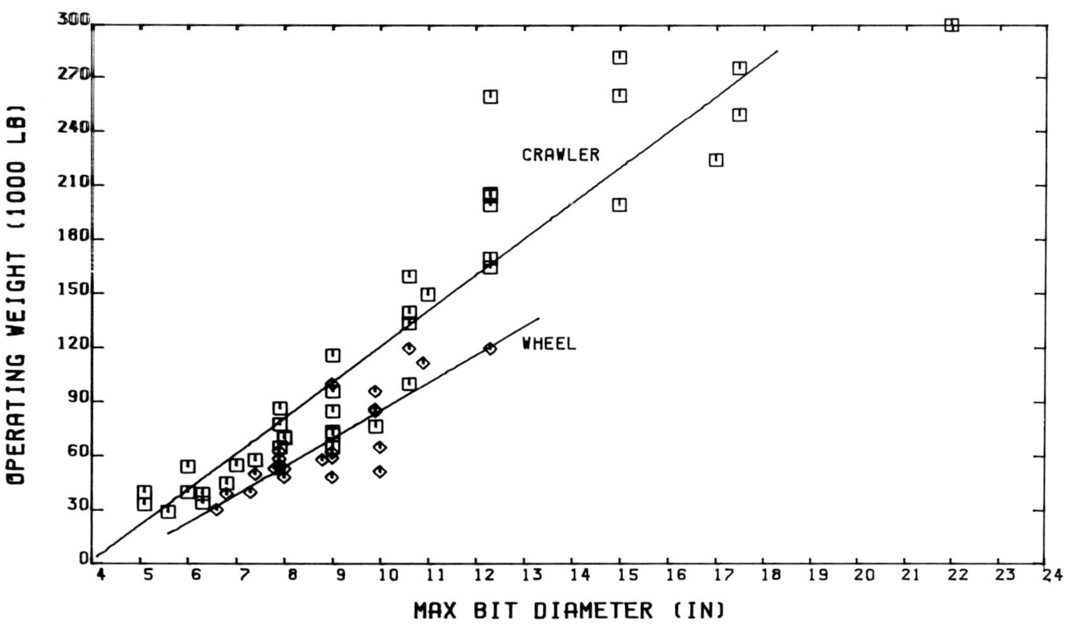

Graph 10.2 Operating weight/max bit diameter

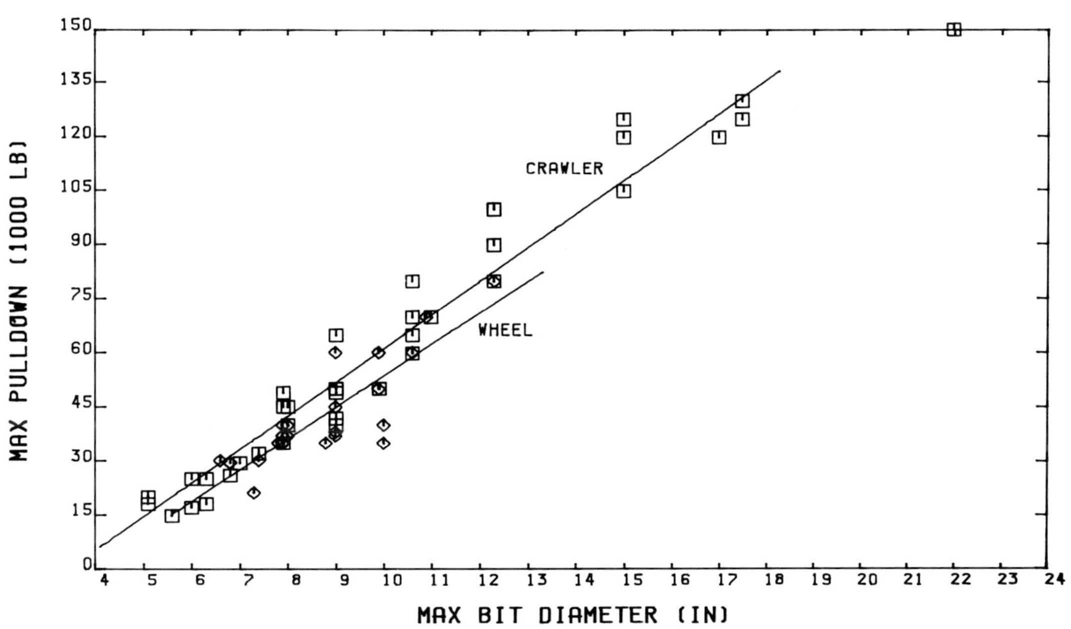

Graph 10.3 Max pulldown/max bit diameter

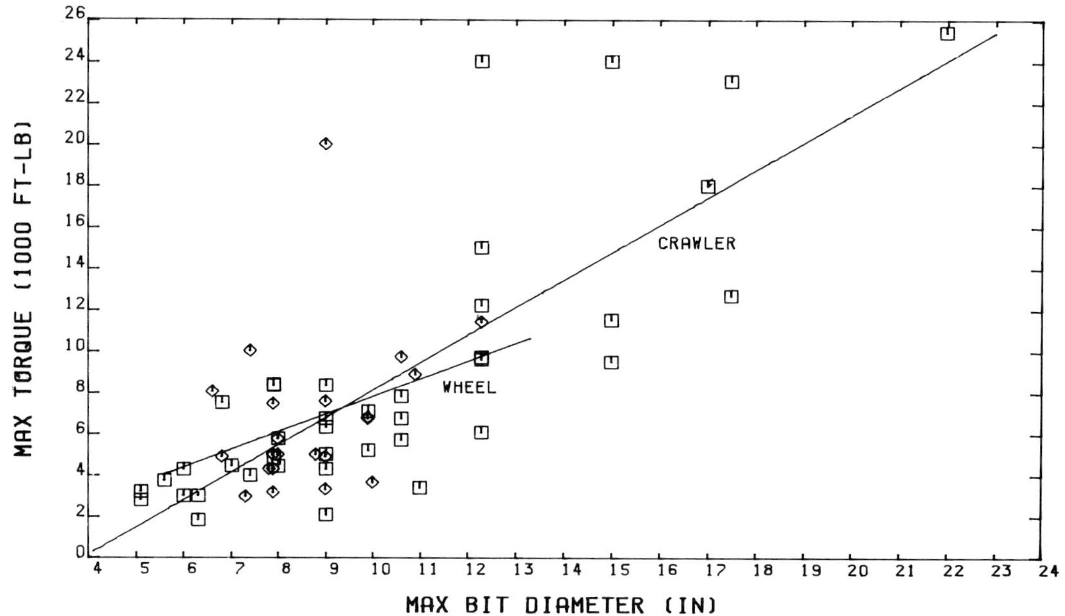

Graph 10.4 Max torque/max bit diameter

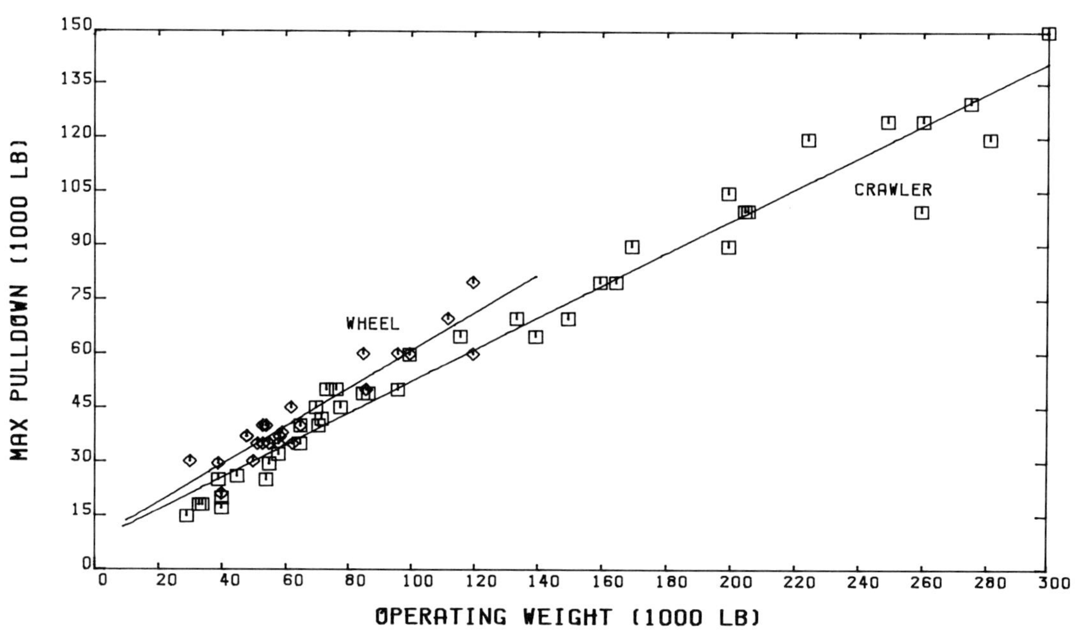

Graph 10.5 Max pulldown/operating weight

diameter. Pipe lengths are assembled with a special box and pin joint, using standardized API (American Petroleum Institute) or BECO (Bucyrus-Erie Company) tapered threads. These special threaded joints have been developed to transmit the very high torques involved while still permitting fast assembly and disassembly.

Mast height is increased to handle longer drill pipes. If the drill pipe length is equal to or greater than that required for the hole depth, no drill pipe additions are required. This optimizes the drilling rate (single pass drilling) because the time lost for pipe changes, as noted above, is eliminated.

If the drilling pattern requires angled holes offset from the vertical, the mast can be tilted. The angles can vary from 0 to 30 degrees, generally in 5 degree increments, with the mast pinned into position during actual drilling. In some cases, additional optional equipment is required, such as deck or lower mast guide bushings.

The machine, after positioning, is raised up on the jacks and leveled. Jacks can be operated simultaneously or individually to correct for severe irregularity in the ground. Leveling normally requires between 0.5 and 1 minute. If the mast must be raised from its horizontal carry position, this will requirė an additional 2 to 5 minutes.

Air compressors on the drills provide a high volume of air at low pressure, which is directed through a swivel down the center of the drill pipe. The air cools the bearings of the cutting bit and, with its high velocity, carries the cuttings generated by the cutting action up the annular ring between the sides of the hole and the drill pipe. The heavier cuttings settle around the top of the hole, generally contained by flexible curtains. The fine particles are drawn up into a deck-mounted dust collection system.

Air transport velocity is dependent on the air volume supplied by the compressor and the area of the annular space. Pipe and bit sizes can be correlated to produce the desired annular area. Consideration must be given to maintaining a minimum gap (3/4 inch) between the pipe and hole, adequate to permit passage of the chips produced. Experience has shown that air velocities of approximately 4500 to 6000 fpm are required for light materials and 7000 to 9000 fpm for heavy materials. Graph 10.6 illustrates an approach to establishing compressor air volume requirements for different combinations of hole and pipe size. The lines labeled 3000 through 7000 represent air velocity with the example (dotted lines) assuming that 5000 fpm would be adequate.

Higher air velocities will remove larger chips and may result in some improvement in bit life and rate of penetration. Higher velocities will also provide an extra margin to offset other factors which might reduce air velocity, including drill pipe wear, and any cavities in the sides of the drill hole. With excessive velocity (above 10,000 fpm), however, there is increased erosion of the drill pipe by the cuttings and increased wear of the dust curtains and other exposed structures.

Air bailing pressures above 40 psi are sufficient to overcome both the losses in the piping and the flow through the orifices in the drill bit. It is possible that higher pressures could result in better bottom hole cleaning, less regrinding of chips, and improved cooling of the bit bearings. Higher pressures reduce the possibility of plugging the air orifices in the bit.

Rotary power requirements are dependent on the formation characterisitcs, bit type and size, down pressure and rotary speed. The rate of penetration is proportional to the rotary speed. Soft formations usually require more than twice the power needed for hard formations.

Crawler mounted units have low propel speeds but good traction, flotation, maneuverability and gradeability. This, together with good stability, reduces the preparation required for access roads and the drilling bench. The truck mounted machines trade off some in-pit mobility for the higher between-pit speeds possible with a roadable unit.

For many years, dust control was achieved primarily with water injection (0.5 to 2.0 gpm) into the bailing air. While this is relatively effective, the water reduces the bit life 15 to 20 % and, of course, introduces additional problems for operations at low temperatures. Dry type systems are available which employ cyclones or filter bags for fine particle separation. These latter units are designed to dump accumulated material on the ground when

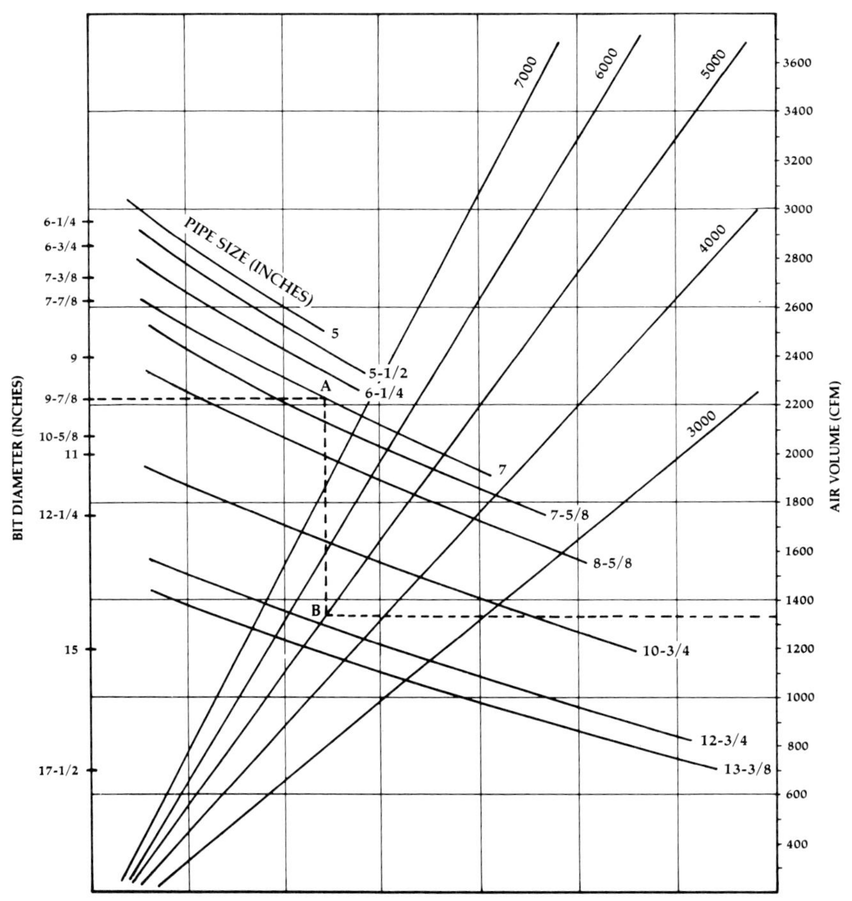

Graph 10.6 Bailing velocity

the drill propels. Dry systems do not always function well where there is ground water or snow which tends to cake the accumulated fines.

The cutting action that occurs while drilling is detailed and illustrated in Graph 10.7. This can be further described as follows:

- With very light down pressure, the bit simply turns and wears itself out.
- With a little more down pressure, the rock begins to fatigue and penetration improves slowly.
- With more down pressure, a rock spalling phase develops and optimum drilling occurs.
- With further increases in down pressure, a foundering phase is initiated and the teeth of the bit are buried with minimal increases in

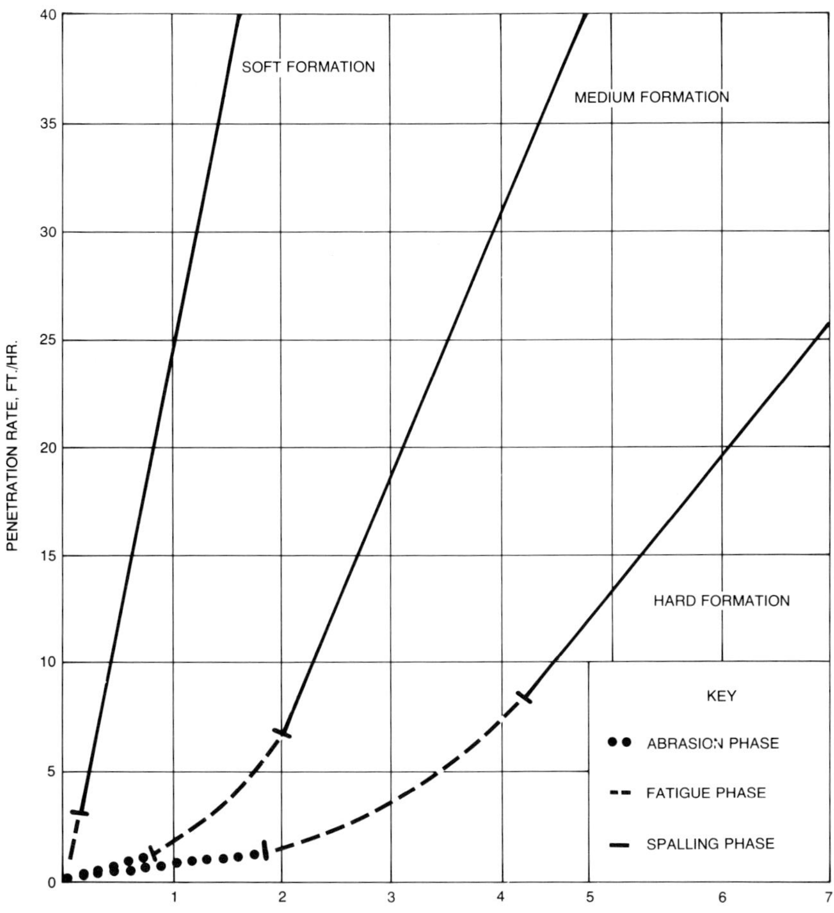

Chart above shows three idealized curves of penetration rate vs. weight in typical soft, medium and hard formations. Note that medium and hard formations have three phases of rock failure during drilling: (1) Abrasion Phase where the formation is worn away; (2) Fatigue Phase where the formation is loaded several times before failure occurs; and (3) Spalling phase where sufficient weight on the bit causes the formation to fracture readily.

When sufficient weight is applied to operate the blast hole bit in the Spalling Phase, the bit drills faster, attains greater overall footage, and requires less energy per foot of formation drilled because chips are larger.

Graph 10.7 Rock drillability

penetration rate and the bit bearings are rapidly destroyed.

Ideally, large chips are produced with minimum energy wasted on regrinding the chips in the hole. The penetration rate increases with bit rpm and down pressure, assuming that there is adequate bailing air for chip removal.

The larger and heavier drills can operate in harder formations. Graphs 10.8 and 10.9 reflect this capability. Pull-down force (down pressure) per inch of width of the drill bit increases with maximum bit diameter or drill size. Maximum torque per inch of maximum bit diameter follows a similar general pattern of increasing with machine size.

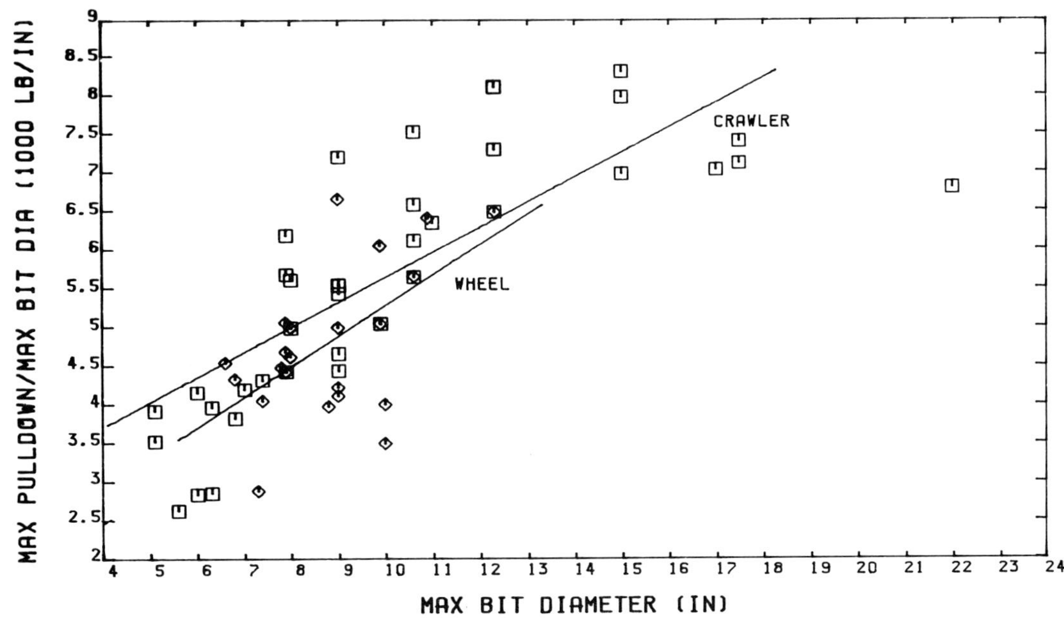

Graph 10.8 Max pulldown per max bit diameter/max bit diameter

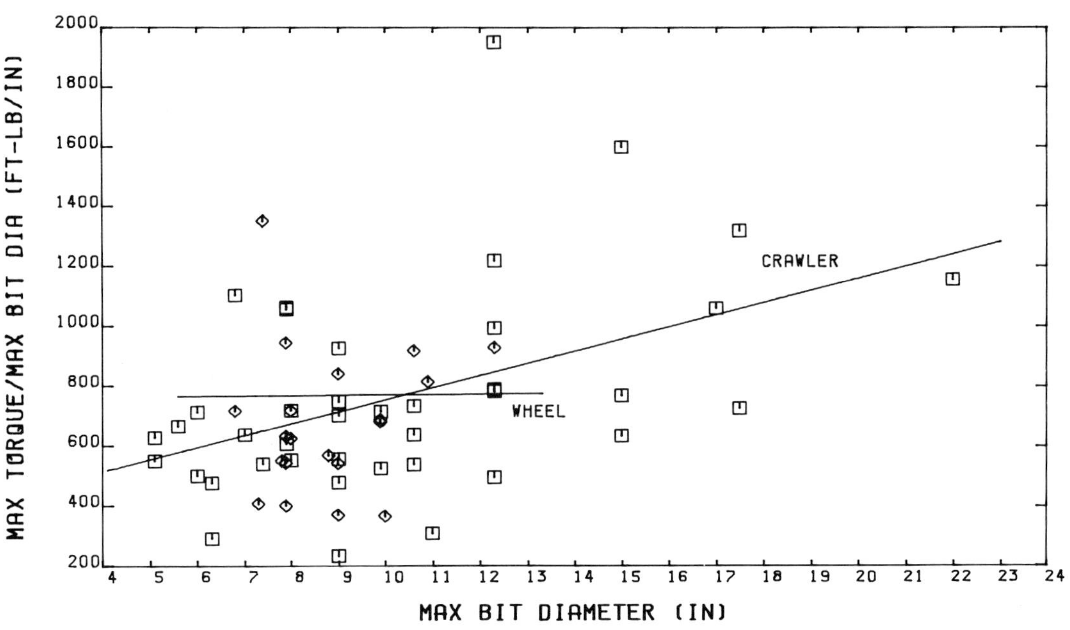

Graph 10.9 Max torque per max bit diameter/max bit diameter

OPERATING PRACTICES

Hole volume and the energy level and detonation characteristics of the explosive employed define the power available for fragmentation. These, together with hole spacing and the set-back from the highwall, determine the degree of fragmentation achieved. The distribution of the explosive down the hole can be varied to match formation characteristics and shape the resulting excavating face. Hole depth is dictated by the mine plan.

Drilling requirements vary with the nature of the formation and the type of excavator utilized. Often, one drill is required for each major piece of face excavating equipment. Topsoil is removed and a drilling bench is prepared in advance of the excavating. Hole locations are surveyed in and marked with stakes as a guide for the drill operator to position the machine. (See Figure 10.4)

The drill bits are threaded to the end of the drill pipe. The type selected varies with formation characteristics. Drag blade or auger type bits are used for smaller holes (3 to 10 inches) in soft materials such as coal. For most overburden drilling, tricone roller bits are used. The basic design is shown in Figure 10.5. Three conical shaped cutting heads rotate on slightly different angled axes. Hardened teeth or shaped tungsten carbide inserts in rows around the circumference of the heads concentrate the applied forces to penetrate and break out chips. The rolling action of the heads, as the bit is continually rotated changes the pattern of tooth contact on the bottom of the hole. The limited intermeshing of the cones provides a self-cleaning action to minimize build-up between the teeth. The diagram shows the air passages which direct roughly 10% of the bailing air to cool the bit bearings. The remaining air is directed through orifices to flush the cutting debris from the bottom of the hole. A combination of roller and ball bearings, plus bushings, are required to withstand the extremely high loads in the limited available space.

Generally, four basic bit configurations are available for soft, medium soft, medium hard and hard formations. These classifications are associated with typical materials:

- Soft formations: soft shales, clay, silt, calcite, unconsolidated formations
- Medium soft formations: firm sandy shale, gypsum, salt, chalk, sandstone
- Medium hard formations: hard shale, siltstone, hard limestone, dolomites, hard sandstone, lime, prophyry copper, basalt
- Hard formations: granite, specular hematite, iron ores, quartizitic materials, taconite

The spacing, shape and size of the hardened steel teeth or inserts on the conical cutting head are matched to the hardness of the formation. (See Photographs 10.7 through 10.10)

- Soft formations: widely spaced, high tooth depths or extended chisel shaped inserts; cutting action produced by gouging and scraping
- Medium soft formations: medium closely spaced, medium short tooth or blunt chisel shaped inserts; cutting action produced by scraping and some chipping and crushing
- Medium hard formations: closely spaced, low inserts, hemispherical or conical shaped inserts; cutting action produced by chipping and crushing
- Hard formations: very closely spaced, very low inserts, hemispherical shaped; cutting produced by grinding

All of the inserts are made of super-hard tungsten carbide. Down pressure must be matched to the bit capabilities to prevent premature failure of the bearings. The larger bits, because of increased bearing size, can drill harder formations. A minimum air pressure drop across the bit of 25 psi is required for proper bearing cooling.

Soft formations typically require:

- High drill torques
- Low down pressures (1500 to 2500 lb/in bit diameter)
- High rotational speeds (70 to 100 rpm)
- High bit penetration rates (50 to 90 ft/hr)
- High rotary horsepower

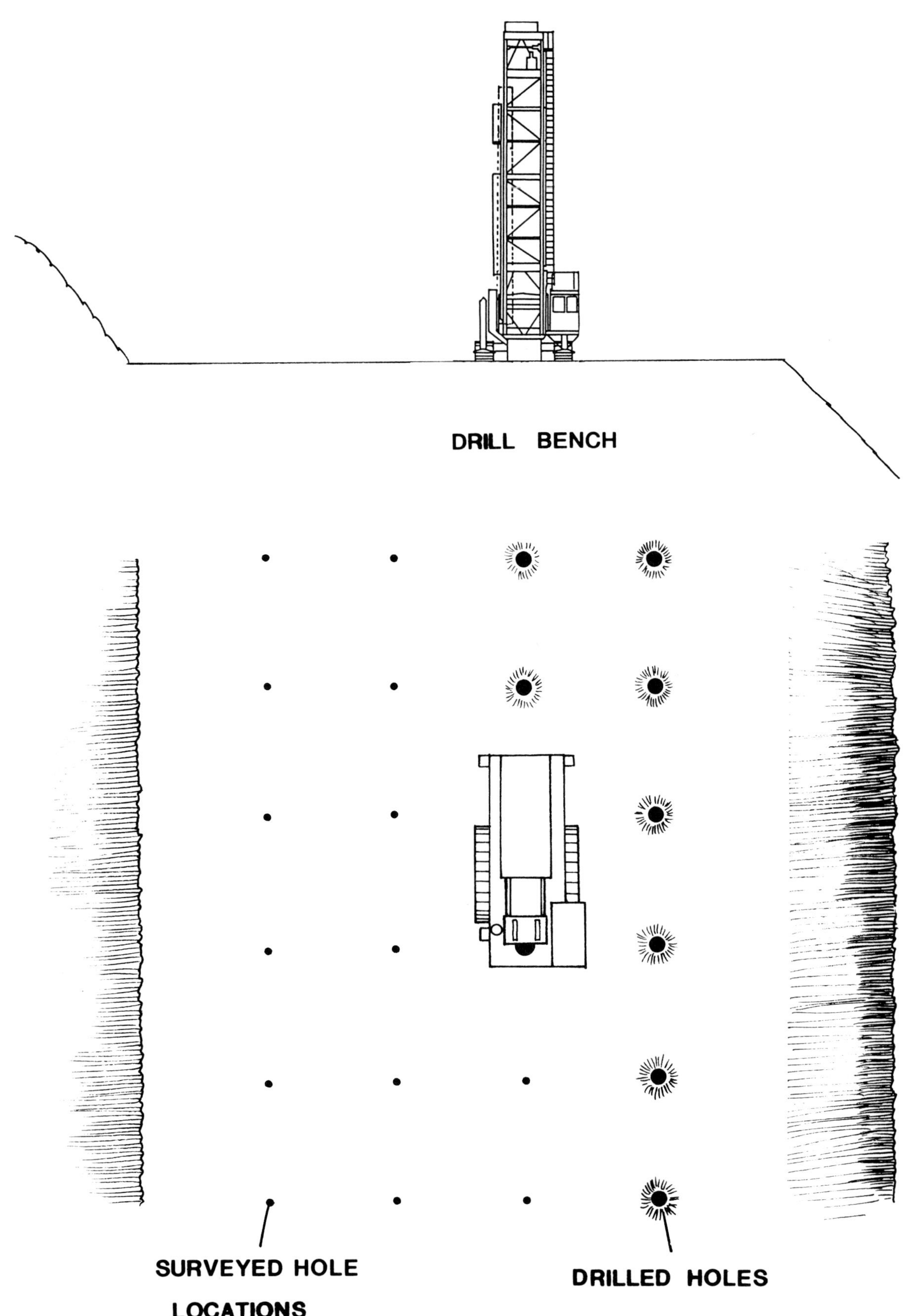

Figure 10.4 Drill Production Cycle

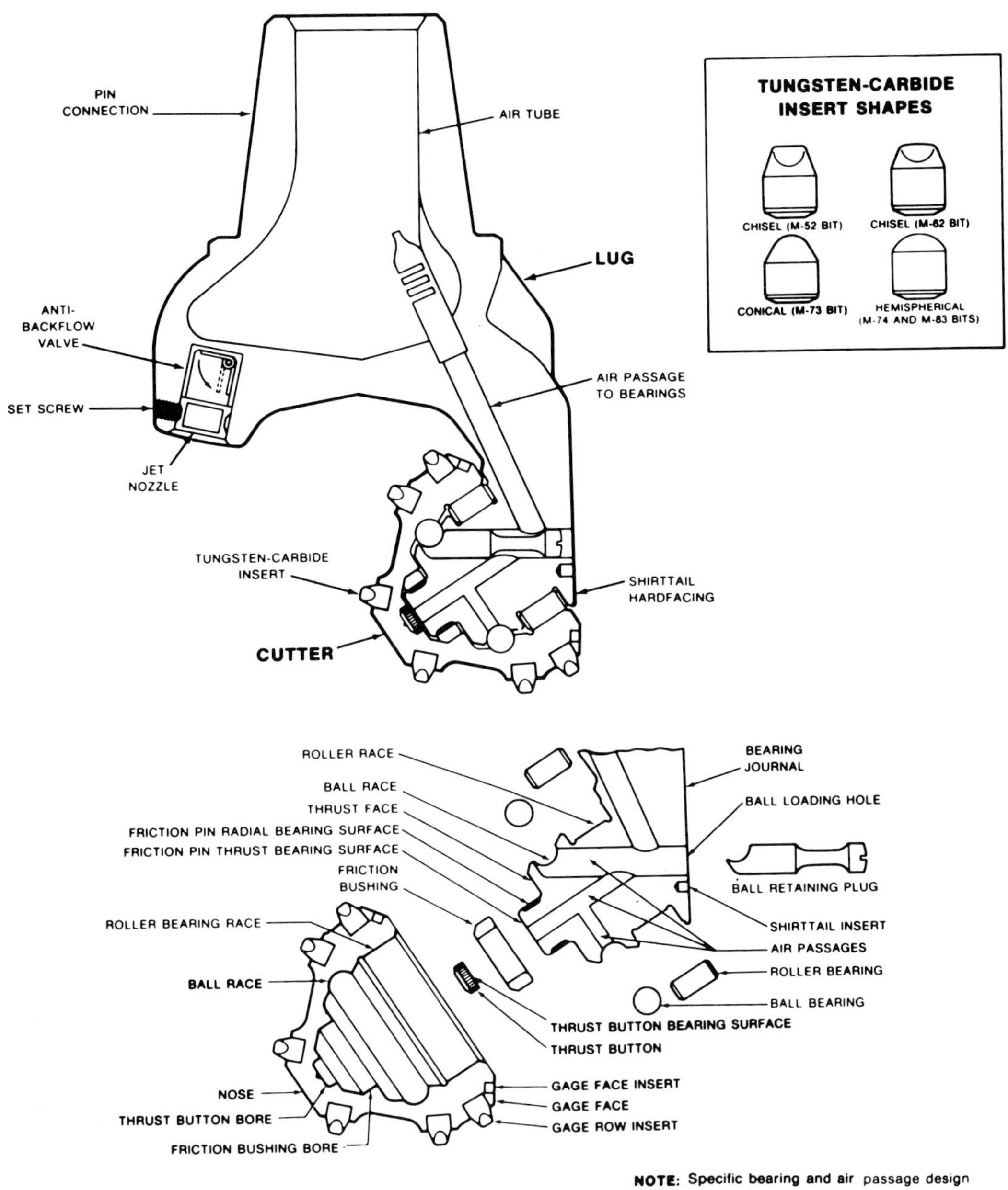

Figure 10.5 Blast Hole Bit Components (Courtesy of Reed Mining Tools, Inc.)

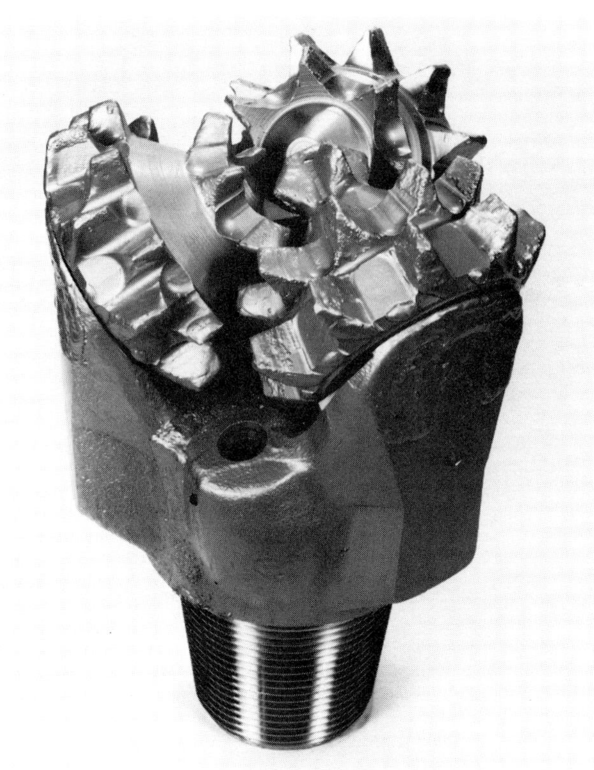

Photograph 10.7 Representative standard steel tooth rock bit for soft formations (Courtesy Hughes Tool Company)

Photograph 10.8 Representative standard steel tooth rock bit for hard formations (Courtesy Hughes Tool Company)

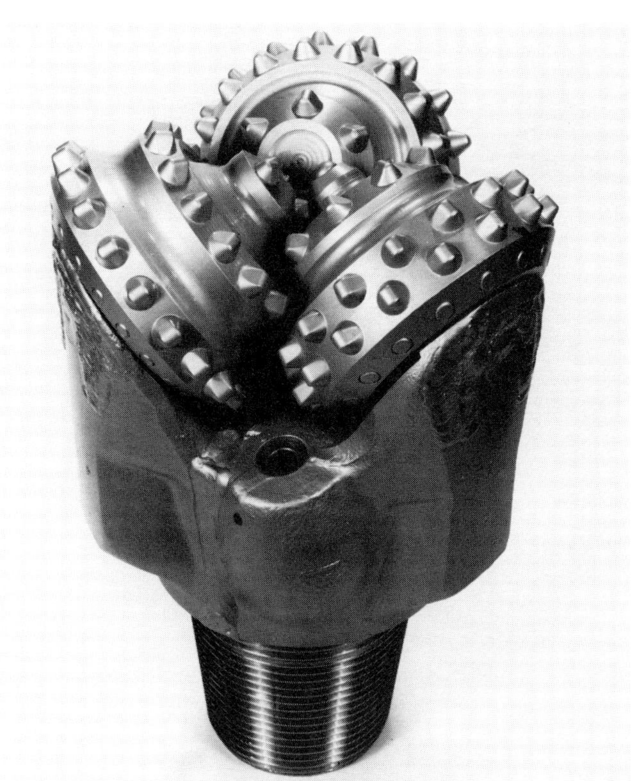

Photograph 10.9 Representative sintered tungsten carbide rock bits for soft formations (Courtesy Hughes Tool Company)

Photograph 10.10 Representative sintered tungsten carbide rock bits for hard formations (Courtesy Hughes Tool Company)

Hard strata typically require:

- Low drill torques
- High down pressures (4000 to 8000 lb/in bit diameter)
- Medium rotational speeds (40 to 80 rpm)
- Low penetration rates (20 to 30 ft/hr)
- Low rotary horsepower

Using improper bits will result in poor penetration and could induce objectionable machine vibrations. Bit selection should be made after consultation with the bit manufacturers and, whenever possible verified by actual site tests. Manufacturers will provide weight (pulldown and dead weight) and speed limitations for specific bits.

Varying geological formations, particularly sloping strata, faults, shifts and voids make drilling difficult. These problem formations cause uneven bit loading, bending of the drill pipe and can induce severe vibrations in the drill. Material which varies from hard to soft can also cause a drill to vibrate and limits the drilling rate.

The machine capability technically limits maximum rpm, down pressure and torques. In practice, these performance areas are generally greater than can be effectively used in the drilling. Excessive bit pressures and speeds result in excessive bit wear and reduced life. The bit should not be fed into soft or loose formations faster than the bailing air can clean out the hole.

Although the carbide insert bits cost considerably more, the footage drilled per bit is increased 4 to 10 times over that of a steel bit. Penetration rates are also improved.

The drill pipe must be straight; otherwise it can lead to early bit failure and will, in addition, cause machine vibrations. The threads should be kept lubricated and shouldered properly when drilling. Care must be exercised to line up pipe accurately for assembly to avoid damage to the threads. Protector caps should be installed for pipe handling on and off the machine.

Air must be available to clear the cuttings off the bottom of the hole. The flow to the bit must be checked regularly to be sure that no obstructions have developed in the bit, drill pipe, swivel or hose. If one cone on the bit is hotter than the others after completing a pass, it indicates a blockage of the air passages to the bearing. If enough air is not flowing to the bottom of the hole, there will be:

- Reduced rate of penetration
- Increased torque and down pressure required
- Increased wear of the drill pipe, bit cutter heads and bit bearings.

Bit air discharge orifices (jets) should be adjusted to maintain a high pressure drop in wet holes or when using water injection for dust control. New drill bits should be operated at low loads and speeds for a short break-in period. A great deal can be learned by examining worn bits. It is often possible to identify operating practices (such as excessive down pressure or rotary speeds) which can be altered to upgrade performance. Bit type selection may also be improved.

When the holes are deep, the drill pipe diameters small, and the down pressures high, stabilizers are frequently installed above the bit. The stabilizer contacts the wall of the hole, centering the bit in the hole. They provide a smoother drilling operation, reducing some of the shock loads and extending bit bearing life. The most popular stabilizer is a roller type with replaceable rollers.

To isolate shock loads on the bit from the rotary drive machinery, cushioned couplings called shock subs can be inserted between the drive spindle and the drill pipe (top drive units). These reduce the transmitted vertical and torsional shock loads, reducing maintenance, noise and improving bit life. Two basic designs are available: one incorporates rubber discs and the other a nitrogen spring.

The drill operator cannot see the bit or the chips coming up from the hole and, therefore, must rely on the machine's gauges recording down pressure, torque and bit rpm for information on drilling performance. Machine vibration can offer another clue to cutting conditions. Optimizing performance and avoiding damage to the bit and drill requires continued attention for prolonged periods; unfortunately, the routine nature of the operation makes it difficult for the operator to maintain his attention.

Recognition of this problem led to the development of optional automated equipment for programmed drilling. This equipment, because of its high cost, is generally only available for the large electric powered crawler drills. The automation is limited to the actual drilling cycle and not the machine positioning. Electronic solid state controls are provided for bit rpm, pulldown force, feed rate and bailing velocity. Limit settings are provided for maximum rpm, drilling depth, maximum torque, maximum air pressure and maximum machine vibration. An experienced operator can effectively increase drill productivity, improve bit life and reduce overall machine maintenance.

It should be noted that the wheel mounted drills which are lighter and, therefore, have lower maximum pulldown forces; are sometimes equipped with higher pressure air compressors. With the higher pressure air, they can use down-the-hole hammers to penetrate hard formations. The down-the-hole units are essentially compact high capacity pneumatic impact (reciprocating) hammers on which the bit is mounted; the hammer is attached at the bottom of the drill pipe. The typical hammers are 4 to 15 inches in diameter. They require air at 125 or 250 psi. Air exhausted from the hammer provides the bailing air. Only low down pressures and rotary speeds are required. The matching bit has a flat face with botton type tungsten carbide inserts. They will penetrate very hard materials but require a high initial investment.

DESIGN FEATURES

See Figure 10.6 for drill nomenclature.

Possibly because of the diversity of the manufacturers, the design componentry of different drill models is extremely varied. An usually large proportion of the currently available machines have been developed within the last ten years, so that the latest technolgy, ranging from static electric to sophisticated hydrostatic drives, is incorporated. Virtually all types of power transfer devices, including hydraulic cylinders, cables, chains, gears, clutches and brakes are applied on most machines. Functions can be powered by A.C. motors, D.C. motors, hydraulic motors (low and high pressure design), or direct from a diesel engine. With this diversity, this discussion on machine design must be restricted to the unique features and characteristics of the drills.

The wheel mounted drills, as noted earlier, utilize diesel or gas engines for the primary drive and direct mechanical or hydraulic drives for the various powered functions. There may be a separate engine for the carrier (truck). Small diesel crawler units use drives of similar configurations. Large crawler machines use either D.C. or A.C. electric systems for the powered functions. In all cases, the machine leveling, mast raising/lowering and pipe handling systems employ hydraulic cylinders. Many of the pulldown systems also use hydraulic cylinders.

Three types of air compressors are used: sliding rotary vane, piston (reciprocating), and oil flooded rotary screw. (See Figure 10.7) Capacities range from 250 to approximately 2500 cubic feet/minute. Compressor size and, in some cases, type are optional. Typical units employed include the following:

- Gardner-Denver: screw — 55 or 100 psi
- Sullair: screw — 100 or 125 psi
- Ingersoll Rand: screw — 110 or 150 psi
- Chicago Pneumatic: screw — 125 psi
- Joy: screw — 50 or 125 psi
- Schramm: piston — 250 psi
- Allis-Chalmers: rotary vane — 40 psi

Note that the vane is essentially a low pressure unit, the screw an intermediate, and the piston a high pressure design. The screw compressor has some appeal because of its simple, compact configuration. They are available in either a single or twin screw configuration, and are oil flooded for sealing and cooling.

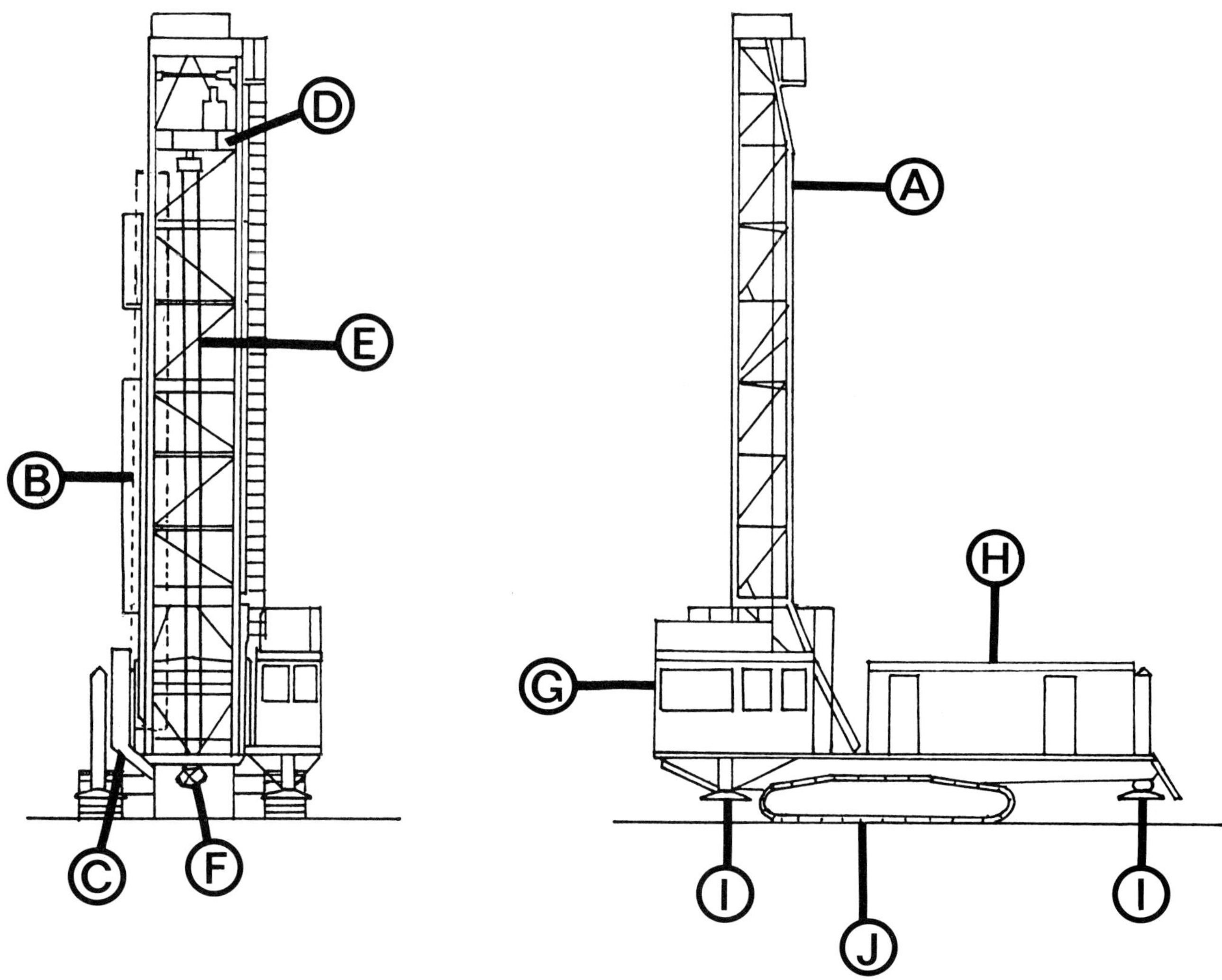

A	Mast	F	Bit
B	Pipe Rack	G	Cab
C	Dust Collection System	H	Machinery House
D	Top Drive	I	Leveling Jacks
E	Drill Pipe	J	Crawlers

Figure 10.6 Crawler Drill Nomenclature

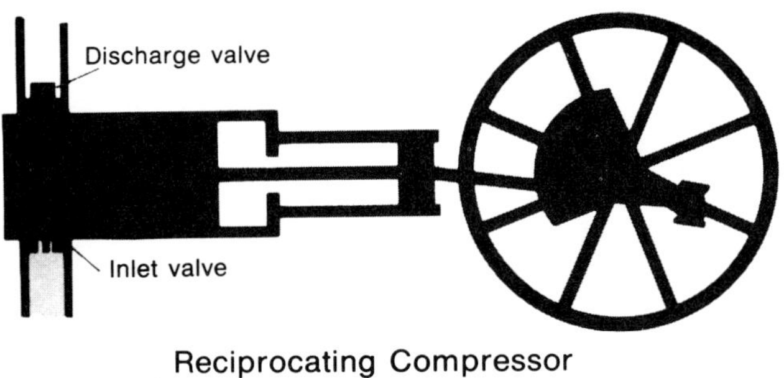

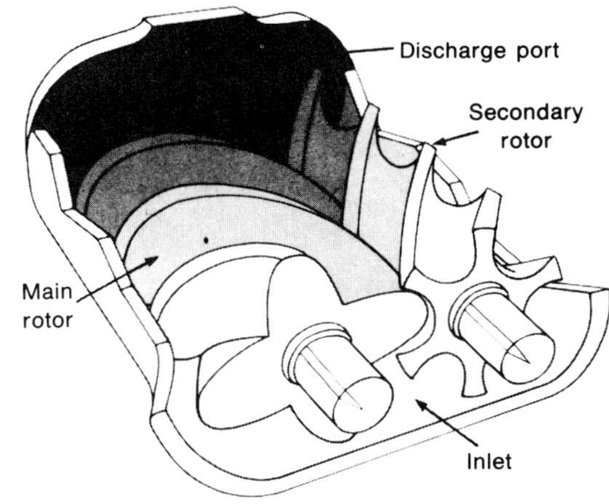

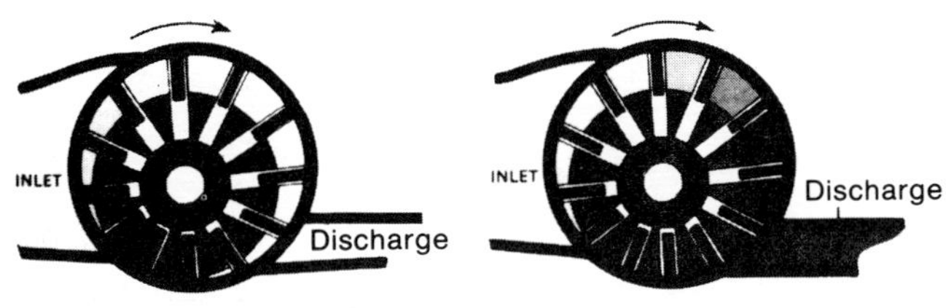

Figure 10.7 Air Compressor Types (Courtesy of Reed Mining Tools, Inc.)

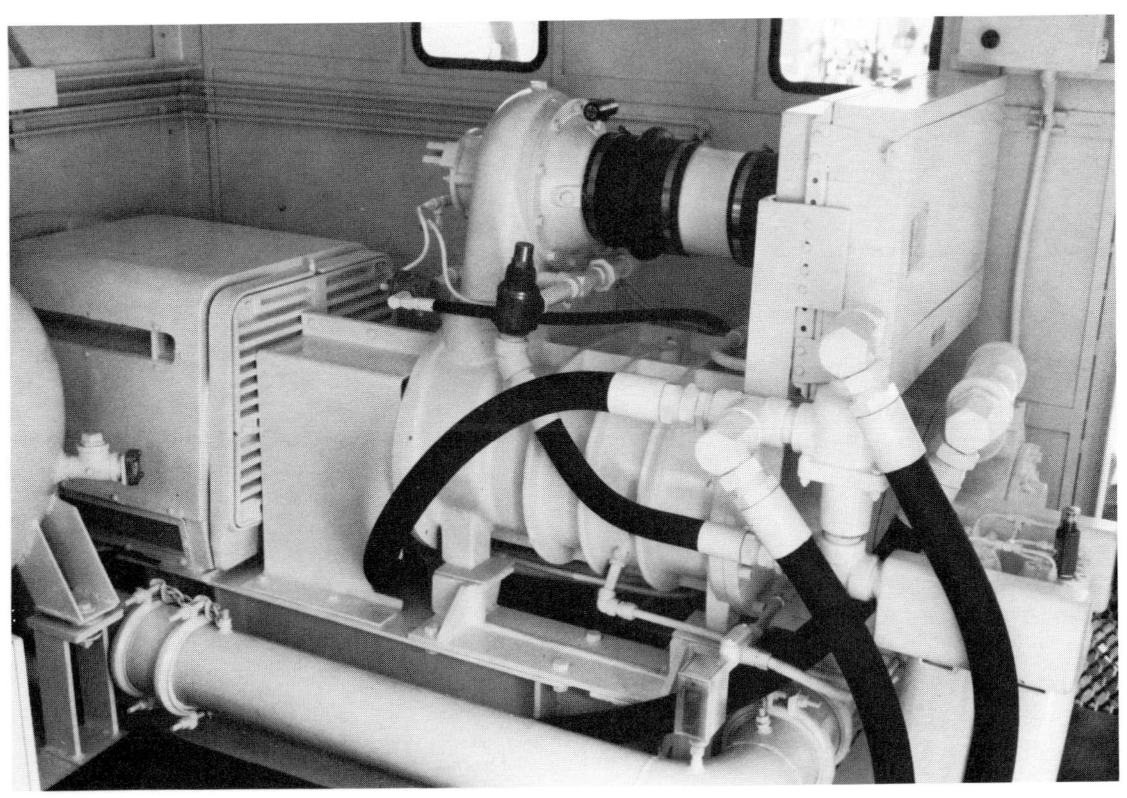

Photograph 10.11 Motor driven screw compressor on a larger crawler drill

Photograph 10.12 Piston compressor on a wheel mounted drill

Because the working environment of the drills is often dusty, special consideration must be given to cleaning the air entering the compressor and the diesel engines. On the open deck machines, two stage dry type air cleaners are sometimes used with an additional cyclone type precleaner.

Rotary torque can be applied to the top end of the drill pipe with either a direct or false kelly system. (See Figure 10.8) In the direct version, electric or hydraulic motors (hydrostatic drive) are mounted on the top of the gear case which drives the drill pipe. During drilling and pipe retraction, the entire assembly moves up and down on guides in the mast. The false kelly design has the drive mounted on the main deck, powering a vertical slotted shaft which drives the drill pipe through a gear reduction. The gear case moves up and down on guides in the mast. A sliding key arrangement permits the transfer of torque from the kelly to the gear reducer. Single or multiple motors of varying sizes and types can be mounted on the gear cases to provide optional rotary power levels.

While the rotary table drive is popular in oil field work and exploration drilling, only a few blast hole drills are so equipped. In this design, deck mounted machinery powers a rotary table which engages a slotted pipe (kelly) to transmit the torque directly to the drilling bit. The kelly, which can slide vertically in the table, is raised and lowered as required. Hydrostatic, infinitely variable drives are common for the table rotation.

The four different arrangement employed to provide the pulldown force (down pressure) for bit penetration are illustrated in Figure 10.9. All of the sketches assume that a top rotary drive system is used. In most cases, these are dual systems; the mechanisms are mounted on the sides of the mast, sharing the loads and moving in unison to raise or lower the rotary drive gear case. The pulldown system must be able to develop high forces, accommodate the flexing of the structure, maintain the drill pipe alignment and produce a travel stroke adequate for the longest drill pipe to be used. High forces and relatively high speeds must be available

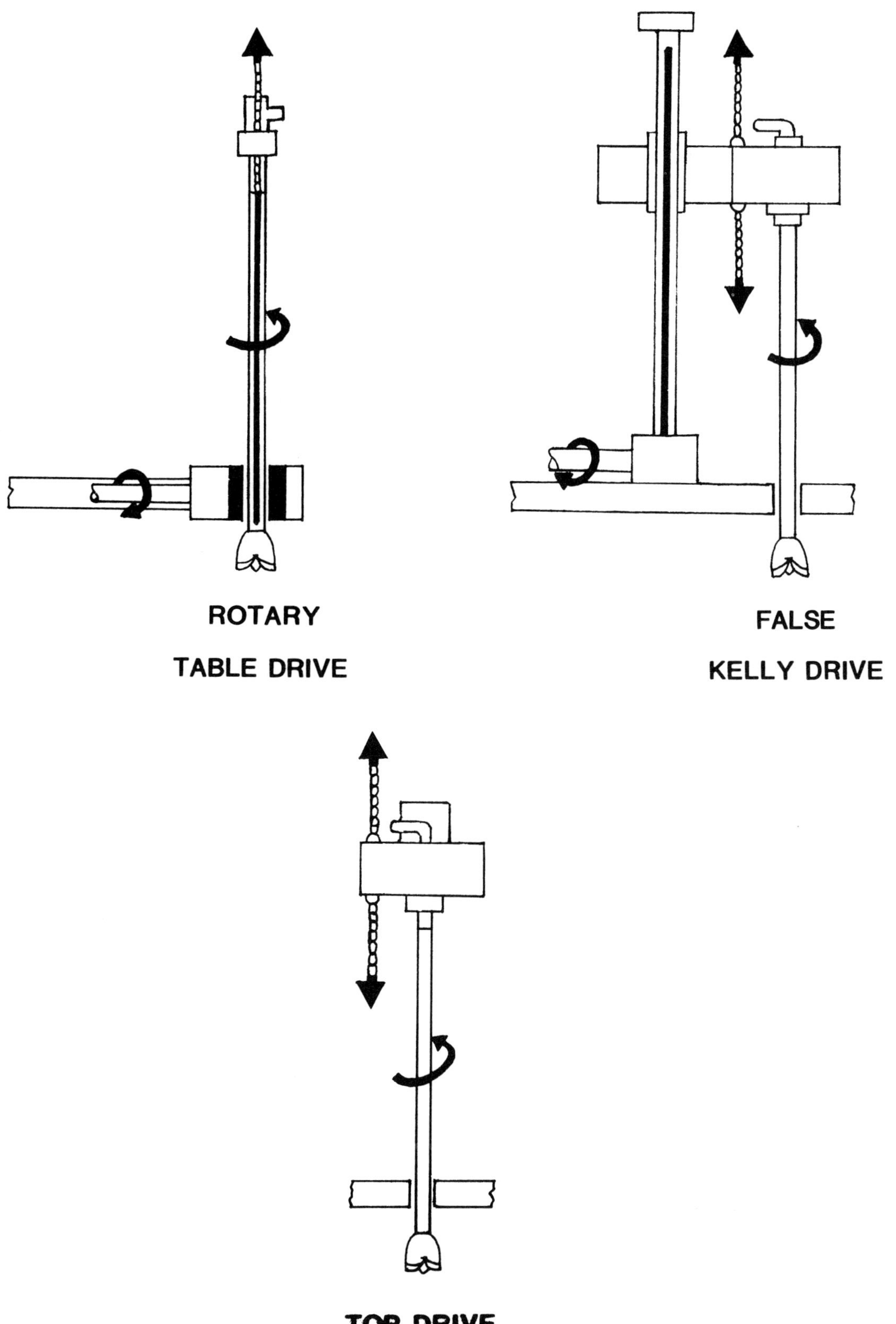

Figure 10.8 Rotary Drive Options

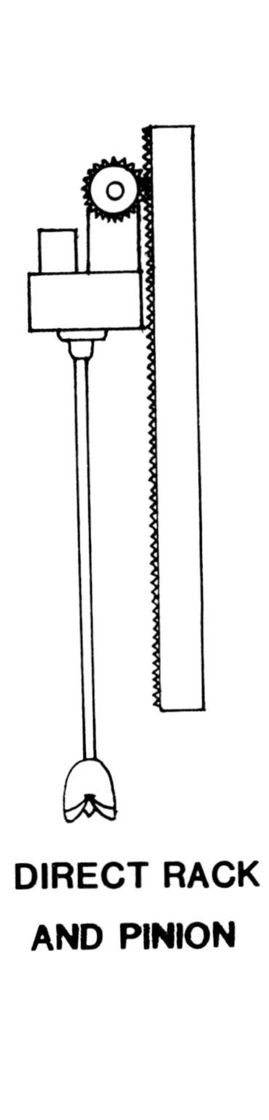

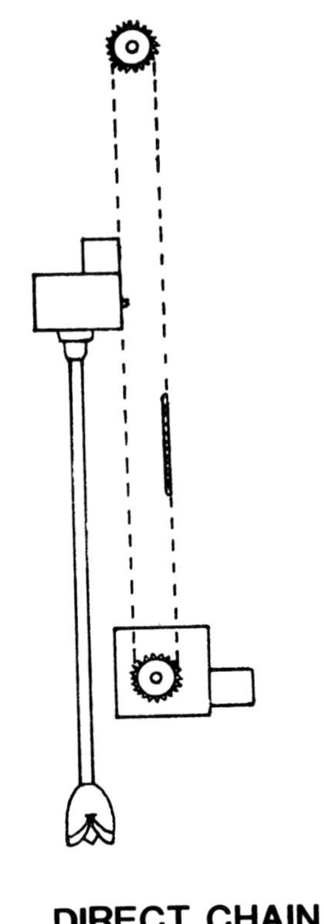

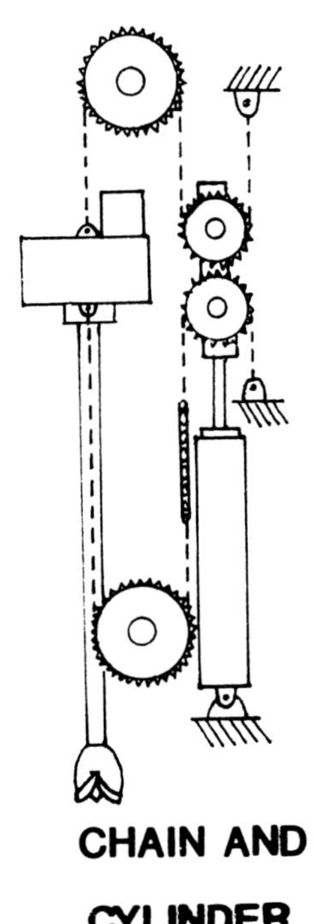

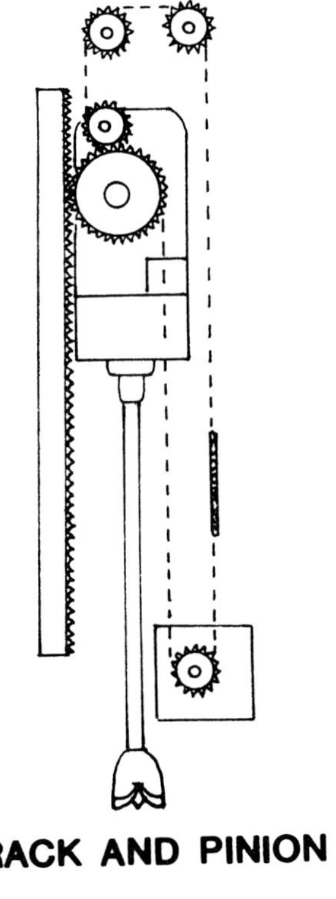

Figure 10.9 Pull Down Options

for both raising and lowering. Generally the actuators are either double acting cylinders or a motor; they are mounted on the main deck to keep the weight low for improved machine stability. Roller chains transmit the motion; small toothed drive and idler spockets provide positive positioning and power transmission. Infinitely variable speed control is common for drilling; high speed ranges are available for raising and lowering during pipe handling. The control system generally permits the automatic maintaining of any set pulldown force.

The mast is the critical structural member on the drill. It is subjected to high pulldown forces and, in the case of the top drive, to rotary drive torques. Vibrations and shock loads from the drilling action are transmitted to the mast with a top drive. The pipe storage racks and the pipe handling equipment are built into the mast. It also supports the guideway for the rotary drive case and maintains the drill pipe alignment. With the pipe rack full, the mast must be capable of being pivoted from a horizontal carry position to a vertical drilling position. These diverse requirements, particularly the high dynamic loads increase the complexity of the design. To overcome field problems, the configuration has been continuously refined. At present there are three designs in service. (See Figure 10.10) The lattice type is the oldest and the most common.

The wheel mounted drills commonly utilize a commercial carrier chassis with a solid suspension, similar to that used for transit cranes. Dependent on machine weight and the axle load limitations, these carriers may have 2, 3, or 4 axles. Two axle drive is employed to improve off-highway mobility.

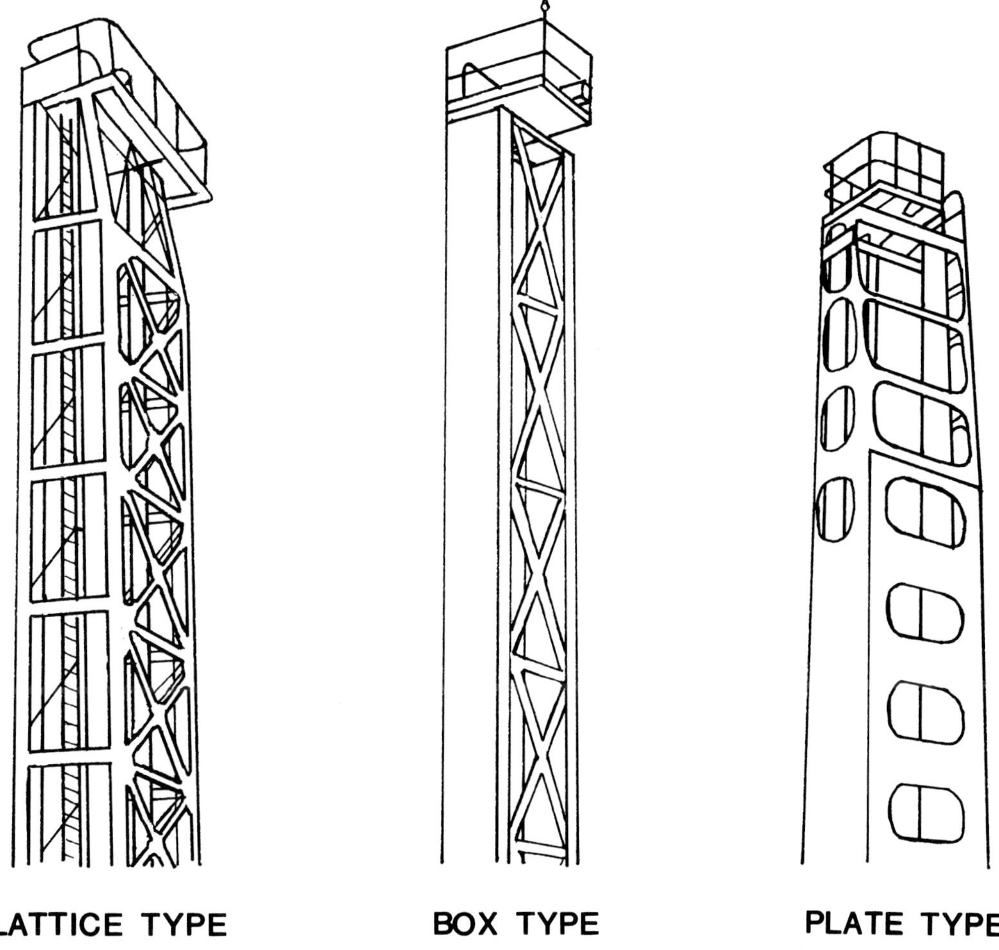

Figure 10.10 Drill Mast Designs

All drilling operations occur with the drill supported on the leveling jacks so that this does not impose any special demand on the mounting. Care must be exercised, however, to be sure there is adequate stability to move across the uneven ground between hole locations with the mast raised. Maneuvering these long wheel based carriers over the hole locations can be difficult. Remote controls at the drilling end of the machine reduce this problem. Some machines can be positioned using the deck engine.

Since most of the large crawler mounted units are electric and have a trailing cable for the power supply, the track system is of a simple, rugged, low speed configuration. On some, the crawlers are driven from the main deck with clutches and brakes, permitting driving or braking of either track and forward or reverse propel with both tracks. Units are also available with hydraulic motors mounted directly on the track side frames to provide independent track drive and spin turns.

The smaller diesel machines have independent track drives, high propel speeds (.7to 1.5 mph); the track is made up of assemblies from dozers or hydraulic excavators of equivalent weight.

SELECTION CONSIDERATIONS

The principal properties of the material that affect drillability are:

- Hardness
- Compressive strength: pressure per unit of area to crush the rock
- Abrasiveness: the capacity of the rock to abrade other materials
- Toughness: the ability of the rock to resist crumbling

The overall effect of these is generally best assessed by actual on-site drilling tests. Alternatively, laboratory tests on samples can provide some insight into future bit performance and life.

Production requirements, in terms of foot of hole to be drilled, is dependent on the mine plan, explosives to be used, type of excavating equipment and the geology of the area. Knowing these, hole size, depths and spacing can be established. Drilling capability must be adequate to keep ahead of the excavating operations.

Factors to be considered in the choice of the drill include the following:

- Machine capabilities (pulldown, rotary torque, etc.) must exceed formation penetration requirements.
- Maximum hole size capability increases with machine size.
- Larger machines are more rugged and can generally drill in harder formations.
- A machine that can handle drill pipe long enough to permit single pass drilling can significantly improve productivity.
- The production rate is dependent both on the actual penentration rate and on the time required for pipe changes and machine repositioning.
- Electric drives have the lowest operating cost, the longest service life and the best track record for reliability.
- Electric drives require an in-pit power distribution system.
- Three levels of pit and area mobility are available; low speed crawlers (electric machines), medium speed crawlers (diesel machines) and roadable high speed carriers (wheel mounted units).
- Dust control requirements are dictated by regulations.
- Optional equipment such as powered cable reels, automatic lubrication, automated controls, etc., can increase the efficiency of the drilling operations.
- Long term productivity is dependent on the ruggedness, reliability and maintainability of the design.

Special consideration must be given to mine life, ruggedness of the terrain, the frequency and distance of moves between drilling locations, the availability of skilled operating and maintenance personnel, and maintenance and service support facilities.

The purchaser of the drill may, based on preference or mine equipment standarization programs, specify some of the major machine elements:

- Compressors: size, type, manufacturer
- Diesel, diesel/electric or electric primary power
- Engines: size, manufacturer
- Operator's cab/station

Optional sizes/type of units are available for some features:

- Dust collector systems
- Hoist/pulldown cylinders
- Rotary drives: type and speed
- Angle drilling features
- Mast length
- Drill pipe lengths
- Oversize crawler shoes
- Pipe racks
- High ambient engine and compressor radiators
- Fuel tanks
- Alternators
- Special paint

Optional equipment is available which can be addded to the basic machine:

- Stabilizers
- Shock subs (anti-shock drill coupling)
- Automation kit to monitor and control drilling parameters such as drill feed rate, torque, etc.
- Automatic centralized lubrication
- Hydraulic drive cable reel (electric machines)
- Propel remote control for positioning
- Depth of drilling indicator
- Tow bar
- Night lighting
- Special mufflers
- Air line lubricator
- Air pressure regulator
- Center guide
- Auxiliary winch
- Ballast kits
- Low temperature starting
- Vandalism protection system
- Break out wrench and special pipe handling accessories
- Ball-bearing deck bushing
- Cold weather options
- Fire extinguishers
- Operator comfort/convenience items such as heater, air conditioner, radio, tinted glass, etc.

NEW DEVELOPMENTS & TRENDS

Over the last years, the number of blast hole drill suppliers has stabilized and the product lines have filled out. Most of the models appear to have been updated to the current level of technology. The largest drills are matched to available bit capacities

Photograph 10.13 Small twin mast tractor mounted auger coal drill (Schroeder)

and there appear to be little demand for larger models. Mining regulations with respect to blasting noise and vibration have dampened the interest in larger holes.

Gardner-Denver has introduced and is continuing to refine two axle, rough terrain wheel-mounted drills. These can provide higher in-pit mobility and improved close quarter maneuverability over the current carrier mounted units and, hence, should appeal to the medium sized mine operators.

Twin mast drills are available (see Photograph 10.13) and have been assembled at some mines for drilling coal where the hole spacing is relatively close and shallow depths are involved. Such units can improve drilling productivity. There seems to be some current interest in expanding this concept into larger units of higher capacity.

Intermittent tests and studies have been made through the years attempting to improve drill penetration rates by superimposing a vibratory motion on the drill bit. While this approach appears effective in soft formations, it has not as yet been developed into a commercial product.

The capabilities of the larger rotary drills, when combined with available tricone bits have reached a point where most hard formations can be drilled at acceptable costs. There is limited future need for special designs, such as jet piercing drills.

The thrust of development in the future would seem to be one of design modification for increased reliability and further improvements in drilling productivity. Modular construction and accessibility for maintenance are being stressed. Emphasis is being placed on speeding up the auxiliary functions to increase the productive time.

Increased power is being provided for the rotary and pulldown drive systems as improvements in bit design and manufacturing extend their capabilities.

Automation of individual functions is progressing. The broad adoption of automated drilling control, however, seems to be making only limited progress.

MACHINE SPECIFICATIONS

See Figure 10.11 — Blast Hole Drill Dimensions
See Table 10.2 — Specifications: Wheel Mounted Blast Hole Drills
See Table 10.3 — Specifications: Crawler Mounted Blast Hole Drills

Drills are listed alphabetically by manufacturer and then in ascending order of drill hole diameter. Operational characteristics and capacities listed are the manufacturer's standard equipment for each machine. For example, drill feed rate and torque (rotary), as listed, are compatable. In most cases, faster feed rates are available with a corresponding decrease in rotary torque. The reverse is also true. Rotation speed, as shown, is the maximum available standard rotation speed. In most cases, other optional speed ranges are available. In addition, for those machines which list direct drive as standard, optional hydraulic drives are usually available. Hoist speed, as shown, is the standard hoist speed; again, optional higher speeds are usually available.

Where optional is shown under Dust Control, it means that more than one system is available and that the user would have a choice. Listed fuel capacity refers to the drill or deck engine and does not include any auxiliary tank capacity. Available drill pipe lengths are listed from shortest to longest. In general, the shortest length listed is the standard length. Pipe rack capacity will often be shown as a range. This is to show that, often, as drill pipe length increases, pipe rack capacity decreases. Machine height with the mast up, and machine length with the mast down, are also given as a range to reflect dimensions changed by varying drill pipe length.

Tire size refers to the tires under the drill itself, i.e., the rear tires.

All specifications, capacities, capabilities and dimensions are based on published manufacturer data. Although the information is believed to be current and the interpretation to be consistent and correct, it is possible that there may be omissions, inaccurancies or out-of-date information in the specification tables. These specifications are not meant to be used either as an in-depth analysis or as a head-to-head comparison of the available equipment, but rather as a general overview of the equipment, available sizes, and approximate operating data. Specific questions relating to performance or purchase should be directed to the manufacturer or authorized distributor/dealer in the user's specific area.

Table 10.4
WHEEL DRILL SPECIFICATIONS

MAKE	MODEL	RANGE OF APPROXIMATE HOLE DIAMETERS (in)	WEIGHT (lbs)	MAX PULLDOWN FORCE (lbs)	MAX DRILL FEED RATE (fpm)	MAX TORQUE (ft-lbs)	MAX ROTATION SPEED (rpm)	MAX PULLBACK FORCE (lbs)	MAX HOIST SPEED (fpm)	MAX DRILL MAST ANGLE (degrees)	AVAILABLE PIPE LENGTHS (ft)
Bucyrus-Erie	2450-R	6.0-10.0	51,300	35,000	—	3,667	150	20,000	120	—	25
Chicago Pneumatic	T-600	to 7.3	40,000	21,000	18	2,975	122	20,800	26	—	20
Chicago Pneumatic	T650-SS	to 7.9	54,000	40,000	25	3,167	152	22,000	46	opt	25
Chicago Pneumatic	T-700	6.8-9.0	62,000	45,000	25	3,333	130	22,500	44	—	25
Davey Rousselle	M8B	4.8-7.4	50,000	30,000	—	10,000	40	35,000	250	—	23,26,32
Davey Rousselle	M10B	6.0-10.0	65,000	40,000	15	5,000	140	22,500	120	—	25
Driltech	D40K-II	8.0	53,000	40,000	72	5,000	106	—	130	opt	25
Driltech	D80K	9.0-12.3	120,000	80,000	10	11,416	100	—	94	opt	30
Gardner-Denver	GD-35T	6.8- 8.8	58,000	35,000	124	5,000	128	—	124	—	25
Ingersoll-Rand	T4-BH	5.0-8.0	48,000	37,000	17	5,750	200	37,000	182	opt	25
Ingersoll-Rand	T-4	5.0-9.0	48,000	37,000	62	7,575	200	37,000	115	opt	25
Ingersoll-Rand	T-5	8.6-9.9	85,000	60,000	14	6,816	79	50,000	88	—	24,30,32
Portadrill	TM6-B	5.6-7.9	58,500	37,000	15	5,000	147	—	105	—	25
Reedrill	SK-35	6.8-7.8	53,000	35,000	10	4,300	135	21,000	92	—	25
Robbins	RRT-35	6.3-7.9	62,500	35,000	15	4,300	150	19,500	100	opt	25
Robbins	RRT-50	6.8-9.9	86,000	50,000	—	6,750	142	33,000	85	opt	20
Robbins	RRT-60	6.8-9.9	96,000	60,000	—	6,750	142	33,000	85	opt	20,25
Robbins	RRT-70	9-10.9	112,000	70,000	—	8,890	120	36,000	100	opt	25,30
Salem Tool	VAD	6.8-9.0	100,000	60,000	—	20,000	127	—	90	—	20
Schramm	T-66H	4.5-6.6	30,000	30,000	78	8,033	24	30,000	65	opt	20
Schramm	T64-HB	to 6.8	39,000	29,450	—	4,883	33	—	87	opt	25
Schramm	T-685	to 7.9	55,000	35,000	63	7,458	85	34,000	85	opt	25
Schramm	T985-HA	7.0-9.0	59,175	38,000	—	4,883	67	—	—	30	25

Table 10.4 (Continued)

WHEEL DRILL SPECIFICATIONS

						COMPRESSOR DATA				HYDRAULIC SYSTEM DATA		
MAKE	MODEL	PULLDOWN TYPE	ROTARY DRIVE TYPE	ROTARY DRIVE POWER	DUST CONTROL SYSTEM	MAKE	TYPE	AIR FLOW (cfm)	AIR PRESSURE (psi)	PUMP FLOW (gpm)	RELIEF VALVE PRESSURE (psi)	PUMP TYPE
Bucyrus-Erie	2450-R	cylinder	top	hyd	wet	Sullair	piston	640	125	—	—	—
Chicago Pneumatic	T-600	cylinder	top	hyd	opt	Chicago Pneumatic	screw	500	250	—	—	—
Chicago Pneumatic	T650-SS	cylinder	top	hyd	dry	Chicago Pneumatic	screw	750	125	—	—	—
Chicago Pneumatic	T-700	cylinder	top	hyd	rotoclone	Chicago Pneumatic	screw	825	125	—	—	—
Davey Rousselle	M8B	hyd motor	table	direct	opt	opt	piston	500	125	—	—	—
Davey Rousselle	M10B	cylinder	top	hyd	rotoclone	opt	opt	600	125	—	—	—
Driltech	D40K-II	cylinder	top	hyd	opt	Sullair	screw	750	350	—	—	piston
Driltech	D80K	cylinder	top	hyd	opt	2/Sullair	screw	1500	100	—	—	piston
Gardner-Denver	GD-35T	hyd motor	top	hyd	dry	Gardner-Denver	screw	770	100	—	—	—
Ingersoll-Rand	T4-BH	cylinder	top	hyd	opt	Ingersoll-Rand	screw	750	150	—	—	piston
Ingersoll-Rand	T-4	cylinder	top	hyd	opt	Ingersoll-Rand	screw	900	150	—	—	piston
Ingersoll-Rand	T-5	cylinder	top	hyd	opt	Ingersoll-Rand	screw	900	150	—	—	piston
Portadrill	TM6-B	cylinder	top	hyd	rotoclone	Sullair	screw	750	125	—	—	—
Reedrill	SK-35	cylinder	top	hyd	opt	Sullair	screw	750	100	—	—	piston
Robbins	RRT-35	cylinder	top	hyd	opt	Joy	screw	800	125	164	—	—
Robbins	RRT-50	cylinder	top	direct	opt	Joy	screw	800	50	—	—	—
Robbins	RRT-60	cylinder	top	direct	opt	Joy	screw	1000	50	—	—	—
Robbins	RRT-70	cylinder	top	hyd	opt	Joy	screw	1170	50	—	—	—
Salem Tool	VAD	—	—	—	—	—	—	900	60	—	—	—
Schramm	T-66H	cylinder	top	hyd	opt	Schramm	screw	600	300	89	2000	vane
Schramm	T64-HB	cylinder	top	hyd	opt	Schramm	screw	425	250	103	2000	vane
Schramm	T-685	cylinder	top	hyd	opt	Schramm	screw	850	350	125	3200	piston
Schramm	T985-HA	cylinder	top	hyd	wet	Schramm	screw	850	250	133	2000	vane

Table 10.4 (Continued)
WHEEL DRILL SPECIFICATIONS

		DRILL ENGINE DATA			TRUCK ENGINE	MAX	CONTINUOUS	FUEL		NO. OF	RANGE OF DRILL PIPE	DRILL OPERATORS
MAKE	MODEL	MAKE	MODEL	RATED HP	RATED HP	SPEED (mph)	GRADEABILITY (%)	TANK (gal)	COOLANT (gal)	LEVELING JACKS	RACK CAPACITIES	CAB (std/opt)
Bucyrus-Erie	2450-R	Cummins	NTC-350	—	drill	—	—	—	—	3	6	—
Chicago Pneumatic	T-600	Detroit	6V-92T	300	220	—	—	150	—	3	4-6	—
Chicago Pneumatic	T650-SS	Cat	D-3406T	325	210	—	—	170	—	3	6	opt
Chicago Pneumatic	T-700	Cat	D3406-TA	375	210	—	—	175	—	3	4	opt
Davey Rousselle	M8B	Detroit	6-71	280	—	—	—	50	—	3	—	n.a.
Davey Rousselle	M10B	GM	12V-71	375	280	—	—	—	—	3	5-6	opt
Driltech	D40K-II	Cat	3406-DITA	400	210	—	—	—	—	3	3	opt
Driltech	D80K	2-Cat	3406	650	325	20	67	—	—	3	3	std
Gardner-Denver	GD-35T	Cat	3406 DIT	325	210	48	—	100	—	3	5-6	std
Ingersoll-Rand	T4-BH	GM	8V-71N	304	220	—	—	—	—	3	5	std
Ingersoll-Rand	T-4	Detroit	12V-71N	360	220	—	—	—	—	3	5-7	std
Ingersoll-Rand	T-5	Detroit	12V-71N	360	228	—	30	—	—	3	3	std
Portadrill	TM6-B	Cat	3406T	325	210	—	—	140	—	3	3	std
Reedrill	SK-35	Cummins	NT-855-P	335	—	—	—	150	—	3	5	std
Robbins	RRT-35	Cummins	NT-855-C	335	225	46	35	—	—	3	4	opt
Robbins	RRT-50	Detroit	8V-71TA	335	drill	32	—	—	—	3	4-8	opt
Robbins	RRT-60	Detroit	8V-71TA	335	drill	32	—	—	—	3	4-8	opt
Robbins	RRT-70	Detroit	8V-71TA	350	drill	20	—	—	—	3	4	opt
Salem Tool	VAD	—	—	—	—	—	—	—	—	—	6	—
Schramm	T-66H	Detroit	8V-92T	400	216	—	—	—	—	3	2-4	std
Schramm	T64-HB	Detroit	8V-71N	304	—	—	—	—	—	3	5	opt
Schramm	T-685	Detroit	8V-92T	400	—	—	—	—	—	3	4	opt
Schramm	T985-HA	Detroit	8V-71N	304	—	—	—	—	—	3	6	opt

Table 10.4 (Continued)

WHEEL DRILL SPECIFICATIONS

MAKE	MODEL	LENGTH MAST UP (ft)	LENGTH RANGE W/MAST DOWN (ft)	HEIGHT RANGE W/MAST UP (ft)	HEIGHT MAST DOWN (ft)	WIDTH (ft)	GROUND CLEARANCE (in)	WHEELBASE (ft)	TREAD (ft)	CLEARANCE CIRCLE (ft)	TOTAL NO. OF AXLES	NO. OF DRIVE AXLES	NO. OF WHEELS	STANDARD TIRES
Bucyrus-Erie	2450-R	30.0	38.3	38.3	12.9	—	—	18.5	—	—	3	2	10	10.00-20
Chicago Pneumatic	T-600	28.5	33.5	35.3	12.5	8.0	—	16.5	—	—	3	2	10	—
Chicago Pneumatic	T650-SS	32.3	37.6	39.5	13.0	8.0	—	17.1	—	—	3	2	10	10.00-20
Chicago Pneumatic	T-700	33.4	35.9	37.8	13.0	8.0	13	18.2	—	—	4	2	—	—
Davey Rousselle	M8B	32.8	34.2-43.2	38.6-47.6	11.0	8.0	—	18.5	7.2	—	3	2	10	10.00-20
Davey Rousselle	M10B	31.4	35.0	37.0	13.3	8.0	—	18.5	—	—	3	2	10	—
Driltech	D40K-II	31.1	35.0	35.7	13.8	8.0	—	—	—	—	3	2	10	—
Driltech	D80K	41.4	46.8	46.4	17.2	15.3	—	—	—	—	4	2	—	14.00-20
Gardner-Denver	GD-35T	33.0	39.3	41.6	13.3	8.0	—	17.1	7.2	—	3	2	10	—
Ingersoll-Rand	T4-BH	28.4	35.0	36.0	12.5	8.0	—	16.8	—	—	3	2	10	—
Ingersoll-Rand	T-4	28.7	35.0	36.0	12.5	8.0	—	16.8	—	—	3	2	10	—
Ingersoll-Rand	T-5	36.6	45.0	48.0	13.5	10.0	—	18.5	—	—	3	opt	10	—
Portadrill	TM6-B	32.0	39.0	40.0	13.7	9.3	—	14.6	—	—	3	2	10	10.00-20
Reedrill	SK-35	34.0	35.0	36.7	13.3	8.0	—	17.1	—	—	3	2	10	—
Robbins	RRT-35	34.4	36.7	36.4	15.0	8.0	—	—	—	98	3	2	10	10.00-20
Robbins	RRT-50	37.5	38.0	33.5	13.3	10.0	—	—	—	82	3	2	10	12.00-20
Robbins	RRT-60	37.5	38.0-38.8	33.5-38.5	13.3	10.0	—	—	—	82	3	2	10	12.00-20
Robbins	RRT-70	38.0	38.6-45.0	38.6-45.0	16.0	11.2	—	—	—	134	3	2	10	14.00-20
Salem Tool	VAD	37.0	39.0	34.0	13.3	10.0	—	—	—	—	—	—	—	—
Schramm	T-66H	28.5	31.5	32.8	11.3	8.0	13	16.5	—	—	3	2	10	—
Schramm	T64-HB	27.1	31.2	32.8	11.1	8.0	13	16.5	—	—	3	2	10	—
Schramm	T-685	31.0	39.4	39.8	13.3	8.0	14	18.2	—	—	3	2	10	—
Schramm	T985-HA	30.6	39.8	40.4	12.8	8.0	16	16.2	—	—	4	2	12	—

Table 10.5
CRAWLER DRILL SPECIFICATIONS

MAKE	MODEL	RANGE OF APPROXIMATE HOLE DIAMETERS (in)	WEIGHT (lbs)	MAX PULLDOWN FORCE (lbs)	MAX DRILL FEED RATE (fpm)	MAX TORQUE (ft-lbs)	MAX ROTATION SPEED (rpm)	MAX PULLBACK FORCE (lbs)	MAX HOIST SPEED (fpm)	RANGE OF DRILL MAST ANGLE (degrees)	AVAILABLE PIPE LENGTHS (ft)
Bucyrus-Erie	40-R	6.8-9.0	96,000	50,000	5	2,100	82	30,000	100	0-30	—
Bucyrus-Erie	45-R	6.8-11.0	150,000	70,000	7	3,400	84	55,000	84	0-30	32,44,55
Bucyrus-Erie	55-R	9.0-12.3	205,000	100,000	16	15,000	95	56,000	74	0-25	50,60
Bucyrus-Erie	60-R	9.0-15.0	261,000	125,000	8	11,525	121	96,000	100	0-30	50,55,60,65
Bucyrus-Erie	61-R	12.3-17.5	276,000	130,000	8	23,050	121	96,000	100	0-30	50,55,60,65
Chicago Pneumatic	C-450	3.5-6.3	34,000	18,000	—	1,834	235	—	150	—	33
Chicago Pneumatic	C-700	6.0-9.0	72,000	42,000	110	6,333	130	26,000	198	0-30	25
Davey Rousselle	M5-CS	3.0-6.0	40,000	17,000	20	3,000	240	17,000	90	0-45	35
Driltech	D25K-II	to 6.0	54,000	25,000	20	4,283	106	—	106	—	25
Driltech	D40K	to 8.0	71,000	40,000	—	4,417	200	—	100	—	25
Driltech	D50K	to 9.0	73,500	50,000	—	5,008	180	—	78	—	25
Gardner-Denver	RDC-16B	4.0-5.1	33,000	18,000	115	2,800	148	—	115	0-15	22,38
Gardner-Denver	GD-25C	5.1-7.4	58,000	32,000	148	4,000	162	—	148	—	25,31
Gardner-Denver	GD-70	to 12.3	200,000	90,000	35	9,725	120	70,000	70	0-20	32,55
Gardner-Denver	GD-100	to 17.5	250,000	125,000	11	12,700	120	50,000	105	0-20	50,55,60,65
Gardner-Denver	GD-120	15.0-22.0	300,000	150,000	3	25,400	120	80,000	130	0-20	50,55,60,66
Ingersoll-Rand	DM20-SP	3.0-5.1	40,000	20,000	100	3,200	385	18,000	100	—	20-40
Ingersoll-Rand	DM25-SP	3.0-6.8	45,000	26,000	110	7,500	165	20,000	110	—	20-50
Ingersoll-Rand	DM35-SP	4.8-7.9	65,000	35,000	15	8,400	145	29,000	85	—	to 65
Ingersoll-Rand	DM45	5.0-8.0	70,000	45,000	50	5,750	200	—	80	0-30	25
Ingersoll-Rand	DM50	7.9-9.9	76,500	50,000	50	7,083	162	—	80	0-30	25
Ingersoll-Rand	DM-100	9.9-12.3	206,000	100,000	—	9,630	125	—	—	—	35

Loftis	Drill King II	6.3-10.6	140,000	65,000	10	—	175	28,000	—	—	30,35
Marion	M-2	9.0-12.3	165,000	80,000	94	6,100	110	60,000	94	0-15	50
Marion	M-3	9.0-12.3	170,000	90,000	94	12,214	110	70,000	94	0-15	50
Marion	M-4	12.3-15.0	200,000	105,000	75	9,500	110	85,000	150	0-30	55
Marion	M-5	15.0-17.0	225,000	120,000	65	18,000	110	100,000	130	0-30	55
Portadrill	CM6-B	6.8-7.9	77,800	45,000	15	4,800	147	—	105	—	25
Reedrill	SK-25	5.4-6.3	39,000	25,000	15	3,000	220	12,000	100	opt	25
Reedrill	SK-40	6.8-9.0	65,000	40,000	10	4,300	135	21,000	92	opt	25
Reedrill	SK-50	6.8-9.9	—	50,000	10	5,200	135	21,000	92	opt	25
Reedrill	SK-60	7.9-10.6	100,000	60,000	10	7,800	135	21,000	92	—	35
Robbins	RR10-S	6.8-9.0	116,000	65,000	60	6,750	145	—	90	—	20,25
Robbins	RR10-HD	6.8-10.6	134,000	70,000	60	6,750	159	28,000	90	—	25
Robbins	RR11-E	9.0-10.6	160,000	80,000	10	5,700	200	41,000	100	0-30	30
Robbins	RR12-E	9.0-12.3	260,500	100,000	60	24,000	120	41,000	155	0-30	32
Robbins	RR15-E	10.6-15.0	282,000	120,000	9	24,000	200	45,000	155	—	40
Schramm	C42H-B	3.0-5.6	29,000	14,700	—	3,733	113	—	130	—	20
Schramm	C64H-C	to 7.0	55,000	29,452	—	4,458	86	—	87	—	20
Schramm	C985-H	to 7.9	86,625	49,000	—	8,341	86	—	—	0-30	—
Schramm	C9120	to 9.0	85,000	49,000	—	8,341	86	—	—	0-30	—

Table 10.5 (Continued)
CRAWLER DRILL SPECIFICATIONS

						COMPRESSOR DATA				HYDRAULIC SYSTEM DATA		
MAKE	MODEL	PULLDOWN TYPE	ROTARY DRIVE TYPE	ROTARY DRIVE POWER	DUST CONTROL SYSTEM	MAKE	TYPE	AIR FLOW (cfm)	AIR PRESSURE (psi)	PUMP FLOW (gpm)	RELIEF VALVE PRESSURE (psi)	PUMP TYPE
Bucyrus-Erie	40-R	hyd motor	top	DC elec	dry, wet	Allis-Chalmers	vane	773	40	—	—	vane
Bucyrus-Erie	45-R	hyd motor	top	DC elec	dry, wet	Allis-Chalmers	vane	982	40	—	—	vane
Bucyrus-Erie	55-R	hyd motor	top	AC elec	dry, wet	Allis-Chalmers	vane	1310	40	—	—	—
Bucyrus-Erie	60-R	hyd motor	top	DC elec	dry, wet	Allis-Chalmers	vane	1310	40	—	—	—
Bucyrus-Erie	61-R	hyd motor	top	DC elec	dry, wet	2/Allis-Chalmers	vane	2620	40	—	—	—
Chicago Pneumatic	C-450	cylinder	top	hyd	dry	Chicago Pneumatic	screw	400	125	—	—	—
Chicago Pneumatic	C-700	cylinder	top	hyd	dry	Chicago Pneumatic	screw	825	125	—	—	—
Davey Rousselle	M5-CS	hyd motor	table	hyd	dry, wet	—	vane	250	125	—	—	—
Driltech	D25K-II	cylinder	top	hyd	all	Sullair	screw	750	250	—	—	piston
Driltech	D40K	cylinder	top	hyd	all	Sullair	screw	750	125	—	—	piston
Driltech	D50K	cylinder	top	hyd	dry, wet	Sullair	screw	900	100	—	—	piston
Gardner-Denver	RDC-16B	hyd motor	top	hyd	all	Gardner-Denver	screw	250	100	—	—	—
Gardner-Denver	GD-25C	hyd motor	top	hyd	dry, wet	Gardner-Denver	screw	650	100	—	—	—
Gardner-Denver	GD-70	hyd motor	top	hyd	dry, wet	Gardner-Denver	screw	1240	55	—	—	piston
Gardner-Denver	GD-100	elec motor	top	DC elec	dry, wet	Gardner-Denver	screw	1480	55	—	—	vane
Gardner-Denver	GD-120	cylinder	top	DC elec	all	Gardner-Denver	screw	1480	55	—	—	vane
Ingersoll-Rand	DM20-SP	hyd motor	table	mech.	all	Ingersoll-Rand	screw	250	125	—	—	—
Ingersoll-Rand	DM25-SP	hyd motor	table	mech.	all	Ingersoll-Rand	screw	450	110	—	—	—
Ingersoll-Rand	DM35-SP	hyd motor	table	mech.	dry, wet	Ingersoll-Rand	screw	600	110	—	—	—
Ingersoll-Rand	DM45	cylinder	top	hyd	dry, wet	Ingersoll-Rand	screw	750	150	—	—	piston
Ingersoll-Rand	DM50	cylinder	top	hyd	dry, wet	Ingersoll-Rand	screw	900	150	—	—	piston
Ingersoll-Rand	DM-100	cylinder	top	hyd	—	Ingersoll-Rand	screw	1420	110	—	—	—

Loftis	Drill King II	cylinder	top	hyd	all	LeRoi	screw	900	100	188	3600	piston
Marion	M-2	hyd motor	top	DC elec	dry,wet	—	—	1000	40	—	—	—
Marion	M-3	hyd motor	top	DC elec	dry,wet	—	—	1000	40	—	—	—
Marion	M-4	hyd motor	top	DC elec	dry,wet	—	—	1310	40	—	—	—
Marion	M-5	hyd motor	top	DC elec	dry,wet	2/	—	2620	40	—	—	—
Portadrill	CM6-B	cylinder	top	hyd	rotoclone	Sullair	screw	750	125	—	—	—
Reedrill	SK-25	cylinder	top	hyd	dry	—	screw	450	100	—	—	piston
Reedrill	SK-40	cylinder	top	hyd	dry	—	screw	900	100	—	—	piston
Reedrill	SK-50	cylinder	top	hyd	dry	—	screw	1200	100	—	—	piston
Reedrill	SK-60	cylinder	top	hyd	dry	—	screw	1250	100	—	—	—
Robbins	RR10-S	cylinder	—	mech	rotoclone	Joy	—	600	125	—	—	—
Robbins	RR10-HD	cylinder	—	hyd	all	Joy	screw	1000	50	—	—	—
Robbins	RR11-E	cylinder	—	DC elec	wet	Joy	screw	1170	50	204	3000	gear
Robbins	RR12-E	cylinder	—	DC elec	all	Joy	screw	1535	125	255	3500	gear
Robbins	RR15-E	cylinder	—	DC elec	all	2/Joy	screw	2340	125	265	—	gear
Schramm	C42H-B	cylinder	top	hyd	wet	Schramm	piston	250	250	72	2000	gear
Schramm	C64H-C	cylinder	top	hyd	wet	Schramm	piston	425	250	90	2000	vane
Schramm	C985-H	cylinder	top	hyd	wet	Schramm	piston	850	250	133	2000	vane
Schramm	C9120	cylinder	top	hyd	wet	Schramm	piston	1200	100	133	2000	vane

Table 10.5 (Continued)

CRAWLER DRILL SPECIFICATIONS

		ENGINE DATA										
MAKE	MODEL	TYPE (diesel or electric)	DIESEL MAKE or ELECTRIC TYPE	DIESEL MODEL or ELECTRIC POWER SUPPLY (volts)	RATED HP	MAX SPEED (mph)	CONTINUOUS GRADEABILITY (%)	FUEL TANK (gal)	COOLANT (gal)	NO. OF LEVELING JACKS	RANGE OF DRILL PIPE RACK CAPACITIES	DRILL OPERATORS CAB (std/opt)
Bucyrus-Erie	40-R	electric	static AC	2400/4160	—	0.9	—	n.a.	n.a.	4	4	std
Bucyrus-Erie	45-R	electric	static AC	2400/4160	—	0.8	—	n.a.	n.a.	4	2-3	std
Bucyrus-Erie	55-R	electric	static AC	2400/4160	—	0.8	—	n.a.	n.a.	4	1-2	std
Bucyrus-Erie	60-R	electric	static AC	2400/4160	—	0.7	—	n.a.	n.a.	4	1-3	std
Bucyrus-Erie	61-R	electric	static AC	2400/4160	—	0.7	—	n.a.	n.a.	4	1-3	std
Chicago Pneumatic	C-450	diesel	—	—	—	2.3	—	—	—	3	2	—
Chicago Pneumatic	C-700	diesel	Cat	3406-TA	375	2.0	50	200	28	3	4	std
Davey Rousselle	M5-CS	diesel	GM	4-71	—	3.5	—	60	—	3	—	—
Driltech	D25K-II	diesel	Cat	3306-DIT	325	1.8	40	150	—	3	3	opt
Driltech	D40K	diesel	Cat	3406	325	1.8	—	—	—	3	3	opt
Driltech	D50K	diesel	Cat	3406	325	1.8	—	—	—	3	3	std
Gardner-Denver	RDC-16B	diesel	Detroit	4-71	131	2.5	30	100	—	4	1-4	opt
Gardner-Denver	GD-25C	diesel	Cat	3306-T	250	1.9	35	200	—	4	1-6	std
Gardner-Denver	GD-70	electric	static AC	4160	500	0.8	30	n.a.	n.a.	4	2-6	std
Gardner-Denver	GD-100	electric	static DC	4160	—	0.6	30	n.a.	n.a.	4	1-4	std
Gardner-Denver	GD-120	electric	static DC	—	—	0.8	21	n.a.	n.a.	4	1-3	std
Ingersoll-Rand	DM20-SP	diesel	Detroit	4-71N	—	2.5	40	100	—	3	0	opt
Ingersoll-Rand	DM25-SP	diesel	Detroit	6-71N	—	2.5	50	100	—	3	0	opt
Ingersoll-Rand	DM35-SP	diesel	Detroit	8V-71N	—	2.0	50	175	—	3	0	opt
Ingersoll-Rand	DM45	diesel	Detroit	8V-71N	—	2.5	—	215	—	3	5	opt
Ingersoll-Rand	DM50	diesel	Detroit	12V-71N	—	2.5	—	215	—	3	3	opt
Ingersoll-Rand	DM-100	electric	—	—	600	—	—	n.a.	n.a.	3	2-4	std

Loftis	Drill King II	diesel	Cat	3306	195	2.4	—	105	12	4	1-3	std
Marion	M-2	electric	static DC	—	—	0.9	—	n.a.	n.a.	4	3	std
Marion	M-3	electric	static DC	—	—	0.9	—	n.a.	n.a.	4	3	std
Marion	M-4	electric	static DC	—	—	1.0	—	n.a.	n.a.	4	3	std
Marion	M-5	electric	static DC	—	—	1.0	—	n.a.	n.a.	4	3	std
Portadrill	CM6-B	diesel	Cat	3406-T	325	2.0	—	140	—	3	3	std
Reedrill	SK-25	diesel	GM	6V-71N	228	—	—	100	—	3	0-3	opt
Reedrill	SK-40	diesel	Cummins	NT-855-P335	335	2.0	—	150	—	3	0-5	opt
Reedrill	SK-50	diesel	Cummins	KT-1150	450	—	—	350	—	3	0-5	std
Reedrill	SK-60	diesel	—	—	430	—	—	350	—	3	0-3	std
Robbins	RR10-S	diesel	opt	opt	opt	3.5	—	—	—	3	5-8	opt
Robbins	RR10-HD	diesel	opt	opt	opt	3.5	—	—	—	4	4-8	opt
Robbins	RR11-E	electric	static AC	7200	—	1.1	—	n.a.	n.a.	4	4	std
Robbins	RR12-E	electric	static AC	—	—	0.8	—	n.a.	n.a.	4	4	std
Robbins	RR15-E	electric	static AC	7200	—	1.1	—	n.a.	n.a.	4	0-3	std
Schramm	C42H-B	diesel	Detroit	6V-71	—	1.5	—	—	—	3	6	std
Schramm	C64H-C	diesel	Detroit	8V-71N	—	1.0	—	—	—	3	5	std
Schramm	C985-H	diesel	Detroit	8V-71N	—	—	—	—	—	3	—	std
Schramm	C9120	diesel	Detroit	8V-71N	—	—	—	—	—	3	—	std

Table 10.5 (Continued)

CRAWLER DRILL SPECIFICATIONS

MAKE	MODEL	LENGTH MAST UP (ft)	LENGTH RANGE W/MAST DOWN (ft)	HEIGHT RANGE W/MAST UP (ft)	HEIGHT MAST DOWN (ft)	WIDTH (ft)	GROUND CLEARANCE (in)	CRAWLER DATA: OVERALL LENGTH (ft)	CRAWLER DATA: OVERALL WIDTH (ft)	CRAWLER DATA: STANDARD SHOE WIDTH (in)	GROUND PRESSURE (psi)
Bucyrus-Erie	40-R	29.7	44.5	47.7	17.0	14.1	—	15.2	—	36	—
Bucyrus-Erie	45-R	36.0	55.2-77.6	57.5-80.0	14.3	19.5	22	18.3	16.1	36	—
Bucyrus-Erie	55-R	38.3	77.1-87.0	80.7-90.7	27.3	24.0	21	21.3	17.3	36	—
Bucyrus-Erie	60-R	42.6	74.8-89.7	75.9-90.9	21.0	20.3	21	24.3	17.8	36	—
Bucyrus-Erie	61-R	42.6	74.8-89.6	75.9-90.9	21.0	20.3	21	24.3	17.8	36	—
Chicago Pneumatic	C-450	20.6	42.7	41.2	9.8	8.0	—	—	—	—	—
Chicago Pneumatic	C-700	34.1	38.0	37.9	14.3	12.0	18	14.5	—	30	8.0
Davey Rousselle	M5-CS	24.0	36.0	35.8	11.5	8.0	—	12.0	8.0	16	—
Driltech	D25K-II	26.2	37.0	36.5	13.6	11.0	—	12.0	10.8	24	9.0
Driltech	D40K	27.8	34.8	36.3	13.0	12.3	—	14.0	12.3	30	—
Driltech	D50K	27.8	34.8	36.3	14.7	12.3	—	14.0	12.3	30	—
Gardner-Denver	RDC-16B	19.2	33.7-44.0	35.9-45.0	11.8	7.9	—	—	7.9	15	10.8
Gardner-Denver	GD-25C	23.8	39.5-44.0	41.8-46.5	12.6	8.0	—	12.1	8.0	18	—
Gardner-Denver	GD-70	39.8	54.8-77.3	58.7-81.2	22.3	19.6	—	22.3	—	36	—
Gardner-Denver	GD-100	44.0	78.0-93.0	81.8-96.8	20.0	18.7	—	22.3	20.7	44	—
Gardner-Denver	GD-120	43.0	78.0-94.0	78.2-94.1	19.8	19.6	—	21.4	18.0	44	—
Ingersoll-Rand	DM20-SP	21.3	37.5-57.2	—	9.4	10.3	—	—	8.0	—	—
Ingersoll-Rand	DM25-SP	23.0	44.0-56.2	—	11.0	10.3	—	—	8.0	—	—
Ingersoll-Rand	DM35-SP	27.7	51.0-86.2	—	15.1	14.0	—	—	11.0	—	—
Ingersoll-Rand	DM45	25.8	35.2	36.0	12.5	12.5	—	—	—	—	—
Ingersoll-Rand	DM50	25.8	35.2	36.0	12.5	12.5	—	—	—	—	—
Ingersoll-Rand	DM-100	38.8	49.8	54.2	21.2	17.5	—	17.5	17.5	48	—

Loftis	Drill King II	33.0	48.0-53.0	52.0-57.0	17.0	11.8	—	16.5	11.3	30	13.6
Marion	M-2	38.5	76.3	75.8	21.7	19.3	—	21.5	17.8	30	12.7
Marion	M-3	38.5	76.3	75.8	21.7	19.3	—	21.5	17.8	30	13.1
Marion	M-4	44.7	83.7	83.3	21.8	20.5	—	23.5	18.8	42	9.9
Marion	M-5	44.7	83.7	83.3	21.8	20.5	—	23.5	18.8	42	11.1
Portadrill	CM6-B	32.0	40.0	42.0	13.6	11.0	—	15.0	10.7	28	—
Reedrill	SK-25	20.8	32.8	—	13.8	8.0	16	11.9	8.8	—	—
Reedrill	SK-40	25.3	34.7	—	13.3	9.8	15	14.5	10.6	—	—
Reedrill	SK-50	—	—	—	—	—	—	—	—	24	—
Reedrill	SK-60	27.3	44.8	47.4	16.1	13.3	—	—	—	30	—
Robbins	RR10-S	30.5	32.8-37.8	32.8-37.8	13.3	11.8	—	—	—	28	—
Robbins	RR10-HD	30.5	37.8	33.8	13.3	11.8	—	—	11.0	—	—
Robbins	RR11-E	38.0	45.0	45.0	16.0	16.0	—	19.0	16.0	42	—
Robbins	RR12-E	42.7	51.0	54.0	20.0	19.0	—	19.0	18.5	42	16.3
Robbins	RR15-E	41.3	57.7	60.0	16.5	19.3	—	19.3	19.3	42	12.6
Schramm	C42H-B	23.5	27.0	28.5	10.4	10.0	12	—	10.0	14	10.6
Schramm	C64H-C	25.7	33.8-43.8	35.0-45.0	10.9	10.0	11	12.0	10.0	24	—
Schramm	C985-H	29.0	38.5	40.0	14.0	13.6	—	14.0	10.0	24	—
Schramm	C9120	29.0	38.5	40.0	14.0	13.6	—	14.0	10.0	24	—

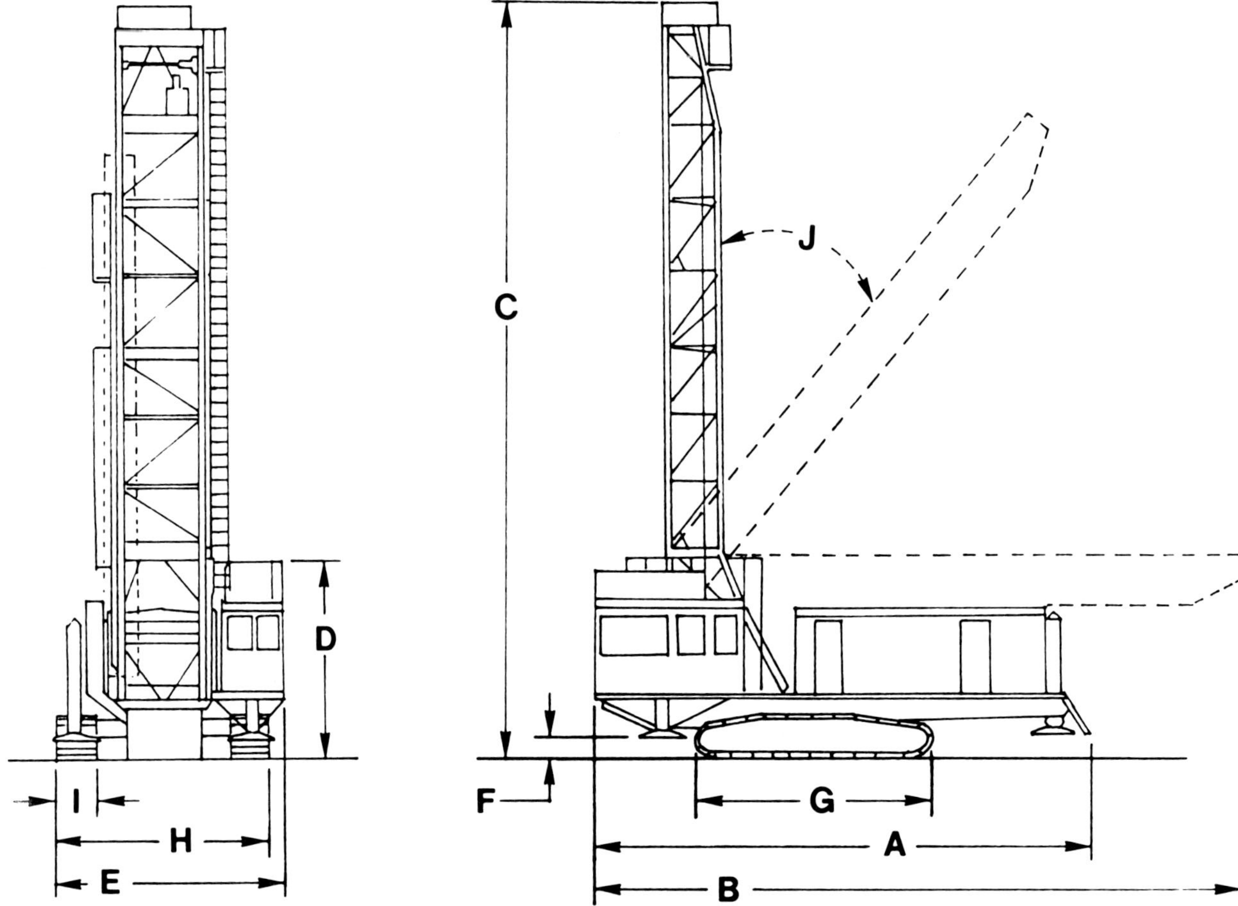

A	Overall Length/Mast Up (ft)	F	Ground Clearance (in)
B	Overall Length/Mast Down (ft)	G	Length of Crawler (ft)
C	Overall Height/Mast Up (ft)	H	Width Over Crawlers (ft)
D	Overall Height/Mast Down (ft)	I	Width of Shoe (in)
E	Overall Width (ft)	J	Drilling Angle (deg) (From Vertical)

Figure 10.11 Drill Dimensions

Appendix A

TIRES

Tire costs constitute one of the major components of operating expenses for wheeled mining equipment. Both machine performance and economics are affected by tire selection, application and maintenance.

The rubber-tired equipment covered in this manual are:

- drills
- dozers
- scrapers
- trucks
- front-end loaders

The tire applications for these machines are quite diverse, ranging from low speed earth moving to high speed material transport. Many of the truck mounted drills have basic road tires. These latter tires are adequately discussed in the general literature. The discussion which follows will concentrate on heavy-duty, off-the-road tires typically found in mining applications.

DESIGN

All tires are designed and built in the same general manner by all tire manufacturers. Principal features and nomenclature of tire construction are:

- Beads: Tire beads are bundles of strong wire running along the inner edge of the tire; the tire carcass is bonded to them and, in turn, the tire beads serve as an anchor to the rim.
- Cord Body: The cord body makes up the carcass of the tire. The term ply rating, as defined in the tire rating system, refers to tire strength and not the actual number of cord plies in the tire.
- Breaker, Tread Plies or Belts: Extra layers of cord, steel breakers or steel cable between the carcass and tread to give the tire more impact resistance
- Sidewalls: The protective rubber covering over the sides of the tire body
- Tread: The part of the tire which contacts the ground

- Inner Liner (tubeless tires): The sealer inside of the tire which serves to retain the compressed air

- Tubes and Flaps (tubed tire): The tube is a separate container for the air between the tire and rim. Flaps are used to protect the tube from being punctured by the tire or rim.

There are two basic tire constructions, the bias ply tire and the radial. (See Figure A.1) The basic difference has to do with the construction of the tire carcass. On bias tires the carcass is constructed of cords (normally made of nylon and rubber cushioned) which cross the tread centerline at an angle, hence bias. Alternating layers of cords are used, called plies. The radial tire has a carcass of only one cord layer (usually steel cable) and it is laid archwise (radially).

Some of the advantages and disadvantages of these two types of construction can be summarized as follows:

- Radial tires cost more than bias.

- Radial tires have higher load and speed capabilities.

- Radial tires have a longer life than bias tires under the same conditions.

- Radial tires have lower rolling resistance and thus better fuel economy.

- Bias tires are less costly to repair.

- Bias tires are less vulnerable to sidewall damage.

- Bias tires are more subject to adverse heat build-up.

Another, relatively new, radically different type of tire is the track type illustrated in Figure A.2. The major features of this type of construction are:

- Rim: Specially designed rim, one or two piece

- Cord Body: Radial steel cord body with a rubber cover

- Belt: A rubber-covered belt made up of steel cables wound around its circumference. It is notched to align itself on the cord body which it fits over. The outside of the belt is ribbed to provide a platform onto which steel shoes are bolted.

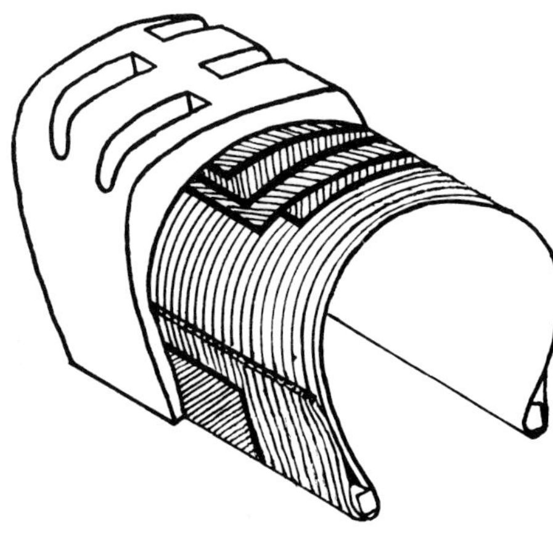

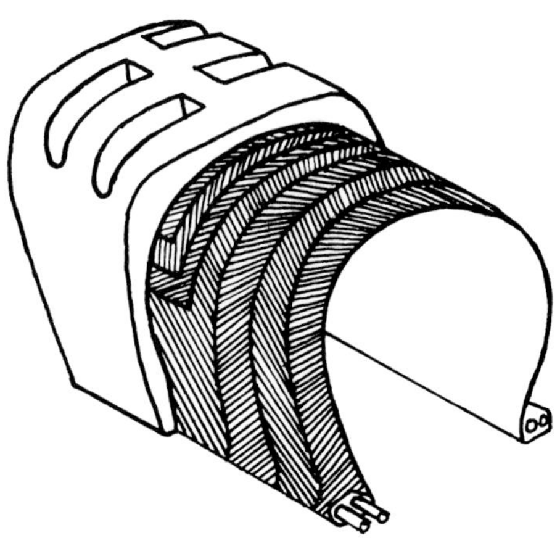

Figure A.1 Tire Construction

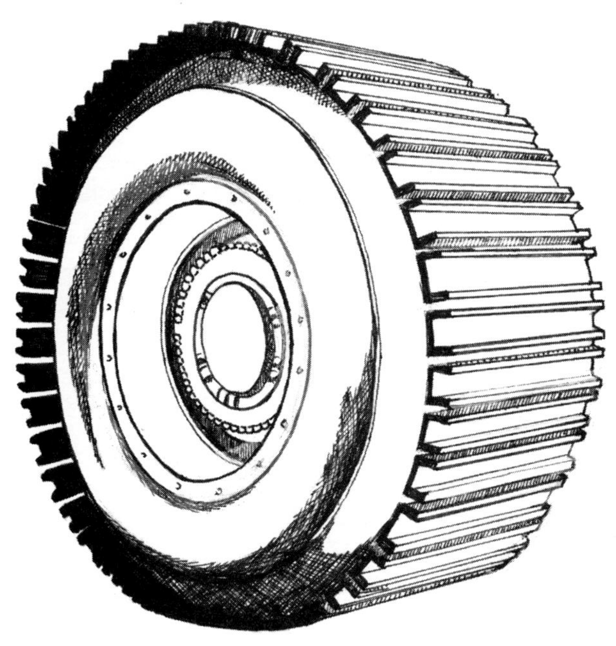

Figure A.2 Track Type Tire

- Steel Track or Shoes (grouser): Bolted to tire assembly forming the ground contact surface

This type of tire has not as yet achieved broad use. It is designed for special applications. Some advantages and disadvantages are:

- Resistant to abrasion and heat
- Provides good traction, stability, and flotation
- Has a very long life
- Each component, i.e., belt, tracks, body, can be individually replaced.
- They give a rougher ride and are noisier than rubber tires.
- Extra weight puts more load on drive train.
- Extra weight increases fuel consumption.
- Very expensive

These tires are most often applied on front-end loaders and wheel dozers. Currently there are only two manufacturers:

- Caterpillar Tractor Company — Beadless steel shoe tire
- Goodyear Tire & Rubber Company — Trak'r Tred, steel track tire

Tire tread designs vary with the manufacturer but there are several basic tread types as illustrated in Figure A.3. These different designs are often indicated by the tire code.

- Traction: Has large deep lugs extending the full width of the tire to maximize traction; lugs are normally about 45 degrees off direction of travel. This tire is used where traction is of prime importance.
- Rock Tire: Normally has thicker cross section than traction tires; often has reduced depth lug or no lug in center of tread pattern to improve wear in abrasive conditions. Lugs are on the outside edges of tire and often perpendicular to direction of travel. This tire is used to improve durability and cut resistance, but it has less traction.
- Smooth Tire: Similar to rock tire except that it has no lugs. This further reduces tractive ability and increases durability in rocky and abrasive situations.
- Rib Tire: Has tire grooves in the same direction as travel. It provides improved steering control.

CLASSIFICATION

Since applications, climates, machines, and operating practices are extremely diverse, there are a variety of tire constructions, tire designs and sizes, as well as tread designs available. Table A.1 gives the commonly accepted classification code system adopted by the tire industry for off-the-road tires.

Tires are designated first by the kind of service, then subdivided by type of tread.

Table A.2 explains the standard tire and rim size designation system.

Table A.1

CODE IDENTIFICATION FOR OFF-THE-ROAD TIRES

SERVICE	APPLICATION
E = Earthmoving	Scrapers, rear and bottom dump trucks. Normal maximum operating speed is 30 mph. Generally, scrapers use wide base tires and haulers use conventional width tires.
L = Loaders or Dozers	Loader and dozer tires have special design features and contain special materials that offer high resistance to abrasion and cutting. Have good pulling power. Normal maximum operating speed is 5 to 10 mph.
G = Graders	Primarily for use on road graders; these tires are also often used on small loaders.
ML = Mining and Logging Trucks	Built for a combination of on and off highway use, such as required in mining and logging; these tires combine heavy duty off-highway and highway service design features. Normal maximum operating speed is about 55 mph.
C = Compaction Machines	Designed for use on self-propelled or tractor drawn pneumatic rollers and compactors.

CODE	READ TYPE	APPLICATION
E-1	Rib	Steering tire
E-2	Traction	Small scrapers where traction is required over soft soil or sand. Regular tread skid depth.
E-3	Rock	Large scrapers and trucks units average TMPH requirements and rock cut exposure. Regular tread skid depth.
E-4	Rock Deep Tread	Applications where severe rock cutting and abrasion protection is needed: TMPH requirements are relatively low. Deep skid depth, approximately 50% deeper than E2 and E3 tread.
E-7	Flotation	For use when the soil requires low tire ground pressures. Shallow tread skid depth.
L-2	Traction	Where traction is required over soft soil or sand. Regular tread skid depth.
L-3	Rock	Resistance to moderate cutting and abrasion; light rock conditions. Regular skid depth
L-4	Rock Deep Tread	Resistance to shot rock or other severe cutting and abrasive applications. Deep tread skid depth about 50% deeper than L2 and L3.
L-5	Rock Extra Deep Tread	Maximum rock cut and abrasion resistance; quarry work and rock drilling. Extra deep tread skid depth approximately 67% deeper than L4.
L-3S	Smooth	
L-4S	Smooth Deep Tread	
L-5S	Smooth Extra Deep Tread	

Note: Combination tread designs are indicated by a combination of the appropriate code numbers. Example: L-5/L-5S

Table A.2

TIRE AND RIM ASSOCIATION SIZE DESIGNATION

Off-highway tires are designated by an alpha-numeric sequence.
Example: 33.25-49 43 PR L4 (L4 refers to the Service Code)

Section Width Reference	—	Nominal Rim Dia.	Carcass Strength Rating
Standard: approximate width (inches) includes 2 zeros following decimal point (ex. 24.00) Tires have a section height over section width ratio of about 0.96. Maximum width 28.53 inches	Hyphen indicates a bias ply construction Replaced by R for a radial construction	Appropriate nominal rim diameter measured through center of tire (excluding flanges)	Carcass strength indicated by ply rating (PR) or load range (LR) These are indices of tire strength and do not necessarily represent number of cord plies in tire
Wide Base: approximate section width (inches) includes digits following decimal point (ex. 33.25) Tires have a section height over a section width ratio of about 0.83. Maximum width 32.75 inches.		Rim is designated by: a nominal rim width, nominal rim diameter, flange height	
65 Series: (low profile) section width includes digits starting with 65 (ex. 65/40) The first number is the ratio of section height to section width which for this series is 0.65. The second number is the approximate section width (inches).			

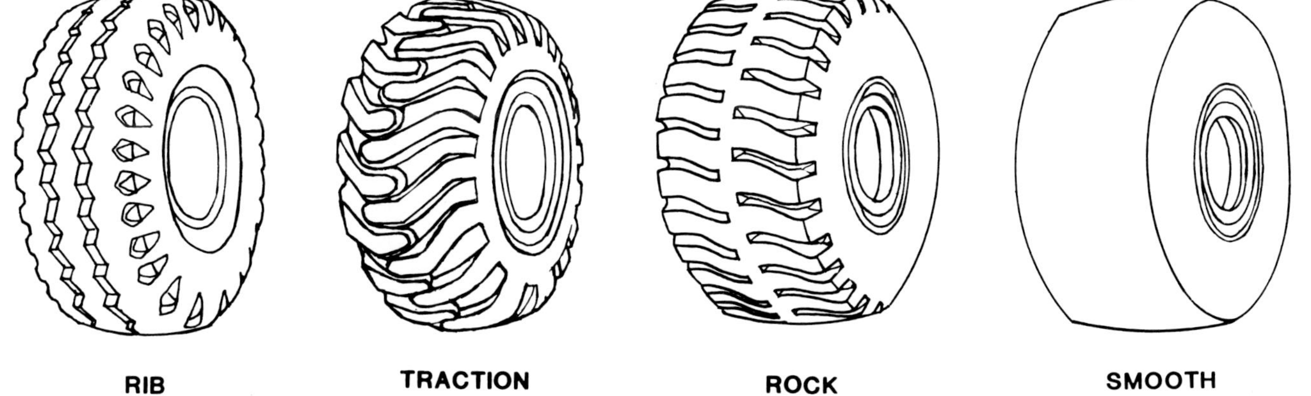

Figure A.3 Typical Tire Treads

GENERAL CHARACTERISTICS

- The load rating of a tire increases with inflation pressure and the number of plies in the tire.

- Narrow tires have less rolling resistance on hard surfaces than wide tires. This could provide some fuel economy.

- Wide tires are used when flotation is a problem and there is high rolling resistance.

- Rock treads will improve traction on smooth surfaces (rock, concrete, compacted earth).

- In general, the service life of tires is considered:
  — Unfavorable in the range of 1,000 to 2,500 operating hours
  — Average from 2,500 to 5,000 operating hours
  — Favorable from 5,000 to 10,000 operating hours

- Rims are an important part of the tire system. Care must be taken to properly match tire and rim, balance, and eliminate any slippage between the two. The rim must be maintained. Rims have a typical service life of about 20,000 hours.

- Chains should be considered where rocky floor conditions cause excessive tire wear. They have an average life of about 2,000 operating hours.

- Tire ballast should be considered on dozers and loaders where additional counter weighting is required (rear wheels) and/or additional weight can be used to increase rimpull or reduce tire spinning.

- Under certain servicing circumstances, off-the-road tires can be dangerous, so recommended safe practices should always be observed.

- Since off-the-road tires can be extremely large and heavy, special handling equipment is often required.

OPERATING CONSIDERATIONS

The most important factor governing tire life is heat. Heat causes deterioration in both the rubber and the tire construction. The primary factors affecting tire performance, wear and life are:

- Overloading causes:
  — High heat build-up
  — Tread and ply separations, flex breaks, radial cracks
  — Rapid irregular tread wear

- Overinflation causes:
  — Harder riding (increased equipment maintenance)
  — Reduced traction and skid resistance
  — Abnormal tire growth
  — Tread separation
  — Decreased cut resistance
  — Increased tendency toward bruises
  — Excessive strain on beads and rim
  — Excessive wear at center of tread due to rounding of tire tread

- Underinflation causes:
  — High heat build up
  — Ply and tread separation
  — Excessive tread wear on edges of tire

- Speed:
  — Increasing speed increases rate of heat generation but not the rate at which it dissipates.
  — Force of impact, i.e. rocks, holes, etc., increase with speed.

- Grades and Curves:
  — Steep grades and sharp curves cause tires to slip, increasing wear.
  — Curves increased the load on outside tires.
  — Grades increased the load on downhill tires.

- Length of haul:
  — Long hauls tend to increase heat build up unless load or speed is reduced.

- Water:
  - — Wet rubber cuts easier than dry rubber.
  - — Water on roadway may hide tire-damaging hazards.

- Floor conditions:
  - — Roads should be maintained (avoid chuck holes).
  - — Loading area should be kept clean.

- Tire storage:
  - — Use first in, first out system.
  - — Keep cool, dark, dry and clean, and preferably indoors.

Correct matching of tire to application, along with proper tire inflation, are the primary components of good maintenance and adequate service life. Tables giving recommended load and inflation limits are readily available; one example of these charts is given in Table A.3. There are other operating practices which can improve tire life:

- Insure clearance between tire and vehicle

- Remove wedged stones between tire and machine, and in tire lugs immediately

- Properly match dual tires

- Repair damaged tires immediately

- Keep tire free of petroleum products (deteriorates rubber)

- Adequately maintain vehicle (axles, wheel alignment, brakes)

Ballast

Machine weight can be increased by substituting heavier materials for the air in the tires. This added weight increases traction and allows the machine to more fully utilize its power. Typically, this approach is utilized only on vehicles in non-load carrying service such as dozers and loaders. Of course, adding weight to a hauling unit will effectively reduce its payload capability.

The most common method of tire ballasting is to replace 75% of the air with a solution of calcium chloride and water. This solution is preferable to pure water since it is heavier and has anti-freeze characteristics. Table A.4 illustrates some of the weight gains achieved through tire ballasting.

Foam Filling

Tires can be filled with cellular foam rubber rather than air. The advantages of this practice of ballasting are: less sensitivity to cuts and punctures, the possibility of longer tread wear, and the elimination of air pressure checks and maintenance. Alternatively, foam filling does increase tire heat build-up.

Chains

Tire chains can be added to tires to improve traction or to provide extra rock protection; the latter reason is the most common in mining applications. Chains are expensive, but can significantly increase tire life. However, they also tend to be noisier and cause a rougher ride. Figure A.4 illustrates a typical tire chain. A variety of chain patterns is available.

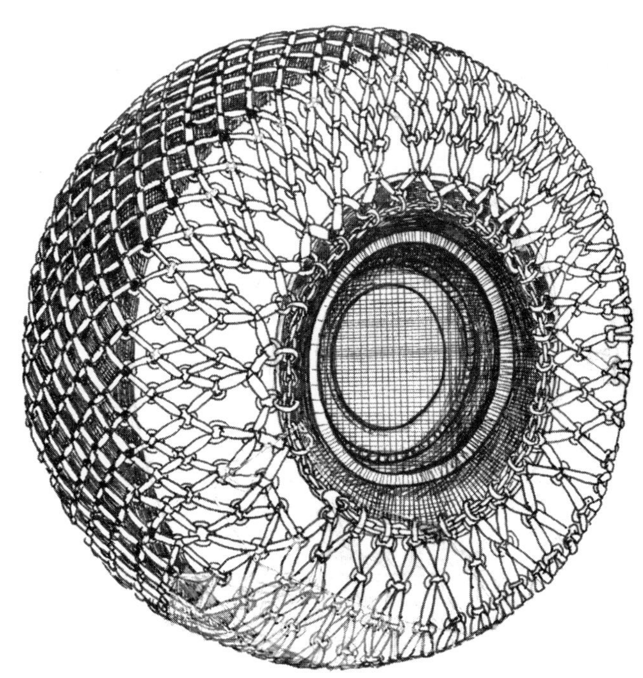

Figure A.4 Tire Chain

TABLE A.3

TYPICAL TIRE LOAD LIMITS
WIDE BASE TIRES*

TIRE SIZE DESIGNATION	TIRE LOAD LIMITS AT VARIOUS COLD INFLATION PRESSURES						
	25	30	35	40	45	50	55
15.5-25	**4740**(8)	5270	**5770**(12)				
17.5-25	**5710**(10)	6360	**6960**(12)	**7520**(14)	**8060**(16)		
20.5-25	7160	**7970**(12)	8720	**9430**(16)	10100	**10750**(20)	
23.5-25	**9250**(12)	10290	**11260**(16)	**12170**(20)	13040	**13870**(24)	
26.5-25	11780	**13110**(16)	**14350**(20)	15510	**16620**(24)	17680	**18690**(28)
26.5-29	12600	**14020**(18)	15340	**16580**(22)	17770	**18900**(26)	
29.5-25	**14790**(16)	16460	**18010**(22)	19480	**20870**(28)		
29.5-29	**15760**(16)	17530	**19180**(22)	20740	**22220**(28)	23630	**24990**(34)
29.5-35	17150	19080	**20880**(22)	22580	**24190**(28)	25720	**27200**(34)
33.25-29	18740	20850	**22820**(26)	24680	**26440**(32)		
33.25-35	**20320**(20)	22600	**24740**(26)	26750	**28650**(32)	30480	**32220**(38)
33.5-33	20920	23270	**25470**(26)	27540	**29500**(32)	31380	**33180**(38)
33.5-39	22530	25070	**27430**(26)	29660	**31770**(32)	33800	**35730**(38)
37.25-35	24760	27550	30150	**32600**(30)	**34930**(36)	37150	**39280**(42)
37.5-33	25510	**28380**(24)	31060	**33590**(30)	**35980**(36)	38270	**40460**(42)
37.5-39	27380	30460	**33330**(28)	36040	**38610**(36)	41070	**43420**(44)
37.5-51	30950	34440	37690	40750	**43660**(36)	46430	**49090**(44)

*Short haul distances under 1.5 miles, maximum speed—40 mph

Notes: Figures in parentheses denote ply ratings for which bold face loads and inflations are maximum
For radial ply tires, increase the inflation pressures shown above by 10 PSI with no increase load ratings

TABLE A.4
TYPICAL WEIGHT GAIN WITH TIRE BALLASTING (75% FULL)

Tire Size	Gallons Water 75%	*3½ Lbs. Calcium Chloride Per Gallon Water			†5 Lbs. Calcium Chloride Per Gallon Water		
		Gallons Water	Lbs. $CaCl_2$	Total Weight	Gallons Water	Lbs. $CaCl_2$	Total Weight
CONVENTIONAL SIZES							
18.00-33	114	98	343	1157	92	460	1231
18.00-49	152	130	455	1543	123	615	1641
21.00-25	131	112	394	1332	106	531	1416
21.00-29	143	123	430	1453	116	580	1547
21.00-35	162	139	485	1641	131	654	1744
21.00-49	206	176	618	2090	167	835	2220
24.00-25	171	146	512	1732	138	690	1841
24.00-29	186	159	577	1885	150	751	2004
24.00-35	211	181	630	2140	170	850	2260
24.00-43	283	199	697	2370	188	940	2510
24.00-49	264	226	790	2680	213	1069	2850
24.00-51	270	232	812	2738	220	1094	2926
27.00-33	284	244	854	2888	230	1151	3071
27.00-49	366	314	1099	3714	296	1480	3949
27.00-51	374	322	1130	3780	306	1515	4069
30.00-33	363	311	1089	3684	294	1470	3917
30.00-51	454	388	1360	4620	366	1830	4880
33.00-51	569	487	1710	5780	459	2295	6120
36.00-51	750	641	2240	7628	605	3025	8060
40.00-57	941	806	2821	9521	761	3806	10121
WIDE BASE AND GRADER SIZES							
15.5-25	46	40	139	470	37	187	500
17.5-25	60	51	180	609	48	243	647
20.5-25	90	77	269	910	72	362	967
23.5-25	118	101	354	1198	95	478	1274
26.5-25	159	136	477	1614	129	643	1716
26.5-29	174	149	521	1764	141	703	1875
29.5-25	207	177	618	2090	167	833	2223
29.5-29	224	192	673	2275	181	907	2419
29.5-35	251	215	753	2547	203	1015	2708
30-33	167	143	500	1690	134	670	1785
33.25-29	296	253	885	2985	238	1190	3170
33.25-35	319	274	958	3242	258	1292	2447
33.5-33	328	281	983	3326	265	1325	3536
33.5-39	363	311	1089	3684	294	1470	3917
37.25-35	355	304	1065	3610	287	1435	3820
37.5-33	423	362	1268	4290	342	1710	4562
37.5-39	466	399	1397	4729	377	1885	5020
37.5-51	552	473	1656	5603	447	2235	5958
43.5-43	291	249	873	2915	235	1175	3126
65 SERIES SIZES							
65/35-33	283	242	847	2880	228	1140	3042
38-39 (special)	389	333	1167	3935	314	1572	4178
65/40-39	418	357	1250	4253	337	1685	4494
65/45-45	586	502	1757	5920	474	2369	6304
65/50-51	741	635	2222	7518	599	2997	7969
67-51 (special)	1349	1156	4046	13641	1091	5456	14511

Note: *3½ lb. solution provides anti-freeze protection to −15° F, slush free
†5 lb. solutiong provides anti-freeze protection to −53°F, slush free

SELECTION CONSIDERATIONS

As indicated, proper tire selection is important for both performance and economics. Manufacturers' representatives are generally both willing and capable of assisting the buyer in tire selection. Following are the major factors to be considered; tire selection is often influenced by prior actual experience.

- The specific piece of equipment will establish the basic classification of the tire and size range.
- Economics will determine the tire type and design.
- Applications must be correlated with service ratings and special requirements, such as ballast, chains, and track type tires.
- Floor material and conditions (abrasiveness, traction and flotation) dictate tread design and section width.
- Gross vehicle weight and load distribution define the tire load rating.
- Travel speeds and loads are limited by accumulated heat damage.

A ton-mile-per-hour (TMPH) rating is used to predict tire temperature build up in tires used for hauling (scrapers and trucks). Tire manufacturers assign TMPH ratings to their tires and the application should not exceed this value. Their ratings are either given in the general literature or are available direct from the manufacturer. The user can calculate TMPH for his job based on the tire load formula: TMPH = average tire load × average speed. Average tire load is the average of loaded and empty loads. Average speed is the number of miles driven in a day divided by the number of operating hours per day. A similar calculation is used for front-end loader tires when the machine is applied for load-and-carry; the rating is called the work capability factor (WCF).

Having determined the basic tire specifications, the next step is to pick the manufacturer. Following in this Appendix is a listing of manufacturers and the various tire designs they offer.

Although tire manufacturers all comply with the basic industry standards, similar tires of different makes will differ somewhat in composition, internal design, etc. They do not all offer a full range of sizes in all classifications. Some, based on experience or specifically at the buyer's request, will make special compositions to meet unique site requirements. All will offer application advice.

Tire costs and delivery are generally established by negotiation and competitive quotations. Because of the high inflation rates associated with petroleum-based products and the very competitive nature of the market, no guideline prices are presented in this manual.

MANUFACTURERS

The manufacturers of the off-road tires available in the U.S. are shown on Table A.5. Distribution is by local dealers. Table A.6 lists these manufacturers (except Michelin) and their tire models. Michelin tires were omitted from the table as their unique tire service code is not easily comparable to the rest of the industry. Michelin does, however, manufacture and market tires suitable for surface mining applications.

Table A.5

TIRE MANUFACTURERS

B. F. Goodrich Company
Tire Division
500 S Main Street
Akron, Ohio 44318

The Firestone Tire & Rubber Company
1200-T Firestone Parkway
Akron, Ohio 44317

The General Tire & Rubber Company
One General Street
Akron, Ohio 44329

The Goodyear Tire & Rubber Company
1144-T E. Market Street
Akron, Ohio 44316

Michelin Tire Corporation
Lake Success, New York 11040

Toyo Tire Corporation
3136 E. Victoria Street
Compton, California 90221

Uniroyal, Inc.
1230 Avenue of the Americas
New York, New York 10020

United Tire & Rubber Co., Ltd.
275 Belfield Road
Rexdale, Ontario, Canada

TABLE A.6

TIRE IDENTIFICATION AND SELECTION CHART

TIRE CODE	CATERPILLAR	FIRESTONE	GENERAL	GOODRICH	GOODYEAR	TOYO	UNIROYAL	UNITED
E-1 Rib		Rib Excavator	Rock Rib LCM	Rib Service	Hard Rock Rib			
E-2 Traction	Traction Earth Mover	Ground Grip "G", Super Ground Grip	All Duty DTL	Super Traction	Earth Mover Sure Grip, Sure Grip Lug		Earth Mover Full Traction, Earth Mover Em-10 Full Traction	
E-3 Rock		Rock Grip Excavator, Super Rock Grip, Grant Steel Radial	ND LCM ND LCM Wide Base, CM 100 SL 100	Rock Service Rock Service II Expediter	Hard Rock Lug Hard Rock Lug 8, Super Hard Rock Lug 8, Super Hard Rock Lug Unisteel RL3	G-18 G-18 Wide Base, G-45	Con-Trak-Tor Con-Trak-Tor SRT, Con-Trak-Tor SRT Wide Base	
E-4 Rock Deep Tread		Super Rock Grip Deep Tread	ND Super LCM	Rock Service Hi Tread	Hard Rock Lug Extra Tread 8	G-18 ET G-36 ET	Super Con-Trak-Tor, Super Con-Trak-Tor SRT	Super Mining Construction Wide Profile, Mining Construction Super Extra Tread Super Mining Construction
E-7 Flotation		All Non-Skid Earthmover Sand Champion	Earth Mover Super Sand Flotation	Earth Mover, Special Sand Service Earthmover	Earthmover All-Weather Rib Sand			
L-2 Traction	Traction Loader Dozer,Grader	Super Groundgrip Conventional & Wide Base	LD, LD All Duty Wide Base	Power Traction, Super Traction Wide Base	Sure Grip Lug D&L	G-15	Design A, Design B	
L-3 Rock	Heavy Duty Rock Loader, Dozer,Grader	Super Rock Grip	LD ND LCM	Rock Service Universal	Super Hard Rock Lug D&L Super Hard Rock Lug 8 D&L Super Hard Rock Loader	G-18	Design C S.R.T. Design D S.R.T.	
L-4 Rock Deep Tread		Super Rock Grip Deep Tread Conventional & Wide Base	LD ND Super LCM, LD 150 Belted	Rock Service High Tread	Super Hard Rock Lug 8 D&L, Super Hard Rock Lug XT D&L, Nylosteel Xtra Tred D&L Belted	G-64 ET G-18 ET	Super Con-Trak-T, S.R.T.	Super Mining Construction Wide Profile
L-5 Rock Extra Deep Tread	Beadless	Super Deep Tread	LD 250 Belted	Rock Service Extra High Tread	Super Hard Rock Lug, Super Xtra Dual Tred, Nylosteel Super Xtra Tred Belted	G-65 EDT, G-25 EDT		Super Extra Mine Haulage, Super Extra Mine Haulage Half-Track
L-4S Smooth Deep Tread					Smooth Xtrared D&L			
L-5 Smooth Extra Deep Tread		Half Tread, Plain Tread	LD 250 Super Smooth		Smooth-5A, AMS-5/8B		Smooth Miner	

Appendix B

DIESEL ENGINES

Diesel engines are the power source for much of the equipment covered in this manual. They are used on:

- Dozers
- Scrapers
- Trucks
- Front-end loaders
- Hydraulic shovels and hoes
- Crawler draglines (some)
- Drills (most)

Gasoline engines are used on some of the smaller sizes of equipment, but they are not common in surface mining applications.

Both gasoline and diesel engines are classified as internal combustion. They burn a fuel to produce mechanical rotary power. A fuel air mixture is compressed by a piston in a cylinder; the mixture is ignited and expands, moving the piston. Multiple pistons, acting on crank arms, produce a smooth output torque. The gasoline engine compresses the air/fuel mixture to a ratio of about 7 or 8 to 1, and requires a spark to ignite the fuel. The diesel engine has a compression ratio of about 20 to 1; this high compression causes the air to heat to such a point that the injected fuel ignites on contact. Diesel engines are generally more fuel efficient, ruggedly built, and have a longer service life.

Diesel engines can be either four cycle or two cycle. The four cycle engine has intake, compression, power, and exhaust strokes; valves at the top of the cylinder are used for both intake and exhaust air. The two cycle engine has an intake/compression and a power/exhaust stroke. Intake air is forced by a rotary pump into the cylinder through ports on the side of the cylinder; exhaust is through valves at the top. Some of the claimed advantages of these two designs are listed in the discussion on selection.

GENERAL CHARACTERISTICS

Although the design fine points of diesel engines vary between manufacturers, within basic parameters, the general characteristics are quite similar. Following is an overview of these characteristics for currently available diesel engines.

- Range of 0.09 to 0.17 horsepower per pound of engine weight
- Displacements from 284 to 3067 cubic inches
- Horsepower from 140 to 1800: maximum horsepower at speeds from 1900 to 2500 rpm
- Torques range from 330 to 4600 ft lbs.; maximum torque at speeds from 1200 to 1800 rpm
- In line 6, V-6, V-8, V-12 and V-16 configurations

Photograph B.1 Turbocharged Detroit Diesel 12V-149T diesel engine

- Base engine dry weights from 1,500 to 10,700 lbs.
- Two or four cycle design
- Nominal torque rise of 15 to 30% for a high rpm engine and 8 to 12% for a low rpm engine
- Individual fuel injectors for each cylinder
- Large capacity gear pumps for bearing lubrication and piston cooling
- Internally drilled oil passages in block and head
- Gear driven accessory drives for water pump, air compressor, and power steering pump
- Spin-on filters
- Replaceable wet (watercooled) cylinder liners
- Aluminum alloy pistons
- Narrow V-type engines, 60 to 65 degrees
- Multiple camshafts for easy replacement
- 7 to 9 main bearings
- Single or dual turbocharging

All diesel engines have fuel injectors for metering and delivering fuel to the cylinders. Depending on the manufacturer, the cylinder may or may not have a precombusion chamber. While one European diesel engine manufacturer (Deutz) does make large air cooled diesels, all the U.S. manufactured engines, are water cooled.

Superchargers

A supercharger is a device which acts as a blower or pump, forcing air into the intake of the engine at higher than atmospheric pressure. More oxygen is compressed into the cylinder so more fuel can be burned and more power produced. The supercharger also helps to overcome friction on the air passages so power is not reduced at high speeds. Since the air is compressed, the engine power is less influenced by the atmospheric air density and engine performance does not fall off as rapidly at higher altitudes.

The turbocharger is probably the most common type of supercharger. It includes a set of turbine vanes driven by the high velocity exhaust gases. These, in turn, drive compressor vanes that force air into the engine. (See Figure B.1) A turbocharger may rotate at up to 100,000 rpm. Essentially, power for the device is supplied by the otherwise wasted energy of the exhaust. Other claimed advantages of turbocharging an engine are a lower combustion temperature resulting in longer engine life, reduced exhaust noise due to dampening by the turbine and increased combustion efficiency resulting in less undesirable exhaust gases and better fuel economy.

Aftercoolers (also called intercooler)

An aftercooler is an auxiliary heat exchanger connected to the engine's cooling system which is used to cool the air between the turbocharger and the intake port of the engine. Cooling the intake air allows it to be compressed to a greater density by the turbocharger. More oxygen is then available for combustion, and engine power is therefore increased.

Low Speed Diesels

Diesels operating at speeds below 100 to 1200 rpm are generally considered as low speed designs. They are substantially larger and heavier than the high speed diesels of equivalent power. They are also significantly more expensive, but provide high reliability with a very long life. Only a few low speed diesels are employed in excavator service for

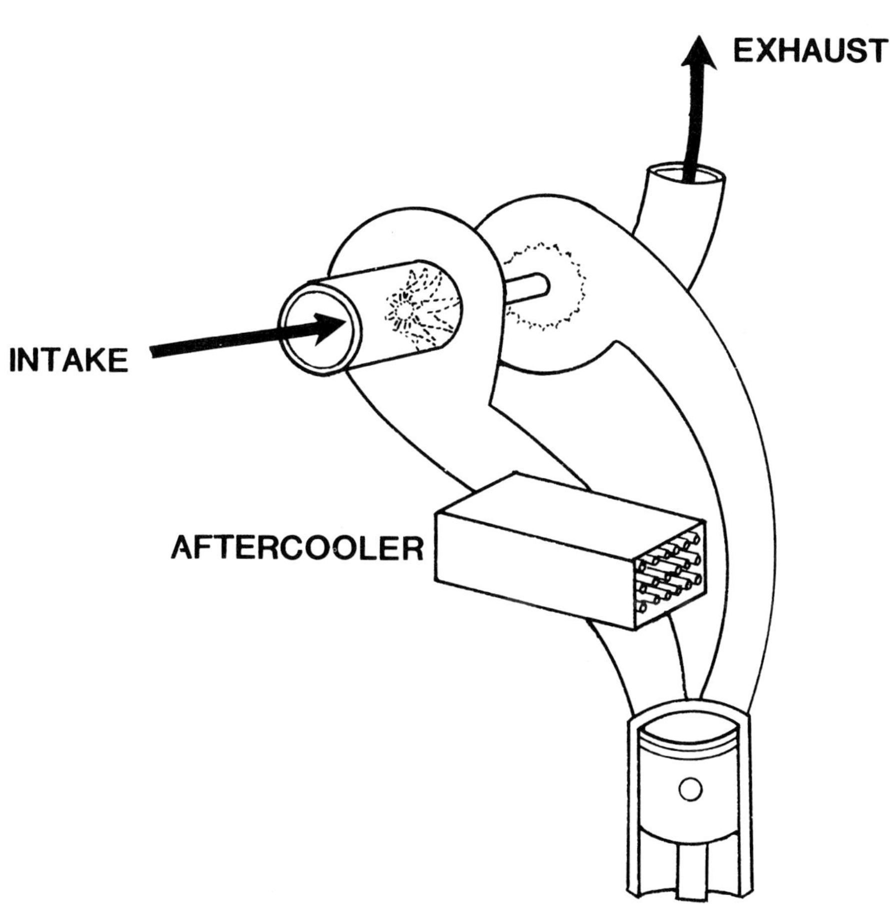

Figure B.1 Turbocharger

powering the larger trucks and draglines where higher speed engines of adequate size are not available. The General Motors Electromotive Division has supplied units for large trucks (above 200 tons). These two cycle engines were developed for locomotive applications and range from 1500 to 3300 horsepower.

Gas Turbine

The gas turbine engine has had limited mining applications; to date, it has only been tried experimentally on large trucks. Basically, the turbine engine has a compressor which supplies air; this air is mixed with fuel in a combustion chamber. It then passes through nozzles and strikes the turbine vanes, providing the driving torque. The big advantage to this type of engine is its very high power to weight ratio; gas turbines are significantly lighter than equivalent powered diesels. Although they are easy to maintain, they are also expensive, noisy, and have relatively high fuel consumption.

OPERATING CONSIDERATIONS

- Since the engine requires air for combustion, its performance is affected by air density and, hence, the altitude. Generally, four cycle engines will loose about 3% power for each 1,000 feet above sea level; two cycles will loose about 1% for each 1,000 feet. The use of a supercharger to force air into the engine will negate most effects of altitude. The specific effect altitude has on an engine can be obtained from its manufacturer.

- As air temperature increases the air density also decreases, so engine power is reduced. Again, use of superchargers and aftercoolers will counteract this effect.

- The operating characteristics of diesel engines are given in performance curves such as those in Graphs B.1 and B.2. These charts are readily

Photograph B.2 Turbocharged Detroit Diesel Allison 12V-92 diesel engine

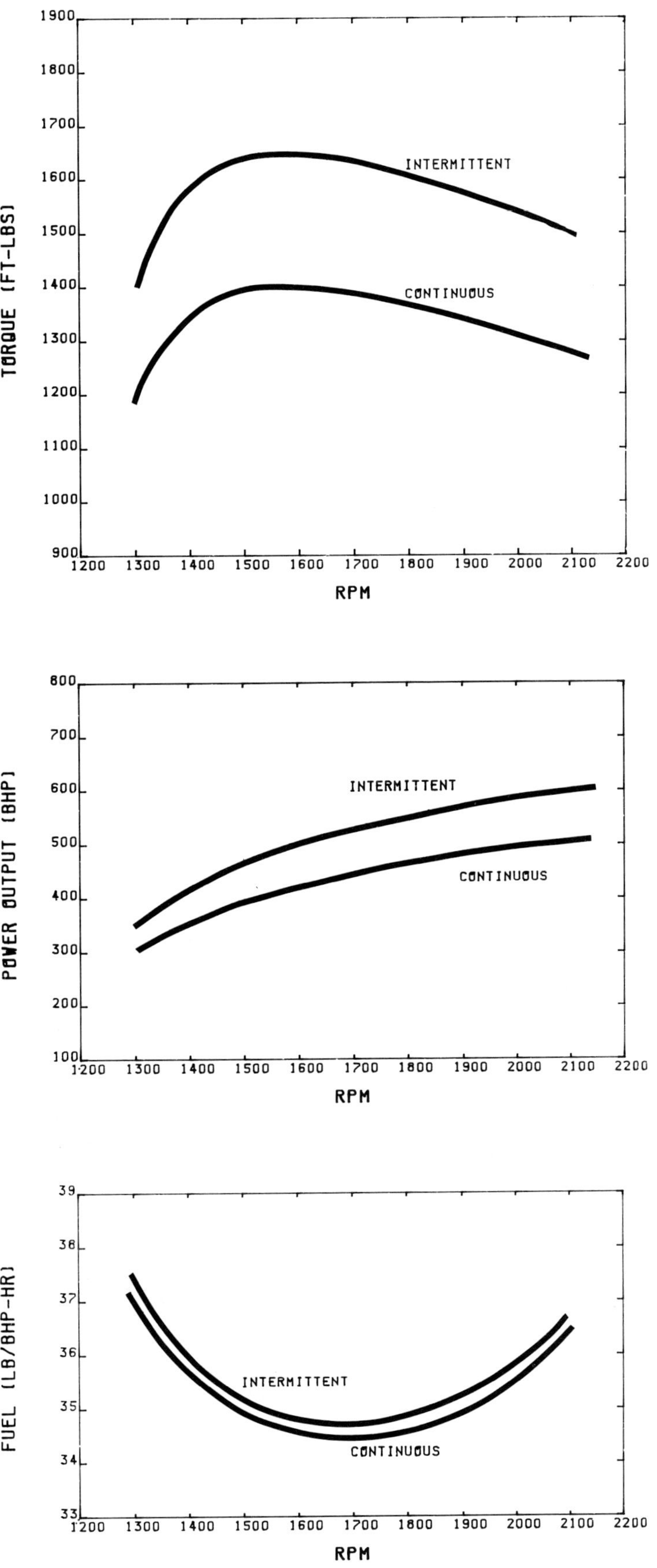

Graph B.1 Typical diesel engine power curves

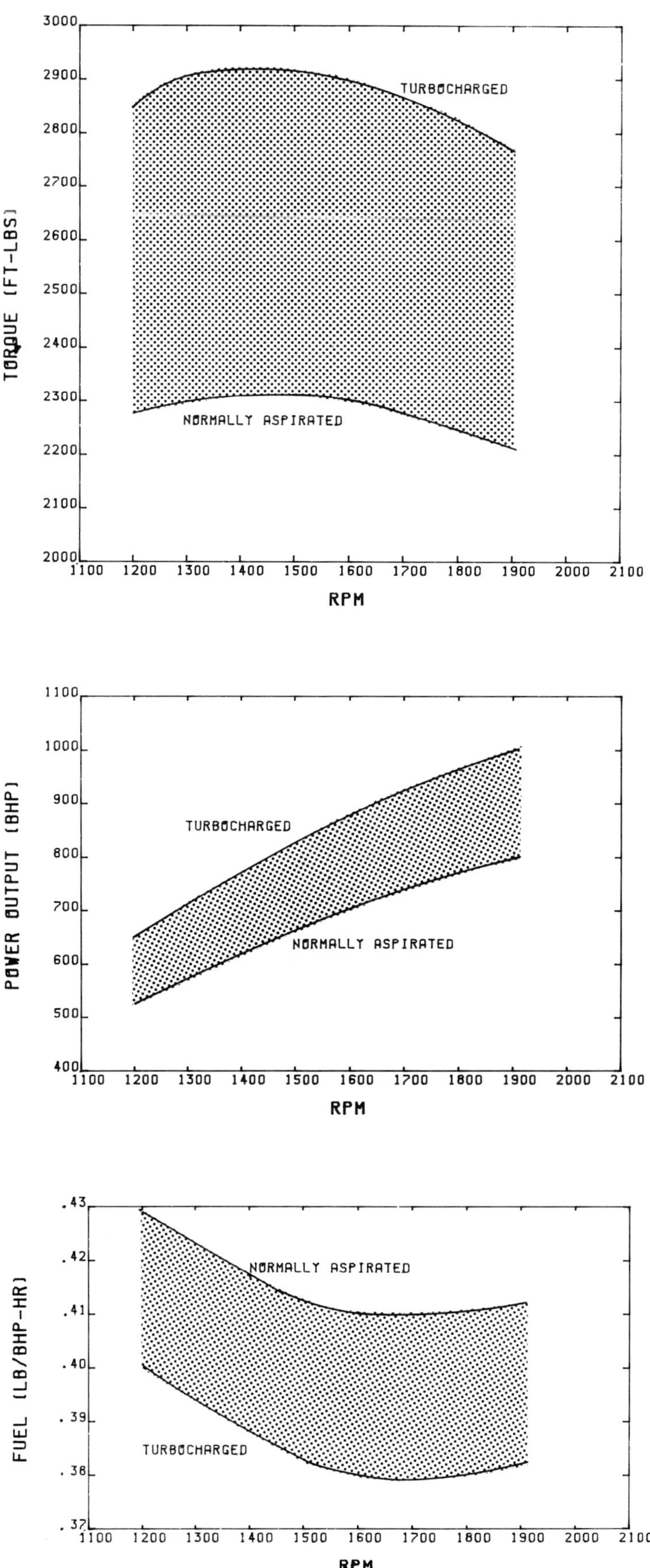

Graph B.2 Comparative power curves — normally aspirated or tubocharged diesel engine

available from engine manufacturers. The continuous duty curve is the suggested rating for uninterrupted service at full throttle for extended periods of time; the intermittent load curve represents higher performance levels that can be sustained only on an intermittent basis.

- Horsepower, torque, and fuel consumption curves represent performance at SAE standard J816b conditions of 500 ft. altitude, 85 degrees F. intake air temperature and 0.38 in. Hg water vapor pressure.

- Manufacturers' curves generally represent performance of the engine with water pump, lubricating oil pump, fuel system, muffler, and air cleaner; not included are alternator, fan, compressor, and optional equipment. Curves represent performance with No. 2 diesel or a fuel corresponding to ASTM D2.

- Fuel consumption is tied to horsepower and service duty cycle; it may increase more than 100% from light to severe duty.

- Diesel engines are typically overhauled after 6,000 to 9,000 hours of service. This range may be extended when the engines are used in conjunction with electric or hydraulic drives designed for constant speed operation.

SELECTION CONSIDERATIONS

As indicated in the discussion and specifications in the previous equipment chapters of this book, there may or may not be an engine choice available on any diesel powered machine. The optional engines offered by the manufacturer generally fall into two categories:

- Relatively equivalent engines from different manufacturers

- Engines with different performance characteristics for differing applications

Manufacturer's representatives will aid in the decision on matching the engine to the job. Primary selection criteria are:

- Two cycle engines have a lower initial cost than four cycle engines; they also have faster acceleration.

- Four cycle engines generally have lower fuel consumption and less engine heat build-up than two cycle engines. They can also use lower grade fuels.

- To prevent loss of power at higher altitudes, engines are equipped with turbochargers.

- Standardizing (same model and/or manufacturer) engines for equipment fleets can allow reduced inventories, maximize part interchangability, and improve maintenance familiarity and efficiency.

MANUFACTURERS

Table B.1 lists the manufacturers of the diesel engines typically found in mining equipment. It also matches equipment manufacturers to various engine options. Such information is readily available from the manufacturers, their representatives, or equipment literature.

Examination of diesel engine product lines shows that manufacturers produce families of engines. Taking a basic engine configuration, attachments are varied to yield a group of engines providing a range of power. Power ratings can be changed by the number of cylinders, size of injectors, RPM, turbocharging, and aftercooling. Within this family of engines, as many as 80% of the parts may be common.

SPECIFICATIONS

Table B.2 lists specifications of commonly used high speed (rpm) diesel engines.

Engines are listed alphabetically by manufacturer, and then in ascending order of displacement (cubic inches) by engine family. Engine families are comprised of those engine models which have identical, individual cylinder displacement and compatible configurations. For example, an in-line 6 can become a V-12 by adding a second bank of

Table B.1

DIESEL ENGINE MANUFACTURERS

Allis-Chalmers Corp. Engine Division P.O. Box 563 Harvey, IL 60426	171–325 FWHP Used by: Fiat-Allis
J.I. Case Company Construction Equip. Div. 700 State Street Racine, WI 53404	140–180 FWHP Used by: J.I. Case
Caterpillar Tractor Co. Engine Division 100 N.E. Adams Street Peoria, IL 61629	140–1200 FWHP Used by: Atlas, Demag, Caterpillar, Chicago-Pneumatic, Driltech, Goodbary, ISCO, Loftis, DJB, Melroe Multiwheel, Portadrill, Unit Rig, Gardner-Denver
Cummins Engine Co., Inc. 1000 Fifth Street Columbus, IN 47201	220–1600 FWHP Used by: Bucyrus-Erie, Big Bud, Clark Erie, Big Bud, Clark, Dart, Demag, Euclid, Fiat-Allis, Goodbary, Harnischfeger, IHC, Hitachi, ISCO, Kress, Kawasaki, Koehring, LeTourneau, Liebherr, MRS, O&K, Poclain, Reedrill, Rimpull, Robbins, Trojan, Unit Rig, Wabco
Deere and Company John Deere Road Moline, IL 61265	145- 260 FWHP Used: Deere and Co.
Deutz Corporation Engine Division 7585 Ponce de Leon Circle Atlanta, GA 30340	280–493 FWHP Used by: Poclain, O&K, Terex
Fiat-Allis Construction Machinery, Inc. 106 Wilmot Road P.O. Box F Deerfield, IL 60015	150–300 FWHP Used by: Fiat-Allis
General Motors Corp. Detroit Diesel Allison Div. 13400 West Outer Drive Detroit, MI 48228	140–1600 FWHP Used by: Amhoist, Bucyrus-Erie, Chicago-Pneumatic, Clark, Dart, Davey-Rousselle, Demag, Drott, Euclid, FMC, Gardner-Denver, General Motors, Goodbary, Hein-Werner, IHC, Ingersoll-Rand, Insley, Koehring, Kress, LeTourneau, Mechanical Excavator, Melroe Multi-Wheel, Northwest, Reedrill, Rimpull, Robbins, Schramm, Terex, Trojan, Unit Rig, Wabco, Warner & Swasey
International Harvester Co. Engine Div. Components Group 401 N. Michigan Avenue Chicago, IL 60611	140–326 FWHP Used by: International Harvester

cylinders. Nominal torque rise indicates the difference between torque at idle and maximum torque.

Most engines have various alternators available. The ratings, as shown, are considered the standard alternators. Many engines are available with more than one starting system, either air or electric. Electric systems can be single or dual units.

Fuel consumption is based on published manufacturer data and graphs. Fuel consumption is taken at 80% of maximum rated RPM. A range is shown indicating different levels of tune and the corresponding effect on fuel consumption. In some cases, the gallons/hour numbers have been converted from lbs/bhp-hour, with #2 diesel fuel assumed to weigh 7.08 lbs/U.S. gallons.

In general, all the engines shown are available with turbocharging. In addition, some form of precombustion air cooling is also available. Aftercooling also known as intercooling, is the typical method of achieving cooler, denser air for combustion.

The manufacturers and models chosen for these specifications represent the diesel engines most often used with mining sized equipment in the United States. These engines fall into the category of high speed engines. Low speed engines (GM Electromotive), and gas turbine engines are not included. All specifications, capacities, capabilities and dimensions are based on published manufacturers' data. Although the information is believed to be current and the interpretation to be consistent and correct, it is possible that omissions and inaccuracies have occurred and that out-of-date information may have been included. These specification charts are not meant to be used either as an indepth analysis, or as a head-to-head comparison of available engines, but rather as a general overview of the equipment, available sizes, and approximate operating data. Specific questions pertaining to performance, or purchase should be directed to the manufacturer or authorized distributor/dealer in the user's specific area.

TABLE B.2

DIESEL ENGINE SPECIFICATIONS

MAKE	MODEL	DISPLACEMENT (cu.in.)	CYLINDERS: NO. AND CONFIGURATION	BORE AND STROKE (in.)	NO. OF CYCLES	RANGE OF MAXIIMUM INTERMITTENT HP AT RATED RPM	RANGE OF MAXIMUM INTERMITTENT TORQUE RATED RPM	NOMINAL TORQUE RISE (%)
Caterpillar	3306	638	inline 6	4.75 × 6.00	four	150-270 at 2200	425@1200-820@1300	16-40
	3406	893	inline 6	5.40 × 6.50	four	280-380 at 2100	1000-1285 at 1200	20-45
	3408	1,099	V-8	5.40 × 6.00	four	400-475 at 2100	1225-1460 at 1200	20-30
	3412	1,649	V-12	5.40 × 6.00	four	575-750 at 2100	1815@1200-2150@1500	14-26
	D-346	1,191	V-8	5.40 × 6.50	four	565-600 at 2000	1640@1500-1745@1425	-
	D-348	1,786	V-12	5.40 × 6.50	four	850@1900-990@2000	2570-2727 at 1400	11-15
	3512	3,158	V-12	6.70 × 7.50	four	1200 at 1800	-	-
Cummins	N-855-C	855	inline 6	5.50 × 6.00	four	220-400 at 2100	644-1150 at 1500	-
	VT-1710-C	1,710	V-12	5.50 × 6.00	four	635-800 at 2100	1747-2200 at 1500	-
	KT-1150-C	1,150	inline 6	6.25 × 6.25	four	450-600 at 2100	1350@1500-1650@1600	-
	KT-2300-C	2,300	V-12	6.25 × 6.25	four	900-1200 at 2100	2475-3300 at 1500	-
	KTA-3067-C	3,067	V-16	6.25 × 6.25	four	1350-1600 at 2100	4050-4400 at 1500	-
Deere & Co.	6619-A	619	inline 6	5.12 × 5.00	four	275	810 at 1400	-
	-	955	V-8	5.50 × 5.00	four	290	858 at 1300	-
General Motors/Detroit Diesel	6V-53	318	V-6	3.88 × 4.50	two	197@2800-233@2500	440 at 1800	-
	4-71	284	inline 4	4.25 × 5.00	two	152-190 at 2100	-	-
	6-71	426	inline 6	4.25 × 5.00	two	210-275 at 2100	577-801 at 1200	-
	8V-71	568	V-8	4.25 × 5.00	two	263-370 at 2100	734-1064 at 1200	29
	12V-71	852	V-12	4.25 × 5.00	two	394-553 at 2100	1100@1200-1200@1600	-
	16V-71	1,136	V-16	4.25 × 5.00	two	720-880 at 2100	-	-
	6V-92	552	V-6	4.84 × 5.00	two	245-330 at 2100	676@1200-890@1400	-
	8V-92	736	V-8	4.84 × 5.00	two	320-430 at 2100	902@1200-1223@1400	-
	12V-92	1,104	V-12	4.84 × 5.00	two	561-665 at 2100	-	-
	16V-92	1,472	V-16	4.84 × 5.00	two	720-860 at 2100	1966-2408 at 1400	-
	12V-149	1,788	V-12	5.75 × 5.75	two	800-1350 at 1900	2310@1500-3445@1600	-
	16V-149	2,384	V-16	5.75 × 5.75	two	1060-1800 at 1900	3080@1500-4595@1600	-
International Harvestor Co.	DT-466	466	inline 6	4.30 × 5.35	four	138@2600-186@2500	334-479 at 1600	-
	DVT-800	800	V-8	5.30 × 4.50	four	300 at 2500	725 at 1825	-
	DTI-817C	817	inline 6	5.38 × 6.00	four	420 at 2200	1208 at 1500	-

TABLE B.2 (Continued)

DIESEL ENGINE SPECIFICATIONS

MAKE	MODEL	ELECTRICAL SYSTEM (volts)	ALTERNATOR (amps)	STARTING SYSTEM TYPE	APPROXIMATE RANGE OF FUEL CONSUMPTION AT 80% LOAD (gal/hr)	APPROXIMATE DRY WEIGHT (lbs)	TURBOCHARGING	AFTERCOOLING
Caterpillar	3306	24	35	-	6-11	1,925	yes	yes
	3406	24	-	-	12-15	2,870	yes	yes
	3408	24	-	-	15-18	3,180	yes	yes
	3412	24	-	-	21-29	4,420	yes	yes
	D-346	-	-	-	27-29	5,700	yes	yes
	D-348	-	-	-	35-41	7,000	yes	yes
	3512	-	-	-	50	11,640	yes	yes
Cummins	N-855-C	24	28	24 volt	10-16	2,590	yes	yes
	VT-1710-C	24	130	24 volt	24-30	5,600	yes	yes
	KT-1150-C	24	90	24 volt	18-22	3,600	yes	yes
	KT-2300-C	24	75	24 volt	34-44	7,950	yes	yes
	KTA-3067-C	24	75	24 volt	55-61	10,700	yes	yes
Deere & Co.	6619-A	24	-	-	-	2,400	yes	yes
	-	24	-	-	-	-	yes	no
General Motors/Detroit Diesel	6V-53	12	-	12 volt	-	1,485	yes	-
	4-71	12	-	12 volt	-	1,780	yes	-
	6-71	12	65	12 volt	8-11	2,190	yes	-
	8V-71	12	65	12 volt	12-15	2,345	yes	yes
	12V-71	-	62	24 volt	15-20	3,300	yes	-
	16V-71	24	-	24 volt	-	4,600	yes	yes
	6V-92	12	75	12 volt	10-14	1,960	yes	yes
	8V-92	24	45	24 volt	14-17	2,345	yes	-
	12V-92	-	-	-	-	3,910	yes	-
	16V-92	24	50	24 volt	29-35	4,600	yes	yes
	12V-149	24	65	24 volt	33-48	8,340	yes	yes
	16V-149	24	65	24 volt	44-61	10,630	yes	yes
International Harvestor Co.	DT-466	12	37	12 volt	-	1,445	yes	no
	DVT-800	24	-	-	-	-	yes	no
	DTI-817C	24	-	-	-	-	yes	yes

Appendix C

SELECTED REFERENCES

Adams, Stanley P. *The Caterpillar 988B Wheel Loader*, Society of Automotive Engineers, #760648, Warrendale, PA, (September, 1976).

Allen, Larry. "Guidelines for Fuel Economy", *Heavy Duty Equipment Management and Maintenance*, (July, 1980), pp. 22-26.

Ajwani, Prem L. *21 Cubic Yard 580 Pay Loader*, Society of Automotive Engineers, #750817, Warrendale, PA, (September, 1975).

Anderson, W.E. and T.M. Brady. "Effects of Maintenance Practices on Wire Rope Life in Dragline Applications", *Mining Engineering*, (November, 1979), pp. 1610-1614.

Application of High Volume Earthmoving Methods to the Reclamation of Area Mined Spoil Banks, The Pittsburgh and Midway Coal Mining Company, U.S. Bureau of Mines, Contract No. H0252012, (February, 1978).

Athans, Arthur N. "Troubleshooting Your Hydraulic System", *Compressed Air*, (March, 1979), pp. 19-22.

Bandy, L.L. *Design and Development of a 425 Flywheel Horsepower Crawler Tractor*, Society of Automotive Engineers, #770502, Warrendale, PA, (April, 1977).

Blast Hole Drilling Economics: Technical Report, Dresser Industries, Inc., Dallas, TX, (March, 1975).

Blast Hole Bit Handbook, Hughes Tool Co., Houston, TX, (August, 1975).

Boersma, Richard F. *Design of a Self-Loading Scraper with Stationary Bowl Floor and Liftable Elevator*, #790537, Warrendale, PA, (April, 1979).

Borquez, Guillermo V. "Estimating Drilling and Blasting Costs — An Analysis and Prediction Model", *Engineering and Mining Journal*, (January, 1981), pp. 83-89.

Brown, D.J.B. and R.J. Heather. *Development of Off-Highway Articulated Dump Trucks*, Society of Automotive Engineers, #790538, Warrendale, PA, (April, 1979).

Bucket Wheel Excavator Study, Skelly and Loy, U.S. Department of Energy, Contract No. ET-77-C01:8914, (1979).

Built In Test Equipment (BITE) To Improve Front End Loader Productivity, McDonnel Douglas Electronics Co., U.S. Department of Energy, Contract No. DE-AC01-80ET14052, (May, 1981).

"Bulldozer Blade Oscillation", *Automotive Engineering*, (November, 1978), pp. 74-80.

Burton, R.T. and P.R. Ukrainetz. *Frozen Soil Cutting Using Vibratory Blades*, Society of Automotive Engineers, #770546, Warrendale, PA, (April, 1977).

"Can a Hydraulic Excavator Increase Your Mine Productivity?", *Coal Mining and Processing*, (March, 1979), pp. 62-64.

Care and Service of Off-the-Highway Tires, Rubber Manufacturers Association, Washington DC, (1977)

Carr, Darrell B. *Dart's 1000 HP Mechanical Drive Off- Highway Trucks*, Society of Automotive Engineers, #770549, Warrendale, PA, (April, 1977).

Carter, Bernie. "Oil Testing Doubles Engine Life and Cuts Maintenance Costs", *Engineering and Mining Journal*, (February, 1981), pp. 111-113.

Caterpillar Performance Handbook, (Edition II), Caterpillar Tractor Co., Peoria, IL, (October, 1980).

"Central Lube Systems Help Sand Producer Halve Repair and Maintenance Costs", *Rock Products*, (November, 1977), pp. 68-70.

Chironis, Nicholas P. "Aluminum Boom Ups Dragline Output", *Coal Age*, (May, 1979), pp. 118-123.

Chironis, Nicholas P. "Draglines: King of the Strippers, Equipment Guide: Draglines", *Coal Age*, (January, 1980), pp. 126-136.

Chironis, Nicholas P. "Electronic Aids Boost Dozer Output, Equipment Guide: Dozers", *Coal Age*, (June, 1981), pp. 110-120.

Chironis, Nicholas P. "Front-End Loaders: Versatile and Productive Machines, Equipment Guide: Front-End Loaders", *Coal Age*, (April, 1980), pp. 127-147.

Chironis, Nicholas P. "Haulage Trucks Still Supreme, Equipment Guide: Off-Highway Trucks", *Coal Age*, (November 1980), pp. 94-109.

Chironis, Nicholas P. "Oscillating Blade Boosts Output", *Coal Age*, (September, 1980), pp. 80-82.

Chironis, Nicholas P. "Scraper Use Picks Up Steam, Equipment Guide: Scrapers", *Coal Age*, (March, 1981), pp. 96-105.

Chironis, Nicholas P. "Sensors Boost Output of Dozers", *Coal Age*, (August, 1978), pp. 60-64.

Chironis, Nicholas P. "Wider Choice of Large Drills Offered, Equipment Guide: Blast Hole Drills", *Coal Age*, (January, 1981), pp. 95-108.

Chitwood, Byron and N.E. Norman. "Blast Hole Drilling Economics: A Look at the Costs Behind the Costs", *Engineering and Mining Journal*, (June, 1977), pp. 168-170.

"Cimarron Puts World's Largest Tractor-Dozer to Work in Kentucky", *Coal Mining and Processing*, (April, 1976), pp. 138-140.

Cochran, Thomas E. *The Caterpillar 980C Wheel Loader*, Society of Automotive Engineers, #790531, Warrendale, Pa., (April, 1979).

Colburn, John W. Jr. *Bulldozing Improvements for Mine Spoil* #780678, Warrendale, PA, (August, 1978).

Constraints Limiting the Availability of Front-End Loaders, Skelly and Loy, U.S. Department of Energy, Contract No. ET-77-Col-8914, (June, 1979).

Crawler Tractor/Loader Safety Manual, Construction Industry Manufacturers Association, Milwaukee, WI, (1973).

Cvengros, David V. and Richard J. Olsen. *Work Capability Factors for Dozer and Loader Tires Operating in Load and Carry Service*, Society of Automotive Engineers, #750574, Warrendale, PA, (April, 1975).

Damon, Alber, Howard W. Stoudt and Ross A. McFarland. *The Human Body in Equipment Design*. Harvard University Press, Cambridge, MA, (1971).

D'Avello, F.S. "Practical Use of Foam in Tires", *Mining Congress Journal*, (July, 1976).

Day, David A. *Construction Equipment Guide*, John Wiley & Sons, New York, NY, (1973).

"Diesel Electronic Fuel Injection Investigated", *Automotive Engineering*, (February, 1981), pp. 62-67.

Dinkelman, G.G. *Controlled Frequency — The Wave of the Future*, Open Pit Mining Association, (June, 1978).

Doidge, William and Bernie Young . "Guidelines for Spec'ing a Coal Hauling Truck", *Heavy Duty Equipment Management and Maintenance*, (November, 1980), pp. 25-29.

Drilling and Reaming; Handbook and Reference Guide for Mining, Quarrying and Construction, Reed Tool Co., Houston, TX, (March, 1976).

Durst, Walter E. *Hydraulic Bucket Wheel Excavators — Design and Performance*, Society of Automotive Engineers, #780489, Warrendale, PA, (April, 1978).

"Earthmoving Machines Require Specialized Cooling Systems", *Automotive Engineering*, (April, 1978), pp. 32-37.

Earthmoving Principles: A Guide to Production and Cost Estimating, International Harvester, Payline Division, Schaumburg, IL, (November, 1975).

Earthmoving Systems: An Analysis of Equipment Selection for Heavy Construction, International Harvester, Payline Division, Schaumburg, Illinois, (April, 1975).

"Elevating Scraper vs. Dozer", *Construction Contracting*, (September, 1980), pp. 26-28.

Farris, John A. and Robert Favor. "Contamination Control in Hydrostatic Transmissions", Parts I and II, *Heavy Duty Equipment Management and Maintenance*, (November, 1979), pp. 38-47, and (December, 1979), pp. 16-23.

"Fire Controls Set for Amax Dragline", *Coal Age*, (April, 1980), pp. 98-100.

Fougerousse, John. "Computerized Scrapers", *Heavy Duty Equipment Management and Maintenance*, (May, 1981), pp. 42-45.

Freebourn, N.E. "Wire Rope Lubrication", *Heavy Duty Equipment Management and Maintenance*, (June, 1981), pp. 20-23.

Fundamentals of Earthmoving, Caterpillar Tractor Co., Peoria, Illinois, (July, 1975).

Goodbary, E.R. and A.E. Hunt. *Design Considerations for Future Off-Highway Haulage Trucks*, National Conference Publication #79/5, The Institution of Engineers, Australia, (July, 1979).

Greaves, Bill. "Bit Selection — the Center of a Good Drilling Program", Dresser Industries, Houston, Texas.

Grenfell, K.P. *The Application of Static Armature Power Supplies for DC Excavator Motion Drives*, Open Pit Mining Association, Electrical Division, St. Louis, Mo., (June, 1970).

Gross, Roger. "Understanding Drive Related Failures", *Heavy Duty Equipment Management and Maintenance*, (November, 1980), pp. 34-41.

Guccione, Eugene. "Donaldson Designs New System for Dustless Blast Hole Drilling", *Coal Mining and Processing*, (April, 1978), pp. 88-91, 126-129.

"Guides to Selecting Scrapers for Earthmoving Operations", *Western Miner*, (September, 1978), pp. 18-20.

Herbaty, Frank. "Essentials of a Maintenance Management System", *Plant Engineering*, (May, 1977), pp. 103-105.

Heusler, Helmut and Udo Reinecke. *RH75 — A 10/13 Cubic Yard Hydraulic Shovel Development and Application*, Society of Automotive Engineers, #770553, (April, 1977).

Hirner, Frank J. "Estimating Production and Costs for Electric Mine Shovels", *Mining Engineering*, (February, 1981), pp. 143-146.

Hirschfeld, Fritz. "Power Trains for Tractors", *Mechanical Engineering*, (April, 1979), pp. 40-41.

Hoeft, Roy J. and Dietrich F. Rikahr. *Why Hydraulic Shovels — Mining*, Society of Automotive Engineers, #780450, Warrendale, PA, (April, 1978).

Hoppe, Richard (Editor). *E/MJ Operating Handbook of Mineral Surface Mining and Exploration*, McGraw-Hill, Inc., New York, NY (1978).

"Hydraulic Excavator Safety", *Heavy Duty Equipment Management and Maintenance*, (July, 1981), pp. 18-21.

Hydraulic Excavator, Users Safety Manual, Construction Industry Manufacturers Association, Milwaukee, WI, (1975).

"Hydraulic Mining Shovels", *Mining Magazine*, (June, 1977), pp. 437-449.

"Hydraulic Systems Safeguard", *Construction Contracting*, (November, 1980), pp. 40-41.

Improved Visibility Systems for Large Haulage Vehicles, MB Associates, U.S. Bureau of Mines, Contract No. PB- 286065, (April, 1978).

Jackson, Dan. "Keeping Equipment On Line And Available", *Engineering and Mining Journal*, (September, 1980), pp. 70-76.

Jackson, Dan. "Miner Strips and Loads at 2800 TPH", *Coal Age*, (March, 1979), pp. 64-69.

Jackson, Dan. "Rip Instead of Drilling and Blasting", *Coal Age*, (August, 1979), pp. 64-70.

Jackson, Robert A. "Cost Per Foot: Key to Rock Bit Selection", *World Oil*, (June, 1972).

Kaufman, Derek and Roger Bowen. "Analyze Haul Truck Costs Wisely", *Coal Age*, (March, 1981), pp. 70-77.

Kaufman, Derek D. "Mechanical vs. Electric Drive", *Engineering and Mining Journal*, (August, 1980), pp. 66-73.

Kay, F.J., J. Gordon and Lester Crow. *Continuous Excavator for Surface Coal Mining*, Society of Automotive Engineers, #770743, Warrendale, PA, (September, 1977).

"Keeping High Drill Availability", *Coal Age*, (July, 1977), pp. 174-178.

Kirtland, J.A., D.C. Andrus and W. Slabiak. *An AC Electric Drive System as Applied to a Tracked Vehicle*, Society of Automotive Engineers, #690442, Warrendale, PA, (May, 1969).

Knapp, Kenneth K. "Hydrostatic Transmissions: How They Work", *Heavy Duty Equipment Management and Maintenance*, (January, 1981), pp. 23-27.

Kulhavey, Joseph A. *Mobile Blast Hole Drilling Equipment*, Society of Automotive Engineers, #770742, Warrendale, PA, (September, 1977).

Lake, David M. and William Brzezniak. *Truck Haulage Using Overhead Electrical Power to Conserve Diesel Fuel and Improve Haulage Economics*, AIME Annual Meeting, Chicago, IL, (February, 1981).

Laswell, Betty J. and Gerald W. "The Pro's and Con's of Rotary Blast Hole Drill Design", *Mining Engineering*, (June, 1978), pp. 625-627.

Learmont, Tom. *Current Developments in Surface Mining Excavators*, International Conference on Mining and Machinery, Brisbane, Australia, (July, 1979).

Li, Ta M. "Rotary Drilling With Automated Controls — New Force in Open Pit Blast Hole Production", *Engineering and Mining Journal*, (August, 1974), pp. 53-60.

"Load and Carry Operations Keep Pit Costs in Line", *Pit and Quarry*, (December, 1978), pp. 108-109.

"Loading Shovels Attain 96% Availability", *Coal Mining and Processing*, (July, 1979), pp. 76-78.

"Lubricating Earthmoving Equipment", *Automotive Engineering*, (May, 1975), pp. 26-30.

Mahoney, J.E. *Design Concepts of Hydraulic Mining Machine — the P&H 1200 Hydraulic Shovel*, International Conference of Mining and Machinery, Brisbane, Australia, (July, 1979), pp. 115-118.

Mance, F.W. "How to Calculate Idler Spacings for More Economical Belt Conveyors", *Coal Mining and Processing*, (January, 1978), pp. 66-70, 80.

Man The Builder: The Functional Design and Job Applications of Power Cranes and Excavators, Power Crane and Shovel Association, Technical Bulletin No. 1, Milwaukee, WI, (January, 1977).

Mason, Richard H. "Front Shovel Works Narrow Benches on Contour Job", *Coal Mining and Processing*, (June, 1978), pp. 88-92.

Matuszak, R.A. "Controlled Frequency — The Brushless Electic Steam Engine", *Mining Engineering:r, (February, 1978), pp. 177-182.*

Meisel, William H. Controls for Hydrostatic Transmissions — Present and Future, Society of Automotive Engineers, #760697, Warrendale, PA, (September, 1976).

Michelin Earthmover Tire Data Book, Michelin Tire Corp., Lake Success, NY, (1980).

Miller, John E. "Development of Automatic Truck Control (ATC)", *Mining Engineering*, (February, 1981), pp. 175-178.

Morgan, W.C. and J.G. Schermerhorn. *Scrapers vs. Draglines — Production and Present-Worth Cost Analysis*, Society of Mining Engineers of AIME, St. Louis, MO, (October, 1977).

Motor Grader Safety Manual, Construction Industry Manufacturers Association, Milwaukee, WI, (1971).

Mueller, Wayne N. *Cooling System Design for Earthmoving Vehicles*, Society of Automotive Engineers, #780453, Warrendale, PA, (April, 1978).

Myntti, D.C. *An Advanced Open Pit Mining Equipment Maintenance Program*, Society of Mining Engineers of AIME, Preprint No. 78-AO-82, Salt Lake City, UT, (February, 1978).

Naft, Manny. *Integration of Component Design for a 170 Ton Off-Highway Truck*, Society of Automotive Engineers, #770741, Warrendale, Pa., (September, 1977).

Nelson, V.A., J. L. Love and A.E. Nordby. *Basic Service Requirements for Hydrostatic Transmissions in Mobile Industrial Equipment*, Society of Automotive Engineers, #760699, Warrendale, PA, (September, 1976).

"New Giant Tractor Features Many Technical Innovations", *Mining Magazine*, (November, 1977), pp. 567-569.

Nichols, H.L. Jr. *Moving The Earth, The Workbook of Excavation*, (Third Edition), North Castle Books, Greenwich, CT, 1976.

Oeller, F. "Tire Protection Chains Reduce Operating Costs", *Rock Products*, (November, 1977), pp. 78-79, 103-105.

Off-Highway Truck Safety Manual, Construction Industry Manufacturers Association, Milwaukee, WI, (1970).

Off The Road Tires: Engineering Data, Goodyear Tire and Rubber Co., Akron, OH, (October, 1976).

Operating Cost Guide: For Estimating Costs of Owning And Operating Cranes And Excavators, Power Crane and Shovel Association, Milwaukee, WI, (1976).

Palmer, W.E. and R.R. Fuehrer. *Noise Control in Planetary Transmissions*, Society of Automotive Engineers, #770561, Warrendale, PA, (April, 1977).

Pellow, Donald W. "How to Assure Safe Use of Wire Rope", *Engineering and Mining Journal*, (May, 1981), pp. 108-111.

Peurifoy, R.L. *Construction Planning, Equipment, and Methods*, (Third Edition), McGraw-Hill Book Co., New York, NY 1979.

Pfleider, Eugene P. (Editor). *Surface Mining*, The American Institute of Mining, Metallurgical and Petroleum Engineers Inc., New York, NY (1972).

Pomerantsev, Alexander. *Preliminary Selection and Estimation of the Basic Parameters of Bucketwheel Excavator*, Society of Mining Engineers of AIME, Preprint No. 79-301, Tucson, AZ, (October, 1979).

Pomproy, William H. *Improved Automatic Fire Protection Systems for Off-Highway Mine Vehicles*, Society of Automotive Engineers, #790880, Warrendale, PA, (September, 1979).

Price, G.C., C.B. Manula and Rajaraman Venkataramani. *Materials Handling Researach: The Bucket-Wheel Excavator*, Bureau of Mines Information Circular #8580, U.S. Department of the Interior, (1973).

Production and Cost Estimating of Material Movement with Earthmoving Equipment, (English Units Version), General Motors Corp., Terex Division, Hudson, OH, (1980).

Pundari, N.B. "Selecting and Using Large Walking Draglines for Overburden Stripping", *Mining Engineering*, (April, 1981), pp. 377-381.

Rasper, Ludwig. *The Bucket Wheel Excavator: Development, Design, Application*, Trans Tech Publications, 1975.

Ritcey, R.M. and W. Taciuk. *Testing and Evaluation of Large, Off-Highway Haulage Truck Brake Systems*, Society of Automotive Engineers, #770547, (April, 1977).

SAE Handbook, Parts 1 and 2, Society of Automotive Engineers, Warrendale, PA, (1979).

"Safety Tips For Hydraulic Excavators", *Engineering and Mining Journal*, (October, 1978), p. 141.

"Safety Tips For Loaders And Dozers", *Heavy Duty Equipment Management and Maintenance*, (August, 1979), pp. 20-25.

"Safety Tips for Off-Highway Trucks", *Heavy Duty Equipment Management and Maintenance*, (July, 1980), pp. 32-38.

Sandberg, C.N. *General Electric's Motorized Wheel Drive For Trucks in Surface Mining*, Open Pit Mining Association, Louisville, KY, (June, 1977).

Schmidt, J.W. and R.L. Kesler. *Power Train Selection For Optimum Performance And Efficiency*, Society of Automotive Engineers, #780961, Warrendale, PA, (November, 1978).

Schweitzer, F.W. and L.G. Dykers. *Belt Conveyor vs. Truck Haulage: Capital vs. Expense*, Society of Mining Engineers of AIME, Preprint #76A0335, Salt Lake City, UT, (September, 1976).

Scott, Koehler S. *Mining Methods and Equipment*, Mining Informational Services, McGraw-Hill Book Co., New York, NY 1980.

Scraper Safety Manual, Construction Industry Manufacturers Association, Milwaukee, WI, (1980).

Smiley, Carl H. "Engine Oil: A High Ticket Item", *Coal Mining and Processing*, (June, 1981), pp. 46-48.

Smiley, Carl H. "How to Fuel-Proof Your Haulers", *Coal Mining and Processing*, (May 1981), pp. 80-81.

Smiley, Carl H. "Managing Tire Costs", *Coal Mining and Processing*, (June, 1981), pp. 53-59.

Smiley, Carl H. "Mechanical Drive Improves Mid-Size Haulers", *Coal Mining and Processing*, (April, 1979), pp. 78-80.

Sprenkel, Roger L. *Ripping — Today and Tomorrow*, Society of Automotive Engineers, #790530, Warrendale, PA, (April, 1979).

Sprouls, Mark. "New Ways to Load Trucks Faster", *Coal Mining And Processing*, (July, 1980), pp. 92-96.

Sprouls, Mark W. "Oak Grove Proves New Loading Method", *Coal Mining and Processing*, (August, 1980), pp. 54-55.

Stafford, Art. "Understanding Turbo Charger Failures", *Heavy Duty Equipment Management and Maintenance*, (May, 1981), pp. 30-34.

Steidle, E. Jr. *Hydraulic And Electric Drive Systems For Heavy Mining Machinery*, International Conference on Mining and Machinery, Brisbane, Australia, (July, 1979).

Steidle, E. Jr. *Modern Design Developments in Dragline Excavators*, International Conference on Mining and Machinery, Brisbane, Australia, (July, 1979).

Surface Coal Mining Reclamation Equipment and Techniques, Bureau of Mines Information Circular #8823, U.S. Department of the Interior, (1980).

"Tank Rig Speeds Drilling Double Seams", *Coal Age*, (May, 1978), pp. 187-190.

"Techniques for Quieting the Diesel", *Automotive Engineering*, (September, 1975), pp. 34-38.

Terai, Akio, Herb Aoki and R.H. Stanage. *New Design Concept For Komatsu D455A Bulldozer and The Actual Results*, Society of Automotive Engineers, #790902, Warrendale, PA, (September, 1979).

Trenary, Bryant. *1500 HP Diesel Electric Tractor*, Society of Automotive Engineers, #760647, Warrendale, PA, (September, 1976).

Utilization of Hydraulic Excavators in Coal Mining Applications, Orenstein & Koppel, Inc., Clifton, NJ, (October, 1978).

Williams, R.L. "The Trend Towards Mechanical Drives For Open Pit Mine Haulage Trucks", *Western Miner*, (December, 1978).

Wood, Stuart Jr. *Heavy Construction Equipment and Methods*, Prentice-Hall, Inc., Englewood Cliffs, NJ, 1977.

LIST OF FIGURES

Chapter 1 Equipment Selection

Chapter 2 Dozers

Chapter 3 Scrapers

Chapter 4 Trucks

Chapter 5 Front-End Loaders

Chapter 6 Hydraulic Excavators

Chapter 7 Electric Shovels

Chapter 8 Draglines

Chapter 9 Bucket Wheel Excavators

Chapter 10 Blast Hole Drills

Appendix A Tires

Appendix B Diesel Engines

LIST OF GRAPHS

Chapter 1 Equipment Selection

Chapter 2 Dozers

Chapter 3 Scrapers

Chapter 4 Trucks

Chapter 5 Front-End Loaders

Chapter 6 Hydraulic Excavators

Chapter 7 Electric Shovels

Chapter 8 Draglines

Chapter 9 Bucket Wheel Excavators

Chapter 10 Blast Hole Drills

Appendix B Diesel Engines

Acknowledgements

The following companies generously permitted the use of their photographs, graphs and/or figures:

Figure 10.5 Reed Mining Tools, Inc.
Figure 10.7 Reed Mining Tools, Inc.
Graph 10.6 Reed Mining Tools, Inc.
Graph 10.7 Reed Mining Tools, Inc.
Photograph 1.5 Caterpillar Tractor Company
Photograph 1.8 JI Case Company
Photograph 1.9 Terex Corporation
Photograph 1.13 Bucyrus-Erie Company
Photograph 2.1 Terex Corporation
Photograph 2.2 Clark Equipment Company
Photograph 2.5 Marathon LeTourneau Company
Photograph 2.7 Caterpillar Tractor Company
Photograph 2.8 Melroe Multi-Wheel
Photograph 3.1 Terex Corporation
Photograph 3.2 Terex Corporation
Photograph 3.3 Caterpillar Tractor Company
Photograph 3.4 Caterpillar Tractor Company
Photograph 3.5 Fiat-Allis
Photograph 3.6 Caterpillar Tractor Company
Photograph 4.1 Euclid Inc.
Photograph 4.2 Wabco Construction & Mining Company
Photograph 4.3 Rimpull Corporation
Photograph 4.4 Wabco Construction & Mining Company
Photograph 4.5 Terex Corporation
Photograph 4.6 Unit Rig and Equipment Company
Photograph 4.7 Kress Corporation
Photograph 5.1 Euclid Inc.
Photograph 5.2 Terex Corporation
Photograph 5.7 Marathon LeTourneau Company
Photograph 6.1 Koehring
Photograph 6.3 Mannesmann Demag Corporation
Photograph 6.4 JI Case Company
Photograph 6.7 O&K (Orenstein and Koppel)
Photograph 6.8 Northwest Engineering Company
Photograph 7.3 Wabco Construction & Mining Company
Photograph 7.6 Bucyrus-Erie Company
Photograph 7.7 Marion Power Shovel Division, Dresser Industries, Inc.
Photograph 7.8 Rimpull Corporation
Photograph 7.9 Bucyrus-Erie Company
Photograph 8.1 American Hoist & Derrick Company
Photograph 8.2 Page Engineering Company
Photograph 8.4 Bucyrus-Erie Company
Photograph 8.5 Marion Power Shovel Division, Dresser Industries, Inc.
Photograph 8.6 Page Engineering Company
Photograph 8.15 Bucyrus-Erie Company
Photograph 10.1 Schramm Inc.
Photograph 10.2 Robbins Division, Joy Manufacturing Company
Photograph 10.3 Gardner-Denver Cooper Mining & Construction

Photograph 10.4 Marion Power Shovel Division, Dresser Industries, Inc.
Photograph 10.10 Hughes Tool Company
Photograph 10.11 Hughes Tool Company
Photograph 10.12 Hughes Tool Company
Photograph 10.13 Hughes Tool Company
Photograph B.1 General Motors Corporation, Detroit Diesel Allison Division
Photograph B.2 General Motors Corporation, Detroit Diesel Allison Division